Newtonian mechanics

A. P. French

PROFESSOR OF PHYSICS, THE MASSACHUSETTS INSTITUTE OF TECHNOLOGY

Newtonian mechanics

THE M.I.T. INTRODUCTORY PHYSICS SERIES

W · W · NORTON & COMPANY
New York · London

W. W. Norton & Company, Inc., 500 Fifth Avenue, New York, N.Y. 10110
www.wwnorton.com

W. W. Norton & Company Ltd., Castle House, 75/76 Wells Street, London W1T 3QT

Copyright © 1971, 1965 by the Massachusetts Institute of Technology

SBN 0-393-09970-9

Library of Congress Catalog Card No. 74-95528

Printed in the United States of America

4 5 6 7 8 9 0

Contents

Preface xi
Prologue 3

PART I   THE APPROACH TO NEWTONIAN DYNAMICS

1 A universe of particles   21

    The particulate view 21
    Electrons and nucleons 24
    Atomic nuclei 25
    Atoms 26
    Molecules; living cells 28
    Sand and dust 31
    Other terrestrial objects 32
    Planets and satellites 33
    Stars 35
    Galaxies 36
        PROBLEMS 38

2 Space, time, and motion   43

    What is motion? 43
    Frames of reference 46
    Coordinate systems 48

Combination of vector displacements 53
The resolution of vectors 56
Vector addition and the properties of space 59
Time 61
Units and standards of length and time 63
Space–time graphs 66
Velocity 67
Instantaneous velocity 68
Relative velocity and relative motion 72
Planetary motions: Ptolemy versus Copernicus 74
PROBLEMS 78

3 Accelerated motions 85

Acceleration 85
The analysis of straight-line motion 87
A comment on extraneous roots 93
Trajectory problems in two dimensions 95
Free fall of individual atoms 98
Other features of motion in free fall 102
Uniform circular motion 105
Velocity and acceleration in polar coordinates 106
PROBLEMS 108

4 Forces and equilibrium 115

Forces in static equilibrium 116
Units of force 118
Equilibrium conditions; forces as vectors 119
Action and reaction in the contact of objects 123
Rotational equilibrium; torque 124
Forces without contact; weight 128
Pulleys and strings 130
PROBLEMS 132

5 The various forces of nature 139

The basic types of forces 139
Gravitational forces 140
Electric and magnetic forces 145
Nuclear forces 147
Forces between neutral atoms 148
Contact forces 150
Frictional contact forces 152
Concluding remarks 154
PROBLEMS 154

6 Force, inertia, and motion 161

The principle of inertia *161*
Force and inertial mass: Newton's law *164*
Some comments on Newton's law *167*
Scales of mass and force *170*
The effect of a continuing force *173*
The invariance of Newton's law; relativity *173*
Invariance with specific force laws *176*
Newton's law and time reversal *178*
Concluding remarks *180*
PROBLEMS *181*

PART II CLASSICAL MECHANICS AT WORK

7 Using Newton's law 187

Some introductory examples *188*
Motion in two dimensions *194*
Motion in a circle *198*
Curvilinear motion with changing speed *200*
Circular paths of charged particles in uniform magnetic fields *202*
Charged particle in a magnetic field *205*
Mass spectrographs *206*
The fracture of rapidly rotating objects *208*
Motion against resistive forces *210*
Detailed analysis of resisted motion *213*
Motion governed by viscosity *218*
Growth and decay of resisted motion *221*
Air resistance and "independence of motions" *225*
Simple harmonic motion *226*
More about simple harmonic motion *231*
PROBLEMS *234*

8 Universal gravitation 245

The discovery of universal gravitation *245*
The orbits of the planets *246*
Planetary periods *249*
Kepler's third law *252*
The moon and the apple *256*
Finding the distance to the moon *259*
The gravitational attraction of a large sphere *261*
Other satellites of the earth *265*
The value of G, *and the mass of the earth* *268*
Local variations of g *270*
The mass of the sun *274*

Finding the distance to the sun 275
Mass and weight 279
Weightlessness 285
Learning about other planets 286
The moons of Jupiter 288
The discovery of Neptune 291
Gravitation outside the solar system 295
Einstein's theory of gravitation 299
    PROBLEMS 301

9 Collisions and conservation laws   307

The laws of impact 308
The conservation of linear momentum 309
Momentum as a vector quantity 310
Action, reaction, and impulse 313
Extending the principle of momentum conservation 318
The force exerted by a stream of particles 321
Reaction from a fluid jet 324
Rocket propulsion 327
Collisions and frames of reference 331
Kinetic energy in collisions 333
The zero-momentum frame 335
Collision processes in two dimensions 339
Elastic nuclear collisions 342
Inelastic and explosive processes 346
What is a collision? 351
Interacting particles subject to external forces 352
The pressure of a gas 354
The neutrino 356
    PROBLEMS 357

10 Energy conservation in dynamics; vibrational motions   367

Introduction 367
Integrals of motion 368
Work, energy, and power 373
Gravitational potential energy 376
More about one-dimensional situations 379
The energy method for one-dimensional motions 381
Some examples of the energy method 384
The harmonic oscillator by the energy method 393
Small oscillations in general 395
The linear oscillator as a two-body problem 397
Collision processes involving energy storage 400
The diatomic molecule 405
    PROBLEMS 411

viii

11 Conservative forces and motion in space   423

Extending the concept of conservative forces   423
Acceleration of two connected masses   425
Object moving in a vertical circle   426
An experiment by Galileo   429
Mass on a parabolic track   431
The simple pendulum   434
The pendulum as a harmonic oscillator   437
The pendulum with larger amplitude   440
Universal gravitation: a conservative central force   442
A gravitating spherical shell   446
A gravitating sphere   450
Escape velocities   453
More about the criteria for conservative forces   457
Fields   461
Equipotential surfaces and the gradient of potential energy   463
Motion in conservative fields   466
The effect of dissipative forces   470
Gauss's law   473
Applications of Gauss's theorem   476
PROBLEMS   478

PART III   SOME SPECIAL TOPICS

12 Inertial forces and non-inertial frames   493

Motion observed from unaccelerated frames   494
Motion observed from an accelerated frame   495
Accelerated frames and inertial forces   497
Accelerometers   501
Accelerating frames and gravity   504
Centrifugal force   507
Centrifuges   511
Coriolis forces   514
Dynamics on a merry-go-round   518
General equation of motion in a rotating frame   519
The earth as a rotating reference frame   524
The tides   531
Tidal heights; effect of the sun   535
The search for a fundamental inertial frame   538
Speculations on the origin of inertia   542
PROBLEMS   546

13 Motion under central forces   555

Basic features of the problem   555
The law of equal areas   557
The conservation of angular momentum   560

Energy conservation in central force motions 563
Use of the effective potential-energy curves 565
Bounded orbits 568
Unbounded orbits 569
Circular orbits in an inverse-square force field 572
Small perturbation of a circular orbit 574
The elliptic orbits of the planets 577
Deducing the inverse-square law from the ellipse 583
Elliptic orbits: analytical treatment 585
Energy in an elliptic orbit 589
Motion near the earth's surface 591
Interplanetary transfer orbits 592
Calculating an orbit from initial conditions 595
A family of related orbits 596
Central force motion as a two-body problem 598
Deducing the orbit from the force law 600
Rutherford scattering 604
Cross sections for scattering 609
Alpha-particle scattering (Geiger and Marsden, Philosophical Magazine *excerpts)* 612
An historical note 615
    PROBLEMS 617

14 Extended systems and rotational dynamics 627

Momentum and kinetic energy of a many-particle system 628
Angular momentum 632
Angular momentum as a fundamental quantity 636
Conservation of angular momentum 639
Moments of inertia of extended objects 643
Two theorems concerning moments of inertia 647
Kinetic energy of rotating objects 651
Angular momentum conservation and kinetic energy 654
Torsional oscillations and rigid pendulums 659
Motion under combined forces and torques 664
Impulsive forces and torques 668
Background to gyroscopic motion 671
Gyroscope in steady precession 677
More about precessional motion 680
Gyroscopes in navigation 683
Atoms and nuclei as gyroscopes 686
Gyroscopic motion in terms of **F** $=$ **ma** 688
Nutation 691
The precession of the equinoxes 694
    PROBLEMS 700

Appendix 709
Bibliography 713
Answers to problems 723
Index 733

Preface

THE WORK of the Education Research Center at M.I.T. (formerly the Science Teaching Center) is concerned with curriculum improvement, with the process of instruction and aids thereto, and with the learning process itself, primarily with respect to students at the college or university undergraduate level. The Center was established by M.I.T. in 1960, with the late Professor Francis L. Friedman as its Director. Since 1961 the Center has been supported mainly by the National Science Foundation; generous support has also been received from the Kettering Foundation, the Shell Companies Foundation, the Victoria Foundation, the W. T. Grant Foundation, and the Bing Foundation.

The M.I.T. Introductory Physics Series, a direct outgrowth of the Center's work, is designed to be a set of short books that, taken collectively, span the main areas of basic physics. The series seeks to emphasize the interaction of experiment and intuition in generating physical theories. The books in the series are intended to provide a variety of possible bases for introductory courses, ranging from those which chiefly emphasize classical physics to those which embody a considerable amount of atomic and quantum physics. The various volumes are intended to be compatible in level and style of treatment but are not conceived as a tightly knit package; on the contrary, each book is designed to be reasonably self-contained and usable as an individual component in many different course structures.

The text material in the present volume is designed to be a more or less self-contained introduction to Newtonian mechanics, such that a student with little or no grounding in the subject can, by beginning at the beginning, be brought gradually to a level of considerable proficiency. A rough guide to the possible use of the book is suggested by its division into three parts. Part I, *The Approach to Newtonian Dynamics*, is intended to serve two purposes. First, it does discuss the basic concepts of kinematics and dynamics, more or less from scratch. Second, it seeks to place the study of mechanics squarely in the context of the world of physical phenomena and of necessarily imperfect physical theories. This is a conscious reaction, on the author's part, against the presentation of mechanics as "applied mathematics," with the divorcement from reality and the misleading impression of rigor that this has engendered in generations of students (especially, alas, those brought up in the British educational system). The student who already has some expertise in using Newton's laws will find little of an analytical or quantitative sort to learn from Part I, but he may still derive some value and interest from reading through it for its broader implications.

Part II, *Classical Mechanics at Work*, is undoubtedly the heart of the book. Some instructors will wish to begin here, and relegate Part I to the status of background reading. The initial emphasis is on Newton's second law applied to individual objects. Later, the emphasis shifts to systems of two or more particles, and to the conservation laws for momentum and energy. A fairly lengthy chapter is devoted to the subject that deserves pride of place in the whole Newtonian scheme—the theory of universal gravitation and its successes, which can still be appreciated as a pinnacle in man's attempts to discover order in the vast universe in which he finds himself.

Part III, *Some Special Topics*, concerns itself with the problems of noninertial frames, central-force motions, and rotational dynamics. Most of this material, except perhaps the fundamental features of rotational motion and angular momentum, could be regarded as optional if this book is used as the basis of a genuinely introductory presentation of mechanics. Undoubtedly the book as a whole contains more material than could in its entirety be covered in a one-term course; one could, however, consider using Parts I and II as a manageable package for beginners, and Parts II and III as a text for students having some prior preparation.

One of the great satisfactions of classical mechanics lies in the vast range and variety of physical systems to which its principles can be applied. The attempt has been made in this book to make explicit reference to such applications and, as in other books in this series, to "document" the presentation with appropriate citations from original sources. Enriched in this way by its own history, classical mechanics has an excitement that is not, in this author's view, surpassed by any of the more recent fields of physical thory.

This book, like the others in the series, owes much to the thoughts, criticisms, and suggestions of many people, both students and instructors. A special acknowledgment in connection with the present volume is due to Prof. A. M. Hudson, of Occidental College, Los Angeles, who worked with the present author in the preparation of the preliminary text from which, five years later, this final version evolved. Grateful thanks are also due to Eva M. Hakala and William H. Ingham for their invaluable help in preparing the manuscript for publication.

A. P. FRENCH

Cambridge, Massachusetts
July 1970

Newtonian mechanics

In the Beginning was Mechanics.
　　　　　MAX VON LAUE, History of Physics (1950)

I offer this work as the mathematical principles of philosophy, for the whole burden of philosophy seems to consist in this—from the phenomena of motions to investigate the forces of nature, and then from these forces to demonstrate the other phenomena.
　　　　　NEWTON, Preface to the Principia (1686)

Prologue

ONE OF THE MOST prominent features of the universe is motion. Galaxies have motions with respect to other galaxies, all stars have motions, the planets have distinctive motions against the background of the stars, the events that capture our attention most quickly in everyday life are those involving motion, and even the apparently inert book that you are now reading is made up of atoms in rapid motion about their equilibrium positions. "Give me matter and motion," said the seventeenth-century French philosopher René Descartes, "and I will construct the universe." There can be no doubt that motion is a phenomenon we must learn to deal with at all levels if we are to understand the world around us.

Isaac Newton developed a precise and powerful theory regarding motion, according to which the *changes* of motion of any object are the result of *forces* acting on it. In so doing he created the subject with which this book is concerned and which is called classical or Newtonian mechanics. It was a landmark in the history of science, because it replaced a merely descriptive account of phenomena with a rational and marvelously successful scheme of cause and effect. Indeed, the strict causal nature of Newtonian mechanics had an impressive influence in the development of Western thought and civilization generally, provoking fundamental questions about the interrelationships of science, philosophy, and religion, with repercussions in social ideas and other areas of human endeavor.

Classical mechanics is a subject with a fascinating dual character. For it starts out from the kinds of everyday experiences that are as old as mankind, yet it brings us face to face with some of the most profound questions about the universe in which we find ourselves. Is it not remarkable that the flight of a thrown pebble, or the fall of an apple, should contain the clue to the mechanics of the heavens and should ultimately involve some of the most basic questions that we are able to formulate about the nature of space and time? Sometimes mechanics is presented as though it consisted merely of the routine application of self-evident or revealed truths. Nothing could be further from the case; it is a superb example of a physical theory, slowly evolved and refined through the continuing interplay between observation and hypothesis.

The richness of our first-hand acquaintance with mechanics is impressive, and through the partnership of mind and eye and hand we solve, by direct action, innumerable dynamical problems without benefit of mathematical analysis. Like Molière's famous character, Monsieur Jourdain, who learned that he had been speaking prose all his life without realizing it, every human being is an expert in the consequences of the laws of mechanics, whether or not he has ever seen these laws written down. The skilled sportsman or athlete has an almost incredible degree of judgment and control of the amount and direction of muscular effort needed to achieve a desired result. It has been estimated, for example, that the World Series baseball championship would have changed hands in 1962 if one crucial swing at the ball had been a mere millimeter lower.[1] But experiencing and controlling the motions of objects in this very personal sense is a far cry from analyzing them in terms of physical laws and equations. It is the task of classical mechanics to discover and formulate the essential principles, so that they can be applied to any situation, particularly to inanimate objects interacting with one another. Our intimate familiarity with our own muscular actions and their consequences, although it represents a kind of understanding (and an important kind, too), does not help us much here.

The greatest triumph of classical mechanics was Newton's own success in analyzing the workings of the solar system—a feat immortalized in the famous couplet of his contemporary and admirer, the poet Alexander Pope:

---

[1]P. Kirkpatrick, *Am. J. Phys.*, **31**, 606 (1963).

> Nature and Nature's Laws lay hid in night
> God said "Let Newton be," and all was light.[1]

Men had observed the motions of the heavenly bodies since time immemorial. They had noticed various regularities and had learned to predict such things as conjunctions of the planets and eclipses of the sun and moon. Then, in the sixteenth century, the Danish astronomer Tycho Brahe amassed meticulous records, of unprecedented accuracy, of the planetary motions. His assistant, Johannes Kepler, after wrestling with this enormous body of information for years, found that all the observations could be summarized as follows:

1. The planets move in ellipses having the sun at one focus.
2. The line joining the sun to a given planet sweeps out equal areas in equal times.
3. The square of a planet's year, divided by the cube of its mean distance from the sun, is the same for all planets.

This represented a magnificent advance in man's knowledge of the mechanics of the heavens, but it was still a description rather than a theory. Why? was the question that still looked for an answer. Then came Newton, with his concept of force as the cause of changes of motion, and with his postulate of a particular law of force—the inverse-square law of gravitation. Using these he demonstrated how Kepler's laws were just one consequence of a scheme of things that also included the falling apple and other terrestrial motions. (Later in this book we shall go into the details of this great achievement of Newton's.)

If universal gravitation had done no more than to relate planetary periods and distances, it would still have been a splendid theory. But, like any other good theory in physics, it had predictive value; that is, it could be applied to situations besides the ones from which it was deduced. Investigating the predictions of a theory may involve looking for hitherto unsuspected phenomena, or it may involve recognizing that an already familiar phenomenon must fit into the new framework. In either case the theory is subjected to searching tests, by which it must

---

[1]To which there is the almost equally famous, although facetious, riposte:
> It did not last; the Devil, howling "Ho,
> Let Einstein be!" restored the status quo.
>
> (Sir John Squire)

stand or fall. With Newton's theory of gravitation, the initial tests resided almost entirely in the analysis of known effects—but what a list! Here are some of the phenomena for which Newton proceeded to give *quantitative* explanations:

1. The bulging of the earth and Jupiter because of their rotation.
2. The variation of the acceleration of gravity with latitude over the earth's surface.
3. The generation of the tides by the combined action of sun and moon.
4. The paths of the comets through the solar system.
5. The slow steady change in direction of the earth's axis of rotation produced by gravitational torques from the sun and moon. (A complete cycle of this variation takes about 25,000 years, and the so-called "precession of the equinoxes" is a manifestation of it.)

This marvelous illumination of the workings of nature represented the last part of Newton's program, as he described it in our opening quotation "...and then from these forces to demonstrate the other phenomena." This modest phrase conceals not only the immensity of the achievement but also the magnitude of the role played by mathematics in this development. Newton had, in the theory of universal gravitation, created what would be called today a mathematical model of the solar system. And having once made the model, he followed out a host of its other implications. The working out was purely mathematical, but the final step—the test of the conclusions—involved a return to the world of physical experience, in the detailed checking of his predictions against the quantitative data of astronomy.

Although Newton's mechanical picture of the universe was amply confirmed in his own time, he did not live to see some of its greatest triumphs. Perhaps the most impressive of these was the use of his laws to identify previously unrecognized members of the solar system. By a painstaking and lengthy analysis of the motions of the known planets, it was inferred that disturbing influences due to other planets must be at work. Thus it was that Neptune was discovered in 1846, and Pluto in 1930. In each case it was a matter of deducing where a telescope should be pointed to reveal a new planet, identifiable through its changing position with respect to the general background of the stars.

What more striking and convincing evidence could there be that the theory works?

Probably everyone who reads this book has some prior acquaintance with classical mechanics and with its expression in mathematically precise statements. And this may make it hard to realize that, as with any other physical theory, its development was not just a matter of mathematical logic applied undiscriminatingly to a mass of data. Was Newton inexorably driven to the inverse-square law? By no means. It was the result of guesswork, intuition, and imagination. In Newton's own words: "I began to think of gravity extending to the orbit of the Moon, and ... from Kepler's Rule of the periodic times of the Planets ... I deduced that the forces which keep the Planets in their orbits must be reciprocally as the squares of their distances from the centers about which they revolve; and thereby compared the force requisite to keep the Moon in her orbit with the force of gravity at the surface of the Earth, and found them to answer pretty nearly." An intellectual leap of this sort—although seldom as great as Newton's—is involved in the creation of any theory or model. It is a process of induction, and it goes beyond the facts immediately at hand. Some facts may even be temporarily brushed aside or ignored in the interests of pursuing the main idea, for a partially correct theory is often better than no theory at all. And at all stages there is a constant interplay between experiment and theory, in the process of which fresh observations are continually suggesting themselves and modification of the theory is an ever-present possibility. The following diagram, the relevance of which goes beyond the realm of classical mechanics, suggests this pattern of man's investigation of matter and motion.

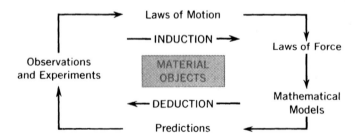

The enormous success of classical mechanics made it seem, at one stage, that nothing more was needed to account for the whole world of physical phenomena. This belief reached a pinnacle toward the end of the nineteenth century, when some optimistic physicists felt that physics was, in principle, complete. They could hardly have chosen a more unfortunate time at which to form such a conclusion, for within the next few decades physics underwent its greatest upheaval since Newton. The discovery of radioactivity, of the electron and the nucleus, and the subtleties of electromagnetism, called for fundamentally new ideas. Thus we know today that Newtonian mechanics, like every physical theory, has its fundamental limitations. The analysis of motions at extremely high speeds requires the use of modified descriptions of space and time, as spelled out by Albert Einstein's special theory of relativity. In the analysis of phenomena on the atomic or subatomic scale, the still more drastic modifications described by quantum theory are required. And Newton's particular version of gravitational theory, for all its success, has had to admit modifications embodied in Einstein's general theory of relativity. But this does not alter the fact that, in an enormous range and variety of situations, Newtonian mechanics provides us with the means to analyze and predict the motions of physical objects, from electrons to galaxies. Its range of validity, and its limits, are indicated very qualitatively in the figure below.

In developing the subject of classical mechanics in this book, we shall try to indicate how the horizons of its application to the physical world, and the horizons of one's own view, can be gradually broadened. Mechanics, as we shall try to present it, is not at all a cut-and-dried subject that would justify its description

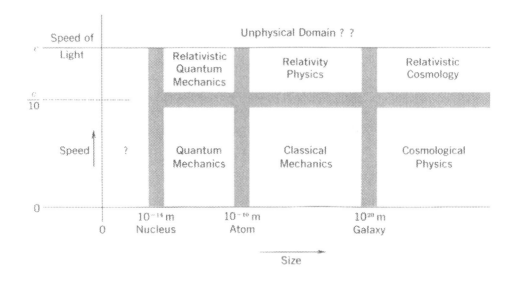

as "applied mathematics," in which the rules of the game are given at the outset and in which one's only concern is with applying the rules to a variety of situations. We wish to offer a different approach, in which at every stage one can be conscious of working with partial or limited data and of making use of assumptions that cannot be rigorously justified. But this is the essence of doing physics. Newton himself said as much. At the beginning of Book III of the *Principia* he propounds four "Rules of Reasoning in Philosophy," of which the last runs as follows:

"In experimental philosophy we are to look upon propositions inferred by general induction from phenomena as accurately or very nearly true, notwithstanding any contrary hypotheses that may be imagined, till such time as other phenomena occur, by which they may either be made more accurate, or liable to exceptions." The person who waits for complete information is on the way to dooming himself never to finish an experiment or to construct a useful theory. Lest this should be taken, however, as an encouragement to slipshod or superficial thinking, we shall end this introduction with a little fable due to George Polya.[1] He writes as a mathematician, but the moral for physicists (and others) is clear.

*The Logician, the Mathematician,
the Physicist, and the Engineer*

"Look at this mathematician," said the logician. "He observes that the first 99 numbers are less than 100 and infers, hence, by what he calls induction, that all numbers are less than a hundred." "A physicist believes," said the mathematician, "that 60 is divisible by 1, 2, 3, 4, 5, and 6. He examines a few more cases, such as 10, 20, and 30, taken at random (as he says). Since 60 is also divisible by these, he considers the experimental evidence sufficient." "Yes, but look at the engineers," said the physicist. "An engineer suspected that all odd numbers are prime numbers. At any rate, 1 can be considered as a prime number, he argued. Then there come 3, 5, and 7, all indubitably primes. Then there comes 9; an awkward case; it does not seem to be a prime num-

---

[1]This cautionary tale is to be found in a book entitled *Induction and Analogy in Mathematics*, Princeton University Press, Princeton, N.J., 1954. This volume and its companion, *Patterns of Plausible Inference*, make delightful reading for any scientist.

ber. Yet 11 and 13 are certainly primes. 'Coming back to 9,' he said, 'I conclude that 9 must be an experimental error.' " But having done his teasing, Polya adds these remarks.

> It is only too obvious that induction can lead to error. Yet it is remarkable that induction sometimes leads to truth, since the chances of error seem so overwhelming. Should we begin with the study of the obvious cases in which induction fails, or with the study of those remarkable cases in which induction succeeds? The study of precious stones is understandably more attractive than that of ordinary pebbles and, moreover, it was much more the precious stones than the pebbles that led the mineralogists to the wonderful science of crystallography.

With that encouragement, we shall, in Chapter 1, begin our approach to the study of classical mechanics, which is one of the most perfect and polished gems in the physicist's treasury. We end this Prologue, however, with some preparatory exercises.

EXERCISES—HORS D'OEUVRES

The literal meaning of the phrase "hors d'oeuvre" is "outside the work." The exercises below correspond exactly to that definition, although it is hoped that they will also whet the appetite as hors d'oeuvres should. They deal mostly with order-of-magnitude estimates (i.e., estimates to the nearest power of 10) and judicious approximations—things that play an important role in a physicist's approach to problems but seldom get emphasized or systematically presented in textbooks. For example, everybody learns the binomial theorem, but how many students think of it as a useful tool for obtaining a quite good value for the hypotenuse of a right triangle, by the approximation

$$(a^2 + b^2)^{1/2} \approx a\left(1 + \frac{b^2}{2a^2}\right)$$

where we assume $b < a$? (Even in the worst possible case, with $b = a$, the result is wrong by only about 6 percent—1.5 instead of 1.414....) Moreover, it takes practice and some conscious effort to develop the habit of assessing, quite crudely, the magnitudes of quantities and the relative importance of various possible effects in a physical system. For example, in dealing with objects moving through liquids, can one quickly decide whether viscosity or turbulence is going to be the chief source of resistance for an object of given speed and linear dimensions? An awareness of the effects of changes of scale can give valuable insights into the properties of systems. [A beautiful example of this is the well-known essay by J. B. S. Haldane, "On Being the Right Size," which is reprinted in *The World of Mathematics*, Vol. II (J. R. Newman, ed.), Simon and Schuster, New York, 1956.] By the use of such methods and ways of thought one can deepen one's appreciation of physical phenomena and can improve one's feeling for what the world is like and how it behaves.

It is surprising how much one can do with the help of a relatively small stock of primary information—which might include such items as the following:

Physical Magnitudes

Gravitational acceleration (g)	10 m/sec^2
Densities of solids and liquids	10^3–10^4 kg/m^3
Density of air at sea level	1 kg/m^3 (approx.)
Length of day	10^5 sec (approx.)
Length of year	3.16×10^7 sec $\approx 10^{7.5}$ sec
Earth's radius	6400 km
Angle subtended by finger thickness at arm's length	1° (approx.)
Thickness of paper	0.1 mm (approx.)
Mass of a paperclip	0.5 g (approx.)
Highest mountains, deepest oceans	10 km (approx.)
Earth–moon separation	3.8×10^5 km
Earth–sun separation	1.5×10^8 km
Atmospheric pressure	Equivalent to weight of 1 kg/cm^2 or a 10-m column of water
Avogadro's number	6.0×10^{23}
Atomic masses	1.6×10^{-27} kg to 4×10^{-25} kg
Linear dimensions of atoms	10^{-10} m (approx.)
Molecules/cm^3 in gas at STP	2.7×10^{19}
Atoms/cm^3 in solids	10^{23} (approx.)
Elementary charge (e)	1.6×10^{-19} C
Electron mass	10^{-30} kg (approx.)
Speed of light	3×10^8 m/sec
Wavelength of light	6×10^{-7} m (approx.)

Mathematical Magnitudes

$$\pi^2 \approx 10 \qquad \log_{10} 4 \approx 0.60$$
$$e \approx 2.7 \qquad \log_{10} e \approx 0.43$$
$$\log_{10} 2 \approx 0.30 \qquad \log_{10} \pi \approx 0.50$$
$$\log_{10} 3 \approx 0.48 \qquad \log_e 10 \approx 2.3$$

Angle (radians) = arc length/radius. Full circle = 2π rad. 1 rad $\approx 0.16 \times$ full circle $\approx 57°$.

Solid angle (steradians) = area/(radius)^2. Full sphere = 4π sr. 1 sr $\approx 0.08 \times$ full sphere.

Approximations

Binomial theorem:

For $x \ll 1$,
$$(1 + x)^n \approx 1 + nx$$
e.g.,
$$(1 + x)^3 \approx 1 + 3x$$
$$(1 - x)^{1/2} \approx 1 - \tfrac{1}{2}x \approx (1 + x)^{-1/2}$$

For $b \ll a$,
$$(a + b)^n = a^n \left(1 + \frac{b}{a}\right)^n \approx a^n \left(1 + n\frac{b}{a}\right)$$

Other expansions:

For $\theta \ll 1$ rad,
$$\sin \theta \approx \theta - \frac{\theta^3}{6} \to \theta$$
$$\cos \theta \approx 1 - \frac{\theta^2}{2} \to 1$$

For $x \ll 1$,
$$\log_e (1 + x) \approx x$$
$$\log_{10} (1 + x) \approx 0.43x$$

No answers are given to the problems that follow. For most of them, you yourself will be the best judge. You may want to turn to an encyclopedia or other reference book to check some of your assumptions or conclusions. If you are not prepared at this point to tackle them all, don't worry; you can always return to them later.

1 What is the order of magnitude of the number of times that the earth has rotated on its axis since the solar system was formed?

2 During the average lifetime of a human being, how many heartbeats are there? How many breaths?

3 Make reasoned estimates of (a) the total number of ancestors you would have (ignoring inbreeding) since the beginning of the human race, and (b) the number of hairs on your head.

4 The present world population (human) is about 3×10^9.
  (a) How many square kilometers of land are there per person?

How many feet long is the side of a square of that area?

(b) If one assumes that the population has been doubling every 50 years throughout the existence of the human race, when did Adam and Eve start it all? If the doubling every 50 years were to continue, how long would it be before people were standing shoulder to shoulder over all the land area of the world?

5 Estimate the order of magnitude of the mass of (a) a speck of dust; (b) a grain of salt (or sugar, or sand); (c) a mouse; (d) an elephant; (e) the water corresponding to 1 in. of rainfall over 1 square mile; (f) a small hill, 500 ft high; and (g) Mount Everest.

6 Estimate the order of magnitude of the number of atoms in (a) a pin's head, (b) a human being, (c) the earth's atmosphere, and (d) the whole earth.

7 Estimate the fraction of the total mass of the earth that is now in the form of living things.

8 Estimate (a) the total volume of ocean water on the earth, and (b) the total mass of salt in all the oceans.

9 It is estimated that there are about 10^{80} protons in the (known) universe. If all these were lumped into a sphere so that they were just touching, what would the radius of the sphere be? Ignore the spaces left when spherical objects are packed and take the radius of a proton to be about 10^{-15} m.

10 The sun is losing mass (in the form of radiant energy) at the rate of about 4 million tons per second. What fraction of its mass has it lost during the lifetime of the solar system?

11 Estimate the time in minutes that it would take for a theatre audience of about 1000 people to use up 10% of the available oxygen if the building were sealed. The average adult absorbs about one sixth of the oxygen that he or she inhales at each breath.

12 Solar energy falls on the earth at the rate of about 2 cal/cm^2/min. Estimate the amount of power, in megawatts or horsepower, represented by the solar energy falling on an area of 100 square miles—about the area of a good-sized city. How would this compare with the total power requirements of such a city? (1 cal = 4.2 J; 1 W = 1 J/sec; 1 hp = 746 W.)

13 Starting from an estimate of the total mileage that an automobile tire will give before wearing out, estimate what thickness of rubber is worn off during one revolution of the wheel. Consider the possible physical significance of the result. (With acknowledgment to E. M. Rogers, *Physics for the Inquiring Mind*, Princeton University Press, Princeton, N.J., 1960.)

14 An inexpensive wristwatch is found to lose 2 min/day.
(a) What is its fractional deviation from the correct rate?

(b) By how much could the length of a ruler (nominally 1 ft long) differ from exactly 12 in. and still be fractionally as accurate as the watch?

15 The astronomer Tycho Brahe made observations on the angular positions of stars and planets by using a quadrant, with one peephole at its center of curvature and another peephole mounted on the arc. One such quadrant had a radius of about 2 m, and Tycho's measurements could usually be trusted to 1 minute of arc ($\frac{1}{60}°$). What diameter of peepholes would have been needed for him to attain this accuracy?

16 Jupiter has a mass about 300 times that of the earth, but its mean density is only about one fifth that of the earth.

(a) What radius would a planet of Jupiter's mass and earth's density have?

(b) What radius would a planet of earth's mass and Jupiter's density have?

17 Identical spheres of material are tightly packed in a given volume of space.

(a) Consider why one does not need to know the radius of the spheres, but only the density of the material, in order to calculate the total mass contained in the volume, provided that the linear dimensions of the volume are large compared to the radius of the individual spheres.

(b) Consider the possibility of packing more material if two sizes of spheres may be chosen and used.

(c) Show that the total surface area of the spheres of part (a) *does* depend on the radius of the spheres (an important consideration in the design of such things as filters, which absorb in proportion to the total exposed surface area within a given volume).

18 Calculate the ratio of surface area to volume for (a) a sphere of radius r, (b) a cube of edge a, and (c) a right circular cylinder of diameter and height both equal to d. For a given value of the volume, which of these shapes has the greatest surface area? The least surface area?

19 How many seconds of arc does the diameter of the earth subtend at the sun? At what distance from an observer should a football be placed to subtend an equal angle?

20 From the time the lower limb of the sun touches the horizon it takes approximately 2 min for the sun to disappear beneath the horizon.

(a) Approximately what angle (expressed both in degrees and in radians) does the diameter of the sun subtend at the earth?

(b) At what distance from your eye does a coin of about $\frac{3}{4}$-in. diameter (e.g., a dime or a nickel) just block out the disk of the sun?

(c) What solid angle (in steradians) does the sun subtend at the earth?

21 How many inches per mile does a terrestrial great circle (e.g., a meridian of longitude) deviate from a straight line?

22 A crude measure of the roughness of a nearly spherical surface could be defined by $\Delta r/r$, where Δr is the height or depth of local irregularities. Estimate this ratio for an orange, a ping-pong ball, and the earth.

23 What is the probability (expressed as 1 chance in 10^n) that a good-sized meteorite falling to earth would strike a man-made structure? A human?

24 Two students want to measure the speed of sound by the following procedure. One of them, positioned some distance away from the other, sets off a firecracker. The second student starts a stopwatch when he sees the flash and stops it when he hears the bang. The speed of sound in air is roughly 300 m/sec, and the students must admit the possibility of an error (of undetermined sign) of perhaps 0.3 sec in the elapsed time recorded. If they wish to keep the error in the measured speed of sound to within 5%, what is the minimum distance over which they can perform the experiment?

25 A right triangle has sides of length 5 m and 1 m adjoining the right angle. Calculate the length of the hypotenuse from the binomial expansion to two terms only, and estimate the fractional error in this approximate result.

26 The radius of a sphere is measured with an uncertainty of 1%. What is the percentage uncertainty in the volume?

27 Construct a piece of semilogarithmic graph paper by using the graduations on your slide rule to mark off the ordinates and a normal ruler to mark off the abscissa. On this piece of paper draw a graph of the function $y = 2^x$.

28 The subjective sensations of loudness or brightness have been judged to be approximately proportional to the logarithm of the intensity, so that equal *multiples* of intensity are associated with equal arithmetic increases in sensation. (For example, intensities proportional to 2, 4, 8, and 16 would correspond to equal increases in sensation.) In acoustics, this has led to the measurement of sound intensities in *decibels*. Taking as a reference value the intensity I_0 of the faintest audible sound, the decibel level of a sound of intensity I is defined by the equation

$$\text{dB} = 10 \log_{10}\left(\frac{I}{I_0}\right)$$

(a) An intolerable noise level is represented by about 120 dB. By what factor does the intensity of such a sound exceed the threshold intensity I_0?

(b) A similar logarithmic scale is used to describe the relative brightness of stars (as seen from the earth) in terms of *magnitudes*. Stars differing by "one magnitude" have a ratio of apparent brightness equal to about 2.5. Thus a "first-magnitude" (very bright) star is 2.5 times brighter than a second-magnitude star, $(2.5)^2$ times brighter than a third-magnitude star, and so on. (These differences are due largely to differences of distance.) The faintest stars detectable with the 200-in. Palomar telescope are of about the twenty-fourth magnitude. By what factor is the amount of light reaching us from such a star less than we receive from a first-magnitude star?

29 The universe appears to be undergoing a general expansion in which the galaxies are receding from us at speeds proportional to their distances. This is described by Hubble's law, $v = \alpha r$, where the constant α corresponds to v becoming equal to the speed of light, c ($= 3 \times 10^8$ m/sec), at $r \approx 10^{26}$ m. This would imply that the mean mass per unit volume in the universe is decreasing with time.

(a) Suppose that the universe is represented by a sphere of volume V at any instant. Show that the fractional increase of volume per unit time is given by

$$\frac{1}{V}\frac{dV}{dt} = 3\alpha$$

(b) Calculate the fractional decrease of mean density per second and per century.

30 The table lists the mean orbit radii of successive planets expressed in terms of the earth's orbit radius. The planets are numbered in order (n):

n	Planet	r/r_E
1	Mercury	0.39
2	Venus	0.72
3	Earth	1.00
4	Mars	1.52
5	Jupiter	5.20
6	Saturn	9.54
7	Uranus	19.2

(a) Make a graph in which $\log(r/r_E)$ is ordinate and the number n is abscissa. (Or, alternatively, plot values of r/r_E against n on semilogarithmic paper.) On this same graph, replot the points for Jupiter, Saturn, and Uranus at values of n increased by unity (i.e., at $n = 6$, 7, and 8). The points representing the seven planets can then be reasonably well fitted by a straight line.

(b) If $n = 5$ in the revised plot is taken to represent the asteroid belt between the orbits of Mars and Jupiter, what value of r/r_E would your graph imply for this? Compare with the actual mean radius of the asteroid belt.

(c) If $n = 9$ is taken to suggest an orbit radius for the next planet (Neptune) beyond Uranus, what value of r/r_E would your graph imply? Compare with the observed value.

(d) Consider whether, in the light of (b) and (c), your graph can be regarded as the expression of a physical law with predictive value. (As a matter of history, it *was* so used. See the account of the discovery of Neptune near the end of Chapter 8.)

PHILOSOPHIÆ
NATURALIS
PRINCIPIA
MATHEMATICA.

Autore *J S. NEWTON*, *Trin. Coll. Cantab. Soc.* Matheseos Professore *Lucasiano*, & Societatis Regalis Sodali.

IMPRIMATUR·
S. PEPYS, *Reg. Soc.* PRÆSES.
Julii 5. 1686.

LONDINI,

Jussu *Societatis Regiæ* ac Typis *Josephi Streater*. Prostat apud plures Bibliopolas. *Anno* MDCLXXXVII.

Facsimile of the title page of the first edition of Newton's Principia *(published 1687). It may be seen that the work was officially accepted by the Royal Society of London in July, 1686, when its president was the famous diarist Samuel Pepys (who was also Secretary to the Admiralty at the time).*

Part I

The approach to Newtonian dynamics

It seems probable to me, that God in the Beginning form'd Matter in solid, massy, hard, impenetrable, moveable Particles . . . .

NEWTON, *Opticks* (1730)

1
A universe of particles

THE PARTICULATE VIEW

THE ESSENCE of the Newtonian approach to mechanics is that the motion of a given object is analyzed in terms of the forces to which it is subjected by its environment. Thus from the very outset we are concerned with discrete objects of various kinds. A special interest attaches to objects that can be treated as if they are point masses; such objects are called particles.[1] In the strictest sense there is nothing in nature that fits this definition. Nevertheless, you have lived for years in a world of particles—electrons, atoms, baseballs, earth satellites, stars, galaxies—and have an excellent idea of what a particle is. If you have read George Orwell's famous political satire *Animal Farm*, you may remember the cynical proclamation: "All animals are equal, but some animals are more equal than others." In somewhat the same way, you may feel that some particles (electrons or protons, for example) are more particulate than others. But in any case the judgement as to whether something is a particle can only be made in terms of specific questions—specific kinds of experiments and observations.

And the answer to the question "Is such and such an object a particle?" is not a clear-cut yes or no, but "It depends." For example, atoms and atomic nuclei will look (i.e., behave) like

---

[1] Actually, Newton himself reserved the word "particle" to denote what we might now call "fundamental particles"—atoms and other such natural building blocks—but the usage has since changed.

Fig. 1-1 Photograph of a portion of the night sky. (Photograph from the Hale Observatories.)

particles if you don't hit them too hard. Planets and stars will look like particles (both visually and in behavior) if you get far enough away from them (see Fig. 1-1). But every one of these objects has spatial extension and an internal structure, and there will always be circumstances in which these features must be taken into account. Very often this will be done by picturing a given object not as a single point particle but as an assemblage of such ideal particles, more or less firmly connected to one another. (If the connections are sufficiently strong, it may be possible to make use of another fiction—the ideal "rigid body"—that further simplifies the analysis of rotational motions, in particular.) For the moment, however, we shall restrict ourselves to a consideration of objects that exist as recognizable, individual entities and behave, in appropriate circumstances, as particles in the idealized dynamical sense.

What sort of information do we need to build up a good description of a particle? Here are a few obvious items, which we write down without any suggestion that the list is exhaustive (or, for that matter, sharply categorized):

1. Mass
2. Size
3. Shape
4. Internal structure
5. Electric charge
6. Magnetic properties
7. Interaction with other particles of the same kind
8. Interaction with different sorts of particles

Partial though that list may be, it is already formidable, and it would not be realistic to tackle it all at once. So we ask a more modest question: What is the *smallest* number of properties that suffices to characterize a particle? If we are concerned with the so-called "elementary" particles (electrons, mesons, etc.), the state of *charge* (positive, negative, or neutral) is an important datum, along with the *mass*, and these two may be sufficient to identify such a particle in many circumstances. Most other objects, composed of large numbers of atoms, are normally electrically neutral, and in any event the mass alone is for many purposes the only property that counts in considering a particle's dynamic behavior—provided we take the forces acting on it as being independently specified.[1] It is, however, useful to know, at least approximately, the size also. Not only is this one of our most informative pieces of data concerning any object, but its magnitude will help to tell us whether, in given circumstances, the particle may reasonably be treated as a point mass.

Recognizing, then, that many of the finer details will have to be filled in later, we shall begin with a minimal description in which particles are objects possessing mass and size. Our survey is not exhaustive or detailed. On the contrary, we have sought

---

[1] Of course, if we want to treat the forces as being derived from characteristic interactions of the body with its surroundings (e.g., by gravitation), then the laws of interaction must also be known. That is the subject of Chapter 5.

to reduce it to a minimum, consistent with illustrating the general scheme of things, by considering only the masses and the linear dimensions of some typical particles. We shall begin with the smallest and least massive particles and go up the scale until we reach what appears to be a fundamental limit. You will appreciate that this account, brief though it is, draws upon the results of a tremendous amount of painstaking observation and research in diverse fields.

A note on units

In this book we shall most frequently employ the meter-kilogram-second (MKS) metric system. You are probably already familiar with it, at least for the basic measures of mass, length, and time. If not, you should learn it at this time. We shall, however, make occasional use of other measures. In mechanics the conversion from one system of measurement to another presents no problem, because it is just a matter of applying simple numerical factors. (This is in contrast to electromagnetism, where the particular choice of primary quantities affects the detailed formulation of the theory.) A tabulation of MKS and other units is given in the Appendix.

ELECTRONS AND NUCLEONS

The principal building blocks of matter from the standpoint of physics and chemistry are electrons, protons, and neutrons. Protons and neutrons are virtually equivalent as constituents of atomic nuclei and are lumped together under the generic title *nucleons*. The vast amount of research on the so-called elementary particles, and on the structure of nucleons, has not brought forth any evidence for particles notably smaller (or notably less massive) than those that were known to science 50 years ago. Thus, although the study of subatomic particles is a field of very great richness and complexity, filled with bizarre and previously unsuspected phenomena, the microscopic limits of the physical world are still well represented by such familiar particles as electrons and protons.

The electron, with a mass of about 10^{-30} kg (9.1×10^{-31} kg to be more precise), is by far the lightest (by more than three powers of 10) of the familiar constituents of matter. (The elusive neutrino, emitted in radioactive beta decay, appears to have no mass at all. This puts it in a rather special category!) The size of the electron is not something that can be unequivocally stated. Indeed, the concept of size is not sharply or uniquely defined for *any* object. If, however, we regard the electron as a sphere of electric charge, its radius can be estimated to be of the order of 10^{-15} m. In our present state of knowledge, the electron can properly be regarded as a fundamental particle, in the sense that there is no evidence that it can be analyzed into other constituents.

The nucleon, with a mass of 1.67×10^{-27} kg, is the other basic ingredient of atoms. In its electrically charged form—the proton—it is (like the electron) completely stable; that is, it survives indefinitely in isolation. In its electrically neutral form— the neutron—it cannot survive in isolation but decays radioactively (with a half-life of about 13 min) into a proton, an electron, and a neutrino. The fact that neutrons spontaneously give birth to the constituents of hydrogen atoms has led some cosmologists to suggest that neutrons represent the true primeval particles of the universe—but that is just a speculation. Nucleons have a diameter of about 3×10^{-15} m—by which we mean that the nuclear matter appears to be confined within a moderately well defined region of this size. Unlike electrons, nucleons seem to have a quite complex internal structure, in which various types of mesons are incorporated. But from the standpoint of atomic physics they can be regarded as primary particles.

ATOMIC NUCLEI

The combination of protons and neutrons to form nuclei provides the basis for the various forms of stable, ordinary matter as we know it. The smallest and lightest nucleus is of course the individual proton. The heaviest naturally occurring nucleus (that of ^{238}U)—contains 238 nucleons and has a mass of 4.0×10^{-25} kg. All nuclei have about the same mass per unit volume, so that their diameters are roughly proportional to the cube roots of the numbers of the nucleons. Thus nuclear diameters cover a range from about 3×10^{-15} to 2×10^{-14} m.

A unit of distance has been defined that is very convenient when dealing with nuclear dimensions. It is named after the

Italian physicist Enrico Fermi[1]:

$$1 \text{ fermi (F)} = 10^{-15} \text{ m} = 10^{-13} \text{ cm}$$

Thus the range of nuclear diameters is from about 3 to 20 F.

The density of nuclear matter is enormous. Given that the uranium nucleus has a mass of about 4×10^{-25} kg and a radius of about 10 F, you can deduce (do it!) that its density is about 10^{17} kg/m^3. This is so vast (it is larger, by a factor of 10^{14}, than the density of water) that we really cannot apprehend it, although we now have evidence that some astronomical objects (neutron stars) are composed of this nuclear matter in bulk.

ATOMS

A great deal was learned about atomic masses long before it was possible to count individual atoms. From the concepts of valence and chemical combinations, chemists established a relative mass scale based on assigning to hydrogen a mass of 1. The *mole* was introduced as that amount of any element or compound whose mass in grams was equal numerically to its relative mass on this scale. Furthermore, from the relative proportions of elements that combined to form compounds, it was known that a mole of any substance must contain the same unique number of atoms (or molecules in the case of compounds)—the number known as Avogadro's constant. But this number was itself unknown. Obviously, if the number could be determined, the mass of an individual atom could be found.

The existence of characteristic mass transfers in electrolysis gave corroborative evidence on relative atomic masses but also pointed the way to absolute mass determinations, for it seemed clear that the electrolytic phenomena stemmed from a characteristic atomic charge unit. All that was necessary was to establish the size of this unit (e)—a feat finally achieved in Millikan's precision measurements in 1909. Some representative atomic mass values are listed in Table 1–1.

---

[1]E. Fermi (1901–1954) was the greatest Italian physicist since Galileo and one of the most distinguished scientists of the twentieth century, gifted in both theoretical and experimental work. He achieved popular fame as the man who produced the first self-sustained nuclear chain reaction, at the University of Chicago in 1942.

TABLE 1-1: ATOMIC MASSES

Element	Electrolytic mass transfer, kg/C	Charge per ion	Approximate relative mass	Atomic mass, kg
H	1.04×10^{-8}	e	1	1.67×10^{-27}
C	6.22×10^{-8}	$2e$	12	2.00×10^{-26}
O	8.29×10^{-8}	$2e$	16	2.66×10^{-26}
Na	2.38×10^{-7}	e	23	3.81×10^{-26}
Al	9.32×10^{-8}	$3e$	27	4.48×10^{-26}
K	4.05×10^{-7}	e	39	6.49×10^{-26}
Zn	3.39×10^{-7}	$2e$	65	1.09×10^{-25}
Ag	1.118×10^{-6}	e	107	1.79×10^{-25}

Modern precision measurements of atomic masses are based on mass spectroscopy (see p. 206 for an account of the principles) and are quoted in terms of an *atomic mass unit* (amu). This is now defined as $\frac{1}{12}$ of the mass of the isotope carbon 12.

$$1 \text{ amu} = 1.66043 \times 10^{-27} \text{ kg}$$

Since almost all the mass of any atom is concentrated in its nucleus (99.95% for hydrogen, rising to 99.98% for uranium), we can say that to a first approximation the mass of an atom is just the mass of its nucleus. But, in terms of size, the atom represents a leap of many orders of magnitude. Nuclear diameters, as we have just seen, are of the order of 10^{-14} m. Atomic diameters are typically about 10^4 times larger than this—i.e., of the order of 10^{-10} m. One way of getting a feeling for what this factor means is to consider that if the dot on a printed letter *i* on this page is taken to represent a medium-weight nucleus, the outer boundary of the atom is about 10 ft away. Think of a grain of fine sand suspended in the middle of your bedroom or study, and you will get a feeling for what that means in three dimensions. (Nuclei are really *very* small.)

It is very convenient to take 10^{-10} m as a unit of distance in describing atomic sizes or interatomic distances in solids and other condensed states in which the atoms are closely packed. The unit is named after the nineteenth-century Swedish physicist, A. J. Ångstrom:

$$1 \text{ angstrom (Å)} = 10^{-10} \text{ m} = 10^{-8} \text{ cm} = 10^5 \text{ F}$$

It is noteworthy that the heaviest atoms are not markedly bigger than the lightest ones, although there are systematic variations,

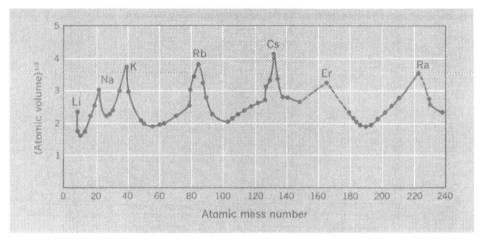

Fig. 1–2   Relative atomic radii (inferred from atomic volumes) versus atomic mass number, A.

with pronounced peaks at the alkali atoms, as one progresses through the periodic table of the elements (see Fig. 1–2).

Atoms are so small that it is hard to develop any real appreciation of the enormous numbers of atoms present in even the tiniest objects. For example, the smallest object that can be seen with a good microscope has a diameter of perhaps a few tenths of a micron and a mass of the order of 10^{-17} to 10^{-16} kg. This minuscule object nevertheless contains something like 1 billion atoms. Or (to take another example) a very good laboratory vacuum may contain residual gas at a pressure of a few times 10^{-11} of atmospheric. One cubic centimeter of such a vacuum would likewise contain about 1 billion atoms.

The atoms or molecules of a gas at normal atmospheric pressure are separated from one another, on the average, by about 10 times their diameter. This justifies (although only barely) the picture of a gas as a collection of particles that move independently of one another most of the time.

MOLECULES; LIVING CELLS

Our first introduction to molecules is likely to be in an elementary chemistry course, which very reasonably limits its attention to simple molecules made up of small numbers of atoms—H_2O, CO_2, Na_2SO_4, C_6H_6, and the like, with molecular weights of the order of 10 or 100 and with diameters of a few angstroms.

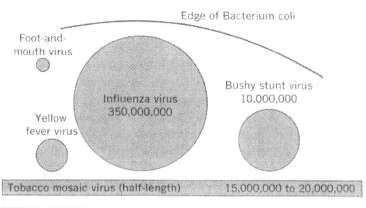

Fig. 1-3 Sizes of microscopic and submicroscopic objects, from bacteria down to molecules. (After J. A. V. Butler, Inside the Living Cell, George Allen & Unwin, London, 1959.)

These then, do not represent much of an advance, either in size or in mass, on the individual atoms we have just been discussing. But through the development of biochemistry and biophysics we have come to know of molecules of remarkable size and complexity. We can feel justified in regarding them as particles on the strength of such features as a unique molecular weight for all molecules of a given type. The biggest objects that are describable as single molecules have molecular weights of the order of 10^7 amu—hence masses of the order of 10^{-20} kg and lengths of the order of 10^{-7} m. Such objects are, however, far more important for their structure, and for their involvement in biological processes, than for their rather precarious status as particles. The particle dynamics of a protein molecule is a pretty slim subject—limited perhaps to the behavior of the molecule in a centrifuge—whereas the elucidation of its structure is a study that requires (and merits) the most intensive efforts of brilliant chemists and crystallographers. It would be both presumptuous and inappropriate to attempt to discuss such matters here, but it is perhaps worth indicating the range of magnitudes of such particles with the help of Fig. 1-3. A convenient unit of length for describing biological systems is the *micron:*

1 micron (μ) = 10^{-6} m = 10^1 Å

The largest object shown in Fig. 1-3 (a bacterium) is about 1μ across and would be visible in a good microscope (for which the limit of resolution is about 0.2μ—rather less than one wavelength of light).

Figure 1-3 includes some viruses, which have a peculiar status between living and nonliving—possessed of a rather definite size and mass, isolatable (perhaps as a crystalline substance), yet able to multiply in a suitable environment. Figure 1-4 is an electron-microscope photograph of some virus particles. These are almost the smallest particles of matter of which we can form a clear image in the ordinary photographic sense. (You have perhaps seen "photographs" of atomic arrangements as observed with the device called a field ion microscope. These are not direct images of individual atoms, although the pattern

Fig. 1-4 Spherical particles of polio virus. [C. E. Schwerdt et al., Proc. Soc. Exptl. Biol. Med., **86**, *310* (1954). Photograph courtesy of Robley C. Williams.]

does reveal their spatial relationships.)[1]

If we go one step further along this biological road, then of course we come to the living cell, which has the kind of significance for a biologist that the atom has for a physical scientist. Certainly it is appropriate to regard biological cells as particles, albeit of such a special kind that the study of most of them lies outside physics. They do, however, provide us with some convenient reference points on our scale of physical magnitudes, and that is our only reason for mentioning them here—except, perhaps, for the matter of reminding ourselves that biological systems also belong within a framework defined by the fundamental atomic interactions.

Although some single cells may be less than 1μ (certain bacteria) or more than 1 cm (e.g., the yolk of a hen's egg), the cells of most living organisms have diameters of the order of 10^{-5} m (10μ) on the average. Thus a human being, with a volume of about 0.1 m^3, contains about 10^{14} cells, each of which (on the average) contains about 10^{14} atoms.

SAND AND DUST

Vast areas of our earth are covered with particles that have come from the breaking down of massive rock formations. These particles, predominantly of quartz (crystalline SiO_2), are chemically very inert and are just the kind of objects to which the word "particle" applies in ordinary speech—small but visible, and inanimate. The earth's surface, and the atmosphere above that surface, are loaded with such particles. The biggest ones (say, of the order of 1 mm across) rest more or less firmly on the ground. Others, orders of magnitude smaller, may be seen as motes of dust dancing in the sunlight, apparently showing no tendency to fall. The fate of a given particle depends on the combined effect of wind (or air resistance) and gravity. Accumulations of windblown sand are found to be made up of particles from a maximum of about 1 mm diameter to a minimum of about 0.01 mm. Below that size the material tends to remain airborne, and the smallest dust particles are of the order of

---

[1] Improvements in electron- and proton-microscope technique reach further and further down into the world of small particles. In 1969 the helical structure of a protein molecule was photographed in an electron microscope.

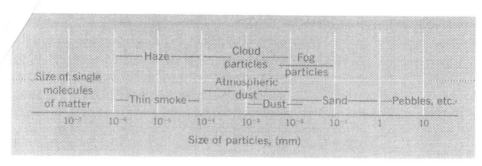

Fig. 1-5 Size distribution of particles found on or above the earth's surface. (After R. A. Bagnold, The Physics of Blown Sand and Desert Dunes, Methuen, London, 1941.)

10^{-4} mm ($=0.1\mu$) diameter. Figure 1-5 shows an approximate classification of particle sizes for dust, sand, and other small particles.

OTHER TERRESTRIAL OBJECTS

In a range from the smallest sand grains to the largest man-made objects we are in the realm of our most immediate experience: things that are large enough to be apprehended by the unaided sight or touch, yet not so big that we cannot achieve a rather direct awareness of them. In ordinary human terms, 100 is already a large number, and a factor of 1000 up or down in linear scale from human dimensions brings us close to the limits of anything that can properly be regarded as a full, first-hand contact with the physical world. Outside that domain we depend chiefly upon indirect evidence, imagination and analogy.

Since the densities of most solid materials (as we find them in the earth's crust) are of the order of magnitude of a few thousand kg/m^3, the range of diameters from a grain of sand (1 mm) to a cliff or a dam (1 km) implies a range of masses from about 1 mg to 10^{12} kg (i.e., 10^9 tons). Actually, if mass itself (or the weight associated with it under the gravitational conditions at the earth's surface) were to be our criterion, then the range we have just stated goes far beyond the span of human perceptions. In terms of weights, it would be fair to say that our direct experience gives us some feeling for objects as light as about 10 mg or as heavy as 1000 kg—i.e., up to a factor of $10^{\pm 4}$ with respect to a central value of the order of 0.1 kg (or a few ounces).

PLANETS AND SATELLITES

For an earthdweller going about his ordinary daily affairs, it is almost impossible—as well as unrealistic—to regard the earth as a particle. We cannot help but be aware, primarily, of the vastness of the earth and of the fact that man's greatest edifices are, at least in terms of physical size, totally insignificant modifications of its surface. Yet if we can imagine ourselves backing off into space for 100 million miles or so, we can arrive at a very different point of view. The earth loses its special status and is placed on the same footing as the other planets. And all of them can, on this scale, be regarded as *particles* moving in their orbits about a much more massive particle, the sun. In Kepler's mathematical description of the orbits of the planets, and subsequently in Newton's dynamical theory of these motions, the planets could legitimately be regarded as point particles—their extent and internal structure were in no way relevant. The reason, of course, was the simple one that on the scale of the solar system the planets *are* little more than mere points, as is obvious whenever we look up into the night sky. The earth's diameter is only about 10^{-4} of the distance between earth and sun—and the sun's own diameter is only 10^{-2} of the same distance. No wonder, then, that a good first approximation to the dynamics of the solar system can be obtained by taking such ratios to be zero.

But we should not rest content with such an approximation; indeed, we cannot. When Newton turned his attention from the planetary orbits to the tides, the physical extent of the earth became a key feature, because it was only this that made possible the existence of tide-producing forces, through a significant change of the moon's gravitational effect over a distance equal to the earth's diameter. In any case, once we free ourselves of the restriction of dealing with terrestrial objects and terrestrial phenomena, the particles that comprise the solar system are among the first to claim our attention.

The planets can be described as roughly spherical particles, somewhat flattened as a result of their own rotation. Their range of diameters is considerable—a factor of nearly 30 between Mercury and Jupiter. Since the mean densities do not differ really drastically (the densest is our earth, at 5.5 times the density of water), the masses cover a very great range indeed. Again the extremes are Mercury and Jupiter, and there is a factor of about 6000 between their masses. Figure 1-6 depicts these principal

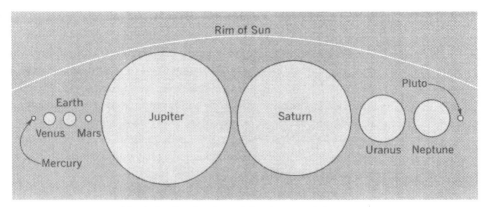

Fig. 1-6 *Relative sizes of the planets and the sun.*

planets in correct relative scale. The size of the sun, which has about the same density as Jupiter but 10 times the diameter and about 1000 times the mass, is indicated for comparison. The same data are presented in Table 1-2.

TABLE 1-2: DATA ON THE PLANETS

	Mean radius, km	R/R_E	M/M_E	Mean density, kg/m^3
Mercury	2.42×10^3	0.38	0.054	5400
Venus	6.10	0.96	0.815	5100
Earth	6.37	1.00	1.000	5520
(Moon)	(1.74)	(0.27)	(0.012)	(3360)
Mars	3.37	0.53	0.108	3970
Jupiter	69.9	10.98	317.8	1330
Saturn	58.5	9.20	95.2	680
Uranus	23.3	3.66	14.5	1600
Neptune	22.1	3.48	17.2	2250
Pluto	3.0	0.47	0.8?	?
(Sun)	(6.96×10^5)	(109.3)	(3.33×10^5)	(1400)

The nine major planets represent almost all the mass of matter around the sun (and Jupiter alone accounts for almost two thirds of it), but the number of other captive objects is enormous. There are, first, the natural satellites. If we ignore man's contributions, there are about 30 known satellites of the planets, most of them extremely tiny in comparison with the planets to which they are tied. (Our own moon is relatively the largest, but even it has only a little over 1% of the mass of the earth.) These satellites have a very special dynamical interest, because from their motions it has been possible to infer the masses of the planets themselves.

The planetary satellites are, however, vastly outnumbered by the minor planets, also called planetoids or asteroids.[1] Tens of thousands of them are orbiting the sun in the region between the orbits of Mars and Jupiter. In size they range from 500 miles in diameter down to 1 mile or less, and they most probably come from the breakup of a larger body. If the astronomer Walter Baade's description of them—"the vermin of the skies"—is at all representative, they are not greatly beloved by professional skywatchers.

STARS

A star is a magnificently complex structure, almost inconceivably gigantic by human standards and with a fascinating interior dynamics that involves nuclear reactions, gravitation, and electromagnetism. Yet when we gaze out into space we see nothing of this, unless we look with the eye—and the instruments—of an astrophysicist. Instead, we see the stars as luminous points, which (in contrast to the planets) continue to appear as point sources when examined through even the most powerful telescopes. In relation to their diameters, in fact, stars are much farther away from each other than are the planets of our solar system.

A convenient unit for specifying astronomical distances is the *light-year*, the distance light travels in a vacuum in 1 year:

$$1 \text{ light-year} = 9.46 \times 10^{15} \text{ m}$$

The *nearest* star to the sun is about 4 light-years away, or about 25,000,000,000,000 miles! A number of other stars that are near neighbors of the sun have mutual separations of the order of 10 light-years or 10^{17} m, which makes the ratio of diameter to separation about 10^{-7} or 10^{-8}. A cluster or galaxy of stars is thus an excellent example of a system of massive point particles, despite the fantastically large size of the particles. The best vacua attainable in the laboratory scarcely approach a corresponding emptiness as given by the ratio of interparticle spacing to particle diameter. One must look to the extreme vacua of interstellar space, where there are regions containing less than one hydrogen atom per cubic centimeter, to find a still emptier kind of system.

[1] The word "asteroid" derives from the star-like appearance of these tiny objects as seen through a telescope.

Most of the objects that are recognizable as stars are within the range of $10^{\pm 2}$ solar masses. In this sense our sun can be regarded as an average or typical star. In terms of size or luminosity (total radiation) the range of variation is very much greater, but a worthwhile account of these features would really call for some discussion of the evolution and interior mechanism of stars; and this is certainly not the place to attempt it.[1] We shall content ourselves, therefore, with remarking that the stars, regarded simply as aggregations of matter, with masses between about 10^{28} and 10^{32} kg (thus containing something of the order of 10^{54} to 10^{58} atoms) can still be regarded as particles when we discuss their motions through space, because of the immense distances that separate them from one another.

GALAXIES

In 1900 the words "galaxy" and "universe" were regarded as being synonymous. Our universe appeared to consist primarily of a huge number of stars—many billions of them—scattered through space (see Fig. 1-1). Here and there, however, could be seen cloudy objects—nebulae—near enough or big enough to have an observable extent and even structure, as contrasted with the pointlike appearance of the stars. By 1900 many thousands of nebulae were known and catalogued. But what were they? To quote the astronomer Allan Sandage: "No one knew before 1900. Very few people knew in 1920. All astronomers knew after 1924."[2] For it was in 1924 that the great astronomer Edwin Hubble produced the conclusive proof that the nebulae were, in the picturesque phrase, "island universes" far outside the region of space occupied by the Milky Way, and that our own Galaxy (distinguished with a capital G) was only one of innumerable systems of the same general kind. The first suggestion for such a picture of the universe was in fact put forward by the philosopher Immanuel Kant as long ago as 1755, but of course at that time it was no more than a pure hypothesis.

---

[1] For extensive discussion, see (for example) F. Hoyle, *Frontiers of Astronomy*, Harper & Row, New York, 1955.

[2] The story of this development is fascinatingly told by Sandage in his introduction to a beautiful book entitled *The Hubble Atlas of Galaxies*, published in 1961 by the Carnegie Institution of Washington, D.C. See also Hubble's own classic work, *The Realm of the Nebulae*, Dover Publications, New York, 1958.

Fig. 1-7 Cluster of galaxies in the constellation Corona Borealis. Distance about 600 million light-years. (Photograph from the Hale Observatories.)

To quote Sandage again: "Galaxies are the largest single aggregates of stars in the universe. They are to astronomy what atoms are to physics." And so far as we can tell at present, they represent the largest particles in the observable scheme of things.

A single galaxy may contain anywhere from about 10^6 to 10^{11} stars. Our own Galaxy appears to be one of the larger ones, with a diameter of about 10^{21} m (10^5 light-years). As we have already seen, the stars within an individual galaxy are very widely spaced indeed, so that the average density of matter in a galaxy is very, very low—only about 10^{-20} kg/m^3. But even so, galaxies represent notable concentrations of matter. On the average the spacing between galaxies is about 100 times their diameter, although there is a tendency for them to exist in clusters with separations perhaps 10 times less than this. Thus even the galaxies may, on an appropriately large scale, be regarded as particles, and the interactions between them may be approximated by treating them as points (see Fig. 1-7).

Astronomical surveys indicate that space contains a roughly uniform concentration of galaxies. If we assume that this continues to be true up to the theoretical limits of observation, we can make some estimate of the content of the universe as a whole. For it would appear that the universe can be represented by a sphere with a radius of about 10^{10} light-years (10^{26} m). The galaxies in general appear to be receding from us with speeds proportional to their distances, and at a distance of 10^{10} light-years the recessional speed would reach the speed of light. At this speed, because of the "galactic red shift" (a form of the wavelength and frequency shift for radiation from a moving source—called in general the Doppler effect), the transmission of energy back to us would fall to zero, thus setting a natural limit to the extent of the knowable universe.

If we take the average density of matter throughout space to be about 10^{-26} kg/m^3 (10^{-6} of the density within an individual galaxy) and assume the total volume of the universe to be of the order of $(10^{26})^3$, i.e., 10^{78} m^3, we arrive at a total mass of about 10^{52} kg (equivalent to about 10^{79} hydrogen atoms). This would then correspond to a total of about 10^{11} galaxies, each containing about 10^{11} stars. Such numbers are of course so stupendous that they defy any attempt to form a mental image of the universe as a whole. But it seems that we can be assured of one thing, at least, which is that the basis of our description of the physical world—what we described at the outset as "the particulate view"—finds some justification over the entire range of our experience, from the nucleus to the cosmos. And the fact that this approach to the description of nature makes sense, while embracing a span of about 10^{40} in distance and 10^{80} in mass, is not merely aesthetically pleasing; it also suggests that something, at least, of the physical description we give to the behavior of atoms will be found applicable to the behavior of galaxies, too. If we want to find a unifying theme for the study of nature, especially for the applications of classical mechanics, this particulate view is a strong candidate.

PROBLEMS

1–1 Make a tabulation of the orders of magnitude of diameter, volume, mass, and density for a wide selection of objects that you regard as being of physical importance—e.g., nucleus, cell, and star. For the diameters and masses, represent the data by labeled points on a straight line marked off logarithmically in successive powers of 10. This should give you a useful overview of the scale of the universe.

1–2 Estimate the number of atoms in:
    (a) The smallest speck of matter you can see with the naked eye.
    (b) The earth.

1-3 Calculate the approximate mass, in tons, of a teaspoonful of nuclear matter—closely packed nuclei or neutrons.

1-4 Sir James Jeans once suggested that each time any one of us draws a breath, there is a good chance that this lungful of air contains at least one molecule from the last breath of Julius Caesar. Make your own calculation to test this hypothesis.

1-5 A vacuum pump evacuates a bottle to a pressure of 10^{-6} mm of mercury (about 10^{-9} of atmospheric pressure). Estimate the magnitudes of the following quantities:
   (a) The number of molecules per cubic centimeter.
   (b) The average distance between molecules.

1-6 In a classic experiment, E. Rutherford and T. Royds [*Phil. Mag.*, **17**, 281 (1909)] showed that the alpha particles emitted in radioactive decay are the nuclei of normal helium atoms. They did this by collecting the gas resulting from the decay and measuring its spectrum. They started with a source of the radioactive gas radon (itself a decay product of radium). The half-life of radon is 3.8 days; i.e., out of any given number of atoms present at any instant, half are left, and half have decayed, 3.8 days later. When the experiment started, the rate of alpha-particle emission by the radon was about 5×10^9 per second. Six days later, enough helium gas had been collected to display a complete helium spectrum when an electric discharge was passed through the tube. What volume of helium gas, as measured at STP, was collected in this experiment? [The number of surviving radioactive atoms as a function of time is given by the exponential decay law, $N(t) = N_0 e^{-\lambda t}$. The number of disintegrations per unit time at any instant is $|dN/dt| = \lambda N_0 e^{-\lambda t}$. Given the value of the half-life, you can deduce the value of the constant, λ.]

1-7 The general expansion of the universe as described by Hubble's law [$v(r) = \alpha r$] implies that the average amount of mass per unit volume in the universe should be decreasing by about 1 part in 10^{17} per second (see Hors d'oeuvres no. 29, p. 16). According to one theory (no longer so strongly held) this loss is being made up by the continuous creation of matter in space.

   (a) Calculate the approximate number of hydrogen atoms per cubic meter per year that would, if produced throughout the volume of the universe, bring in the necessary amount of new mass.

   (b) It has been hypothesized (by J. G. King) that the creation of new matter, if it occurs, might well take place, not uniformly throughout space, but at a rate proportional to the amount of old matter present within a given volume. On this hypothesis, calculate the number of hydrogen atoms equivalent to the new mass created per day in a vessel containing 5 kg of mercury. (J. G. King concluded on this basis that a test of the hypothesis was feasible.) Convert your result to an equivalent volume of hydrogen gas as measured at a pressure of 1 Torr (1 mm of mercury).

1-8 Theodore Rosebury, in his book *Life on Man* (Viking, New York, 1969), remarks that the total bulk of all the microorganisms that live on the surface of a human being (excluding a far larger quantity on the inner surfaces of the intestinal system) could easily be accommodated in the bottom of a thimble. If the mean radius of these microorganisms is taken to be $5\,\mu$ ($1\,\mu = 10^{-6}$ m), what is the order of magnitude of the population of such organisms that each of us carries around all the time? Compare the result with the total human population of the globe.

1-9 It seems probable that the planets were formed by condensation from a nebula, surrounding the sun, whose outer diameter corresponded roughly to the orbit of Pluto.

(a) If this nebula were assumed to be in the form of a disk with a thickness equal to about one tenth of its radius, what would have been its mean density in kg/m^3? (Do not include the mass of the sun in the calculation.)

(b) A better picture appears to be of a gas cloud whose thickness increases roughly in proportion to the radius. Suppose that Jupiter were formed from a ring-shaped portion of the cloud, extending radially for half the distance from the asteroid belt to the orbit of Saturn and with a thickness equal to this radial extension. What would have been the mean density of this portion of the nebula?

(In terms of the radius of the earth's orbit, the orbit of Pluto has a radius of about 40 units, the asteroid belt is at about 3 units, and the orbit of Saturn has a radius of about 9.5 units.)

1-10 The core of a large globular cluster of stars may typically contain about 30,000 stars within a radius of about 5 light-years.

(a) Estimate the ratio of the mean separation of stars to the stellar diameter in such a core.

(b) At approximately what degree of vacuum would the same ratio of separation to particle diameter be obtained for the molecules of a gas?

I do not define time, space, place and motion, since they are well known to all.

NEWTON, *Principia* (1686)

2 Space, time, and motion

WHAT IS MOTION?

YOU ARE undoubtedly familiar with motion in all kinds of manifestations, but what would you say if you were asked to *define* it? The chances are that you would find yourself formulating a statement in which the phrase "a change of position with time," or something equivalent to that, expressed the central thought. For it seems that our ability to give any precise account of motion depends in an essential way on the use of the separate concepts of space and time. We say that an object is moving if it occupies different positions at different instants, and any stroboscopic photograph, such as that shown in Fig. 2–1, gives vivid expression to this mental picture.

All of us grow up to be good Newtonians in the sense that our intuitive ideas about space and time are closely in harmony with those of Newton himself. The following paragraphs are a deliberate attempt to express these ideas in simple terms. The description may appear natural and plausible, but it embodies many notions which, on closer scrutiny, will turn out to be naive, and difficult or impossible to defend. So the account below (set apart with square brackets to emphasize its provisional status) should not be accepted at its face value but should be read with a healthy touch of skepticism.

[*Space*, in Newton's view, is absolute, in the sense that it exists permanently and independently of whether there is any matter in the space or moving through it. To quote Newton's own words in the *Principia:* "Absolute space, in its own nature,

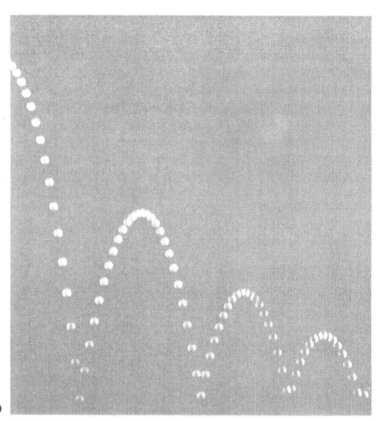

Fig. 2-1 Stroboscopic photograph of a motion. (From PSSC Physics, D. C. Heath, Lexington, Massachusetts, 1965.)

without relation to anything external, remains always similar and immovable."

[Space is thus a sort of stationary three-dimensional matrix into which one can place objects or through which objects can move without producing any interaction between the object and the space. Each object in the universe exists at a particular point in space and time. An object in motion undergoes a continuous change of its position with time. And although it would not be practicable, one can imagine the charting of positions with the help of a vast network of meter sticks, laid out end to end in a three-dimensional, cubical array throughout space. One can conceive of extending such measurements to any point in the universe. In other words, the space is there, and we simply have a practical task of attaching markers to it. Moreover, our physical measurements agree with the theorems of Euclidean geometry, and space is thus assumed to be Euclidean.

[*Time*, in Newton's view, is also absolute and flows on without regard to any physical object or event. Again quoting from the *Principia:* "Absolute, true, and mathematical time, of itself, and from its own nature, flows equably without relation to anything external, and by another name is called duration." The language is elegant but delightfully uninformative. As Newton said in the remark quoted at the beginning of this chapter, he did not attempt to *define* either space or time.

[One can neither speed up time nor slow down its rate, and this flow of time exists uniformly throughout the universe. If we imagine the instant "now" as it occurs simultaneously on every planet and star in the universe, and an hour later mark the end of this 60-minute interval, we assume that such a time interval has been identical for every object in the universe, as could (in principle) be verified by observations of physical, chemical, or biological processes at various locations. As an aid to measuring time intervals, it would be possible, in principle, to place identical clocks at each intersection of a meter-stick framework and to synchronize these clocks so that they indicate the same time at a common, simultaneous instant. Being identical clocks, they would thereafter correctly mark off the flow of absolute time and remain synchronized with each other.

[Space and time, although completely independent of each other, are in a sense interrelated insofar as we find it impossible to conceive of objects existing in space for no time at all, or existing for a finite time interval but "nowhere" in space. Both space and time are assumed to be infinitely divisible—to have no ultimate structure.]

The preceding five paragraphs describe in everyday language some commonsense notions about the nature of space and time. Embedded in these notions are many assumptions that we adopt, either knowingly or unconsciously, in developing our picture of the universe. It is fascinating therefore that, however intuitively correct they seem, many of these ideas have consequences that are inconsistent with experience. This first became apparent (as we mentioned in the Prologue) in connection with motions at very high speeds, approaching or equaling the speed of light, and with the phenomena of electromagnetism; and it was Einstein, in his development of special relativity theory, who exposed some of the most important limitations of classical ideas, including Newton's own ideas about relativity, and then showed how they needed to be modified, especially with regard to the concept of time.

The crux of the matter is that it is one thing to have abstract concepts of absolute space and time, and it is another thing to have a way of describing the actual motion of an object in terms of measured changes of position during measured intervals of time. Newton himself understood this very well, at least as far as spatial measurements were concerned. Thus in the *Principia* we find him remarking: "But because the parts of space cannot be seen, or distinguished from one another by our senses, therefore in their stead we use sensible [i.e., observable] measures of them . . . And so, instead of absolute places and motions, we use relative ones . . . For it may be that there is no body really at rest, to which the places and measures of others may be referred." If there is any knowledge to be gained about absolute space, it can only be by inference from these relative measurements. Thus our attention turns to the only basis we have for describing motion—observation of what a given object does in relation to other objects.

FRAMES OF REFERENCE

If you should hear somebody say "That car is moving," you would be quite certain that what is being described is a change of position of the car with respect to the earth's surface and any buildings and the like that may be nearby. Anybody who announced "There is relative motion between that car and the earth" would be rightly regarded as a tiresome pedant. But this does not alter the fact that it takes the pedantic statement to express the true content of the colloquial one. We accept the local surroundings—a collection of objects attached to the earth and therefore at rest relative to one another—as defining a *frame of reference* with respect to which the changes of position of other objects can be observed and measured.

It is clear that the choice of a particular frame of reference to which to refer the motion of an object is entirely a matter of taste and convenience, but it is often advantageous to use a reference frame in which the description of the motion is simplest. A ship, for example, is for many purposes a self-contained world within which the position or path of any person is most efficiently described in terms of three perpendicular axes based on the directions fore and aft, port and starboard, and up and down (according to deck number). The uniform motion of the ship itself with respect to a frame of reference attached to the earth may be unnoticed or even ignored by the passengers, as long as they can rely on the navigation officers.

Is the earth itself at rest? We would not say so. We have become accustomed to the fact that the earth, like the other planets, is continually changing its position with respect to a greater frame of reference represented by the stars. And since the stars constitute an almost completely unchanging array of

Fig. 2-2 Apparent circular motions of the stars. (Carillon Tower of Wellesley College, Wellesley, Mass., from J. C. Duncan, Astronomy, *5th edition, Harper & Row, 1954, p. 19.)*

reference points in the sky, we regard the totality of them as representing a fixed frame of reference, within which the earth both rotates and moves bodily. It remains true, however, that our primary data are only of *relative* positions and displacements; the belief that it makes more sense to assume that the earth is really rotating on its axis once every 24 hours, rather than that the system of stars is going around *us*, is one that cannot be justified by primary observations alone (see Fig. 2–2). With our present knowledge of the great masses and enormous numbers of the stars, it does, to be sure, simply seem more *reasonable* to attribute the motion to our puny earth. But it would be hard to elevate that subjective judgment into a physical law.

Later we shall see that there are powerful *theoretical* reasons for preferring some reference frames to others. The "best" choice of reference frame becomes ultimately a question of dynamics—i.e., dependent on the actual laws of motion and force. But the choice of a particular reference frame is often made without regard to the dynamics, and for the present we shall just concern ourselves with the purely kinematic problems of analyzing positions and motions with respect to any given frame.

COORDINATE SYSTEMS

A frame of reference, as we have said, is defined by some array of physical objects that remain at rest relative to one another. Within any such frame, we make measurements of position and displacement by setting up a *coordinate system* of some kind. In doing this we have a free choice of origin and of the kind of coordinate system that is best suited to the purpose at hand. Since the space of our experience has three dimensions, we must in general specify three separate quantities in order to fix uniquely the position of a point. However, most of the problems that we shall consider will be of motion confined to a single plane, so let us first consider the specification of positions and displacements in two dimensions only.

As you are doubtless well aware, the position of a point in a plane is most often designated with respect to two mutually perpendicular straight lines, which we call the x and y axes of a coordinate system, intersecting at an origin O. The position of the point P [Fig. 2–3(a)] relative to O is then described by a *position vector* $\mathbf{r}$, as shown, characterized by a specific length and

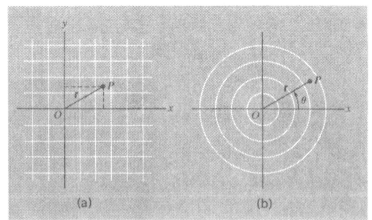

Fig. 2-3 (a) Square grid; the basis of Cartesian coordinates in a plane. (b) Plane polar coordinate grid.

a specific direction. Using our perpendicular axes, we can uniquely define **r** by the pair of rectangular (Cartesian) coordinates (x, y), which are projections of **r** onto the x and y axes, respectively. However, another important way of specifying **r** is in terms of polar coordinates (r, θ), as shown in Fig. 2-3(b). Here r is the distance of P from O and θ is the angle that **r** makes with the positive x axis, as measured in a counterclockwise (conventionally positive) direction. The two schemes of designating the position of P are related as follows:

$$\text{(In two dimensions)} \begin{cases} r^2 = x^2 + y^2 \\ \tan \theta = \dfrac{y}{x} \\ x = r \cos \theta \qquad y = r \sin \theta \end{cases} \tag{2-1}$$

Figure 2-4 shows examples of the use of these coordinate systems.

We shall on various occasions be making use of *unit vectors* that represent displacements of unit length along the basic coordinate directions. In the rectangular (Cartesian) system we shall denote the unit vectors in the x and y directions by **i** and **j**, respectively. The position vector **r** can be written as the sum of its two vector components:

$$\text{(In two dimensions)} \qquad \mathbf{r} = x\mathbf{i} + y\mathbf{j} \tag{2-2}$$

In the polar coordinate system, we shall use the symbol $\mathbf{e}_r$ to denote a unit vector in the direction of increasing r at constant θ and the symbol $\mathbf{e}_\theta$ to denote a unit vector at right angles to **r** in the direction of increasing θ. (The use of the symbol **e** for this purpose comes from the German word for unit, which is

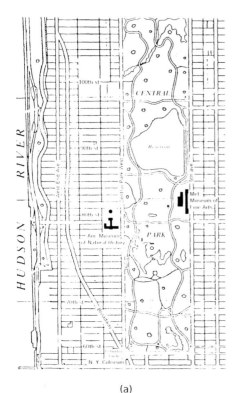
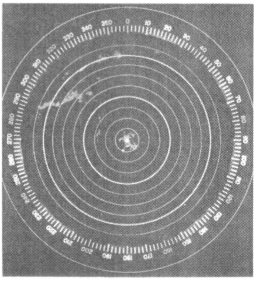

(a)   (b)

Fig. 2-4 (a) Example of Cartesian coordinates in use—a portion of midtown Manhattan, New York City. (b) Example of plane polar coordinates in use—a radar scope face with a few incipient thunderstorms (June 3, 1970). North is shown as 0° azimuth. The heavy circles of r = const are at 100 km, 200 km; the lighter circles are spaced by 25 km. (Photograph courtesy of Department of Meteorology, M.I.T.)

"Einheit.") In this polar-coordinate system, the vector **r** is simply equal to $r\mathbf{e}_r$, and one might wonder why the unit vector $\mathbf{e}_\theta$ is introduced at all. As we shall see, however, it becomes very important as soon as we consider motions rather than static displacements, for motions will often have a component perpendicular to **r**.

Although the above coordinate schemes are the most familiar ones—and are the only ones we shall be using in this book for two-dimensional problems—it is worth noting that *any* mapping of the surface that uniquely fixes the position of a point is a possible system. Figure 2-5 shows two examples—one a nonorthogonal system based on straight axes, and the other an orthogonal system based on two sets of intersecting curves. Such systems are introduced to capitalize on the kinds of symmetry that particular physical systems may possess.

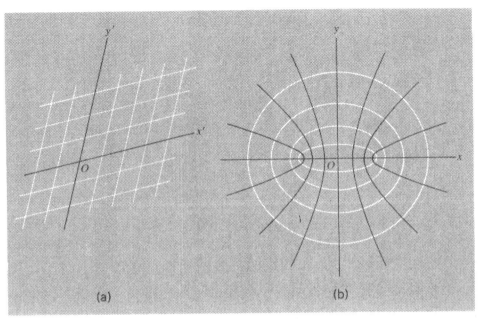

Fig. 2-5 (a) Oblique coordinate system, of the kind that is frequently used in Minkowski diagrams (space-time diagrams) in special relativity. (b) Orthogonal curvilinear coordinate system, made of intersecting sets of confocal ellipses and hyperbolas.

If it is necessary to specify all three of the spatial coordinates of a point, the most generally useful coordinate systems are the three-dimensional rectangular (Cartesian) coordinates (x, y, z), and spherical polar coordinates (r, θ, φ). These are both illustrated in Fig. 2–6(a). The Cartesian system is almost always chosen to be right-handed, by which we mean that the positive z direction is chosen so that, looking upward along it, the process of rotating from the positive x direction toward the positive y direction corresponds to that of a right-handed screw. It then follows that the cyclic permutations of this operation are also right-handed—from $+y$ to $+z$, looking along $+x$, and from $+z$ to $+x$, looking along $+y$. You may note that the two-dimensional coordinate system, as shown in Fig. 2–3(a), would on this convention be associated with a positive z axis sticking up toward you out of the plane of the paper. Introducing a unit vector **k** in the $+z$ direction, analogous to the **i** and **j** of Eq. (2-2), we can write the vector **r** as the sum of its three Cartesian vector components:

$$\mathbf{r} = x\mathbf{i} + y\mathbf{j} + z\mathbf{k} \qquad (2\text{-}3)$$

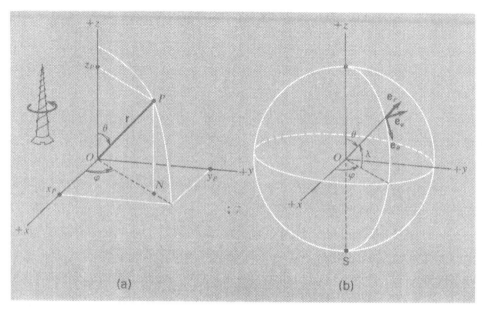

Fig. 2-6 (a) Coordinates of a point in three dimensions, showing both spherical polar and right-handed Cartesian coordinates. (b) Point located by angular coordinates (latitude and longitude) on a sphere, and the unit vectors of a local Cartesian coordinate system at the point in question.

The description of the position or displacement in spherical polar coordinates makes use of one distance and two angles. (Notice that three dimensions requires three independent coordinates, whatever particular form they may take.) The distance is, as with plane polar coordinates, the distance r from the chosen origin. One of the angles [the one shown as θ in Fig. 2-6(a)] is simply the angle between the vector **r** and the positive z axis; it is known as the polar angle. The other angle represents the angle between the zx plane and the plane defined by the z axis and **r**. It can be found by drawing a perpendicular PN from the end point P of **r** onto the xy plane and measuring the angle between the positive x axis and the projection ON. This angle (φ) is called the azimuth. The geometry of the figure shows that the rectangular and spherical polar coordinates are related as follows:

$$x = r \sin \theta \cos \varphi$$
$$y = r \sin \theta \sin \varphi \quad (2\text{-}4)$$
$$z = r \cos \theta$$

If we set $\theta = \pi/2$, we make $z = 0$ and so get back to the two-dimensional world of the xy plane. The first two equations of (2–4) then give us

$$x = r \cos \varphi$$
$$y = r \sin \varphi$$

It is very unfortunate that a long-established tradition uses the symbol θ, as we ourselves did earlier, to denote the angle between the vector **r** and the x axis in this special two-dimensional case. This need not become a cause for confusion, but one does need to be on the alert for the inconsistency of these conventions.

We have all grown up with one important use of spherical polar coordinates, the mapping of the earth's surface. This is indicated in Fig. 2–6(b). The longitude of a given point is just the angle φ, and the latitude is an angle, λ, equal to $\pi/2 - \theta$. (This entails calling north latitudes positive and south latitudes negative.) At any given point on the earth's surface a set of three mutually orthogonal unit vectors defines for us a local coordinate system; the unit vector $\mathbf{e}_r$ points vertically upward, the vector $\mathbf{e}_\theta$ points due south, parallel to the surface, and the third unit vector, $\mathbf{e}_\varphi$, points due east, also parallel to the surface. As with the plane polar coordinates, the vector **r** is given simply by $r\mathbf{e}_r$.

COMBINATION OF VECTOR DISPLACEMENTS

Suppose we were at a point P_1 on a flat horizontal plane [Fig. 2–7(a)] and wished to go to another point P_2. Imagine that we chose to make the trip by moving only east and north (represented by $+x$ and $+y$ in the figure). We know there are two particularly

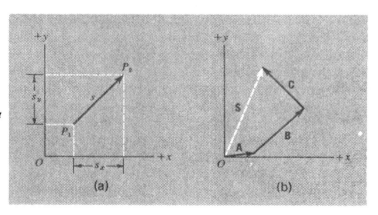

Fig. 2–7 (a) Successive displacements on a plane; the final position is independent of the order in which the displacements are made. (b) Addition of several displacement vectors in a plane.

*Fig. 2-8 Scalar multiples of a given vector **r**, including negative multiples.*

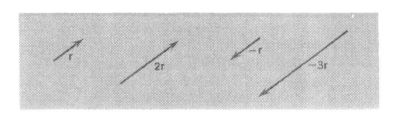

straightforward ways of doing this: (1) travel a certain distance s_x due east and then a certain distance s_y due north, or (2) travel s_y due north, followed by s_x due east. The order in which we take these two component displacements does not matter; we reach the same point P_2 in either case. Our representation of the vector **r** in Eq. (2-2) as the sum of the individual vectors $x\mathbf{i}$ and $y\mathbf{j}$ is an example of such a combination. This simple and familiar property of linear displacements exemplifies an essential feature of all those quantities we call vectors and is not confined to combinations at right angles. Thus, for example, in Fig. 2-7(b) we illustrate how three vector displacements, **A**, **B**, and **C**, placed head to tail, can be combined into a single vector displacement **S** drawn from the original starting point to the final end point. This is what we mean by *adding* the vectors **A**, **B**, and **C**. The order in which vectors are added is of no consequence: thus successive displacements of an object can be combined according to the vector addition law, without regard to the sequence in which the displacements are made. What we mean by a vector quantity, in general, is that it is a directed quantity obeying the same laws of combination as positional displacements.

We shall often be concerned with forming a numerical multiple of a given vector. A positive multiplier, n, means that we change the length of the vector by the factor n without changing its direction. The negative of a vector (multiplication by -1) is defined to mean a vector of equal magnitude but in the opposite direction, so that added to the original vector it gives zero. A negative multiplier, $-n$, then defines a vector reversed in direction and changed in length by the factor n. These operations are illustrated in Fig. 2-8.

Subtracting one vector from another is accomplished by noting that subtraction basically involves the addition of a negative quantity. Thus if vector **B** is to be subtracted from vector **A**, we form the vector $-\mathbf{B}$ and add it to **A**:

$$\mathbf{A} - \mathbf{B} = \mathbf{A} + (-\mathbf{B})$$

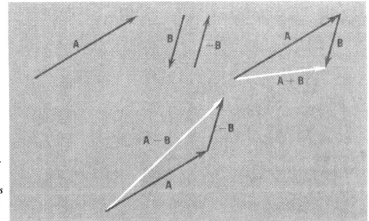

Fig. 2-9 Addition and subtraction of two given vectors. Note that the magnitude of the vector difference may be (as here) larger than the sum.

In Fig. 2-9 we show both the sum and the difference of two given vectors. We have deliberately chosen the directions of **A** and **B** to be such that the vector $\mathbf{A} - \mathbf{B}$ is *longer* than the vector $\mathbf{A} + \mathbf{B}$; this will help to emphasize the fact that vector combination is something rather different from simple arithmetical combination.

The evaluation of the vector distance from a point P_1 to a point P_2, when originally the positions of these points are given separately with respect to an origin O [see Fig. (2-10)], is a direct application of vector subtraction. The position of P_2 relative to P_1 is given by the vector $\mathbf{r}_{12}$ such that

$$\mathbf{r}_{12} = \mathbf{r}_2 - \mathbf{r}_1$$

(The subscript "12" is to be read as "one-two" and is a common notation in the description of two-particle systems.) Similarly, the position of P_1 relative to P_2 is given by the vector $\mathbf{r}_{21} = \mathbf{r}_1 - \mathbf{r}_2$. Clearly $\mathbf{r}_{21} = -\mathbf{r}_{12}$.

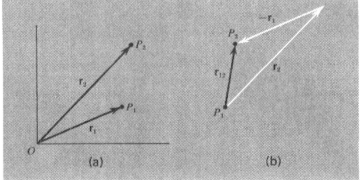

Fig. 2-10 Construction of the relative position vector of one point (P_2) with respect to another point (P_1).

Combination of vector displacements

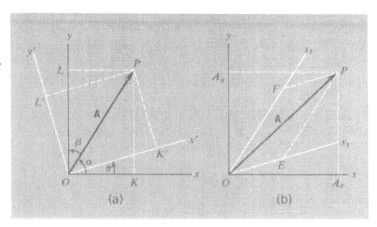

Fig. 2-11 (a) Components of a given vector in two different rectangular coordinate systems related by an angular displacement θ in the xy plane. (b) Components of a given vector in a rectangular coordinate system and in an oblique coordinate system.

THE RESOLUTION OF VECTORS[1]

In discussing the description of a given vector in terms of its components, we have indicated that this is a process that can be operated in either direction. There is the analysis (resolution) of the vector into its components in a given coordinate system, or there is the synthesis (addition) of the vector components to reconstitute the original vector. There is, however, an important difference between these two operations. The vector sum of the components is unique—it is the particular vector that we are considering—but the process of resolving the vector into components can be done in an infinity of ways, depending on the choice of coordinate system.

If we are using a coordinate system based on orthogonal axes (whether Cartesian or polar or anything else), the components of the vector are easily found by multiplying the length of the vector by the cosine of the angle that the vector makes with each of the coordinate axes in turn. Thus, for example, if we had a vector **A** confined to the xy plane [see Fig. 2-11(a)] its components in the xy coordinate system are given by

$$A_x = A \cos \alpha \qquad A_y = A \cos \beta$$

We know that $\beta = (\pi/2) - \alpha$, but by introducing the two separate angles we have a formalism that lends itself to being

---

[1] It is hoped that this section may be helpful in a general way, but the only feature that will be specifically needed later is the scalar product of two arbitrary vectors.

extended to the case of three dimensions, making use of the three separate angles α, β, and γ that the vector makes with the three axes. In this more general case there is no simple connection between the angles themselves, but we have the relationship

$$\cos^2\alpha + \cos^2\beta + \cos^2\gamma = 1$$

Our two-dimensional case corresponds to putting $\gamma = \pi/2$.

The total vector **A** in Fig. 2-11(a) can of course be written

$$\begin{aligned}\mathbf{A} &= A_x\mathbf{i} + A_y\mathbf{j} \\ &= (A\cos\alpha)\mathbf{i} + (A\cos\beta)\mathbf{j}\end{aligned}$$

A very convenient way of expressing such results is made possible by introducing what is called the *scalar product* of two vectors. This is defined in general in the following way: If the angle between any two vectors, **A** and **B**, is θ, then the scalar product, S, is equal to the product of the lengths of the two vectors and the cosine of the angle θ. This product is also called the *dot product* because it is conventionally written as $\mathbf{A} \cdot \mathbf{B}$. Thus we have

$$\text{scalar product } (S) = \mathbf{A} \cdot \mathbf{B} = AB\cos\theta$$

If for the vector **B** we now choose one or other of the unit vectors of an orthogonal coordinate system, the scalar product of **A** with the unit vector is just the component of **A** along the direction characterized by the unit vector:

$$A_x = \mathbf{A} \cdot \mathbf{i} \qquad A_y = \mathbf{A} \cdot \mathbf{j}$$

Thus the vector **A** can be written as follows:

$$\mathbf{A} = (\mathbf{A} \cdot \mathbf{i})\mathbf{i} + (\mathbf{A} \cdot \mathbf{j})\mathbf{j}$$

This result can, in fact, be developed directly from the basic statement that **A** can be written as a vector sum of components along x and y:

$$\mathbf{A} = A_x\mathbf{i} + A_y\mathbf{j}$$

Forming the scalar product of both sides of this equation with the unit vector **i**, we have

$$\mathbf{A} \cdot \mathbf{i} = A_x(\mathbf{i} \cdot \mathbf{i}) + A_y(\mathbf{j} \cdot \mathbf{i})$$

Now $(\mathbf{i} \cdot \mathbf{i}) = 1$ and $(\mathbf{j} \cdot \mathbf{i}) = 0$, because these vectors are all of unit length and the values of θ are 0 and $\pi/2$, respectively. Thus we have a more or less automatic procedure for selecting out and evaluating each component in turn.

If one were to take this no further, the above development would seem perhaps pointlessly complicated. Its value becomes more apparent if one is interested in relating the components of a given vector in different coordinate systems. Consider, for example, the second set of axes (x', y') shown in Fig. 2-11(a); they are obtained by a positive (counterclockwise) rotation from the original (x, y) system. The vector **A** then has two equally valid representations:

$$\mathbf{A} = A_x\mathbf{i} + A_y\mathbf{j} = A_x'\mathbf{i}' + A_y'\mathbf{j}'$$

If we want to find A_x' in terms of A_x and A_y, we just form the scalar product with $\mathbf{i}'$ throughout. This gives us

$$A_x' = A_x(\mathbf{i} \cdot \mathbf{i}') + A_y(\mathbf{j} \cdot \mathbf{i}')$$

Looking at Fig. 2-11(a), we see the following relationships:

$$\mathbf{i} \cdot \mathbf{i}' = \cos\theta \quad \mathbf{j} \cdot \mathbf{i}' = \cos\left(\frac{\pi}{2} - \theta\right) = \sin\theta$$

Hence

$$A_x' = A_x \cos\theta + A_y \sin\theta$$

Similarly,

$$A_y' = A_x(\mathbf{i} \cdot \mathbf{j}') + A_y(\mathbf{j} \cdot \mathbf{j}')$$
$$= A_x \cos\left(\frac{\pi}{2} + \theta\right) + A_y \cos\theta$$

Therefore,

$$A_y' = -A_x \sin\theta + A_y \cos\theta$$

This procedure avoids the need for tiresome and sometimes awkward considerations of geometrical projections of the vector **A** onto various axes of coordinates.

The same approach can be useful if a vector is to be resolved into nonorthogonal components. Consider, for example, the situation shown in Fig. 2-11(b). The axes are the lines Os_1 and Os_2, and the components of the vector **A** in this system are OE and OF. If we denote unit vectors along the coordinate directions by $\mathbf{e}_1$ and $\mathbf{e}_2$, we have

$$\mathbf{A} = s_1\mathbf{e}_1 + s_2\mathbf{e}_2 = A_x\mathbf{i} + A_y\mathbf{j}$$

If we form the scalar product throughout with, let us say, $\mathbf{e}_1$, we have

$$s_1 + s_2(\mathbf{e}_2 \cdot \mathbf{e}_1) = A_x(\mathbf{i} \cdot \mathbf{e}_1) + A_y(\mathbf{j} \cdot \mathbf{e}_1)$$

Given a knowledge of the angles between the various axes, this is a linear equation involving the two unknowns s_1 and s_2. A second equation can be obtained by forming the scalar product throughout with $\mathbf{e}_2$ instead of $\mathbf{e}_1$, and it then becomes possible to solve for s_1 and s_2 separately. It is important, in this case, to recognize that the base vectors $\mathbf{e}_1$ and $\mathbf{e}_2$ are no longer orthogonal, so that the scalar product $(\mathbf{e}_1 \cdot \mathbf{e}_2)$ does not vanish. The use of oblique coordinate systems of this kind is, however, rather special, and as a rule the resolution of a vector along the three independent directions of an orthogonal coordinate system is the reasonable and useful thing to do.

VECTOR ADDITION AND THE PROPERTIES OF SPACE[1]

You may be tempted to think that the basic law of vector addition, and the fact that the final result is independent of the order in which the combining vectors are taken, is more or less obvious. Let us therefore point out that it depends crucially on having our space obey the rules of Euclidean geometry. If we are dealing with displacements confined to a two-dimensional world as represented by a surface, it is essential that this surface be *flat*. This is not just a pedantic consideration, because one of our most important two-dimensional reference frames—the earth's surface—is curved. As long as displacements are small compared to the radius of the curvature, our surface is for practical purposes flat, our observations conform to Euclidean plane geometry, and all is well. But if the displacements along the surface are sufficiently great, this idealization cannot be used. For example, a displacement 1000 miles eastward from a point on the equator, followed by a displacement 1000 miles northward, does *not* bring one to the same place as two equivalent displacements (i.e., again on great circles intersecting at right angles) taken in

---

[1]This section can be omitted without loss of continuity.

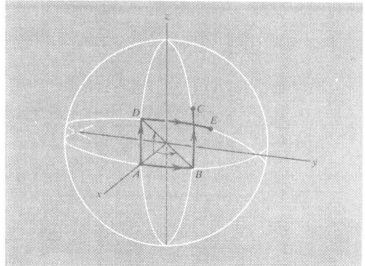

Fig. 2-12 *Successive displacements on a sphere are not commutative if the sizes of the displacements are not small compared to the radius of the sphere.*

the opposite order; it misses by about 40 miles! (See Fig. 2-12.) This means, in effect, that the correctness of vector addition is a matter for experiment and that tests for departures from it can be used to make deductions about the geometrical properties of the space in which we operate. For example, we could take the results of sensitive measurements on the difference between the two possible ways of making two successive displacements on a sphere and use the data to *deduce* the radius of the sphere.

When we acknowledge that the space of our ordinary experience is really three-dimensional, then, of course, we look at the foregoing analysis from a different point of view. We recognize that displacements on the surface of a sphere can actually be seen as displacements in a three-dimensional world that obeys Euclidean geometry rather than in a merely two-dimensional world that appears, within itself, to be non-Euclidean. But at this point a very interesting speculation suggests itself: Can we say that our space of three dimensions is rigorously Euclidean? Is it possible that the result of adding displacements along the three basic coordinate directions is dependent to some minute extent on the order of addition? If this were discovered to be the case, then we might proceed by analogy and introduce a fourth spatial dimension, associated with some characteristic radius of curvature, such that our non-Euclidean space of three dimensions could be described as Euclidean in a "hyperspace" of four dimensions.

Some distinguished scientists, beginning with the great Karl Friedrich Gauss ("the Prince of mathematicians")[1] have sought by direct observation to test the validity of Euclidean geometry, by measuring whether the angles of a closed triangle add up to exactly 180°. (For example, in the non-Euclidean space represented by the surface of a sphere, the angles of a triangle add up to more than 180°.) No departure from a Euclidean character for three-dimensional space has been detected through such observations. The concept that space may, however, be "curved," and that this curvature might be revealed if one could only carry out observations over sufficiently great distances, occupies an important place in theoretical cosmology.

You may have read about the curvature of space in another connection—Einstein's theory of gravitation—which describes local gravitational effects in terms of a modification of geometry in the space surrounding a massive object such as the sun. We shall not pursue this topic here, although we shall touch on it very briefly at the end of our account of gravitation in general (Chapter 8).

TIME

In the preceding sections we have developed the basic analysis of spatial displacements. To describe motion we must link such displacements to the time intervals during which they occur. Before considering this as a quantitative problem, let us very briefly supplement the remarks that we made at the beginning of the chapter concerning the actual nature of time. This is, of course, a huge subject that has engaged the thoughts and speculations of men—philosophers, scientists, and humanists alike—throughout history, and continues to do so. We shall not presume to do more than to examine one or two aspects of the problem from the standpoint of physical science.

The sense of the passage of time is deeply embedded in every one of us. We know, in some elemental sense, what time is.

---

[1]Karl Friedrich Gauss (1777–1855) was one of the outstanding mathematicians of all time. In the originality and range of his work he has never been surpassed, and probably never equaled. He delved deeply into astronomy and geodesy and was perhaps the first to recognize the possibility of a non-Euclidean geometry for space. See E. T. Bell's essay about him in *The World of Mathematics* (J. R. Newman, ed.), Simon and Schuster, New York, 1956.

But can we *say* what it is? The distinguished Dutch physicist H. A. Kramers once remarked: "My own pet notion is that in the world of human thought generally, and in physical science particularly, the most important and most fruitful concepts are those to which it is impossible to attach a well-defined meaning."[1] To nothing, perhaps, does this apply more cogently than to time. Nevertheless, if one tries to analyze the problem, one can perhaps begin to see that it is not entirely elusive. Even though a definition of time may be hard to come by, one can recognize that our concept of the passage of time is tied very directly to the fact that things *change*. In particular we are aware of certain *recurrent* events or situations—the beats of our pulse, the daily passage of the sun, the seasons, and so on. We almost subconsciously treat these as though they are markers on some continuous line that already exists—rather like milestones along a road. But all we have in terms of direct knowledge is the set of markers; the rest is an intellectual construction. Thus, although it may be valuable to have an abstract concept of time continuously flowing, our first-hand experience is only the observed behavior of a device called "a clock." In order to assign a quantitative measure to the duration of some process or the interval between two events, we simply associate the beginning and end points with readings on a clock. It is not essential that the clock make use of a recurrent phenomenon—one need only consider ancient devices such as water clocks and graduated candles, or modern parallels such as continuously weakening radioactive sources—but we do ask that it provide us with a means of marking off successive intervals in some recognizable way, and most such devices do, in fact, make use of repetitive phenomena of some kind.

How do we know that the successive time intervals defined by our chosen clock are truly equal? The fact is that we don't; it is ultimately a matter of faith. No clock is perfect, but we have learned to recognize that some clocks are better than others—better in the sense that the segments into which they divide our experience are more nearly equal. A doctor observes his wristwatch and tells us that our pulse is irregular; the wristwatch, however, is found to be itself irregular when compared

---

[1]*Physical Sciences and Human Values* (a symposium), Princeton University Press, Princeton, N.J., 1947.

more critically against a crystal-controlled oscillator; the oscillator wanders noticeably when checked against a clock based on atomic vibrations. Whether or not the uniform flow of time, as an ultimate abstraction, has any physical meaning, the remarkable fact is that we approach the measurement of time *as if* this steady flow existed. We evaluate the behavior of any given clock by observing its consistency and reproducibility, so that in using it we can quote its measure of a time or a time interval with a specified range of possible error. And then, when we proceed from individual measurements to general equations involving time, we introduce the symbol t and treat t as a continuous variable in the mathematical sense.

UNITS AND STANDARDS OF LENGTH AND TIME

Most of our discussion of motion will be in terms of unspecified positions and times, represented symbolically by **r**, t, and so on. It should never be forgotten, however, that the description of actual motions involves the numerical measures of such quantities and the use of universally accepted units and standards. Our choice of acceptable standards of both distance and time is the result of a continuing search for the highest degree of consistency and reproducibility in such measurements. The evolution and present state of this process is briefly summarized below.

Length

The current standard of length—the meter—was introduced, along with the rest of the metric system, in the drive for scientific and cultural order that developed in France in the latter half of the 18th century. The meter was originally intended to represent 1 ten-millionth (10^{-7}) of the distance from pole to equator of the earth along a meridian of longitude. But it proved impossible to construct any sufficiently precise standard on the basis of this definition, and the meter was then defined as the length of a particular metal bar kept in Sèvres, near Paris. Much later (in 1960) it was redefined as a specific multiple of the wavelength of a characteristic orange-red spectral line emitted by the isotope krypton-86.

This was not, however, the end of the story. It seemed at one time that a new definition of the meter in terms of wavelengths would be based on some chosen spectral line in the exceedingly sharp spectral lines emitted by lasers. But in 1983 a completely different approach was adopted. The speed of light in vacuum was accepted as a universal constant, and instead of being stated with a certain error, based on uncertainties of distance and time measurements, it was defined by international agreement to have the absolute value of 299,792,458 m/s, without any associated error. (The assignment of a completely defined numerical value to the speed of light was recognized as having significant advantages in astronomy and geodesy.) The meter is therefore now defined as the distance traveled by light in a vacuum in 1/299,792,458 seconds. This then rests on a definition of the second, based on atomic clocks, as described below.

Time

The process of defining a standard of time involves a feature that sets it significantly apart from the establishment of a standard of length. This is that, as Allen Astin has remarked: "We cannot choose a particular sample of time and keep it on hand for reference."[1] We depend upon identifying some recurring phenomenon and *assuming* that it always supplies us with time intervals of the same length.

The standard of time—the second—was originally based on the assumed constancy of the earth's rotation. It was first defined as being equal to 1/86,400 of a mean solar day—i.e., the average, over 1 year, of the time from noon to noon or midnight to midnight at a given place on the earth's surface. This is an awkward definition, because the length of the day, as measured from noon to noon, is not a constant; it varies because the earth's speed and its distance from the sun are continuously changing during one complete orbit. A logically more satisfactory definition of the second can be based on the *sidereal day*—the time for any given star to return to the same position overhead. If the earth's rotation were truly uniform, the length of every sidereal day would be the same.

In fact, it has gradually come to be recognized, thanks to

---

[1]Allen V. Astin, "Standards of Measurement," *Sci. Am.*, **218** (6), 50 (1968). Some people would, however, argue that even a solid, tangible bar as a length standard is equally vulnerable on philosophical and logical grounds.

the extraordinary precision of astronomical measurements extended over thousands of years, that neither the length of the solar year nor the earth's rate of rotation on its axis is exactly constant, the latter in particular being subject to minute but abrupt variations. The year has been found to be lengthening at the rate of about $\frac{1}{2}$ sec per century, so that in 1956 the second was redefined as being equal to 1/31,556,925.9747 of the tropical year 1900. (A tropical year is defined as the interval of time between two successive passages of the sun through the vernal equinox. We shall not attempt here to describe just how one "latches on" to a second of the year 1900 for calibration purposes.) It should be recognized that the variations being discussed here are fantastically small, as is implied by the ability to define the year in terms of the second to 12 significant figures.

Finally, in 1967, the use of atomic vibrations to specify a time standard was adopted by international agreement; it defines the second as corresponding to 9,192,631,770 cycles of vibration in an atomic clock controlled by one of the characteristic frequencies associated with atoms of the isotope cesium 133.

Quite apart from the practical challenge of defining units and standards of time with the maximum attainable precision, there are some questions of fundamental interest involved. Does time as defined by celestial motions, controlled by gravitation, keep step at all epochs with time as defined by atomic vibrations, controlled by electric forces within the atom? It was suggested in 1938 by P. A. M. Dirac[1] that the constant of universal gravitation might be slowly changing with time—with a "time constant" on the order of the age of the universe itself, i.e., about 10^{10} years. If this were true, our astronomical and atomic standards would, in the long run, be found to reveal discrepancies.

This matter of units and standards is one that most of us do not bother our heads with. We think we know well enough what is meant by a meter and a second; and a ruler or a watch is usually near at hand. But perhaps the above discussion may help to suggest that the detailed story of how these basic measures are defined, redefined, and made more and more precise is a quite fascinating business—especially, perhaps, for time, to

---

[1] Dirac, a British theoretical physicist, was one of the leaders in the development of quantum theory around 1926–1930. He was awarded the Nobel prize for this work.

which astronomical observations over the centuries have contributed data of a refinement that almost passes belief.[1]

SPACE-TIME GRAPHS

The primary data in the description of any motion will be a set of associated measures of position and time. Such data might, for example, be a tabulation in an astronomer's logbook, or a single stroboscopic photograph such as Fig. 2-1. In general the statement of position at any instant will require the use of three coordinates, corresponding to the three independent dimensions of space. In many circumstances, however, the motion may be confined to a plane, requiring two coordinates only, or to a single line, so that a single positional coordinate suffices. In this last case, and especially if the motion is along a *straight* line, it is often extremely convenient to display the motion in

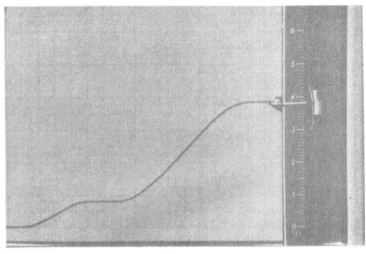

Fig. 2-13 Example of a space-time graph for a one-dimensional motion. (From the PSSC film, "Straight Line Kinematics," by E. M. Hafner, Education Development Center Film Studio, Newton, Mass., 1959.)

[1] For further reading see, for instance, *An Introduction to the Physics of Mass, Length and Time*, by N. Feather, Edinburgh University Press, Edinburgh, 1959. On the matter of time in particular, which probably holds the greatest interest, the following books and articles are recommended: J. T. Fraser (ed.), *The Voices of Time*, George Braziller, New York, 1966; T. Gold and D. L. Schumacher (eds.), *The Nature of Time*, Cornell University Press, Ithaca, N.Y., 1967; G. M. Clemence, "Standards of Time and Frequency," *Science*, **123**, 567 (1956); Lee Coe, "The Nature of Time," *Am. J. Phys.*, **37**, 810 (1969); Richard Schlegel, *Time and the Physical World*, Dover, New York, 1968.

terms of a space–time graph in which, as a rule, the time is regarded as an independent variable, plotted along the abscissa, and the position is plotted along the ordinate. Figure 2–13 shows such a graph. It has the great merit that it conveys directly, in a way that a numerical table cannot, a complete picture of a given motion. One can immediately identify points of maximum and minimum distance from the origin, regions of time in which the motion temporarily ceases altogether, and so on.[1]

VELOCITY

The central concept in the quantitative description of motion is that of *velocity*. It is a vector. The quantitative measure of velocity, in the case of one-dimensional motion, is one of the first pieces of information that we can extract from a space–time graph such as Fig. 2–13. Our way of designating velocities—miles per hour, meters per second, and so on—is a constant reminder of the fact that velocity is a *derived* quantity, based on these separate measures of space and time. Has it ever struck you that although physicists have invented names for the units of measurement of all sorts of physical quantities, they have never introduced special names for units of velocity? (Seafaring men have done it, though, with their unit the *knot*, equal to 1 nautical mile per hour.) In nature itself, things seem to be quite otherwise, for although we have not as yet identified anything that is directly recognizable as a fundamental natural unit of length or time, we do find a fundamental unit of velocity—the magnitude (c) of the velocity of light in empty space:

$$c = (2.997925 \pm 0.000001) \times 10^8 \text{ m/sec}$$

It has become customary in high-energy particle physics to express velocities as fractions of c. And in a comparable way, in connection with high-speed flight, the Mach number is used to express the speed of an aircraft as a fraction or multiple of the speed of sound in air. But this does not alter the fact that

---

[1]Such graphical representations of correlated quantities often provide a much more immediate and vivid insight into a situation than do numerical tabulations or algebraic formulas, and even a rough graph, sketched freehand, can be a great aid to thinking about a situation. A facility in drawing and interpreting graphs as expressions of physical relationships is well worth developing.

our basic description of velocities is in terms of the number of units of distance per unit of time.

The measurement of a velocity requires at least two measurements of the position of an object and the two corresponding measurements of time. Let us denote these measurements by $(\mathbf{r}_1, t_1)$ and $(\mathbf{r}_2, t_2)$. Using these we can deduce the magnitude and direction of what we can loosely call the average velocity between those points:

$$\mathbf{v}_{av} = \frac{\mathbf{r}_2 - \mathbf{r}_1}{t_2 - t_1}$$

However, this average velocity is not, in most cases, a very interesting quantity. Sometimes we may find that a graph of s versus t (let us assume a one-dimensional motion) is a straight line, so that the value of v deduced from any two pairs of values s and t is the same. But there is a much more basic and general problem: What can we do about defining and evaluating the velocity at an arbitrary instant in a nonuniform motion such as that represented by Fig. 2-13? The next section is devoted to this question, which is of fundamental importance to the whole of the mathematical analysis of motion.

INSTANTANEOUS VELOCITY

Richard P. Feynman tells the story of the lady who is caught for speeding at 60 miles per hour and says to the police officer: "That's impossible, sir, I was traveling for only seven minutes."[1] The lady's objection does not convince us (or the police officer); we understand that what is at issue is not the persistence of a uniform motion for a long time but the property of the motion as measured over a time interval that might be arbitrarily short. In order to talk about this in specific terms, imagine that alongside a straight section of a road in each direction from a chosen point, P, we have placed a set of equally spaced poles, say at

[1] This story, further embellished by Feynman with entertaining and instructive details, can be found in *The Feynman Lectures on Physics*, Vol. I (R. P. Feynman, R. B. Leighton, and M. Sands, eds.), Addison-Wesley, Reading, Mass., 1963. These lectures, of wonderful freshness and originality, range over the whole of physics and provide rich and exciting fare for everyone, whether a beginner or a veteran in the subject. Feynman, one of the most outstanding physicists of our time, was awarded the Nobel prize in 1966 for his fundamental contributions to quantum field theory.

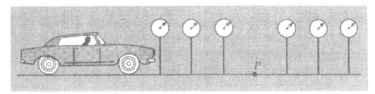

Fig. 2-14 Arrangement for inferring the instantaneous velocity of a car as it passes the point P.

5-meter intervals (see Fig. 2–14). On each pole is an electric clock with a sweep second hand. The clocks are synchronized and run continuously. A movie camera is placed opposite each pole, so as to photograph the pole, the clock face, and the road. A car comes along the road; the cameras are set running and the photographic records are developed. It is like the photofinish of a race. Each film will contain one frame in which, let us say, the front bumper of the car is more or less exactly in line with the pole. We now assemble the records. Take the data for the poles in pairs, equal distances before and after the central point P. For a given pair let the separation in distance be called Δs (a multiple of 10 m) and let the difference of time readings be called Δt. Then the ratio $\Delta s/\Delta t$ is the average velocity over the range of distance Δx centered on P. Now construct a graph, as shown in Fig. 2–15(a), of $\Delta s/\Delta t$ as a function of Δs. Unless the motion of the car has been extraordinarily erratic, the points can be fitted by a smooth curve that flattens out for the smaller values of Δs. We extrapolate it backward, and the value of $\Delta s/\Delta t$ for $\Delta s = 0$ is our measure of the instantaneous velocity at the central point P. We could equally well plot the values of $\Delta s/\Delta t$ against the time intervals Δt, as shown in Fig. 2–15(b); the value of $\Delta s/\Delta t$ at $\Delta t = 0$ is again the same, even though the

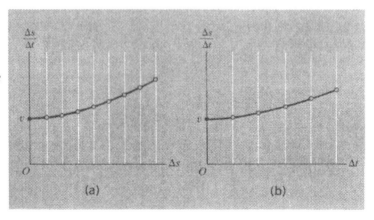

Fig. 2-15 (a) Evaluation of the instantaneous velocity ds/dt by extrapolation to $\Delta s = 0$ of a graph of $\Delta s/\Delta t$ against Δs. (b) Evaluation of the instantaneous velocity ds/dt by extrapolation to $\Delta t = 0$ of a graph of $\Delta s/\Delta t$ against Δt.

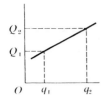

Fig. 2-16 The "slope" of a graph of one physical quantity against another is not primarily geometrical; it is defined by the ratio of the changes ΔQ and Δq, each measured in whatever units are appropriate.

graph itself looks somewhat different. And here, in physical terms, is what we mean by evaluating the limit of the average velocity, when the range of position or time over which $\Delta s/\Delta t$ is evaluated is shrunk to zero.

The process described above corresponds to the mathematical process of determining a derivative—in this case, of displacement with respect to time. Using the standard calculus notation, which in the context of the above discussion almost speaks for itself, we write

$$\text{instantaneous velocity } v = \lim_{\Delta t \to 0} \frac{\Delta s}{\Delta t} = \frac{ds}{dt} \quad (2\text{-}5)$$

The quantity ds/dt, a synonym for the limit of $\Delta s/\Delta t$, is, in mathematical parlance, the first derivative of s with respect to t. In geometrical terms, it represents the slope of (a tangent to) the graph of s versus t at a particular value of t.

[A note concerning the meaning of the word "slope" in graphs of physical data is in order. The graph—e.g., as in Fig. 2-16— is a display of the numerical measure of one physical quantity plotted against the numerical measure of another, using scales that are entirely arbitrary and dictated by convenience alone. Thus there is, in general, no physical significance to the inclination of such a line as measured by a protractor. By "slope" we mean simply the ratio of the change in the quantity represented on the ordinate to the corresponding change in the quantity represented on the abscissa—e.g., $(Q_2 - Q_1)/(q_2 - q_1)$. In this ratio, numerator and denominator are each a pure number times a unit, and the slope is then expressible as the quotient of these numbers, labeled with its own characteristic units—e.g., meters per second.]

If the same ideas as above are applied to changes with time of the displacement in space, we arrive at a general definition of instantaneous velocity as a vector:

$$\mathbf{v} = \lim_{\Delta t \to 0} \frac{\Delta \mathbf{r}}{\Delta t} = \frac{d\mathbf{r}}{dt} \quad (2\text{-}6)$$

In Fig. 2-17(a) we indicate what the evaluation of $\Delta \mathbf{r}$ entails. Clearly, if we display the path of an object as a continuous curve, as in Fig. 2-17(b), the instantaneous velocity vector is tangent

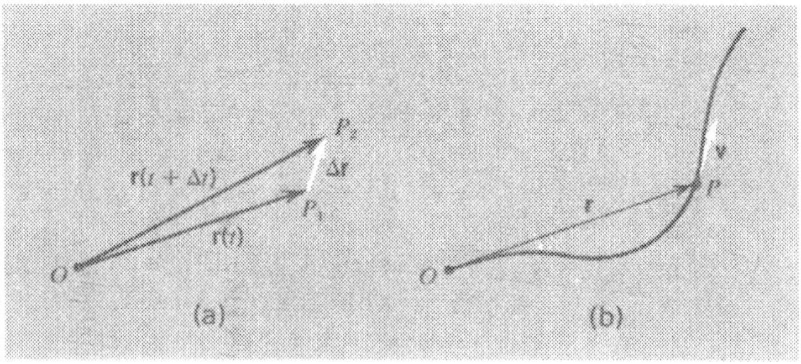

Fig. 2-17 (a) Vector diagram to define a small change of position. (b) Instantaneous velocity vector, tangent to the path.

to this curve; this is implicit in the definition.

With this departure from straight-line motion, we shall draw attention to the distinction that is made technically (but not always faithfully observed) between the words *velocity* and *speed*. The speed is the *magnitude* of the vector velocity **v**. The speed is thus, by definition, a positive, scalar quantity. The v that figures in our analysis of straight-line motion has, in fact, represented the velocity; it may take on negative as well as positive values, which of course is all the information needed to specify the direction of motion in the one-dimensional case.

Even if we are dealing with motion along a single straight line, the use of the vectorial description of position and velocity will be necessary if displacements are referred to an origin not

Fig. 2-18 (a) Vector positions in a straight-line motion, referred to an origin not on the line. (b) Corresponding diagram for a curvilinear path.

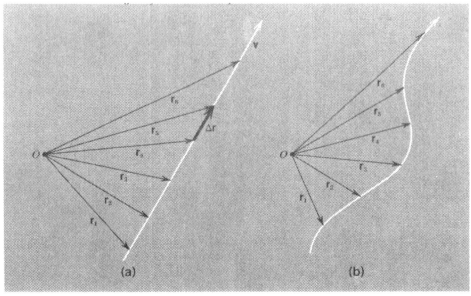

on the line itself. It is, in fact, a very instructive thing to consider a straight-line motion from this point of view [see Fig. 2-18(a)], because it helps us to see the one-dimensional situation in its larger context. We can become accustomed to the idea that the displacement vector $\Delta \mathbf{r}$ is, in general, in a different direction from the position vector $\mathbf{r}$, when the latter is referred to an arbitrary origin O. This then makes the transition from the description of a rectilinear path to the description of an arbitrary curved path [Fig. 2-18(b)] seem less abrupt. It also emphasizes the fact that although it may be very convenient, in the case of straight-line motion, to choose an origin on the line itself, this is certainly not necessary and may not always be possible.

RELATIVE VELOCITY AND RELATIVE MOTION

Since the vector velocity is the time derivative of the vector displacement, the velocity of one object relative to another is just the vector difference of the individual velocities. Thus if one object is at $\mathbf{r}_1$ and another object is at $\mathbf{r}_2$, the vector distance $\mathbf{R}$ from object 1 to object 2 is given by

$$\mathbf{R} = \mathbf{r}_2 - \mathbf{r}_1$$

The rate of change of $\mathbf{R}$ is then the velocity, $\mathbf{V}$, of object 2 relative to object 1, and we have

$$\mathbf{V} = \frac{d\mathbf{R}}{dt} = \frac{d\mathbf{r}_2}{dt} - \frac{d\mathbf{r}_1}{dt}$$

i.e.,

$$\mathbf{V} = \mathbf{v}_2 - \mathbf{v}_1 \tag{2-7}$$

This relative velocity $\mathbf{V}$ is the velocity of object 2 in a frame of reference attached to object 1.

In discussing frames of reference earlier in this chapter, we pointed out how the choice of some particular frame of reference may be advantageous because it gives us the clearest picture of what is going on. Nothing could illustrate this better than the practical problems of navigation and the avoidance of collisions at sea or in the air. Imagine, for example, two ships that at some instant are in the situation shown in Fig. 2-19(a). The vectors $\mathbf{v}_1$ and $\mathbf{v}_2$ represent their velocities (which we take to be constant) with respect to the body of water in which they both

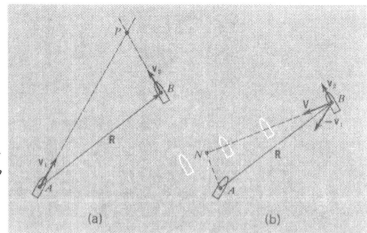

Fig. 2-19 (a) Paths of two ships moving at constant velocity along courses that intersect. (b) Path of ship B relative to ship A, showing that they do not collide even though their paths cross.

move. The paths of the ships, extended along the directions of motion from the initial points A and B, intersect at a point P. Will the ships collide, or will they pass one another at a safe distance? The answer to this question is not at all clear if we stick to the ocean frame, but if we describe things from the standpoint of one of the two ships the analysis becomes very straightforward. Let us imagine that we are standing on the deck of the ship marked A. Putting ourselves in that frame of reference means giving ourselves the velocity v_1 with respect to the water. But from *our* standpoint it is as if the water, and everything else, were given a velocity equal and opposite to v_1. Thus to every motion as observed in the ocean frame we add the vector $-v_1$, as implied by Eq. (2-7). This automatically, and by definition, brings A to rest, as it were, and shows us that the velocity of the ship B, relative to A, is obtained by combining the vectors v_2 and $-v_1$, as shown in Fig. 2-19(b). The vector *distance* between the ships is unaffected by this change of viewpoint. So now we can see the whole picture. B follows the straight line shown, as indicated by several successive positions in the diagram. It will miss A by the distance AN, the perpendicular distance from A to the line of V. The time at which this closest approach occurs is equal to the distance BN divided by the magnitude of V. Thus B seems to sweep across A's bow, more or less sideways. If you have had occasion to observe a close encounter of this sort, especially if it is out on the open water with no landmarks in sight, you will know that it can be a curious experience, quite disturbing to the intuitions, because the observed motion of the other ship seems to be unrelated to the direction in which it is pointing.

PLANETARY MOTIONS: PTOLEMY VERSUS COPERNICUS

Some of the most fascinating problems in the study of motion, and in particular of its relative character, have arisen in man's attempts to elucidate the motions of the heavenly bodies, including our own earth, through space. Observational astronomy, the first of the exact sciences, has yielded data of marvelous accuracy for several thousand years. But the question has always been how to interpret these data. Let us consider some of the main features of the problem.

The first thing to recognize is that naked-eye astronomy is,

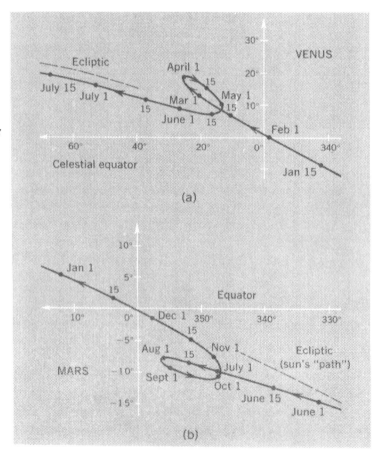

Fig. 2-20 (a) Path of Venus among the stars during a 6-month period, showing reversed (retrograde) motion at one stage.

(b) Similar set of observations on Mars. (Both diagrams after E. M. Rogers, Physics for the Inquiring Mind, Princeton University Press, Princeton, N.J., 1960.)

almost exclusively, the study of directions rather than distances. These unaided observations reveal nothing about the distances of the stars. The great Greek astronomers Aristarchus (third century B.C.) and Hipparchus (~150 B.C.) did make reasoned estimates of the distances of sun and moon (the latter very successfully),[1] but the only direct clues to the distances of the planets are through such quantitative evidence as changes of apparent brightness with time, suggesting that whatever the distances of the planets from the earth may be, they undergo systematic variations. Thus, as is still the practice, the positions of astronomical objects are defined in the first place in terms of their directions only (this being all we need to find them with a telescope) and can be described as if they were points on the surface of a sphere of large but arbitrary radius—the celestial sphere—with its center at the earth, and with a polar axis and an equator defined by the earth's own axis of rotation (cf. Fig. 2–2). In these terms the primary data on the motions of the planets are of the type shown in Fig. 2–20.

It is already a tribute to the genius of the early astronomers that they were able to visualize such strange-looking paths as the projections, on the celestial sphere, of orbital motions of various kinds. In particular, the belief took hold that the orbits must be combinations of circular motions. This is not the place to go into a detailed account of the problem; some outstandingly fine accounts exist elsewhere.[2] Instead, we shall simply focus on an idealized presentation of the two main models: an earth-centered (geocentric) or a sun-centered (heliocentric) solar system.

The most intuitively reasonable picture of the universe, in terms of everyday experience, is undoubtedly one that places the earth at the center of everything. Not one of us, without benefit of hindsight, could interpret his first impressions in any other way, and the ancient descriptions, such as the biblical one in Genesis, are entirely justifiable in these terms. It was the Alexandrian astronomer Ptolemy (~150 A.D.) who built this picture into a quantitative model of planetary motions and

---

[1] See the problems at the end of this chapter, and also Chapter 8.
[2] See, for example, E. M. Rogers, *Physics for the Inquiring Mind*, Princeton University Press, Princeton, N.J., 1960, Chaps. 12–18, or E. C. Kemble, *Physical Science, Its Structure and Development*, MIT Press, Cambridge, Mass., 1966, Chaps. 1–5; or any of a number of excellent books on elementary astronomy.

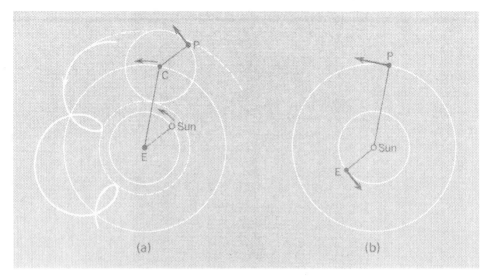

Fig. 2-21 (a) Apparent motion of a planet as explained in the Ptolemaic system. The planet P moves on the epicycle whose center C follows a circular path around the earth, E. (b) Copernican explanation of observed motion. The epicycle of (a) is seen as being a reflection of the earth's own motion around the sun.

described it in his great work, the *Almagest*. Figure 2-21(a) illustrates the essential features. Setting aside the effects of the earth's daily rotation, the motion of the sun is an approximately circular path, with a period of 1 year, around the earth, E, as center.[1] The motion of a planet, however, is compound. It can be fairly closely approximated by imagining that a point C travels uniformly around a circular path, and that the planet, P, travels in another circular path with respect to C as center. This extra circle is called an *epicycle;* the combination of these two motions, if they are in the same plane, gives rise to a complicated path that can have backward loops as shown. Precisely the same result would be obtained if we interchanged the roles of the two circles. (Verify this.)

If we now imagine viewing this motion from the earth, and projecting it onto the celestial sphere or to any other constant distance, we obtain almost the kind of variation of angular position that is shown in Fig. 2-20. We can come even closer by tilting the plane of the epicycle out of the plane of the primary

[1] This path, which carries the sun eastward through the constellations around the celestial sphere, is known as the *ecliptic*.

circle a little. To fit the motion of a particular planet it is necessary to choose appropriate values for the ratio of the radii of the two circles and also for the times to make one complete circuit of each. It is a noteworthy fact that in two cases (Mercury and Venus) the period of the primary circle is exactly 1 sidereal year (i.e., the period for one complete orbit of the sun around the ecliptic), and in the other three cases (Mars, Jupiter, and Saturn) the period of the epicycle is 1 sidereal year. Indeed, this can be taken as the crucial clue to what we now regard as the truer picture.

Suppose that we now place the sun at the center, as in Fig. 2-21(b), and make the earth travel around a circle that has the same radius as the epicycle in Fig. 2-21(a). Then if the planet, P, travels around another simple circle, of radius equal to that of the primary circle in Fig. 2-21(a), the *relative* positions and velocities of E and P can be made precisely the same as before. We have drawn the positions of E and P in the two diagrams to display this exact correspondence. The reason for the appearance of the sidereal year in one or other of the two component motions of a planet in the Ptolemaic model is now very clear, and much of the arbitrariness of the whole description disappears. It is this new, heliocentric, description with which we associate the name of Copernicus. He presented it in great detail in his principal work, *De Revolutionibus Orbium Celestium* ("On the Revolutions of the Celestial Spheres") published in 1543, the year of his death. Actually, the first suggestion that the sun occupied the central position had been made about 1800 years earlier by Aristarchus, who from other observations knew that the sun was much bigger than the earth (although he seriously underestimated just how huge and distant it is). There is no record, however, that the heliocentric theory was developed in quantitative detail before Copernicus. It is interesting, by the way, to note his clear understanding of the relativity of motion. Here is a translation of his own statement of the principle: "For all change in position which is seen is due to a motion either of the observer or of the thing looked at, or to changes in the position of both, provided that these are different. For when things are moved equally relatively to the same things, no motion is perceived, as between the object seen and the observer."[1]

[1]From Arthur Berry, *A Short History of Astronomy*, Dover Publications, New York, 1961.

Given the persistent intrusion of the sidereal year into the Ptolemaic scheme, it may seem surprising that the heliocentric picture did not prevail at a much earlier stage, especially when its possibility had been recognized by Aristarchus about 400 years before Ptolemy's day. It must be remembered, however, that we have been presenting a greatly oversimplified model of the solar system, and the ancient astronomers were legitimately worried over discrepancies between these idealized models and the precise, hard facts of observation. Both Ptolemy and Copernicus were driven to introduce numerous auxiliary circular motions to obtain even approximate agreement between theory and observation. The Copernican scheme, in the form in which Copernicus himself developed it, was not in fact notably less arbitrary or less complex than the Ptolemaic. Not until the theory could free itself of the circle as the basis of all celestial motion was a fundamental solution to be finally attained, although the introduction of dynamical considerations—the laws of force and the laws of motion—transformed the context within which the observations were interpreted. We shall come back to these questions in Chapter 8 and, more fully, in Chapter 13.

PROBLEMS

2–1 Starting from a point that can be taken as the origin, a ship travels 30 miles northeast in a straight line, and then 40 miles on a course that heads SSW (a direction making a counterclockwise angle of $247\frac{1}{2}°$ with a reference line drawn eastward). Find the x and y coordinates of its final position (x eastward, y northward) and its distance from the starting point.

2–2 The scalar (dot) product of two vectors, $\mathbf{A} \cdot \mathbf{B}$, is equal to $AB \cos \theta_{AB}$, where θ_{AB} is the angle between the vectors.

(a) By expressing the vectors in terms of their Cartesian components, show that

$$\cos \theta_{AB} = \frac{A_x B_x + A_y B_y + A_z B_z}{AB}$$

(b) By using the relation between rectangular and spherical polar coordinates [Eq. (2–4)], show that the angle θ_{12} between the radii to two points (R, θ_1, φ_1) and (R, θ_2, φ_2) on a sphere is given by

$$\cos \theta_{12} = \cos \theta_1 \cos \theta_2 + \sin \theta_1 \sin \theta_2 \cos (\varphi_2 - \varphi_1)$$

(Note that the distance between the two points as measured along the great circle that passes through them is equal to $R\theta_{12}$, where θ_{12} is expressed in radians. This can be used, for example, to calculate mileages between points on the earth's surface.)

2-3 (a) Calculate the Cartesian coordinates of New York, U.S.A. (41° N, 74° W) and Sydney, Australia (34° S, 151° E). Take an origin of coordinates at the earth's center, with a z axis through the north pole and an x axis passing through the equator at the zero of longitude. The earth's mean radius is 6370 km.

(b) Find the distance along an imaginary straight tunnel bored through the earth between New York and Sydney.

(c) Compare the result of (b) with the shortest practicable route between these points by a great-circle flight. You can either calculate this, using the result of Problem 2-2(b), or measure it directly on a globe with the help of a piece of string.

2-4 (a) Starting from a point on the equator of a sphere of radius R, a particle travels through an angle α eastward and then through an angle β along a great circle toward the north pole. If the initial position of the point is taken to correspond to $x = R$, $y = 0$, $z = 0$, show that its final coordinates are $R \cos \alpha \cos \beta$, $R \sin \alpha \cos \beta$, and $R \sin \beta$. Verify that $x^2 + y^2 + z^2 = R^2$.

(b) Find the coordinates of the final position of the same particle if it first travels through an angle α northward, then changes course by 90° and travels through an angle β along a great circle that starts out eastward.

(c) Show that the straight-line distance Δs between the end points of the displacements in (a) and (b) is given by
$$\Delta s^2 = 2R^2(\sin \beta - \sin \alpha \cos \beta)^2$$

(d) Using the above result, check the statement in the text (p. 60) that there is a difference of about 40 miles in the end points of (1) a displacement of 1000 miles eastward on the earth along the equator, followed by a displacement 1000 miles north, and (2) a displacement of 1000 miles north, followed by a displacement of 1000 miles starting out eastward on another great circle. (Put $R\alpha = R\beta = 1000$ miles. The approximation $\cos \theta \approx 1 - \theta^2/2$ will be found useful.)

2-5 If you found yourself transported to an unfamiliar planet, what methods could you suggest

(a) To verify that the planet is spherical?

(b) To find the value of its radius?

2-6 The radius of the earth was found more than 2000 years ago by Eratosthenes through a brilliant piece of analysis. He lived at Alexandria, at the mouth of the Nile, and observed that on midsummer day at noon, the sun's rays were at 7.2° to the vertical (see the figure). He also knew that the people living at a place 500 miles south of

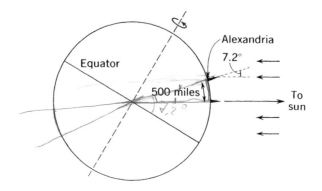

Alexandria saw the sun as being directly overhead at the same date and time. From this information, Eratosthenes deduced the value of the earth's radius. What was his answer?

2-7 It has been suggested that a fundamental unit of length is represented by a distance about equal to a nucleon diameter, and that a fundamental unit of time is represented by the time it would take a light signal (i.e., the fastest kind of signal achievable) to travel across a nucleon diameter. Express the radius of the universe and the age of the universe in terms of these units, and ponder the results.

2-8 A particle is confined to motion along the x axis between reflecting walls at $x = 0$ and $x = a$. Between these two limits it moves freely at constant velocity. Construct a space–time graph of its motion

(a) If the walls are perfectly reflecting, so that upon reaching either wall the particle's velocity changes sign but not magnitude.

(b) If upon each reflection the magnitude of the velocity is reduced by a factor f (i.e., $v_2 = -fv_1$).

2-9 A particle that starts at $x = 0$ at $t = 0$ with velocity $+v$ (along x) collides with an identical particle that starts at $x = x_0$ at $t = 0$ with velocity $-v/2$. Construct a space–time graph of the motion before and after collision

(a) For the case that the particles collide elastically, exchanging velocities.

(b) For the case that the particles stick together upon impact.

2-10 A particle moves along the curve $y = Ax^2$ such that its x position is given by $x = Bt$.

(a) Express the vector position of the particle in the form $\mathbf{r}(t) = x\mathbf{i} + y\mathbf{j}$.

(b) Calculate the speed v ($= ds/dt$) of the particle along this path at an arbitrary instant t.

2-11 The refraction of light may be understood by purely kinematic considerations. We need to assume that light takes the shortest

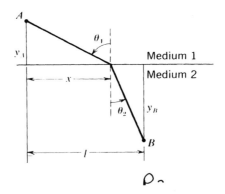

(timewise) path between two points (Fermat's principle of least time). Referring to the figure, let the speed of light in medium 1 be v_1 and in medium 2, v_2. Calculate the time it takes light to go from point A to point B as a function of the variable x. Minimize with respect to x. Given that

$$v_1 = c/n_1 \quad \text{and} \quad v_2 = c/n_2$$

where the n's are known as indices of refraction, prove Snell's law of refraction:

$$n_1 \sin \theta_1 = n_2 \sin \theta_2$$

2–12 At 12:00 hours ship A is 10 km east and 20 km north of a certain port. It is steaming at 40 km/hr in a direction 30° east of north. At the same time ship B is 50 km east and 40 km north of the port, and is steaming at 20 km/hr in a direction 30° west of north.

(a) Draw a diagram of this situation, and find the velocity of B relative to A.

(b) If the ships continue to move with the above velocities, what is their closest distance to one another and when does it occur?

2–13 The distance from A to B is l. A plane flies a straight course from A to B and back again with a constant speed V relative to the air. Calculate the total time taken for this round trip if a wind of speed v is blowing in the following directions:

(a) Along the line from A to B.
(b) Perpendicular to this line.
(c) At an angle θ to this line.

Show that the time of the round trip is always increased by the existence of the wind.

2–14 A ship is steaming parallel to a straight coastline, distance D offshore, at speed V. A coastguard cutter, whose speed is v ($<V$) sets out from a port to intercept the ship.

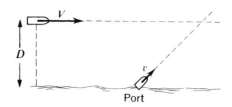

(a) Show that the cutter must start out before the ship passes a point a distance $D(V^2 - v^2)^{1/2}/v$ back along the coast. (*Hint:* Draw a vector diagram to show the velocity of the cutter as seen from the ship.)

(b) If the cutter starts out at the latest possible moment, where and when does it reach the ship?

Bulge rotates to keep pace with moon

Earth's daily rotation

2-15 With respect to the "fixed stars," the earth rotates once on its axis in one sidereal day—that is how the sidereal day is defined.

(a) The length of the year is about 366 sidereal days. By what amount is the mean *solar* day (from noon to noon) longer than the sidereal day?

(b) The moon completes one orbit with respect to the stars in 27.3 *sidereal* days. That is, in this time the line from earth to moon turns through 360° with respect to the stars. The time between corresponding high tides on successive days is longer than 1 solar day (24 hr) because of this motion of the moon. (The high tide is an ocean bulge at a fixed direction with respect to the moon—see the figure.) Show that the daily lag is close to 50.5 min (60 min = $\frac{1}{24}$ solar day).

2-16 (a) The orbital radii of Venus and Mars are 0.72 and 1.52 times the radius of the earth's orbit. Their periods are about 0.62 and 1.88 times the earth's year. Using these data, construct diagrams by which to find how the apparent angular positions of Venus and Mars change with time as seen from the earth, assuming that the orbits of all three planets lie in the same plane. Compare your results with Fig. 2–20.

(b) With respect to the ecliptic (the plane of the earth's own orbit) the planes of the orbits of Venus and Mars are tilted by about 3.5° and 2°, respectively. Consider how the apparent paths of Venus and Mars are affected by this additional feature.

2-17 (a) What methods can you suggest for finding the distance from the earth to the moon (without using radar or space flight)?

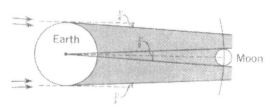

(b) The astronomer Hipparchus, more than 2000 years ago, found the distance of the moon as a multiple of the earth's radius by observing the duration of a total eclipse of the moon by the earth (see the figure). The rays from the sun have a spread of directions of about $\pm\frac{1}{4}°$, and the moon itself subtends an angle of just about $\frac{1}{2}°$ (the same as the sun within about 2%). The moon takes about 29 days to circle the earth, and the duration of the total eclipse is about $1\frac{3}{4}$ hr. Use these data to obtain the moon's distance.

2-18 The astronomer Aristarchus had the idea of comparing the distances of the sun and the moon from the earth by measuring the angular separation θ between them when the moon was exactly half full (see the figure). Using our present knowledge of what these distances are, criticize the feasibility of the method. Aristarchus found $\theta = 87°$. What result would this imply? Calculate what the angle really is and what error would be introduced in the distance if this angle were uncertain by $\pm 0.1°$.

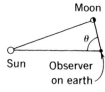

At present it is the purpose of our Author merely to investigate and to demonstrate some of the properties of accelerated motion (whatever the cause of this acceleration may be).

GALILEO, *Dialogues Concerning Two New Sciences* (1638)

3
Accelerated motions

ACCELERATION

FROM THE PURELY descriptive point of view, the central feature of motion is velocity—the instantaneous rate of change of position with time. But we must dig a little deeper to get to the quantity that proves to be the crucial one in relating motion per se (kinematics) to motion as governed by forces (dynamics). This is *acceleration*, the rate of change of velocity with time. Again we shall develop the basic ideas in the first instance in the context of straight-line motion. If the instantaneous velocity is found to be a linear function of time, as in Fig. 3–1(a), then we conclude that the instantaneous acceleration is the same at all times and equal to the slope of the graph as evaluated from any two points on it.[1] If, on the other hand, the values of the instantaneous velocity define some sort of curve, as in Fig. 3–1(b), then we must obtain the instantaneous acceleration at a particular value of t by a limiting process:

$$\text{instantaneous acceleration } a = \lim_{\Delta t \to 0} \frac{\Delta v}{\Delta t} = \frac{dv}{dt} \qquad (3\text{-}1)$$

Thus the acceleration is the first derivative of the velocity with

---

[1]The units in which this is expressed will be defined by our chosen unit of velocity divided by whatever unit of time is found convenient. For example, the acceleration of a car is expressed most effectively and vividly in m.p.h. per second. But in the more analytical treatment of motions it is almost essential to use the same time units throughout, e.g., expressing velocities in m/sec and accelerations in m/sec^2. Otherwise, when we see a symbol such as Δt, we have to stop and ask ourselves which of the different units it is measured in, and that makes for confusion and error.

Fig. 3-1 (a) Graph of v versus t for a uniformly accelerated motion. (b) Motion with varying acceleration. In the indicated time interval Δt the acceleration is negative.

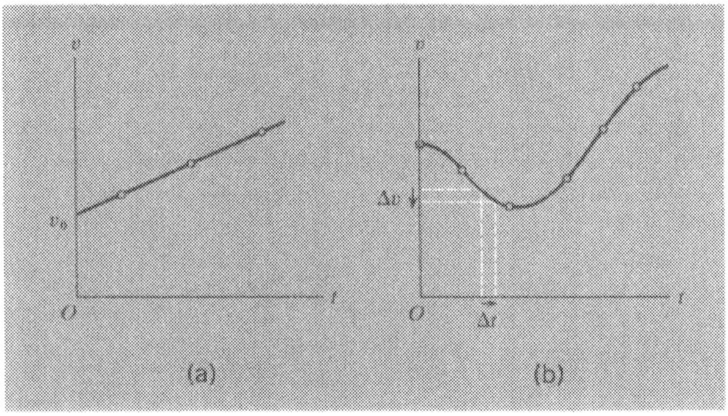

respect to time. If, however, we wish to tie our definition of acceleration to a primary record of position against time, then we can write it as the second derivative of s with respect to t:

$$a = \frac{d^2s}{dt^2} \quad (3\text{-}2)$$

In general, we must be ready to take into account a variation of velocity in direction as well as magnitude. This then requires us to consider the acceleration explicitly as a vector quantity. Just as we previously considered vectorial changes of position, so now we can show the instantaneous velocity vectors at two neighboring instants, as in Fig. 3–2, and can proceed to a statement of the instantaneous vector acceleration, **a**:

$$\mathbf{a} = \lim_{\Delta t \to 0} \frac{\Delta \mathbf{v}}{\Delta t} = \frac{d\mathbf{v}}{dt} = \frac{d^2\mathbf{r}}{dt^2}$$

As far as kinematics by itself is concerned, there is no good reason why we should stop here. We could define and evaluate the rate of change of acceleration, but in general this does not represent information of any basic physical interest, and so our discussion of mechanics is based almost exclusively on the three quantities displacement, velocity, and acceleration.

You may feel, especially if you have some prior familiarity with calculus, that we have gone to excessive lengths in our dis-

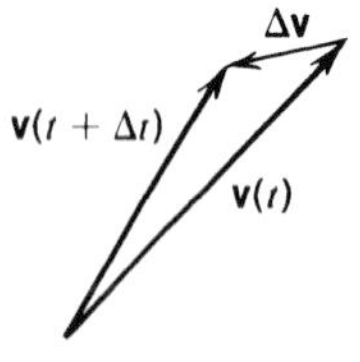

Fig. 3-2 Small change of a velocity that is changing in both magnitude and direction.

86   Accelerated motions

cussion of instantaneous velocity and acceleration. But make no mistake about it; these are very subtle concepts. The notion that an object could both be *at* a certain point and *moving* past that point was one that perplexed some of the best minds of antiquity. Indeed, it is the subject of one of the famous paradoxes of the Greek philosopher, Zeno, who contended that if an object was moving it could not be said to *be* anywhere.[1]

If you want to test your own mastery of these ideas, try explaining to someone how an object that is at a certain point with zero velocity (i.e., instantaneously at rest) can nonetheless move away from that point by virtue of having an acceleration. It really isn't trivial.

THE ANALYSIS OF STRAIGHT-LINE MOTION

Given a detailed record of position versus time in a straight-line motion, the procedures that we have described enable us to find the associated variations of velocity and acceleration. The sequence of diagrams in Fig. 3-3, going from top to bottom, shows an example of this. But how about the converse of this process: given the acceleration as a function of time, to infer the graphs of velocity and displacement? The basic definitions of velocity and acceleration suggest the appropriate procedure for doing this. From Eq. (3-1) we see that the change of velocity Δv in a short time Δt is given, at least approximately, by the equation

$$\Delta v = a \, \Delta t$$

Of course, "approximately" is not good enough, but we recognize that the smaller we choose Δt to be, the more nearly is Δv an accurate statement of the change of v. Again we resort to a graphical presentation. In Fig. 3-4(a) we show a graph of acceleration versus time; it then becomes apparent that $a \, \Delta t$ can be read as the area of a narrow rectangular strip, the top of which cuts across the curve of a versus t. From there it is a short step to concluding that the over-all change of v between two given values of t is obtained by summing all such rectangular contributions. (We must, of course, recognize that wherever a is negative, the area represented by $a \, \Delta t$, and hence the change of v, is also negative.) We then imagine that the widths Δt are made

---

[1]On a quite different basis, the uncertainty principle of quantum mechanics expresses our inability to measure both the position and the velocity of a moving object with arbitrarily high precision.

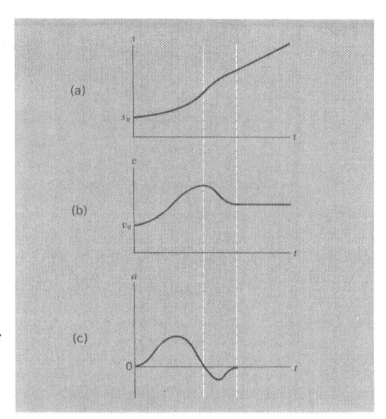

Fig. 3-3 Set of related graphs showing the time dependence of (a) position, (b) velocity, and (c) acceleration.

vanishingly small so that the sum of all the strips coincides, in this limit, with the area under the smooth curve of a versus t. This is then written mathematically as a definite integral:

$$v_2 - v_1 = \int_{t_1}^{t_2} a(t)\,dt \tag{3-3}$$

where we write $a(t)$ to show that the acceleration is to be considered as a specific function of time. Most often this integral is evaluated up to some indefinite time t, starting from some chosen zero of time at which the velocity is v_0. Thus we put

$$v - v_0 = \int_0^t a(t)\,dt \tag{3-4}$$

Notice, then, that our integral, starting from knowledge of the acceleration as a function of time, gives us only the *change* of velocity during the time t. Information about the value of v at $t = 0$ (or at some other specific time) must be supplied sep-

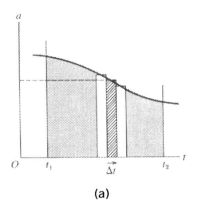 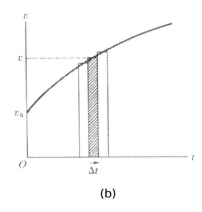

(a) (b)

Fig. 3–4 (a) Graphical integration of a graph of acceleration versus time to find change of velocity.
(b) Graphical integration of velocity–time graph to find displacement.

arately; v_0 is a typical example of a *constant of integration* that requires some knowledge of the *initial conditions*—or of the specific value of v at any one value of t.

In like manner, given the curve such as that of Fig. 3–4(b) of v against t (which may represent the initial data or may itself have come from the above integration of the acceleration function) we can proceed to find the distance traveled. It is represented by the area contained between the velocity–time curve, the t axis, and the ordinates at two given values of t:

$$s_2 - s_1 = \int_{t_1}^{t_2} v(t)\, dt \tag{3-5}$$

Again it is most usual to evaluate the integral from $t = 0$ up to an arbitrary time, and again a constant of integration—the position s_0 at $t = 0$—must be supplied:

$$s - s_0 = \int_0^t v(t)\, dt \tag{3-6}$$

Very often, of course, it is possible to choose $s_0 = 0$, but one should never forget that the area included under the velocity–time graph gives us only the *change* of position.

The simplest applications of these kinematic equations, for $a = 0$ or $a = $ constant, are undoubtedly familiar to you. In the first case the velocity–time graph is simply a rectangle, as shown in Fig. 3–5(a). If the acceleration is constant (but not zero),

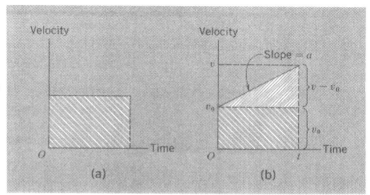

Fig. 3-5 (a) Velocity–time graph for the special case of zero acceleration. (b) Velocity–time graph for a constant (positive) acceleration.

Fig. 3-5(b) is appropriate. The magnitude of the constant acceleration is given by

$$a = \text{slope} = \frac{v - v_0}{t}$$

This conforms to the definition of acceleration in Eq. (3-1). If we took a as given, then we would obtain this same result by integration, according to Eq. (3-4),

$$v - v_0 = a \int_0^t dt = at \tag{3-7}$$

The area in Fig. 3-5(b) that represents the distance traveled can be thought of as made up of the two shaded regions as shown. Hence

$$s - s_0 = v_0 t + \tfrac{1}{2}(v - v_0)t$$

Combining the last two equations, we get

$$s - s_0 = v_0 t + \tfrac{1}{2}at^2$$

which we can recognize also as the result of evaluating the integral in Eq. (3-6) with $v(t) = v_0 + at$:

$$s - s_0 = \int_0^t (v_0 + at)\, dt = v_0 t + \tfrac{1}{2}at^2 \tag{3-8}$$

It is sometimes convenient to remove all explicit reference to the time, by combining Eqs. (3-7) and (3-8). This gives us

$$v^2 = v_0^2 + 2a(s - s_0) \tag{3-9}$$

For ease of reference, we repeat these equations below as a group:

Kinematic equations (valid only for constant a)
$$v = v_0 + at \qquad (3\text{-}10a)$$
$$v^2 = v_0^2 + 2a(s - s_0) \qquad (3\text{-}10b)$$
$$s = s_0 + v_0 t + \tfrac{1}{2}at^2 \qquad (3\text{-}10c)$$

Although these mathematical expressions for accelerated motion are tidy and extremely useful, it should be remembered that a truly constant acceleration is never maintained indefinitely. For example, the problems that everyone learns to solve on free fall under gravity, using a constant acceleration g, really do not correspond to the facts, because air resistance causes the acceleration to become less as the velocity increases. For low velocities the error may not be big enough to worry about, but it is there. Later we shall be dealing with situations in which the acceleration varies in some mathematically well-defined way with position or time. Thus the emphasis will shift away from Eq. (3–10) and toward the more general statements expressed in Eqs. (3–4) and (3–6). Another very important factor in solving real problems in kinematics is the digital computer. Whether or not the acceleration is described by a mathematically convenient function, the actual technique of getting numerical answers to problems on motion—e.g., the path of a rocket or a satellite—will be the summation of small but finite contributions, corresponding to the strips of Fig. 3–4. The program for solving problems in motion is then represented not by mathematical integrals, but by equations such as the following:

$$\begin{aligned} v(t + \Delta t) &= v(t) + a(t)\Delta t \\ v(t) &= v_0 + \sum a(t)\Delta t \\ s(t + \Delta t) &= s(t) + v(t)\Delta t \\ s(t) &= s_0 + \sum v(t)\Delta t \end{aligned} \qquad (3\text{-}11)$$

There are many problems in motion that can be handled as one-dimensional problems, even though the space in which dynamical processes go on is a space of three dimensions. A prime reason for this is that it is often feasible to resolve the vectors of position, velocity and acceleration into their components in a rectangular coordinate system and then proceed to work with the separate components. Under these conditions, as we have mentioned before, it is not necessary to make use of vector notation as such, even though we know that we are dealing with directional quantities. It becomes sufficient to choose an axis of reference along the given line of the motion and adopt a convention that selects one direction along this axis as positive and the opposite direction as negative. Which we choose as positive is arbitrary, but having made a choice for the purpose of a particular problem we must stick to it. Directed quantities that are to be found as a result of a calculation—e.g., the unknown final coordinate, x, of a particle moving along a straight line—will always be taken as measured along the positive direction of the axis. If the answer comes out negative, this automatically tells us that the final position is on the negative side of the origin. In other words, it is not necessary (and may be actually inadvisable, because it can lead to confusion) to inject preconceived ideas as to which sign or direction a quantity will be found to have; the mathematics will do it for you.

Example. A particle starts out at $t = 0$ from the point $x_0 = 10$ m with an initial velocity $v_0 = 15$ m/sec and a constant acceleration $a = -5$ m/sec^2 in the x direction. Find its velocity and position at $t = 8$ sec.

We have

$$v(t) = v_0 + at$$
$$v(8) = 15 + (-5)(8) = -25 \text{ m/sec}$$

Also

$$x(t) = x_0 + v_0 t + \tfrac{1}{2}at^2$$
$$x(8) = 10 + (15)(8) + \tfrac{1}{2}(-5)(8)^2 = -30 \text{ m}$$

Thus at $t = 8$ sec the particle has passed back through the origin to a point on the far side and is traveling in the direction of increasingly negative x. The whole progress of the motion up to $t = 8$ sec is shown in the two graphs of Fig. 3-6. Notice how much information is provided at a glance in these diagrams. One sees at once how the velocity falls to zero at $t = 3$ sec and at $t = 6$ sec becomes equal and opposite to its initial value. One sees how the maximum value of x corresponds to the instant at which v reaches zero and reverses sign, and that the particle returns to its original displacement (x_0) at $t = 6$ sec when v has reached the negative of its initial value v_0, so that the total area under the velocity–time curve up to that instant is zero. Even though this is a very simple and straightforward example, it dis-

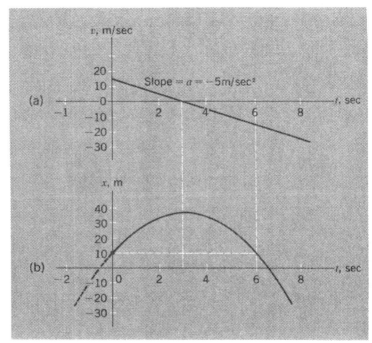

Fig. 3-6 (a) Velocity–time graph for specific values of initial velocity and constant (negative) acceleration. (b) Position versus time for the motion represented in (a).

plays many features that are worthy of note. And seeing how the various details fit together will greatly strengthen one's grip of basic kinematics.

A COMMENT ON EXTRANEOUS ROOTS

Occasionally, in turning the mathematical handle in the solution of the kinematic equations, one cranks out extraneous roots that are contrary to the physical situation. How does one recognize these "incorrect" answers and when can they be discarded?

Many of these extraneous answers have their origin in the fact that in solving a problem we always state initial conditions which specify the situation at the moment we first begin to follow the motion of the particle at $t = 0$. Specifying the position and velocity at $t = 0$ does not tell us anything whatsoever about the past dynamical history of the particle. Indeed, it is of no consequence whatsoever how the particle attained these initial values. For example, if we think of the motion of a body falling freely from rest at an initial height h above the earth's surface, we have

$v_0 = 0$ and $y_0 = h$. We may have held the body at its initial position and released it from rest, or we may have thrown it upward so that it rises to a maximum height h. The motion of the particle subsequent to the instant of time ($t = 0$) when it was at the height h with zero velocity is identical in both cases.

The most frequent source of extraneous roots is the equation relating displacement to time in motion with constant acceleration —i.e., Eq. (3-10c). Mathematically, this is a quadratic equation for t and must be solved as such if the displacement is given and the time is to be found.

Because initial conditions alone do not give us information about the earlier motion of the particle (unless additional relevant data are given), a root of this equation corresponding to a negative value of t may not be valid. To illustrate this, consider the problem used as an example in the last section. Suppose we ask for the value of t at which $x = 0$. We have

$$x(t) = 10 + 15t - \tfrac{5}{2}t^2$$

Putting $x = 0$, we get the following quadratic equation:

$$t^2 - 6t - 4 = 0$$

with the roots $t = 3 \pm \sqrt{13} = 6.6$ sec or -0.6 sec. Figure 3-6(b) makes quite apparent the origin of these two roots and shows how the negative root follows from an extrapolation of the graph backward into the region of times prior to $t = 0$. But that may be quite unjustified. We might, if asked, say: "Oh yes, I held the particle at $+10$ m from the origin until $t = 0$ and then fired it off in the positive x direction with its initial velocity of 15 m/sec." If that were the case, the solution $x = 0$ at $t = -0.6$ sec would be a complete fabrication; one would be forced to recognize that it simply did not correspond to reality. One should not, however, discard extraneous roots without first asking, in the way we have just done, whether there is a clear physical reason for doing so.

The matter of extraneous roots has had an interesting consequence in the history of physics. In quantum mechanics, developing a relativistically correct equation led to two values for the total energy of an electron: positive and negative. Negative values were initially rejected outright as having no physical significance. After all, what meaning can one attach to a kinetic energy less than zero? However, at a later time, P. A. M. Dirac investigated more carefully the nature of these negative energy

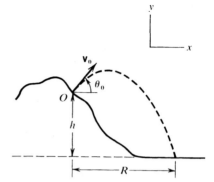

Fig. 3–7 Idealized parabolic trajectory for motion under gravity in the absence of air resistance.

states and was led to a highly successful theory of electrons which predicted the existence of positrons and other "antimatter" particles.

TRAJECTORY PROBLEMS IN TWO DIMENSIONS

One of the most famous and widely studied problems in motion is that of free fall near the earth's surface. It provides an illustration of the fact that, in the rectangular-coordinate system represented by horizontal and vertical directions, the two orthogonal components of the motion are completely independent (provided that air resistance is negligible—see p. 225). The path of an object may be treated as two, separate motions occurring simultaneously, and each may be analyzed as if the other were not present. Galileo, in his *Dialogue on the Two Chief World Systems* (1632), first recognized this fact.[1]

We shall consider the motion of an object hurled with initial speed v_0 at an angle θ as shown in Fig. 3–7 from a height h

---

[1]The accomplishments of Galileo Galilei, born in Pisa in 1564, the year of Shakespeare's birth and Michaelangelo's death, are often cited as the beginning of modern science. Galileo's publication on astronomy, *Dialogue on The Two Chief World Systems*, incorporated the Copernican model and led to conflicts with church authorities. While technically a prisoner of the Inquisition, Galileo turned to the studies of mechanics and published (1638) surreptitiously in Holland the results of his investigations "Discourses and Mathematical Demonstrations Concerning Two New Sciences Pertaining to Mechanics and Local Motion," commonly referred to as *Two New Sciences* (translated, Dover Publications, New York). These books, written largely in the form of imaginary conversations, have a surprisingly modern flavor, and impress one with Galileo's insight and intellectual sophistication. They are well worth reading.

above a level plain. As you know, an approximation to the actual motion is obtained by assuming that the horizontal component of velocity, v_x, remains constant and that the vertical component of velocity, v_y, is subject to a constant acceleration of magnitude g ($= 9.8$ m/sec^2) downward. This approximate description of the motion works well provided that the effects of air resistance are unimportant, which generally speaking requires compact (dense) objects and fairly low velocities. Later we shall consider cases in which the resistive effects are important and our present idealized picture of the motion becomes seriously inadequate.

Let us now see how to answer the following questions:

1. How long is the object in flight?
2. What is the range R (the horizontal distance traveled)?
3. What is the velocity upon striking the ground?

We shall choose our origin of coordinates at the starting point of the particle, and we shall take the *positive* coordinate directions to be *upward* (y) and to the *right* (x). The values of the initial and final coordinates, the initial velocity components, and the acceleration components are then as shown in Fig. 3-8.

Fig. 3-8 Analysis of the motion of Fig. 3-7 in terms of separate horizontal and vertical components.

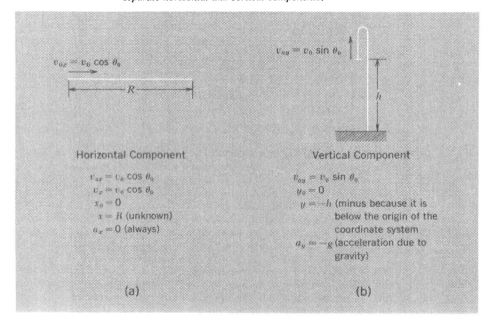

[A note about signs is in order here. The acceleration due to gravity is represented by a vector, **g**. In this book we shall use the symbol g to represent the scalar magnitude of this vector. That is, g denotes the *positive* quantity equal to 9.8 m/sec^2 or 32 ft/sec^2. This is in accord with the usual convention in which the symbol A represents the positive scalar magnitude of any vector **A**. *After* a coordinate system has been chosen, and if the *upward* vertical direction is taken to be positive, then the y component of the vector **g** becomes $-g$ (i.e., $a_y = -9.8$ m/sec^2). In some situations it may be convenient to choose the downward direction as positive. In this case the y component of **g** must be set equal to $+g$. We must of course be consistent, within a given calculation, about what we mean by the positive coordinate direction, but we are completely free to take whichever choice we please; the actual content of the final answers cannot depend on this.]

We return now to the trajectory problem, as depicted in Figs. 3–7 and 3–8:

1. How long is the object in flight?

We know the initial velocity, the initial and final values of y, and the vertical acceleration. Therefore, if we take Eq. (3–10c), as applied to motion in the y direction, we know everything except the time of flight t:

$$y = v_{0y}t + \tfrac{1}{2}a_y t^2$$

i.e.,

$$-h = (v_0 \sin \theta_0)t + \tfrac{1}{2}(-g)t^2$$

We solve this equation and take the positive root as the physically relevant one.

2. What is the range R?

Apply Eq. (3–10c) to the horizontal-component problem:

$$x = v_{0x}t + \tfrac{1}{2}a_x t^2$$

i.e.,

$$R = (v_0 \cos \theta_0)t + 0 = (v_0 \cos \theta_0)t$$

In this we substitute the specific value of t obtained in calculation 1.

3. What is the velocity upon striking the ground?

The *components* of **v** may be found from Eq. (3–10b) using the value of t obtained from calculation 1.

$$v_x = v_{0x} + a_x t \qquad v_y = v_{0y} + a_y t$$
$$v_x = v_0 \cos \theta_0 \qquad v_y = v_0 \sin \theta_0 + (-g)t$$

Then from these components we have

Magnitude of **v**: $\qquad v = (v_x^2 + v_y^2)^{1/2}$

Direction of **v**: $\qquad \tan \theta = \dfrac{v_y}{v_x}$

(In this case $\tan \theta$ will be negative, because θ represents a direction pointing downward below the horizontal.) Alternatively, we can calculate the magnitude of v^2 directly from Eq. (3–10b) as follows:

$$v_x^2 = v_{0x}^2 = v_0^2 \cos^2 \theta_0$$
$$v_y^2 = v_{0y}^2 + 2a_y y = v_0^2 \sin^2 \theta_0 + 2(-g)(-h)$$

Therefore,

$$v^2 = v_x^2 + v_y^2 = v_0^2 + 2gh$$

The *direction* of **v** can then be calculated from the relation

$$\sin \theta = \dfrac{v_y}{v}$$

FREE FALL OF INDIVIDUAL ATOMS

Atoms moving in a vacuum cast sharp shadows and give evidence of traveling in straight lines. On the other hand, it must surely be true that atoms and molecules, as samples of ordinary matter, are subject to the usual free fall under gravity at the earth's surface. The effect is not very noticeable because evaporated atoms have high average speeds—about the same as a rifle bullet—but it is measurable. How far would a beam of atoms (initially moving horizontally) fall vertically under gravity while traveling a horizontal distance L at such speeds? Figure 3–9(a) illustrates the problem. It would be just like the trajectory problem of the last section, except that this time we take the horizontal distance as given and then solve for the vertical distance y. If an origin is taken at the point O as shown, from which the atoms

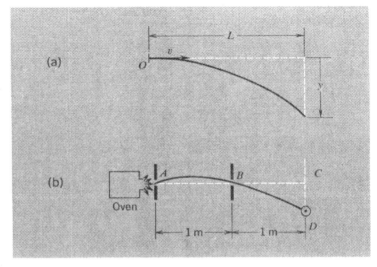

Fig. 3-9 (a) Trajectory of atoms in vacuum with an initial horizontal velocity. The vertical displacement is greatly exaggerated. (b) Parabolic trajectory of atoms in an atomic beam that must pass through the slits A and B to reach the detector D.

start out horizontally with a speed v, we have

Horizontal component:

$$x = v_{0x}t + \tfrac{1}{2}a_x t^2$$

Therefore,

$$L = vt$$

Vertical component:

$$y = v_{0y}t + \tfrac{1}{2}a_y t^2$$
$$y = 0 + \tfrac{1}{2}(-g)t^2$$

which upon substitution for t gives us

$$y = \frac{-g}{2}\left(\frac{L}{v}\right)^2$$

Suppose that we apply this result to a beam of atoms with a speed of about 500 m/sec. In traveling a horizontal distance of 1 m, the time of flight would be 1/500 sec (2 msec) and we should have

$$y = -\frac{9.8}{2}\left(\frac{1}{500}\right)^2 \approx -2 \times 10^{-5} \text{ m}$$

The deviation from a straight-line path is thus extremely small, only a few hundredths of a millimeter.

Despite the small size of the effect, it has been studied with precision in an experiment by Estermann et al.[1] Their arrangement was as shown schematically in Fig. 3-9(b). Atoms of cesium or potassium were evaporated out of an "oven" at about 450°K inside a vacuum system. Since the atoms emerge from the oven in a variety of directions, the beam was collimated by two slits A and B on the same horizontal level, as shown. The slits were about 0.02 mm wide. The beam of atoms was detected by a horizontal hot-wire detector D, also about 0.02 mm across.[2] As Fig. 3-9(b) shows, any atom that reaches the detector must have a small initial upward component velocity at A in order to negotiate the slit system. However, from a point midway between A and B where the beam is horizontal, the trajectory is just like that shown in Fig. 3-9(a).

The detector is moved vertically across the beam and the ion current, produced by atoms that strike the wire, is recorded. In the absence of any gravitational deflection, the intensity distribution across the beam should be trapezoidal as in Fig. 3-10(a) because the central region would be bordered by "penumbra" regions that are a consequence of the two-slit system. Some results are shown in Fig. 3-10(b).

The most obvious feature of these graphs is that they reveal a wide spread of speeds in the atoms of a beam. Some atoms are moving so fast that they are scarcely deflected at all; others are moving so slowly that their deflection is many times greater than the most probable deflection (which corresponds to the maximum of the intensity distribution). The complete curve must reflect a characteristic distribution of speeds of the atoms in the oven at a particular temperature.

We shall not consider the detailed shape of the intensity pattern but will fix attention on the deflection of the peak. A comparison of the graphs for cesium and potassium makes it obvious that potassium atoms (atomic mass = 39) move on the

---

[1] I. Estermann, O. C. Simpson, and O. Stern, *Phys. Rev.*, **71**, 238 (1947). Similar experiments were reported at about the same time on the free fall of thermal neutrons—see L. J. Rainwater and W. W. Havens, *Phys. Rev.*, **70**, 136 (1946).

[2] A neutral Cs or K atom, striking the hot wire, becomes ionized by losing an electron. A nearby electrode at a negative potential with respect to the wire will collect these positive ions and the resultant current flow can be detected with a sensitive electrometer or galvanometer. If a thin straight wire is used, it acts as a detector of width equal to its own diameter. Some atoms (e.g., the halogens) tend to capture electrons and form negative ions; the sign of the potential of the electrode can be adjusted accordingly.

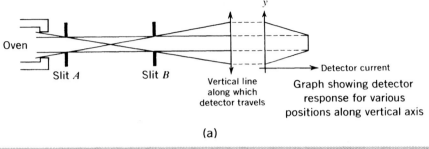
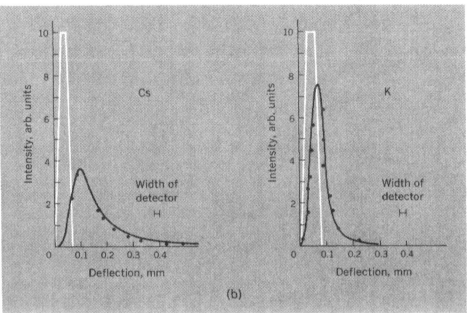

*Fig. 3-10 (a) Magnified detail of geometrical image formed by atoms moving in straight lines through two slits. The intensity is proportional to the area of the source that can be "seen" by the detector in any given position. (b) Actual data on the deflection of beams of cesium and potassium atoms. [After I. Estermann, O. C. Simpson, and O. Stern, Phys. Rev., **71**, 238 (1947).]*

average much faster than cesium atoms (atomic mass = 133) at comparable temperatures. This is an expression of the fact that the molecules of different gases at the same temperature have equal *average* kinetic energies—a result that we quote without further discussion at this point. As for the actual magnitude of the thermal velocities, let us look at the peak of the cesium curve. It is displaced relative to the center of a gravitation-free beam by about 0.11 mm. Now in the arrangement shown in Fig.

3-9(b), if $AB = BC = L$, then an analysis of the trajectory will show (see Problem 3-10) that atoms of speed v will be displaced downward by a distance y given approximately by

$$y \approx \frac{gL^2}{v^2}$$

Hence

$$v \approx L \left(\frac{g}{y}\right)^{1/2}$$

Substituting the values $L = 1$m and $y = 1.1 \times 10^{-4}$m we get, for the approximate speed, $v = 300$ m/sec.

This very beautiful and delicate experiment was used as a test of the theoretical velocity distribution of atoms at a given temperature.[1] We cite it here as a nice illustration that atoms, like baseballs or earth satellites, follow curved paths under the action of gravitational forces. Since the motion takes place in a vacuum it is, in fact, a more justifiable application of the idealized laws of free fall than are the more usual problems of objects moving through the air. You may wonder if it is possible to demonstrate the free fall of individual electrons in a similar way. This is an immensely more difficult problem, because whereas atoms, being electrically neutral, experience only the gravitational acceleration, electrons are exposed to stray electric forces that completely swamp all gravitational effects unless extraordinary precautions are taken. Nonetheless, some experiments have been attempted on this problem, although the interpretation of the results is a rather complicated affair.

OTHER FEATURES OF MOTION IN FREE FALL

In the idealized description of motion of a freely moving object near the earth's surface, the horizontal component of velocity always remains constant and the vertical acceleration always has the same value, g (downward). It may be interesting to point out that, under these assumptions, every trajectory associated with the same value of the constant horizontal velocity, v_{0x}, forms part of a single parabola—a kind of universal curve (see Fig. 3-11) on which one can mark in the beginning and end points of any particular trajectory, as shown. There is nothing profound

---

[1] A similar experiment, using velocity-selected atoms, has been reported by N. B. Johnson and J. C. Zorn, *Am. J. Phys.*, **37**, 554 (1969).

Fig. 3-11 "Universal parabola" that embodies all possible parabolic trajectories for a given downward acceleration and a given horizontal component of velocity. The heavily marked part of the curve between points C and D corresponds to the atomic beam trajectory of Fig. 3-9(b).

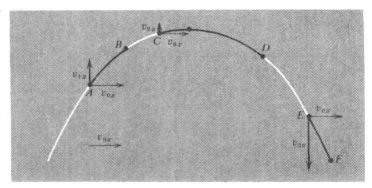

about this, but it can be useful in helping one to see any individual trajectory as part of a larger scheme. For example, it makes very clear the relation between the two atomic-beam paths shown in Fig. 3–9.

A closely related feature is the way in which the total velocity vector changes during the course of the motion. This is illustrated in Fig. 3–12. Suppose that the initial velocity is represented by the vector v_1. Then the velocity v_2, at a time Δt later, is obtained by adding to v_1 a vertical (downward) vector $a\,\Delta t$ as shown. Similarly, every vector representing the instantaneous velocity at a subsequent stage in the motion has its end point on a vertical line drawn from the end of v_1. This result embodies the fact that the horizontal component of every such vector has the same value.

Yet another aspect of this same free-fall problem is illustrated by the venerable demonstration of the hunter and the

Fig. 3–12 Array of successive velocity vectors for a motion in which the acceleration is constant and vertically downward, as in free fall under gravity in the absence of air resistance. (This type of diagram is known as a hodograph.)

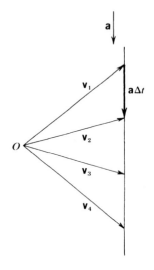

103 Other features of motion in free fall

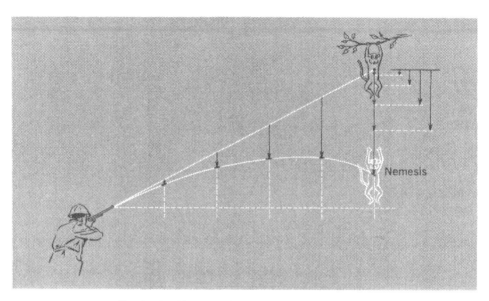

Fig. 3-13 Classic monkey-shooting demonstration. The bullet and the monkey undergo equal gravitational displacements in equal times and are doomed to meet if the monkey lets go as soon as he sees the gun fired.

monkey. The hunter aims directly at the monkey as it hangs from a limb (see Fig. 3-13). This is really a mistake, because it makes no allowance for the fact that the bullet follows a parabolic path as shown. But the monkey makes a compensating mistake. Seeing the gun aimed directly at him, he lets go of the limb as soon as he sees the flash of the gun. Thus the bullet and the monkey begin falling at the same instant (ignoring any delays due to the time of transit of the light flash and—less justifiably—the reaction time of the monkey). It then follows that, in whatever time it takes for the bullet to travel the horizontal distance from the gun to the vertical line of the monkey's descent, both bullet and monkey receive the same contributions, $\frac{1}{2}gt^2$, to their displacement as a result of the gravitational acceleration alone. Thus the bullet's trajectory crosses the line of the monkey's fall at a point that is bound to be reached at the same time by the monkey—with dire consequences to himself. Note that this result is independent of both the speed of the bullet and the value of g; it requires only that the bullet would, in the absence of gravity, go straight to the monkey's original position. Quite remarkable, on the face of it, yet easily understood in terms of the basic analysis of accelerated motion.

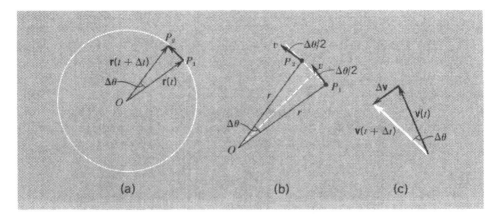

Fig. 3-14 (a) Small displacement (P_1P_2) in a uniform circular motion. (b) Velocity vectors at the beginning and end of the short element of path. (c) Vector diagram for the evaluation of the change of velocity, $\Delta \mathbf{v}$.

UNIFORM CIRCULAR MOTION

Probably the most interesting direct application of the vector definitions of velocity and acceleration is to the problem of motion in a circular path at some constant speed. In this case, if the center of the circle is chosen as an origin, the vector **r** always has the same length and simply changes its direction at a uniform rate. The instantaneous velocity is always at right angles to **r**, and its magnitude v is constant. From this we can readily calculate the acceleration. For during a short time, Δt, the distance traveled is $v \Delta t$, from P_1 to P_2 along a circular arc [Fig. 3–14(a)]. The angle $\Delta \theta$ between the two corresponding directions of **r** is therefore given by

$$\Delta \theta = \frac{v \Delta t}{r}$$

Imagine that the bisector of this angle is drawn [Fig. 3–14(b)] and consider the changes in velocity parallel and perpendicular to this bisector. Initially the velocity has a component $v \sin(\Delta \theta/2)$ away from O, and $v \cos(\Delta \theta/2)$ transversely. Subsequently it has a component $v \sin(\Delta \theta/2)$ *toward* O, and again $v \cos(\Delta \theta/2)$ transversely in the same direction as before. Thus the change of velocity is of magnitude $2v \sin(\Delta \theta/2)$ *toward* O. Figure 3–14(c) shows how this same result comes from considering a vector diagram in which $\Delta \mathbf{v}$ is defined as that vector which, added to $\mathbf{v}(t)$, gives $\mathbf{v}(t + \Delta t)$.

As $\Delta\theta$ is made vanishingly small, $\sin(\Delta\theta/2)$ becomes indistinguishable from $\Delta\theta/2$ itself (in radian measure).[1] Thus we can put

$$|\Delta \mathbf{v}| = 2v \sin(\Delta\theta/2) \to v\,\Delta\theta$$

But $\Delta\theta = v\,\Delta t/r$, so we have

$$|\Delta \mathbf{v}| = v^2\,\Delta t/r$$

Hence the magnitude of the acceleration is given by

(Uniform circular motion) $\quad |\mathbf{a}| = \dfrac{v^2}{r} \quad$ (3-12)

and its direction is radially inward, regardless of whether the circular path is being traced out clockwise or counterclockwise. This is called the *centripetal acceleration* (literally, "center-seeking") associated with any circular motion. The need for a dynamical means of supplying this acceleration to an object is an essential feature of any motion that is not strictly straight, because any change in the direction of the path implies a component of $\Delta \mathbf{v}$ perpendicular to $\mathbf{v}$ itself.

VELOCITY AND ACCELERATION IN POLAR COORDINATES

The result of the last section, and other results of more general application, are very nicely developed with the help of polar coordinates in the plane. The use of this type of analysis is particularly appropriate if the origin represents a center of force of some kind—e.g., the sun, acting on an orbiting planet. The starting point is to write the position vector $\mathbf{r}$ as the product of the scalar distance r and the unit vector $\mathbf{e}_r$:

$$\mathbf{r} = r\mathbf{e}_r \qquad (3\text{-}13)$$

We now consider the change of $\mathbf{r}$ with time. This can arise from a change of its length, or from a change of its direction, or from

---

[1]Mathematically, this approximation is equivalent to $(\sin\theta)/\theta \to 1$ for $\theta \to 0$. We shall be using this approximation often. For a discussion of it, see, for example, G. B. Thomas Jr., *Calculus and Analytic Geometry*, 3rd ed., Addison-Wesley, Reading, Mass., 1960, p. 172.

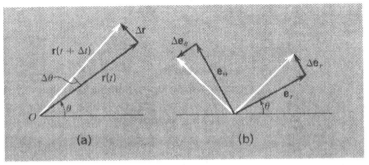

Fig. 3-15 (a) Vector change of displacement, Δr, during a short time Δt in a uniform circular motion. (b) Changes in the unit vectors e_r and e_θ during Δt, showing how Δe_r is parallel to e_θ and Δe_θ is parallel (but opposite) to e_r.

a combination of both. For the present we shall limit ourselves to circular motion, in which the length of **r** remains constant. The change of **r** in a short time Δt is then as shown in Fig. 3-15(a), which is almost the same as Fig. 3-14(a). The *direction* of this change (Δ**r**) is in the direction of the unit vector e_θ drawn at right angles to e_r as shown in Fig. 3-15(b). Its magnitude, as is clear from Fig. 3-15(a), is equal to $r\,\Delta\theta$. Thus we can put

$$\Delta\mathbf{r} = r\,\Delta\theta\,\mathbf{e}_\theta$$

Dividing by Δt, and letting Δt tend to zero, we then have the result

$$\text{(Circular motion)} \qquad \mathbf{v} = \frac{d\mathbf{r}}{dt} = r\frac{d\theta}{dt}\mathbf{e}_\theta \qquad (3\text{-}14a)$$

If we designate $d\theta/dt$ by the single symbol ω, for angular velocity (measured in rad/sec), we have

$$\text{(Circular motion)} \qquad \mathbf{v} = \omega r \mathbf{e}_\theta = v \mathbf{e}_\theta \qquad (3\text{-}14b)$$

The derivation of the above result embodies the important fact that the unit vector $\mathbf{e}_r$ is changing with time. Although its length is by definition constant, its direction changes in accord with the direction of **r** itself. In fact, we can obtain the explicit expression of its rate of change as a special case of Eq. (3-14a), with $r = 1$:

$$\frac{d}{dt}(\mathbf{e}_r) = \frac{d\theta}{dt}\mathbf{e}_\theta = \omega \mathbf{e}_\theta \qquad (3\text{-}15a)$$

In an exactly similar way, as Fig. 3-15(b) shows, a change of θ implies a change of the other unit vector, $\mathbf{e}_\theta$. If the change of θ is positive, as shown, it can be seen that the change of $\mathbf{e}_\theta$ is in the direction of $-\mathbf{e}_r$; it is given by the equation

$$\frac{d}{dt}(\mathbf{e}_\theta) = -\frac{d\theta}{dt}\mathbf{e}_r = -\omega\mathbf{e}_r \qquad (3\text{-}15b)$$

This possible time dependence of the unit vectors in a polar-coordinate system is a feature that has no counterpart in rectangular coordinates, where the unit vectors **i**, **j**, and **k** are defined to have the same directions for all values of the position vector **r**.

Once we have Eqs. (3–14a) and (3–14b) we can proceed to calculate the acceleration by taking the next time derivative. If we limit ourselves to the case of *uniform* circular motion, both r and ω are constant, so we have

(Uniform circular motion) $\qquad \mathbf{a} = \omega r \dfrac{d}{dt}(\mathbf{e}_\theta) = -\omega^2 r \mathbf{e}_r = -\dfrac{v^2}{r}\mathbf{e}_r$

$$(3\text{-}16)$$

Thus the result expressed by Eq. (3–12) falls out automatically, together with its correct direction. If we label this acceleration specifically as a *radial* acceleration of magnitude a_r, we can put

$$a_r = -\omega^2 r$$

If, still restricting ourselves to motion in a circle, we remove the condition that the motion be uniform, then the acceleration vector **a** has a transverse component also. Starting from Eq. (3–14b), we have

(Arbitrary circular motion) $\qquad \mathbf{a} = \dfrac{dv}{dt}\mathbf{e}_\theta - v\dfrac{d\theta}{dt}\mathbf{e}_r \qquad (3\text{-}17)$

The radial component of **a** is the same as we obtained for uniform circular motion (since $d\theta/dt = v/r = \omega$), but it is now joined by a transverse component, a_θ. Thus we have

(Arbitrary circular motion) $\qquad \begin{cases} a_r = -\dfrac{v^2}{r} = -\omega v = -\omega^2 r \\ a_\theta = \dfrac{dv}{dt} = r\dfrac{d\omega}{dt} = r\dfrac{d^2\theta}{dt^2} \end{cases} \qquad (3\text{-}18)$

(where $\omega = d\theta/dt$)

PROBLEMS

3–1 At $t = 0$ an object is released from rest at the top of a tall building. At the time t_0 a second object is dropped from the same point.

(a) Ignoring air resistance, show that the time at which the objects have a vertical separation l is given by

$$t = \frac{l}{gt_0} + \frac{t_0}{2}$$

How do you interpret this result for $l < \frac{1}{2}gt_0^2$?

(b) The above formula implies that there is an optimum value of t_0 so that the separation l reaches some specified value at the earliest possible value of t. Calculate this optimum value of t_0, and interpret the result.

3-2 Below are some careful measurements taken on a stroboscopic photograph of a particle undergoing accelerated motion. The distance is measured from the starting point, but the zero of time is set at the first position that could be separately identified:

Time (in strobe flashes)	Distance (cm) in photo
0	0.56
1	0.84
2	1.17
3	1.57
4	2.00
5	2.53
6	3.08
7	3.71
8	4.39

Plot a straight-line graph, based on these data, to show that they are fitted by the equation $s = \frac{1}{2}a(t - t_0)^2$, and find t_0.

3-3 A child's toy car rolling across a sloping floor is known to have a constant acceleration. Taking $x = 0$ at $t = 0$, it is observed that the car is at $x = 3$ m at $t = 1$ sec, and at $x = 4$ m at $t = 2$ sec.

(a) What are the acceleration and initial velocity of the car?

(b) Plot the position of the car as a function of time up to $t = 4$ sec.

(c) When is the car at $x = 2$ m?

3-4 The faculty resident of a dormitory sees an illegal water-filled balloon fall vertically past his window. Having lightning reflexes, he observes that the balloon took 0.15 sec to pass from top to bottom of his window—a distance of 2 m. Assuming that the balloon was released from rest, how high above the bottom of his window was the guilty party?

3-5 The graph on the next page is an actual record of distance versus time in a straight-line motion.

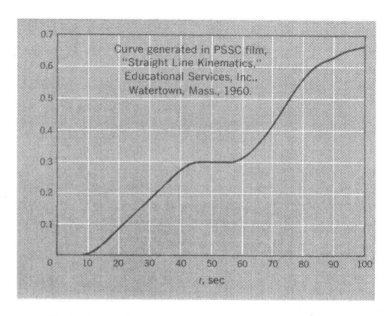

(a) Find the values of the instantaneous velocity at $t = 25$, 45, and 65 sec.

(b) Sketch a graph of v as a function of t for the whole trip.

(c) From (b) estimate very roughly the times at which the acceleration had its greatest positive and negative values.

3-6 In 1965 the world records for women's sprint races over different distances were as follows:

60 m	7.2 sec
100 yd	10.3 sec
100 m	11.2 sec

(a) Make an accurate graph of distance in meters versus time in seconds.

(b) The graph will show you that the data can be well fitted by assuming that a sprinter has a certain acceleration a for a short time τ and then continues with a constant speed v. Set up the equation for distance x in terms of a, τ, and t.

(c) Find the numerical values of v, a, and τ that fit the description given in (b). If this description is correct, what is the distance in meters traveled by the sprinters before they reach their steady velocity?

3-7 Two cars are traveling, one behind the other, on a straight road. Each has a speed of 70 ft/sec (about 50 mph) and the distance between them is 90 ft. The driver of the rear car decides to overtake the car ahead and does so by accelerating at 6 ft/sec^2 up to 100 ft/sec (about 70 mph) after which he continues at this speed until he is 90 ft ahead of the other car. How far does the overtaking car travel along the road between the beginning and end of this operation? If a third car were in sight, coming in the opposite direction at 88 ft/sec (60 mph), what would be the minimum safe distance between the third car and the overtaking car at the beginning of the overtaking operation? (If you are a driver, take note of how large this distance is.)

3-8 In *Paradise Lost*, Book I, John Milton describes the fall of Vulcan from Heaven to earth in the following words:

> ... from Morn
> To Noon he fell, from Noon to dewy Eve,
> A Summer's day; and with the setting Sun
> Dropt from the Zenith like a falling Star . . . .

(It was this nasty fall that gave Vulcan his limp, as a result of his being thrown out of Heaven by Jove.)

(a) Clearly air resistance can be ignored in this trip, which was mostly through outer space. If we assume that the acceleration had the value g (9.8 m/sec^2) throughout, how high would Heaven be according to Milton's data? What would have been Vulcan's velocity upon entering the top of the atmosphere?

(b) (Much harder) One really should take account of the fact that the acceleration varies inversely as the square of the distance from the earth's center. Obtain revised values for the altitude of Heaven and the atmospheric entry speed.

3-9 A particle moves in a vertical plane with constant acceleration. Below are values of its x (horizontal) and y (vertical) coordinates at three successive instants of time:

t, sec	x, m	y, m
0	4.914	4.054
2×10^{-2}	5.000	4.000
4×10^{-2}	5.098	3.958

Using the basic definitions of velocity and acceleration ($v_x = \Delta x/\Delta t$, etc.), calculate

(a) The x and y components of the average velocity vector during the time intervals 0 to 2×10^{-2} sec and 2×10^{-2} to 4×10^{-2} sec.

(b) The acceleration vector.

3-10 (a) The figure [similar to Fig. 3-9(b)] shows a parabolic atomic-beam trajectory in vacuum, passing through two narrow slits, a distance L apart on the same horizontal level, and traveling an additional horizontal distance L to the detector. Verify that the atoms arrive at the detector at a vertical distance y below the first slit, such

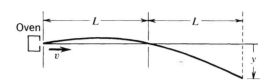

that $y \approx gL^2/v^2$, where v is the speed of the atoms. (You can assume $y \ll L$.)

(b) A beam of rubidium atoms (atomic weight 85) passes through two slits at the same level, 1 m apart, and travels an additional distance of 2 m to a detector. The maximum intensity is recorded when the detector is 0.2 mm below the level of the other slits. What is the speed of the atoms detected under these conditions? Compare with the results for K and Cs shown in Fig. 3–10(b). What was the initial vertical component of velocity at the first slit?

3–11 (a) Galileo, in his book *Two New Sciences* (1638), stated that the theoretical maximum range of a projectile of given initial speed over level ground is obtained at a firing angle of 45° to the horizontal, and furthermore that the ranges for angles $45° \pm \delta$ (where δ can be any angle < 45°) are equal to one another. Verify these results if you have not been through such calculations previously.

(b) Show that for any angle of projection θ (to the horizontal) the maximum height reached by a projectile is half what it would be at the same instant if gravity were absent.

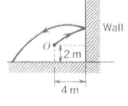

3–12 A perfectly elastic ball is thrown against a house and bounces back over the head of the thrower, as shown in the figure. When it leaves the thrower's hand, the ball is 2 m above the ground and 4 m from the wall, and has $v_{0x} = v_{0y} = 10$ m/sec. How far behind the thrower does the ball hit the ground? (Assume that $g = 10$ m/sec^2.)

3–13 A man stands on a smooth hillside that makes a constant angle α with the horizontal. He throws a pebble with an initial speed v_0 at an angle θ above the horizontal (see the figure).

(a) Show that, if air resistance can be ignored, the pebble lands at a distance s down the slope, such that

$$s = \frac{2v_0^2 \sin(\theta + \alpha) \cos\theta}{g \cos^2\alpha}$$

(b) Hence show that, for given values of v_0 and α, the biggest value of s is obtained with $\theta = 45° - \alpha/2$ and is given by

$$s_{max} = \frac{v_0^2(1 + \sin\alpha)}{g \cos^2\alpha}$$

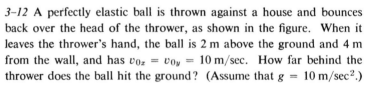

(Use calculus if you like, but it is not necessary.)

3–14 A baseball is hit out of the stadium and is observed to pass over the stands, 400 ft from home plate, at a height of 50 ft. The ball leaves the bat at an angle of 45° to the horizontal, and is 4 ft above the ground when struck. If air resistance can be ignored (which it actually cannot), what magnitude of the ball's initial speed would be implied by these numbers? ($g = 32$ ft/sec^2.)

3–15 A stopwatch has a hand of length 2.5 cm that makes one complete revolution in 10 sec.

(a) What is the vector displacement of the tip of the hand between the points marked 6 sec and 8 sec? (Take an origin of rectangular coordinates at the center of the watch-face, with a y axis passing upward through $t = 0$.)

(b) What are the velocity and acceleration of the tip as it passes the point marked 4 sec on the dial?

3–16 Calculate the following centripetal accelerations as fractions or multiples of g (≈ 10 m/sec^2):

(a) The acceleration toward the earth's axis of a person standing on the earth at 45° latitude.

(b) The acceleration of the moon toward the earth.

(c) The acceleration of an electron moving around a proton at a speed of about 2×10^6 m/sec in an orbit of radius 0.5 Å (the first orbit of the Bohr atomic model).

(d) The acceleration of a point on the rim of a bicycle wheel of 26 in. diameter, traveling at 25 mph.

3–17 A particle moves in a plane; its position can be described by rectangular coordinates (x, y) or by polar coordinates (r, θ), where $x = r \cos \theta$ and $y = r \sin \theta$.

(a) Calculate a_x and a_y as the time derivatives of $r \cos \theta$ and $r \sin \theta$, respectively, where both r and θ are assumed to depend on t.

(b) Verify that the acceleration components in polar coordinates are given by

$$a_r = a_x \cos \theta + a_y \sin \theta$$
$$a_\theta = -a_x \sin \theta + a_y \cos \theta$$

Substitute the values of a_x and a_y from (a) and thus obtain the general expressions for a_r and a_θ in polar coordinates.

3–18 A particle oscillates along the x axis according to the equation $x = 0.05 \sin(5t - \pi/6)$, where x is in meters and t in sec.

(a) What are its velocity and acceleration at $t = 0$?

(b) Make a drawing to show this motion as the projection of a uniform circular motion.

(c) Using (b), find how long it is, after the particle passes through the position $x = 0.04$ m with a negative velocity, before it passes again through the same point, this time with positive velocity.

It seems clear to me that no one ever does mean or ever has meant by "force" rate of change of momentum.
                                      C. D. BROAD, *Scientific Thought* (1923)

4
Forces and equilibrium

AS WE SAID at the beginning of this book, Newton's great achievement, in creating the science of mechanics, was to develop quantitative relationships between the forces acting on an object and the changes in the object's motion. More than that, he declared that the main task of mechanics was to learn about forces from observed motions. But this does not alter the fact that the idea of *force* exists independently of the quantitative laws of motion and comes initially from very subjective experiences—the muscular effort involved in applying a push or a pull. We shall begin from this point of view, and rather than plunge at once into dynamics, we shall first take a look at forces in balance. It has become rather unfashionable to do this because, as you probably already know, the accepted units for the absolute measurement of force are defined in terms of the motions that unbalanced forces produce. We are bound to come to that, and in doing so we shall come to the heart of mechanics. Nevertheless, the quantitative notion of force can be (and was) developed in another context—the study of objects at rest, in *static equilibrium*. Indeed, our basic knowledge of the two most important forces in mechanics—the gravitational force and the Coulomb force between electric charges—was obtained largely through laboratory observations of static equilibrium situations, using techniques that are still important and broadly applied.

In discussing forces in equilibrium, we shall begin with experiences that are familiar and seemingly quite straightforward. Later, after considering some of the problems of motion, we shall recognize that tacit assumptions and unsuspected subtleties are involved; we shall then be equipped to return to the problem of equilibrium with deeper insights and a broader view. But, as we said earlier, that kind of development is very good physics; we do not try to handle everything at once, but proceed by easy stages to extend the range and sophistication of our ideas.

FORCES IN STATIC EQUILIBRIUM

Let us consider a very simple physical system—an archer's bow. In Fig. 4-1(a) we show the bow in its resting state; the string is straight and taut. We know that to bring the bow into the situation shown in Fig. 4-1(b), and to hold it there, a force must be supplied. Remove that force and the bowstring snaps back toward its straight condition, perhaps launching an arrow in the process [Fig. 4-1(c)]. What conditions have to be satisfied to hold the bow in the shape shown in Fig. 4-1(b)? Putting aside the human, subjective aspects of the situation, one can say that the forces at the point C are in balance. But what does this mean? A force is not disembodied; it is applied to *something*.

Fig. 4-1 (a) Schematic diagram of an archer's bow in resting state. (b) Bowstring drawn back at midpoint, requiring applied force along the bisector of the angle between the two segments of string. (c) Arrow in the process of being launched. (d) Forces applied at the center of the bowstring in situation (b).

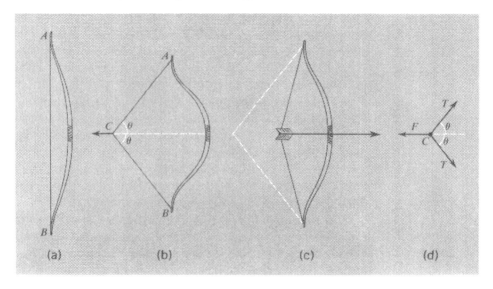

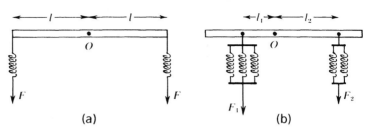

Fig. 4-2 (a) Balancing of equal forces in simple equal-arm arrangement. (b) Balancing of unequal forces in accordance with the law of the lever.

One cannot imagine a force in the absence of a physical object on which it is exerted. In this case the object in question is some small part of the bowstring in the immediate neighborhood of point C. This little piece of string is exposed to pulls along directions CA and CB from the adjoining portions of the string, and to a third force, supplied by the archer, along the bisector of angle ACB. We can draw a separate diagram, as in Fig. 4-1(d), showing the piece of string at C in equilibrium under the action of three forces—a force F from the archer, and two forces, which by symmetry are of equal magnitude T, from the string. These latter combine to give a force, equal and opposite to F, along the line of symmetry. But how do we *know* that the equilibrium requires that the bowstring supply a net force equal and opposite to F? You might say that this is obvious, but we can back up this intuition with a real experiment. It will involve one assumption: that identical objects, equally deformed, supply pulls or pushes of equal size. For example, a small loop of string can remain at rest if pulled in opposite directions by two identical coiled springs extended by equal amounts. And this balancing of equal and opposite forces is the most elementary of all equilibrium situations.

Problems such as that of the archer's bow will not be new to you. But let us point out that the analysis of them involves our ability to assign numerical magnitudes to individual forces and to compare one force with another. How can we do that? Archimedes showed the way, when he discovered the law of the lever over 2000 years ago. Equal forces balance when applied at equal distances on either side of the pivot O [Fig. 4-2(a)]. Unequal forces balance [Fig. 4-2(b)] if the condition $F_1 l_1 = F_2 l_2$ is satisfied. Considerations of symmetry alone would suggest the first result, but the second has to be based on experiment.[1] To establish it we need to have a way of obtaining

---

[1] Archimedes believed that he could obtain the result by pure logic, but this is not so. See Problem 4-5.

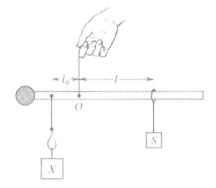

Fig. 4–3 Basic arrangement of a "steelyard" for weighing an unknown (X) with the help of a standard weight (S) that can be moved along the horizontal arm.

multiples of a given force. One method would be to construct a number of identical coiled springs. We could first verify that, when stretched by equal amounts, they would balance one another individually in an equal-arm arrangement. Then we could attach several, all at the same distance on one side of the pivot, and balance them with a single spring, or some different multiple, at the appropriate distance on the other side, as shown in Fig. 4–2(b). (Note that such a procedure entails no assumptions about the way in which the force varies with extension for an individual spring.) Once the law of the lever has been justified in this way, we can use it to measure an unknown force in terms of an arbitrary standard. The technique has been used since Archimedes' time, at the very least, for the purpose of weighing objects (see Fig. 4–3), and its basic principle, which allows us to balance a small force against a large one, using suitably chosen lever arms, is exploited in many familiar mechanical devices.

UNITS OF FORCE

For the purposes of this introductory study of forces in equilibrium, we do not strictly need to worry about the absolute magnitudes of forces. The introduction of an arbitrary unit of force, and the specification of other forces as multiples of this unit, would really suffice. However, it will often be convenient to express forces in terms of their customary measures, so before going further we shall state what these measures are. (You have probably met them all before, in any case.)

The *unit* of force that we shall most frequently employ, here and throughout the book, is the *newton* (N). This is a force of such a size that it can give an acceleration of 1 m/sec² to a mass of 1 kg; it represents the basic unit of force in the MKS system. The detailed basis of this definition will be discussed in Chapter 6; all that we need for the moment is the recognition that this does uniquely define a force of a certain size. The magnitude of this unit is such that the gravitational pull exerted on a mass of 1 kg near the earth's surface is about 9.8 N. A force of about 1 N is represented by the earth's pull on a medium-sized apple—a most appropriate result in view of the old tradition (which may well be true) that the fall of an apple provided the starting point of Newton's profound thoughts about gravitation.

In the CGS (centimeter-gram-second) system, the unit of force is the *dyne* (dyn), defined as the force that can give an acceleration of 1 cm/sec² to a mass of 1 g. Since 1 cm = 10^{-2} m, and 1 g = 10^{-3} kg, we have the relationship

$$1 \text{ dyn} = 10^{-5} \text{ N}$$

(In saying this we take the relation $F = ma$ as already established, so that changing m and a by given factors implies changing F by the product of those factors.) The dyne is thus an exceedingly small force, about equal to the earth's gravitational pull on a mass of 1 mg—as represented, for example, by roughly a $\frac{1}{8}$-in.-square piece of this page.

The only other force that we shall have occasion to mention is the *pound*. Unlike the newton and the dyne, the pound is (or at least was, originally) based directly on the measure of the earth's gravitational pull on a standard object. As soon as one recognizes that the gravitational pull on any given object changes from one place to another, this definition loses its exactness and has to be adjusted. However, setting aside such difficulties of detail, we can still say that the pound is a force equal in magnitude to about 4.5 N. Every time we buy something that is weighed out on a spring scale, we are accepting the use of gravitational units of force such as the pound. We shall consider the practical implications of that in more detail later (Chapter 8).

EQUILIBRIUM CONDITIONS; FORCES AS VECTORS

The static equilibrium of a given object entails two distinct conditions:

1. The object shall not be subject to any net force tending to

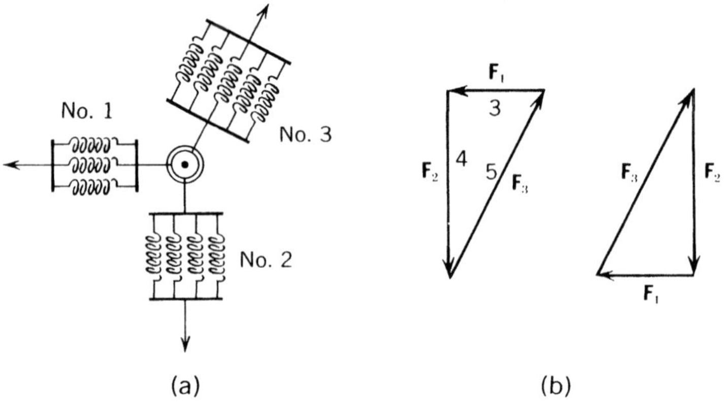

Fig. 4-4 (a) An object in equilibrium under the action of three nonparallel forces. (b) Two equally acceptable vector diagrams showing the equilibrium condition $\Sigma \mathbf{F} = 0$.

move it bodily; it is in what we call *translational equilibrium*.

2. The object shall not be subject to any net influence tending to twist or rotate it; it is in what we call *rotational equilibrium*.

The first condition involves, in general, the combination of forces acting in different directions, as in our initial example of the bowstring. It has been known since long before Newton's time that forces are *vectors*. This says much more than that they have characteristic directions; it says that they combine with one another in the same way as the prototype vector quantity, positional displacement. Imagine, for example, a ring that is a loose fit over a vertical peg. Suppose that it has three strings attached to it and that these strings are pulled by forces of relative magnitudes 3, 4, and 5, as defined by corresponding numbers of identical springs equally extended [see Fig. 4-4(a)]. Then experiment will show that the ring remains in equilibrium, even when the peg is removed, if the directions of the forces correspond exactly to those of a 3-4-5 triangle. This means that the forces, represented as vectors with lengths proportional to their magnitudes, form a closed triangle, which can be drawn in two different ways, as shown in Fig. 4-4(b). In formal terms, we say that the three force vectors add up to zero:

$$\mathbf{F}_1 + \mathbf{F}_2 + \mathbf{F}_3 = 0$$

It is a basic property of vectors that the order of addition is immaterial (Chapter 2). Thus, if we have a large number of forces applied to the same object, it is possible to represent their addition in many different ways; Fig. 4-5 gives an example. The one essential feature is that, in every case, the force vectors form

Fig. 4-5 (a) Several forces acting at the same point. (b) The force vectors form a closed polygon, showing equilibrium. (c) Equivalent vector diagram to (b).

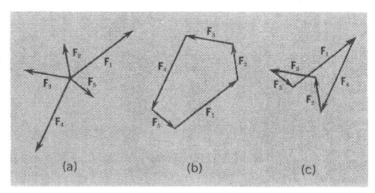

a closed polygon (i.e., they add up to zero) if equilibrium exists. Since forces may be applied in any direction in three-dimensional space, the force polygon is not necessarily confined to a plane, and the single statement that the force vectors add up to zero will in general be analyzable into three separate statements pertaining to three independently chosen directions—usually, although not necessarily, the mutually orthogonal axes of a rectangular coordinate system. Geometrically, one can think of this as the projection of the closed vector polygon onto different planes; regardless of the distortions of shape, the projected polygon remains a closed figure. When written out in algebraic terms, the projection involves a statement of the analysis of an individual force vector into components, or resolved parts, along the chosen directions. Thus the first condition of equilibrium—equilibrium with respect to bodily translation—can be written as follows:

Vector statement:

$$\mathbf{F} = \mathbf{F}_1 + \mathbf{F}_2 + \mathbf{F}_3 + \cdots = 0 \tag{4-1a}$$

Component statement:

$$F_x = F_{1x} + F_{2x} + F_{3x} + \cdots = 0$$
$$F_y = F_{1y} + F_{2y} + F_{3y} + \cdots = 0 \tag{4-1b}$$
$$F_z = F_{1z} + F_{2z} + F_{3z} + \cdots = 0$$

It is worth noticing the relationship between the process of adding force vectors and the process of resolving an individual force vector into its components. In Fig. 4-6(a) we show a force **F**, in the xy plane, resolved into the usual orthogonal components:

$$\mathbf{F} = F_x\mathbf{i} + F_y\mathbf{j}$$

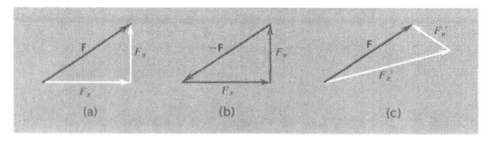

Fig. 4–6 (a) *Resolution of a force into orthogonal components.* (b) *The components of* **F**, *added vectorially to the force* −**F**, *give zero.* (c) *Resolution of* **F** *into nonorthogonal components.*

We represent **F** by a full line and its components by dashed lines, to emphasize that the components are a replacement—a substitute—for **F** itself. In Fig. 4–6(b) we show a closed vector force triangle, in which the vectors $F_x\mathbf{i}$ and $F_y\mathbf{j}$, *added to* the force vector −**F**, represent an equilibrium situation:

$$F_x\mathbf{i} + F_y\mathbf{j} + (-\mathbf{F}) = 0$$

We can then recognize that any other pair of vectors that give zero when added to −**F** are a complete equivalent to **F** itself. Thus we have at our disposal an infinite variety of ways of resolving a given force into components. For example, the two forces $F_{x'}$ and $F_{y'}$ in Fig. 4–6(c) could represent the way of analyzing **F** into components in an oblique coordinate system; the fact that $F_{x'}$ may happen to be larger than **F** itself does not invalidate this way of resolving the original force.

In a similar vein, if any set of forces combine to give zero, as in Fig. 4–7(a), we can regard one of them (which we have

Fig. 4–7 (a) *The combination of forces* $\mathbf{F}_1$, $\mathbf{F}_2$, *and* $\mathbf{F}_3$ *is balanced by the equilibrant* $\mathbf{F}_E$. (b) *The same combination of forces* $\mathbf{F}_1$, $\mathbf{F}_2$, *and* $\mathbf{F}_3$ *has the resultant* $\mathbf{F}_R$.

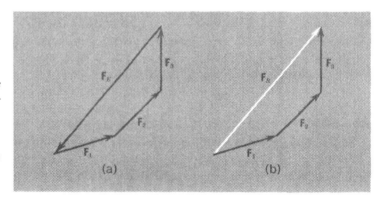

labeled here as F_E) as the *equilibrant* (i.e., the balancer) of all the others. A force equal and opposite to F_E is then a force completely equivalent to the vector sum of these other forces: it is their *resultant*, shown as F_R in Fig. 4–7(b). A familiarity with these simple relationships can be a considerable help in problems of combining or resolving forces and can strengthen one's understanding of equilibrium situations in general.

ACTION AND REACTION IN THE CONTACT OF OBJECTS

We have already used a kind of principle of uniformity to prescribe a way of reproducing a force of a given magnitude or making a known multiple of the force. We assume that identical springs, stretched or compressed by equal amounts, define equal forces. This assumption leads to consistent conclusions in the analysis of equilibrium situations. Picture now a simple experiment that any two people can do with a pair of identical springs. They stand facing each other, and one of them, *A*, agrees to be the active agent. His goal is to push (via the spring that he holds) on the end of the spring held by *B*, the passive partner, who tries to avoid pushing back. The size of the push that each person exerts is measured by the amount by which his spring is compressed.

You know the result of the experiment, of course. Regardless of the maneuvers made by *A* and *B*, the compressions of the two springs are always the same; only the amount of both compressions together can be varied. This displays, in a particularly direct manner, the phenomenon that is expressed by the familiar statement: Action and reaction are equal and opposite. In its general form this says that, regardless of the detailed form of the contact or of the relative hardness or softness of the two objects involved, the magnitudes of the forces that each object exerts on the other are always exactly equal. Note particularly that, from the very way they are defined, these two forces cannot both act on the same object. This may seem a trivial and obvious remark, but many calculations in elementary mechanics have come to grief through a failure to recognize it.

The production of a force of reaction in response to an applied force always involves deformation to some extent. You push on a wall, for example, and that is a conscious muscular act; but how does the wall know to push back? The answer is that it yields, however imperceptibly, and it is as a result of such elastic deformations that a contact force exerted by the wall comes into existence. No matter how rigid a surface may seem, it always gives a little under a push or a pull and cannot supply a contrary force until it has done so. There is no basic difference between what happens when we sit down in a comfortably upholstered chair and what happens when we sit down on a concrete floor. But in the one case the springs are soft and yield visibly by distances of an inch or more, whereas in the other case the springs are essentially the individual atoms in a tightly packed solid structure, and a deformation by a small fraction of an atomic diameter is enough to produce the support required.

ROTATIONAL EQUILIBRIUM; TORQUE

We shall now consider the second condition for static equilibrium of an object, assuming that the first condition, $\sum \mathbf{F} = 0$, has been satisfied. Whether the object is, in fact, in equilibrium will now depend on whether or not the forces are applied in such a way as to produce a resultant twist. Figure 4–8 illustrates the problem with the simplest possible example. An object is acted on by two equal and opposite forces. If, as in Fig. 4–8(a), the forces are along the line joining the points A and B at which the forces are applied, the object is truly in equilibrium; it has no tendency to rotate. In any other circumstances [e.g., as shown in Fig. 4–8(b)] the object is bound to twist. If the directions of the forces remain unchanged as the object turns, equilibrium orientation is finally reached, as shown in Fig. 4–8(c). How do we con-

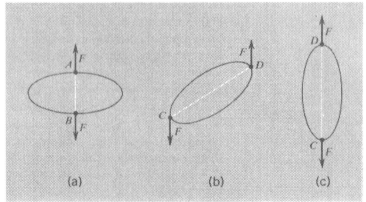

Fig. 4–8 (a) Rotational equilibrium with equal, opposite forces applied at different points of an extended object. (b) Equal, opposite forces applied in a way that does not give rotational equilibrium. (c) If free to rotate, the object moves from orientation (b) to an equilibrium orientation.

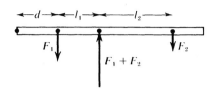

Fig. 4-9 All the forces acting on a pivoted bar (of negligible weight) in rotational equilibrium with one force applied on each side of the pivot.

struct out of such familiar knowledge a quantitative criterion for rotational equilibrium?

The law of the lever provides the clue. Look again at the situations shown in Fig. 4–2. In particular, consider situation (b). The balancing of the forces F_1 and F_2 with respect to the pivot at O requires the condition $F_1 l_1 = F_2 l_2$. The product of the force and its lever arm describes its "leverage," or twisting ability; the technical term for this is *torque*. The torques of F_1 and F_2 with respect to O are equal in magnitude but opposite in direction—that due to F_2 is clockwise and that due to F_1 is counterclockwise. Let us call one of them positive and the other one negative; then the condition of balance can be expressed in another way: *The total torque is equal to zero.*

Although the situation as described above is extremely simple, there is more to it than meets the eye, because a further force is exerted on the bar at the position of the pivot; it must be of magnitude $F_1 + F_2$ if the first condition of equilibrium is to be satisfied. Thus the complete array of forces is as shown in Fig. 4-9. Now, to be sure, this third force exerts no torque about the pivot point O itself. However, what if we choose to consider the torques about, let us say, the left-hand end of the bar, some distance d to the left of the point of application of F_1? Then clearly, with respect to this new origin, the force at O is supplying a counterclockwise torque—but it turns out that this is exactly balanced by the sum of the torques due to F_1 and F_2 (both clockwise, notice, with respect to the new, hypothetical pivot), provided that the condition $F_1 l_1 = F_2 l_2$ is satisfied. If this result is new to you, take a moment to convince yourself of its correctness. What it says is that, if the vector sum of the forces on an object is zero, and if the sum of the torques about any one point is zero, then the sum of the torques about any other point is also zero.

So far we have limited ourselves to the balancing of torques of parallel forces. Now let us make things more general. Suppose that a force **F** is applied at a point P, somewhere on or in an

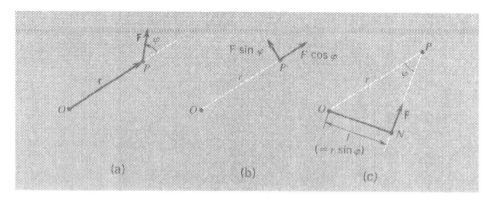

*Fig. 4–10 (a) Force **F** applied at a vector distance **r** from a pivot point. (b) Resolution of **F** into components along and perpendicular to **r**. (c) Evaluation of torque of **F** by finding its effective lever arm, l.*

object [Fig. 4–10(a)]. Consider the torque produced by **F** about some other point O, which might be the position of a real pivot, or just an arbitrary point. Let the vector distance from O to P be **r**. The first thing to notice is that **r** and **F** between them define a plane, which we have chosen to be the plane of the diagram. Experience, so familiar that it has become second nature, tells us that if O were indeed a real pivot point, the effect of **F** would be to produce rotation about an axis perpendicular to the plane in which **r** and **F** lie. It therefore makes excellent sense to associate this direction with the torque itself, regarded as a vector of some sort. Now, what about the magnitude of the torque? We can calculate this in two ways. The first, indicated in Fig. 4–10(b), is to resolve the force into components along and perpendicular to **r**. If the angle between **r** and **F** is φ, these components are $F\cos\varphi$ and $F\sin\varphi$, respectively. The radial component represents a force directed straight through O and hence contributes nothing to the torque. The transverse component, perpendicular to **r**, gives a torque of magnitude $rF\sin\varphi$. Another way of seeing this result is, as suggested in Fig. 4–10(c), to recognize that the effective lever arm of the total force **F** can be formed by drawing the perpendicular ON from O to the line along which **F** acts. Then the torque due to **F** is just the same as if it were actually applied at the point N, at right angles to a lever arm of length l equal to $r\sin\varphi$.[1]

We shall introduce the single symbol M for the magnitude

[1] Leonardo da Vinci envisaged the effective lever arm of a force in this way and called it "the spiritual distance of the force."

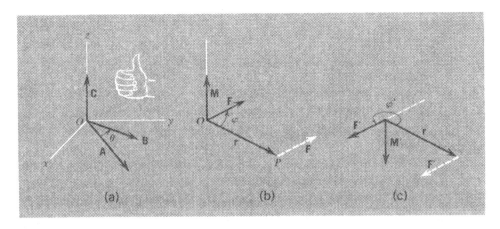

Fig. 4-11 (a) Cross product, C, of two arbitrary vectors, A and B. (b) Torque vector, M, as the cross product of r and F. (c) Situation resulting in a torque vector opposite in direction to that in (b).

of the torque. Then we have

$$M = rF \sin \varphi \qquad (4\text{-}2a)$$

This equation does not contain the necessary information about the *direction* of the torque, but a compact statement in vector algebra, invented specifically for such purposes, is at hand. This is the so-called cross product or vector product of two vectors.

Given two vectors **A** and **B**, the *cross product* **C** is defined to be a vector perpendicular to the plane of **A** and **B** and of magnitude given by

$$C = AB \sin \theta$$

where θ is the lesser of the two angles between A and B (the other being $2\pi - \theta$). There are, of course, two opposite vector directions normal to the plane of **A** and **B**. To establish a unique convention we proceed as follows. Imagine rotating the vector **A** through the (smaller) angle θ until it lies along the direction of **B** [see Fig. 4-11(a)]. This establishes a sense of rotation. If the fingers of the *right* hand are curled around in the sense of rotation, keeping the thumb extended (do it!), the direction of the cross product is along the direction of the pointing thumb. The shorthand mathematical statement, which is understood to embody all these properties, is then written as

$$\mathbf{C} = \mathbf{A} \times \mathbf{B}$$

Note carefully that the order of the factors is crucial; reversing the order reverses the sign:

B × **A** = −(**A** × **B**)

Using this vector notation, the torque as a vector quantity is completely specified by the following equation:

M = **r** × **F** (4-2b)

Figure 4-11(b) and (c) illustrates this for two different values of φ; in each case a right-handed rotation about the direction in which **M** points represents, as you can verify, the direction in which **F** would cause rotation to occur. Then, finally, we can write down the vector sum of all the torques acting on an object, and the second condition of equilibrium—equilibrium with respect to rotation—can be written as follows:

$$\sum \mathbf{M} = \mathbf{r}_1 \times \mathbf{F}_1 + \mathbf{r}_2 \times \mathbf{F}_2 + \mathbf{r}_3 \times \mathbf{F}_3 + \cdots = 0 \quad (4\text{-}3)$$

If the object on which a set of forces acts can be regarded as an ideal particle (i.e., a point object), then the condition of rotational equilibrium becomes superfluous. Since all the forces are applied at the same point, they cannot exert a net torque about this point; and if the condition $\sum \mathbf{F} = 0$ is also satisfied, they cannot exert a net torque about any other point either. If one wants to put this in more formal terms, one can say that the same value of **r** applies to every term in Eq. (4-3), so that the condition $\sum \mathbf{M} = 0$ reduces to the condition

r × ($\sum$**F**) = 0

and so embodies the condition $\sum \mathbf{F} = 0$ for translational equilibrium; the equation for rotational equilibrium adds no new information.

FORCES WITHOUT CONTACT; WEIGHT

Figure 4-12 depicts a pair of situations that look so simple, and embody results that are so familiar, that you may never have paused to wonder about the relation between them. In Fig. 4-12(a) an object is shown attached to two spring balances that pull on it horizontally with equal and opposite forces—a clear case of static equilibrium. In Fig. 4-12(b) the same object hangs

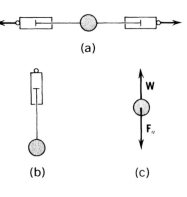

Fig. 4-12 (a) Object in equilibrium under equal and opposite horizontal forces (from spring balances). (b) Object in equilibrium under the action of a single vertical contact force (from a spring balance) and an invisible influence (gravity). (c) The reading on the spring balance in (b) defines the measured weight (W) of the object. By inference the gravitational force F_g is equal and opposite to W.

from a single spring balance that pulls vertically upward on it. Experience tells us that this is also a case of static equilibrium, yet as far as visible connections are concerned, it is quite clearly an unbalanced situation. How do we resolve the dilemma? This may seem like a trite or even trivial question. In fact, it is a very profound one. At this time we shall examine it only from the standpoint of our knowledge of static equilibrium. Later we shall see that it has much wider ramifications.

Try to imagine coming to situation (b) of Fig. 4–12 for the first time, with a background of experience in other kinds of static equilibrium. You have learned from this previous experience that equilibrium always corresponds to having the vector sum of the applied forces equal to zero. You are used to having tangible and visible evidence of these forces being applied, via strings, springs, and so on. Now you see an equilibrium situation in which only one force is in full view, so to speak. But your confidence in the general validity of the equilibrium conditions is so strong that you say: "Even though there is no other contact, there must be another force acting on the object, equal and opposite to the force supplied by the spring." And so you postulate the force of gravity, pulling downward on the object. But your only measure of it is the reading on the spring balance that counteracts this gravitational force.

It is in terms such as these that we have come to postulate and accept the existence of a downward gravitational force exerted on every object at or near the earth's surface. In order to hold an object at rest relative to the earth, we must apply a certain upward force on it. What we shall call the *weight* of an object is the downward force equal and opposite to this equilibrating force. This is an important definition; we shall restate it: *The weight of* an object, **W**, will be defined as the downward force whose magnitude W is equal to that of the upward force that must be applied to the object to hold it at rest relative to the earth. This is an example of what is called an *operational* definition; we describe the actual process that is to be followed to get a practical measurement of the quantity in question. Notice especially that we are *not* defining the weight of an object as the gravitational force on it. If the connection with the supporting spring is broken, so that the object begins to fall freely, then by our definition the object is now *weightless*, although there is no suggestion that the gravitational force on the object has been changed in any way in the process of breaking the connection. (We shall discuss weightlessness further in Chapter 8.)

If we return to the situation in which the object is held stationary relative to the earth by the pull of the spring, then our picture of the forces acting on the object is as shown in Fig. 4–12(c). On the assumption that this is indeed a static equilibrium situation, then we can say that the spring force **W** and the gravitational force $\mathbf{F}_g$ are equal and opposite, so that the magnitude of **W** does provide, under these circumstances, a measure of the magnitude of $\mathbf{F}_g$. But keep it firmly in mind that, in our use of the terms, weight and gravitational force are not synonymous. By maintaining this distinction we shall be much better equipped to handle dynamical problems involving gravity later on.

PULLEYS AND STRINGS

The use of pulleys and strings to transmit forces has more physics in it than one might think and contains some nice applications of the principles of static equilibrium. Consider a string passing over a circular pulley of radius R [Fig. 4–13(a)]. Let the string be pulled by forces $\mathbf{F}_1$ and $\mathbf{F}_2$ as shown. If this is to be a situation of static equilibrium, the pulley must be in both translational and rotational equilibrium. Let us consider the rotational equilibrium first. If the pivot is effectively frictionless, the only torques on the pulley are supplied by the forces $\mathbf{F}_1$ and $\mathbf{F}_2$. Since both of these are applied tangentially, they have equal lever arms of length R (assuming that the pulley is pivoted exactly at its center). Therefore, to give zero net torque, the magnitudes of $\mathbf{F}_1$ and $\mathbf{F}_2$ must both be equal to the same value, T, which we can thus call *the* tension in the string—the strength

Fig. 4-13 (a) Tensions in the segments of a string where it meets a stationary circular pulley must be in rotational equilibrium. (b) Set of three forces involved in the static equilibrium of a pulley. (c) Pulley as a device for applying a force of given magnitude in any desired direction.

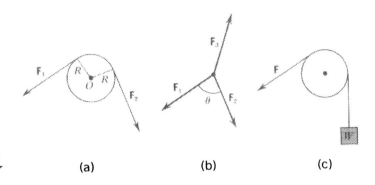

of the pull that would be found to be exerted by the string at any point along its length. The pulley thus enables us to change the direction of an applied force without changing its magnitude. However, to satisfy the condition of translational equilibrium, the pivot must be able to supply a force $\mathbf{F}_3$ equal and opposite to the vector sum of $\mathbf{F}_1$ and $\mathbf{F}_2$. This force must therefore lie along the bisector of the angle θ between the two straight segments of the string, and its magnitude must be equal to $2T \cos(\theta/2)$ [see Fig. 4-13(b)].

If the tension at one end of a string passing over a pulley is supplied by a suspended object, then a force $\mathbf{F}$ of magnitude equal to the weight, W, of the object is needed to maintain equilibrium [Fig. 4-13(c)]. This means that a pulley-string system can be used to supply a force of magnitude W in any desired direction. Figure 4-14 illustrates the typical kind of arrangement that exploits these features in a simple experiment to study the equilibrium of three concurrent forces. To the extent that the pulleys can be treated as ideal, the magnitudes of the forces $\mathbf{F}_1$,

Fig. 4-14 (a) Simple static equilibrium arrangement involving three nonparallel forces (string tensions) applied at the point P. (b) Vector diagram showing the equilibrium condition.

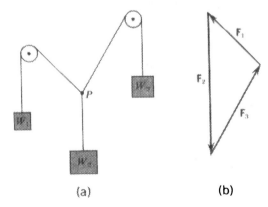

F_2, and F_3 are equal to the respective weights W_1, W_2, and W_3 of the suspended objects.

PROBLEMS

4-1 The ends of a rope are held by two men who pull on it with equal and opposite forces of magnitude F. Construct a clear argument to show why the tension in the rope is F, not $2F$.

4-2 It is a well-known fact that the total gravitational force on an object may be represented as a single force acting through a uniquely defined point—the "center of gravity"—regardless of the orientation of the object.

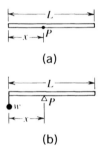

(a) For a uniform bar, the center of gravity (CG) coincides with the geometrical center. Use this fact to show that the total gravitational torque about the point P [see part (a) of the figure] may be considered as arising from a single force W at the bar's center, or from two individual forces of magnitudes Wx/L and $W(L - x)/L$ acting at the midpoints of the two segments defined by P.

(b) If a bar or rod has a weight W, and a small weight w is hung at one end [see part (b) of the figure], use the simpler of the above two methods to show that the system balances on a fulcrum placed at P if $x = LW/2(W + w)$.

4-3 Diagram (a) represents a rectangular board, of negligible weight, with individual concentrated weights mounted at its corners.

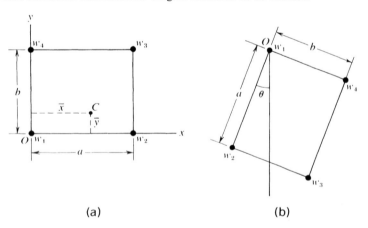

(a)   (b)

(a) To find the position of the CG of this system, one can proceed as follows: Choose an origin at the corner O, and introduce x and y axes as shown. Imagine the board to be pivoted about a horizontal axis along y, and calculate the distance $\bar{x}$ from this axis at which an upward force W ($= w_1 + w_2 + w_3 + w_4$) will keep the system in rotational equilibrium. Next imagine the board to be pivoted about a horizontal axis along x, and calculate the corresponding distance $\bar{y}$. Then the center of gravity, C, is at the point $(\bar{x}, \bar{y})$.

(b) An experimental method of locating the CG is to hang the board from two corners in succession (or any other two points, for that matter) and mark the direction of a plumbline across the board in each case. To verify that this is consistent with (a), imagine the board to be suspended from O in a vertical plane [diagram (b)] and show by direct consideration of the balancing of torques due to w_2, w_3, and w_4 that the board hangs in such a way that the vertical line from O passes through C (so that $\tan \theta = \bar{y}/\bar{x}$).

4-4 You have just finished a 20-page letter to your girl friend (or boyfriend) and you want to mail it at once, so that it will be collected first thing in the morning. You have a supply of stamps, but unfortunately it is 2 A.M., the local post office is closed, and you haven't a letter scale of your own. However, you do have a 12-in. ruler and a nickel, and you happen to have learned somewhere that the density of nickel is about 9 g/cm^3. The ruler itself balances at its midpoint. When the nickel is placed on the ruler at the 1-in. mark, the balance point is at the 5-in. mark. When the letter is placed on the ruler, centered at the 2-in. mark, the balance point is at $3\frac{1}{4}$ in. The postal rate is 20 cents per half-ounce (international airmail). How much postage should you put on? (This problem is drawn from real life. You might like to do a similar experiment for yourself, perhaps using a different coin as your standard of weight.)

4-5 (a) As mentioned in the text (p. 117), Archimedes gave what he believed to be a theoretical proof of the law of the lever. Starting from the necessity that equal forces, F, at equal distances, l, from a fulcrum must balance (by symmetry), he argued that one of these forces could, again by symmetry, be replaced by a force $F/2$ at the fulcrum and another force $F/2$ at $2l$. Show that this argument depends on the truth of what it is purporting to prove.

(b) A less vulnerable argument is based on the experimental

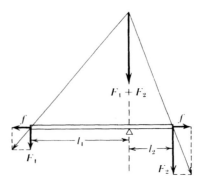

knowledge that forces combine as vectors. Suppose that parallel forces F_1 and F_2 are applied to a bar as shown. Imagine that equal and opposite forces of magnitude f are introduced as shown. This gives us two resultant force vectors that intersect at a point that defines the line of action of their resultant (of magnitude $F_1 + F_2$). Show that this resultant intersects the bar at the pivot point for which $F_1 l_1 = F_2 l_2$.

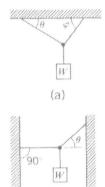

4-6 Find the tensions in the ropes in the two configurations shown. The weight W is in static equilibrium in each case.

4-7 (a) A clothesline is tied between two poles, 10 m apart, in such a way that the sag is negligible. When a wet shirt with a mass of 0.5 kg is hung at the middle of the line, the midpoint is pulled down by 8 cm. What is the tension in the clothesline?

(b) (With acknowledgements to F. W. Sears.) A car is stranded in a ditch, but the driver has a length of rope. The driver knows that he is not strong enough to pull the car out directly. Instead, he ties the rope tightly between the car and a tree that happens to be 50 ft away; he then pushes transversely on the rope at its midpoint. If the midpoint of the rope is displaced transversely by 3 ft when the man pushes with a force of 500 N (= 50 kg), what force does this exert on the car? If this were sufficient to begin to move the car, and the man pushed the midpoint of the rope another 2 ft, how far would the car be shifted, assuming that the rope does not stretch any further? Does this seem like a practical method for dealing with the situation?

4-8 Prove that if three forces act on an object in equilibrium, they must be coplanar and their lines of action must meet at one point (unless all three forces are parallel).

4-9 Painters sometimes work on a plank supported at its ends by long ropes that pass over fixed pulleys, as shown in the figure. One end of each rope is attached to the plank. On the other side of the pulley the rope is looped around a hook on the plank, thus holding the plank at any desired height. A painter weighing 175 lb works on such a plank, of weight 50 lb.

(a) Keeping in mind that he must be able to move from side to side, what is the maximum tension in the ropes?

(b) Suppose that he uses a rope that supports no more than 150 lb. One day he finds a firm nail on the wall and loops the rope around this instead of around the hook on the plank. But as soon as he lets go of the rope, it breaks and he falls to the ground. Why?

4-10 An inn sign weighing 100 lb is hung as shown in the figure. The supporting arm, freely pivoted at the wall, weighs 50 lb, and the system is held by a guywire that should not be subjected to a tension of more than 250 lb.

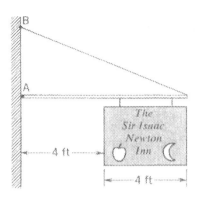

(a) What is the minimum safe distance of the point B above point A?

(b) What is the magnitude and direction of the force exerted on the supporting arm at A under these conditions?

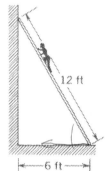

4-11 A man begins to climb up a 12-ft ladder (see the figure). The man weighs 180 lb, the ladder 20 lb. The wall against which the ladder rests is very smooth, which means that the tangential (vertical) component of force at the contact between ladder and wall is negligible. The foot of the ladder is placed 6 ft from the wall. The ladder, with the man's weight on it, will slip if the tangential (horizontal) force at the contact between ladder and ground exceeds 80 lb. How far up the ladder can the man safely climb?

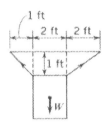

4-12 You want to hang a picture at a certain place on a wall, but the only available nails are at points 1 ft to the left and 2 ft to the right of the edges of the picture (see the figure). You attach strings of the appropriate lengths from these nails to the top corners of the picture, as shown, but the picture will not hang straight unless you add a balancing weight of some kind.

(a) If the picture with its frame weighs 10 lb, what is the least possible balancing weight, and where would you put it? (*Hint:* Find the point of intersection of the two tension forces in the strings.)

(b) In the absence of the balancing weight, how would the picture hang? (If you want to go beyond a qualitative discussion, be prepared for some rather messy trigonometry.)

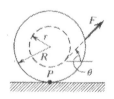

4-13 A yo-yo rests on a table (see the figure) and the free end of its string is gently pulled at an angle θ to the horizontal as shown.

(a) What is the critical value of θ such that the yo-yo remains stationary, even though it is free to roll? (This problem may be solved geometrically if you consider the torques about P, the point of contact with the table.)

(b) What happens for greater or lesser values of θ? (If you have a yo-yo, test your conclusions experimentally.)

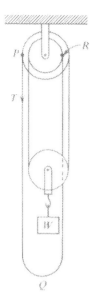

4-14 A simple and widely used chain hoist is based on what is called a differential pulley. In this arrangement two pulleys of slightly different diameter are rigidly connected with a common, fixed axis of rotation. An endless chain passes over these pulleys and around a free pulley from which the load W is suspended (see the figure). If the components of the differential pulley have radii a and $0.9a$, respectively, what downward pull applied to one side of the freely hanging part of the chain will (ignoring friction) suffice to prevent the load from descending if

(a) the weight of the chain itself can be neglected?

(b) (more realistic) one takes account of the fact that the freely hanging portion of the chain (PQR) has a total weight w?

4-15 A force **F** with components $F_x = 3N$, $F_y = 4N$, and $F_z = 0$ is applied at the point $x = 0$, $y = 5\,\text{m}$, and $z = 4\,\text{m}$. Find the magnitude and direction of the torque, **M**, of **F** about the origin. (Express the direction in terms of the direction cosines—i.e., the cosines of the angles that **M** makes with the axes.)

4-16 Analyze in qualitative but careful terms how the act of pushing vertically downward on the pedal of a bicycle results in the production of a horizontal force that can accelerate the bicycle forward. (Clearly the contact of the rear wheel with the ground plays an essential role in this situation.)

And thus Nature will be very conformable to herself and very simple, performing all the great Motions of the heavenly Bodies by the attraction of gravity . . . and almost all the small ones of their Particles by some other attracting and repelling Powers . . . .

NEWTON, *Opticks* (1730)

5
The various forces of nature

THE BASIC TYPES OF FORCES[1]

ALL FORCES ARISE from the interactions between different objects. Once upon a time it must have seemed that these interactions were bewilderingly diverse, and one of the most remarkable features in the development of modern science has been the growing realization that only a very few basically distinct kinds of interaction are at work. The following are the only forces that we know of at present:

1. Gravitational forces, which arise between objects because of their masses.

2. Electromagnetic forces, due to electric charges at rest or in motion.

3. Nuclear forces, which dominate the interaction between subatomic particles if they are separated by distances less than about 10^{-15} m.

It may be that even this degree of categorization will prove

---

[1] An excellent background to this topic is the PSSC film, "Forces," by J. R. Zacharias, produced by Education Development Center, Inc., Newton, Mass., 1959. The title of this chapter is borrowed from that of a set of popular lectures delivered in London by the great scientist, Michael Faraday, just over 100 years ago and available in paperback (Viking, New York, 1960). They make easy and rather delightful reading.

to be unnecessarily great; the theoretical physicist's dream is to find a unifying idea that would allow us to recognize *all* these forces as aspects of one and the same thing. Albert Einstein spent most of his later years on this problem but to no avail, and at the present time the assumption of several different kinds of forces seems meaningful as well as convenient.

In the following sections we shall briefly consider these three primary types of forces, with examples of physical systems in which they are significant. It will be useful, and from the standpoint of classical mechanics very important, to add to our classification what we shall call "contact forces"—the forces manifested in the mechanical contact of ordinary objects. Although these forces are merely the gross, large-scale manifestation of the basic electromagnetic forces between large numbers of atoms, they serve so well to describe most of the familiar interactions in mechanical phenomena that they merit a category of their own.

GRAVITATIONAL FORCES

All our experience suggests that a gravitational interaction between material objects is a universal phenomenon. It is always an attractive interaction. The gravitational forces exerted by the earth on different objects near its surface can be compared, by using a spring balance for example, and these gravitational attractions are proportional, in every case, to the property of the attracted object that we call its mass. (The content and the implications of this familiar and seemingly simple statement will be a matter for detailed discussion in later chapters.)

The general law of gravitational interaction arrived at by Newton states that the force **F** with which any particle attracts any other is proportional to the product of the masses of the particles, inversely proportional to the square of their separation, and directed along the line separating the two particles. It is found experimentally that this force has no measurable dependence on the velocity of the particle on which it acts.[1] The magnitude of the force $\mathbf{F}_{12}$ that a particle of mass m_1 exerts on a

---

[1] Departures from this velocity independence are analyzed in the general theory of relativity. They are discernible only if the effect of quantities of the order of v^2/c^2 (where c is the speed of light) can be detected in the gravitational interaction.

particle of mass m_2 can be written as

$$F_{12} = G\frac{m_1 m_2}{r_{12}^2} \tag{5-1}$$

where r_{12} is the distance from the center of m_1 to the center of m_2 and G is a constant of proportionality called the universal gravitational constant. This law of force holds for point masses and for uniform spheres of finite size. The value of the constant G is 0.667×10^{-10} m^3/kg-sec^2.

Equation (5-1), and the preceding verbal expression of it, is the first example in this book of the quantitative expression of a physical law. It is worth spending a few words to discuss what it really says. Every mathematical statement of experimental relationships in physics is no more than a statement of a relationship between *numbers*. By m_1 and m_2 we simply mean the *numerical measures* of the masses of particles 1 and 2 in terms of some arbitrarily chosen unit. The concept of mass is one that has been developed to aid us in our description of nature—likewise the concepts of force and distance. But it is always the numbers that we deal with. Thus the full verbal equivalent of Eq. (5-1) would be:

The numerical measure of **the force F with which any particle attracts any other is proportional to the product of the** *numerical measures of* **their masses and inversely proportional to the square of the** *numerical measure of* **their separation.**

Keep those italicized phrases in mind whenever you read the mathematical statement of a physical relationship. They are almost never included explicitly, yet without them there is a danger of reading more into such a statement than it really contains. Thus the customary type of abbreviated colloquial statement of the law of gravitation begins: "The force of gravitational attraction between two particles is proportional to the product of their masses . . . ." What, one may ask, is meant by the *product* of two masses? What sort of a physical quantity is that? Even very good scientists have been drawn, at times, into almost metaphysical arguments through the effort to read some special significance into the "dimensions" of such combinations. By reminding oneself what an equation such as Eq. (5-1) actually represents, one can avoid any such confusion.

The classic experiment to measure G was performed by the British physicist Henry Cavendish in 1798. It involved a

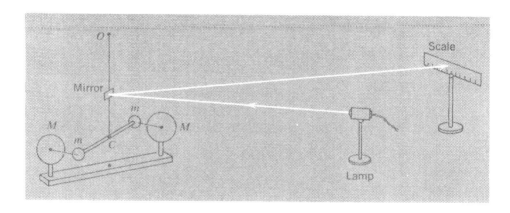

Fig. 5-1 Schematic diagram of a gravity torsion-balance experiment.

measurement of the gravitational force between solid lead spheres of modest dimensions and made use of an ingenious mechanical arrangement—a torsion balance—to detect this tiny attractive force. Figure 5-1 shows the essential features of the arrangement. Two small lead spheres are placed at the ends of a light rod. The rod is suspended horizontally by a very thin metal fiber attached to its midpoint. Two larger lead spheres are positioned close to the smaller ones, in symmetrical fashion, so that the gravitational attraction between these pairs of large and small masses tends to rotate the rod in a horizontal plane. An equilibrium orientation is reached when the twist of the supporting fiber provides a restoring effect that just balances the gravitational attraction. A beam of light reflected from a small mirror attached to the rod allows the tiny angular deflection to be amplified into a substantial movement of a spot formed by the reflected light on a distant wall (the "optical lever" effect). A torsion-balance arrangement of this type, incorporating an optical lever, is one of the most sensitive of all mechanical devices.

The gravitational force is astonishingly weak under the conditions of a laboratory experiment involving the interaction of relatively small objects, and the detection and measurement of it is an extremely delicate operation. For example, a modern version of the Cavendish apparatus[1] uses two small suspended lead spheres, each of 15-g mass, and large spheres of 1.5 kg each. The center-to-center distance between a small sphere and a

[1] Manufactured by the Leybold Co.

Fig. 5-2 Globular cluster of stars held together by gravitational forces (globular cluster M 13 in the constellation Hercules). (Photograph from the Hale Observatories.)

large one is about 5 cm. Under these conditions the gravitational force of attraction is only about 6×10^{-10} N. The weight of a single human hair is about 10,000 times greater than this!

Although the gravitational interaction is intrinsically so very feeble, it plays the prime role in most astronomical systems, because (1) the interacting objects are extremely massive and (2) other forces are almost absent. An interesting example is provided by globular star clusters. These are collections of stars in a spherically symmetric distribution; more than 120 such clusters have been identified in our galaxy. Figure 5-2 shows a globular cluster containing perhaps more than a million stars. One can infer from the symmetry of this system that there is probably no net rotational motion to the cluster as a whole though individual stars undoubtedly travel in all directions. Direct experimental evidence on the actual motions of individual stars is very meager. If we consider a star "at rest" near the outer edge of the cluster, the net force due to all other stars should be (by symmetry) toward the center of the cluster and the star will accelerate, reaching its maximum speed at the center. The net force on the star (again by symmetry considerations) will approach zero as the star reaches the center of the cluster. Having gained considerable speed, the star will pass through the center and gradually slow down due to the resultant force of attraction

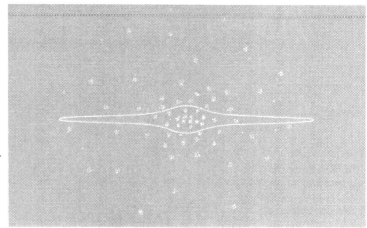

Fig. 5-3 Schematic diagram of a group of globular clusters associated with our Galaxy.

of all the other stars (which at all times is a net force directed toward the center of the cluster). The star will finally reach a point diametrically opposite its starting position. Thus a star could perform oscillations along a diameter passing through the center of cluster. Because the separation between stars is so very much greater than the diameter of any star, the chance of a stellar collision is very small, even at the center of the cluster (although the photograph would not suggest this).

There is clear astronomical evidence for other systems of a similar type but on a much larger scale. Our Galaxy is surrounded by a spherically symmetric "halo" of globular clusters (see the sketch shown in Fig. 5-3). Since the individual clusters maintain their compact identity, each can itself be considered a "mass point," and together they form a sort of supercluster of globular clusters. Direct observational evidence shows that clusters are traveling in all directions and there is no reason to doubt that a single cluster can perform oscillations through the heart of our Galaxy (just as single stars can oscillate within a cluster), again with almost negligible chance of individual stars colliding with each other.

These two similar examples illustrate cases where gravitational attraction between masses is the sole force that governs the motion. Usually gravitational forces are important only where at least one body of astronomical size is involved—note that it is solely because the earth is in this category that gravity exercises a major influence on our everyday lives. One can think of certain exceptions to this general rule, as for example the initial stages of the aggregation of neutral hydrogen atoms under their mutual gravitational attractions to form a protogalaxy.

ELECTRIC AND MAGNETIC FORCES

The forces that electrically charged particles exert on one another are of fundamental importance in nature. The basic law of electric force is that found by the nineteenth-century French physicist C. A. Coulomb, and known by his name. Like Cavendish, Coulomb used a torsion balance for his measurements.[1] But whereas Cavendish simply measured the gravitational constant, taking the basic form of the force law as already known, Coulomb explored the actual form of the law of electric force. Coulomb's law states that a charged particle at rest will attract or repel another charged particle at rest with a force proportional to the product of the charges, inversely proportional to the square of their separation and directed along the line separating the two particles.[2] The force is attractive when the charges are unlike and repulsive when they are alike in sign. If one denotes by q_1 and q_2 the charges carried by the particles, the magnitude of the force that particle 1 exerts on particle 2 is given by

$$F_{12} = k \frac{q_1 q_2}{r_{12}^2} \qquad (5-2)$$

identical in form with the law of universal gravitation. The metric unit of electric charge is the coulomb (C). The constant k in Coulomb's law has the value 9×10^9 N-m^2/C^2.

A coulomb is a huge amount of electric charge—vastly more than is usually found in isolation in nature. One coulomb separated from a similar charge 1 mile distant would experience a repulsive force of about 3500 N (approximately 800 lb). Yet an electrically neutral droplet of water $\frac{1}{8}$ in. in diameter contains nearly 1000 C of positive charge in the nuclei of the hydrogen and oxygen atoms, balanced by an equal amount of negatively charged electrons. (Check this for yourself.)

---

[1] Their experiments were performed at almost the same time, but apparently they acted quite independently of one another in their choice and development of the torsion-balance technique.

[2] Note that insertion, where appropriate, of the phrase "numerical measure of" is tacitly assumed.

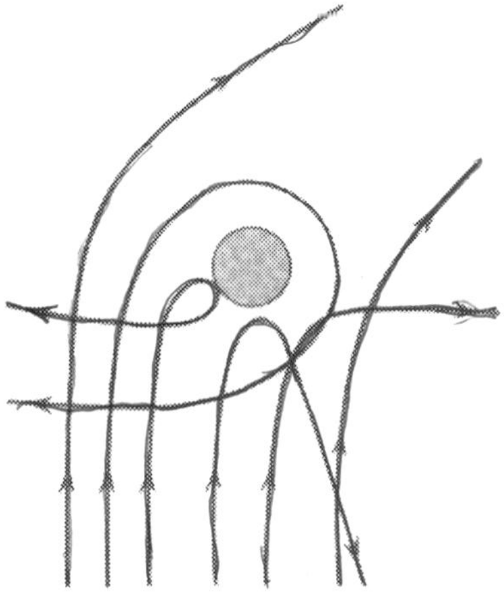

Fig. 5-4 Calculated trajectories of protons of about 14 Gev kinetic energy approaching the earth in its magnetic equatorial place and deflected by its magnetic field. (After D. J. X. Montgomery, Cosmic Ray Physics, *Princeton University Press, 1949.)*

It is interesting to compare the sizes of electrical forces with those of gravity. Consider, for example, the gravitational torsion balance mentioned earlier. If only one out of every 10^{18} electrons in each lead sphere were missing, the resultant imbalance of electric charge on the masses would produce an electrical force comparable to the gravitational force. Through such examples one can appreciate the immense strength of the electrical force compared to the gravitational force.

Although gravity is always present, the electrical force is overwhelmingly the most significant agent in all chemical and biological processes and in the interactions between physical objects of everyday size (i.e., below those of the astronomical domain). It holds atoms together, provides the rigidity and tensile strength of material objects, and is the only force involved in chemical reactions.

We have been discussing the electrical forces between stationary charged particles. *Moving* charged particles also exert electrical forces on each other. But an additional force arises in this case which we call the magnetic force. It has the interesting property that it depends on the velocity of the charges and always acts on a given charged particle at right angles to the particle's motion. We shall discuss these magnetic forces more precisely in Chapter 7.

An illustration of the magnetic force is provided by the trajectories of protons (positively charged) ejected from the sun and approaching the earth. Figure 5-4 shows possible paths of protons approaching in the equatorial plane, when they pass in the vicinity of the earth's magnetic field. It is confidently believed that the earth's magnetic field itself arises as a result of charged particles in motion inside the earth, so basically this is an example of the electromagnetic interaction between charges in motion.

Actually, from the standpoint of relativity theory, the magnetic force is not something new and different. Charges that are moving with respect to one observer can be stationary with respect to another. Thus, if one accepts the basic idea of relativity, one may expect to be able to relate a magnetic force, as observed in one reference frame, to a Coulomb force, as observed in another frame. For the detailed working out of this idea, see for example the volume *Special Relativity* in this series.

NUCLEAR FORCES

Although electric forces are responsible for holding atoms together, they would, by themselves, prevent the existence of atomic nuclei. For we all know that nuclei contain protons, electrically repelling one another and not stabilized by a compensating negative charge. But nature has supplied another force, known as the *strong interaction*, which binds together the nucleons (protons and neutrons) in a nucleus. Although much stronger than the Coulomb force at sufficiently short distances, its properties were relatively unknown until recently because of its extremely short range. For distances greater than about 10^{-13} cm ($= 1$ F) this nuclear force quickly becomes negligibly small, but it dominates over all other interactions between nucleons at shorter distances. It is an exceedingly complex type of interaction, attractive down to about 0.4 F and strongly repulsive for still smaller separations. It is, in part, a noncentral force—that is, in contrast to the gravitational and Coulomb interactions, it is not directed along the line joining the centers of the interacting particles. Somewhat analogously to the Coulomb force, which exists only between electrically charged particles, the strong nuclear interaction exists only among a certain class of particles, known as *hadrons* (Gr. *hadros:* heavy, bulky), which besides the nucleons themselves includes a number of lighter particles (mesons) and heavier particles (baryons), all of which are unstable and very short-lived (10^{-8} sec or less).

Another type of force associated with nuclear interactions

is also known to exist but is not very well understood as yet. It is called the "weak interaction." The range of this force is even less than that of the strong interactions, and its strength is estimated to be only about 10^{-15} as great as the other. This weak interaction is important only for certain types of nuclear process, such as radioactive beta decay.

It is instructive to compare the magnitudes of the different types of forces exerted between two protons 10^{-15} m apart—i.e., separated by about one nucleon diameter:

Type of interaction	Approximate magnitude of the force, N
Gravitational	2×10^{-34}
Coulomb	2×10^2
Nuclear (strong)	2×10^3

The magnitude of the weak nuclear interaction, within the range in which it is effective, is of the order of 10^{-13} of the electrical force, but even so it is greater, by a factor of the order of 10^{23}, than the calculated gravitational force at the same distance between two particles.

The nuclear forces have really no place in a discussion of Newtonian mechanics. Indeed, the very use of the word "force" in connection with nuclear interactions is open to question, for any statement about the force between two nuclear particles is at best a remote inference. We cannot cite any direct observations comparable to those of Cavendish and Coulomb for gravitational and electrical forces. Moreover, the subject of nuclear forces, like everything else in subatomic physics, requires the ideas and techniques of quantum mechanics, which is a theory that has almost no use for the concept of force as such. Thus, when we talk of classical or Newtonian mechanics, we are really concerned with situations in which the only relevant kinds of interactions are electromagnetic or gravitational.

FORCES BETWEEN NEUTRAL ATOMS

It is a fact of profound importance that, in contrast to gravitation, the electric forces between individual particles may be repulsive as well as attractive, because of the existence of two different signs of electric charge. This duality of charge makes possible the existence of matter that is electrically neutral in bulk and of atoms that are individually neutral by virtue of having equal numbers of protons and electrons.[1]

On the face of it, therefore, two neutral atoms would exert no forces on one another at all when separated (if one excepts the usually negligible gravitational force). This, however, is not quite true. It *would* be the case if the positive and negative charges in an atom were located at a single point. We know, however, that the electrons and the nucleus are separated by a certain small amount. Also, the atom is not a rigid structure; and although the "center of gravity" of the negative charge distribution due to the electrons coincides with the positive nucleus when the atom is isolated, the approach of another particle can disturb this situation. One manifestation of this is a characteristic force of attraction between neutral atoms—a force named after the great Dutch physicist J. van der Waals. This force increases much more rapidly than $1/r^2$ as two atoms approach one another—or (to put it in a way that carries a subtly different message) the force falls off with increasing distance much more rapidly than the attraction between two unbalanced charges of opposite sign.

The basis of the van der Waals force is still, of course, the inverse-square law of force between point charges, but its detailed character cannot be calculated without the use of quantum mechanics. The final result is a force varying as $1/r^7$ between neutral atoms of the same kind.

There is, however, another kind of force that comes into play between neutral atoms when the attempt is made to squash them together. This is a positive (repulsive) force that increases with decreasing separation even more rapidly than the van der Waals force. The result is that the *net* force exerted by one neutral atom on another, as a function of the separation r between their centers, has the kind of variation shown in Fig. 5–5. The force passes through zero at a certain value of r, and this value (r_0) can be thought of as the sum of the two atomic radii—i.e., one atomic diameter if the atoms are identical. The repulsive component of F grows so sharply with further decrease of r below r_0 that atoms behave to some approximation as hard spheres; one manifestation of this is the highly incompressible

---

[1]J. G. King has shown by direct investigation of the degree of neutrality of a volume of gas that the basic units of positive and negative charge in ordinary matter cannot differ by more than about 1 part in 10^{22} (private communication).

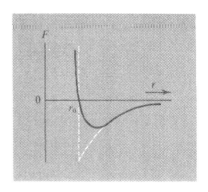

Fig. 5-5 Qualitative graph of the force between two neutral atoms as a function of the distance between their centers. The dashed line represents the nonphysical idealization of atoms that act as completely hard spheres that attract one another.

nature of condensed matter. The dashed line in Fig. 5-5 indicates the result of idealizing the interatomic force to correspond to complete impenetrability for $r < r_0$ and the van der Waals attraction for $r > r_0$. This model can be used quite effectively to analyze the deviations of an actual gas from the ideal gas laws.

Since almost all the objects we deal with in classical mechanics are electrically neutral, this basic atomic interaction—the electric force between neutral particles—is of fundamental importance, as we shall discuss next.

CONTACT FORCES

Many of the physical systems we shall deal with are the ordinary objects of everyday experience, acted upon by such forces as friction, the push and pull of struts and beams, the tension of strings and cables, and so on. Each of these forces involves what we naively call physical "contact" with the object under observation.

Consider, for example, a book resting on a horizontal table. The book is supported by the sum total of countless electromagnetic interactions between atoms in the adjacent surface layers of the book and the table. A submicroscopic analysis of these interactions would be prohibitively complex. For most purposes, however, we can ignore this complexity and can lump all these interactions together into a single force that we shall call a *contact* force. This is a rather artificial category but a useful one. Broadly speaking, all the familiar forces of a mechanical nature, including the force that a liquid or a gas exerts on a surface, are contact forces in this sense. Our discussion in the previous section makes it clear that they are electric forces

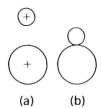

Fig. 5-6 (a) Two charged spheres, obviously not in contact. The weight of the upper sphere is balanced by the electrical repulsion. (b) Two uncharged spheres, apparently in contact. On a sufficiently magnified scale, this appearance of a contact of sharp geometrical boundaries would disappear.

exerted between electrically *neutral* objects. The development of such forces when one smooth, hard object is pressed against another comes about through a distortion of the distributions of positive and negative electric charges. It is characteristic of such forces that their variation with the distance between the objects is much more rapid than the inverse-square dependence that holds for objects carrying a net charge. Thus the contact forces are, in effect, forces of very short range; they fall to negligible size when objects are more than about one atomic diameter apart. The fact, however, that they do have a systematic dependence on the distance of separation means that the notion of what we ultimately mean by "contact" is not clear-cut. There is no *fundamental* distinction between the situation represented by two charged spheres, visibly held apart by their electrical repulsion, and the same two spheres, uncharged, apparently in contact, as shown in Fig. 5-6. In each case the upper sphere assumes a final position of equilibrium in which the net electrical force on it just balances its weight. For any smaller separation it exceeds the weight. But if we display these variations of force with distance, as in Fig. 5-7, we see a drastic difference. In the case of uncharged objects, the transition from negligible force to very large force is so abrupt that it supports our impression of a completely rigid object with a geometrically sharp

Fig. 5-7 (a) Qualitative graph of force versus separation for the charged spheres of Fig. 5-6(a). The contact is "soft"— i.e., the difference between F and W varies slowly with r. (b) Comparable graph for the uncharged spheres of Fig. 5-6(b). The contact is "hard" —the near equality of F and W occurs only over an extremely narrow range of values of r.

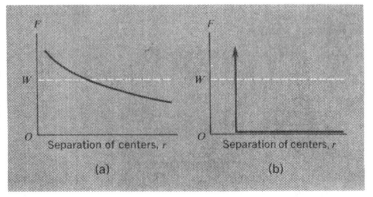

boundary and no interaction outside that boundary. Do not forget, however, that this *is* an idealization, and that sufficiently refined measurements would always reveal a continuous variation of force with separation.

FRICTIONAL CONTACT FORCES

The discussion in the previous section has concentrated on contact forces called into play by the simple pushing together of two objects. Such forces are then at right angles to the surface of contact—we call them *normal* forces in the geometrical sense of that word. But much importance and interest attaches to the tangential forces of friction that appear when the attempt is made to drag an object sideways along a surface. Figure 5–8(a) depicts a block resting on a horizontal surface; its weight is supported by a normal contact force N. We now apply a horizontal force P. Suppose that the magnitude of P is gradually increased from zero. At first nothing seems to happen; the block remains still. We know, from our analysis of equilibrium situations, that this means that a force equal and opposite to P is being supplied via the contact between the block and the surface. This is the frictional force, $\mathfrak{F}$. It automatically adjusts itself to balance P, just as the normal force N would automatically increase if we deliberately pushed down harder, in a vertical direction, on the top of the block. In both cases we can imagine minute deformations of the electric charge distributions along the interface, sufficient to develop the requisite forces. But then, when P is increased beyond a certain value, the frictional force $\mathfrak{F}$ is no longer able to keep step with it. The equilibrium is broken down, and motion ensues. A graph of $\mathfrak{F}$ against the applied force P might look like Fig. 5–8(b). Once $\mathfrak{F}$ has been brought to its maximum limiting value, it may even drop at first as P is

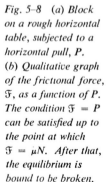
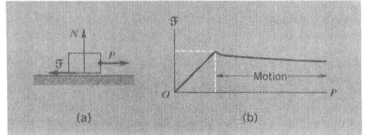

Fig. 5–8 (a) Block on a rough horizontal table, subjected to a horizontal pull, P. (b) Qualitative graph of the frictional force, $\mathfrak{F}$, as a function of P. The condition $\mathfrak{F} = P$ can be satisfied up to the point at which $\mathfrak{F} = \mu N$. After that, the equilibrium is bound to be broken.

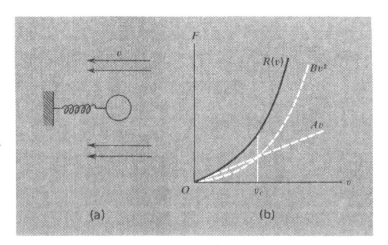

Fig. 5-9 (a) Force on a sphere in a flowing fluid. (b) Total force of fluid friction is made up of separate terms that are respectively linear and quadratic in the relative flow velocity, v.

further increased, although it tends to remain at a fairly constant value thereafter. However, the whole regime $P > \mathcal{F}$ corresponds to motion, and $\mathcal{F}$ may depend in detail on the velocity. The only uniquely defined region is that of static equilibrium, throughout which we can put $\mathcal{F} = P$, as represented by the 45° line on the graph of Fig. 5-8(b). The other feature of interest is the empirical fact that the limiting value of $\mathcal{F}$ is roughly proportional to the normal force N, so that their quotient—the coefficient of friction (μ)—is a property of the two surfaces in contact:

$$\mathcal{F}_{\max} = \mu N \tag{5-3}$$

The above discussion applies to the contact of two solid surfaces. If the contact is lubricated, the behavior is very different. One is then dealing, in effect, with the properties of the contact between a fluid (liquid or gas) and a solid. The basic properties of so-called *fluid friction* can be studied by measuring the force exerted on a fixed solid object as a stream of fluid is driven past it at a given speed v [see Fig. 5-9(a)]. Over a wide range of values of v, this fluid frictional resistance is well described by the following formula:

$$R(v) = Av + Bv^2 \tag{5-4}$$

where A and B are constants for a given object in a given fluid. The first term depends on the viscosity of the fluid, and the second is associated with the production of turbulence. Since the ratio of the second term to the first is proportional to v, one knows that at sufficiently high speeds the fluid friction is dominated by turbulence, however small the ratio B/A may be. The same consideration guarantees that at sufficiently *low* speeds the resistance will be dominated by the viscous term, directly proportional to v [see Fig. 5-9(b)].

CONCLUDING REMARKS

In this chapter we have given a brief account of the three major types of physical interactions and have indicated the general areas in which they are dominant. To recapitulate: Nuclear forces are significant only for nuclear distances, the gravitational force is important only if objects of astronomical scale are involved, and nearly everything else ultimately depends on electromagnetic interactions. The study of physics is essentially the attempt to understand these interactions and all their consequences. In mechanics we have, for the most part, the more modest goal of taking the forces as given and considering various dynamical situations in which they enter. We shall, however, be discussing two classic cases—gravitation and alpha-particle scattering—in which Newtonian mechanics provided the key to the basic laws of force. The present chapter has provided a kind of preview, because it summarizes the state of our current knowledge without entering into any detailed discussion of how we have come to know it. The real work lies ahead!

PROBLEMS

5-1 At what distance from the earth, on the line from the earth to the sun, do the gravitational forces exerted on a mass by the earth and the sun become equal and opposite? Compare the result with the radius of the moon's orbit around the earth.

5-2 By what angle, in seconds of arc, will a plumbline be pulled out of its normal vertical direction by the gravitational attraction of a 10-ton truck that parks 20 ft away? Do you think that this effect could be detected?

5-3 In a Cavendish-type apparatus (see the figure) the large spheres are each 2 kg, the small spheres each 20 g. The length of the arm connecting the small spheres is 20 cm, and the distance between the centers of a small sphere and the big sphere close to it is 5 cm. The torsion constant of the suspending fiber is 5×10^{-8} m-N/rad. The angular deflection of the suspended system is deduced from the displacement

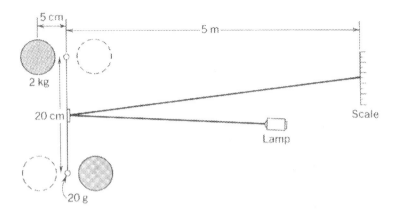

of a reflected spot of light on a scale 5 m away. (Remember that the change in direction of the reflected light is twice the angle through which the mirror is turned.)

It is observed that when the large spheres are moved from their initial positions to equivalent positions on the other side (dashed lines) the mean position of the spot of light is shifted by 8 cm.

(a) Deduce the value of G according to these data, ignoring the effect on each small sphere of the force due to the more distant of the larger spheres.

(b) Estimate the percentage correction on the result of (a) required to allow for the effect of the more distant spheres.

5-4 The original Cavendish experiment was done with a large-size apparatus, as is natural if one wants to make the gravitational forces and torques as big as possible. However, this requires a strong, stiff wire to support the suspended masses. Much later (1895) C. V. Boys made a miniature apparatus, using thin fibers of fused quartz for the suspensions. It is an interesting exercise to see how the attainable sensitivity of the apparatus depends on its size. Imagine two versions of the Cavendish apparatus, A and B, both using solid lead spheres, in which the radii and separations of all the masses in B, together with the length of the torsion fiber, are scaled down by a certain factor L with respect to A. We then design for maximum sensitivity in each apparatus by using the thinnest possible torsion fiber that will take the weight of the suspended masses without breaking. Now for a torsion fiber of given material and of circular cross section, the maximum supportable load is proportional to d^2, where d is its diameter, and its torsion constant is proportional to d^4/l, where l is its length. Using this information, compare the maximum angular deflections obtainable with the two different sizes of apparatus. (Remember, the *lengths* of the torsion fibers also differ by the scaling factor L.)

5-5 The radius of the hydrogen atom according to the original Bohr theory is 0.5 Å.

(a) What is the Coulomb force between the proton and the electron at this distance? What is the gravitational force?

(b) How far apart must the proton and electron be for the Coulomb force to be equal to the value that the gravitational attraction has at 0.5 Å? What familiar astronomical distance is this comparable to?

5-6 Suppose electrons could be added to earth and moon until the Coulomb repulsion thus developed was of just the size to balance the gravitational attraction. What would be the smallest *total* mass of electrons that would achieve this?

5-7 For a person living at 45° latitude, what is the approximate fractional difference, during the day, between the maximum and minimum gravitational forces due to the moon—the change resulting from the fact that the earth's rotation causes the person's distance from the moon to vary? What is a manifestation of this kind of force effect in nature?

5-8 You know that the Coulomb force and the gravitational force both obey an inverse-square law. Suppose that it were put to you that the origin of the gravitational force is a minute difference between the natural unit of positive charge, as carried by a proton, and the natural unit of negative charge, as carried by an electron. Thus "neutral" matter, containing equal numbers of protons and electrons, would not be quite neutral in fact.

(a) What fractional difference between the positive and negative elementary charges would lead to "gravitational" forces of the right magnitude between lumps of ordinary "neutral" matter? How could such a difference be looked for by laboratory experiments?

(b) Is the theory tenable?

5-9 (a) The text (p. 148) quotes a value of the nuclear force for two nucleons close together but also suggests that to describe the nuclear interactions in terms of forces is not very practical. Can you suggest any way in which a nuclear force as such could be measured?

(b) According to one of the earliest and simplest theoretical descriptions of the nuclear interaction (by H. Yukawa) the force of attraction between two nucleons at large separation would be given by

$$F(r) = -\frac{A}{r} e^{-r/r_0}$$

where the distance r_0 is about 10^{-15} m and the constant A is about 10^{-11} N-m. At about what separation between a proton and a neutron would the nuclear force be equal to the gravitational force between these two particles?

5-10 Can you think of any systems or processes in which gravitational, electromagnetic, and nuclear forces all play an important role?

5-11 As mentioned in the text, the attractive (long-range) part of the force between neutral molecules varies as $1/r^7$. For a number of molecules, the order of magnitude of this van der Waals force is well represented by the equation

$$F_{VW}(r) \approx -10^{-76}/r^7$$

where F_{VW} is in newtons and r in meters. Compare the magnitude of the van der Waals force with the Coulomb force between two elementary charges [Eq. (5-2), with $q_1 = q_2 = e = 1.6 \times 10^{-19}$ C]:

(a) For $r = 4$ Å. (This is a distance about equal to the diameter of a molecule of oxygen or nitrogen and hence barely exceeding the closest approach of the centers of two such molecules in a collision.)

(b) For a value of r corresponding to the mean distance between molecules in a gas at STP.

5-12 One of the seemingly weakest forms of contact force is the surface tension of a liquid film. One of the seemingly strongest is the tensile force of a stretched metal wire. However, when expressed in terms of a force between individual atoms in contact, they do not look so different. Use the following data to evaluate them in these terms:

(a) If a water film is formed between a rectangular wire frame, 3 in. wide, and a freely sliding transverse wire (see the figure), it takes the weight of about 1 g to prevent the film from contracting. This contractile force can be ascribed to the contact of the atoms lying within a monomolecular layer along each side of the film. Supposing that the molecules are 3 A across and closely packed, calculate the force per molecule.

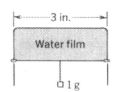

(b) A copper wire of 0.025-in. diameter was found to break when a weight of about 10 kg was hung from its lower end. First, calculate this breaking force in tons per square inch. If the fracture is assumed to involve the rupturing of the contacts between the atoms on the upper and lower sides of a horizontal section right across the wire, calculate the force per atom, assuming an atomic diameter of about 3 A.

5-13 A time-honored trick method for approximately locating the midpoint of a long uniform rod or bar is to support it horizontally at any two arbitrary points on one's index fingers and then move the fingers together. (Of course, just finding its balance point on one finger alone works very well, too!) Explain the workings of the trick method, using your knowledge of the basic principles of static equilibrium and a property of frictional forces: that they have a maximum value equal to a constant μ (the coefficient of friction) times the component of force normal to the surface of contact between two objects.

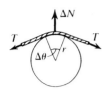

5-14 (a) A string in tension is in contact with a circular rod (radius *r*) over an arc subtending a small angle $\Delta\theta$ (see the figure). Show that the force with which the string presses radially inward on the pulley (and hence the normal force ΔN with which the pulley pushes on the string) is equal to $T\,\Delta\theta$.

(b) Hence show that the normal force per unit length is equal to T/r. This is a sort of pressure which, for a given value of T, gets bigger as r decreases. (This helps to explain why, when a string is tightly tied around a package, it cuts into the package most deeply as it passes around corners, where r is least.)

(c) If the contact is not perfectly smooth, the values of the tension at the two ends of the arc can differ by a certain amount ΔT before slipping occurs. The value of ΔT is equal to $\mu\,\Delta N$, where μ is the coefficient of friction between string and rod. Deduce from this the exponential relation

$$T(\theta) = T_0 e^{\mu\theta}$$

where T_0 is the tension applied at one end of an arbitrary arc (θ) of string and $T(\theta)$ is the tension at the other end.

(d) The above result expresses the possibility of withstanding a large tension T in a rope by wrapping the rope around a cylinder, a phenomenon that has been exploited since time immemorial by sailors. Suppose, for example, that the value of μ in the contact between a rope and a bollard on a dock is 0.2. For $T_0 = 100$ lb, calculate the values of T corresponding to one, two, three and four complete turns of rope around the bollard.

(It is interesting to note that T is proportional to T_0. This allows sailors to produce a big pull or not, at will, by having a rope passing around a continuously rotating motor-driven drum. The arrangement can be described as a force amplifier.)

5-15 In a very delicate torsion-balance experiment, such as the Cavendish experiment, the stray forces due to the fluid friction of slow air currents pushing on the suspended system may be quite significant. To make this quantitative, consider the gravity torsion-balance experiment described in Problem 5-3. For suspended spheres of the size stated ($r \approx 0.8$ cm), the force due to a flow of air of speed v is given approximately by the formula [Eg. (5-4)]

$$R \text{ (newtons)} = 2.5 \times 10^{-6} v + 5 \times 10^{-5} v^2$$

where v is in m/sec. Calculate the value of v that would cause a force due to air currents that equaled the gravitational force exerted on the sphere in this experiment (i.e., the force exerted on a 20-g sphere by a 2-kg sphere with their centers 5 cm apart). (*Hint:* Do not bother to solve a quadratic equation for v. Just find the values of v for which the contributions to R, *taken separately*, would equal the gravitational force. The smaller of the two values of v so obtained is clearly already enough to spoil the experiment.)

To tell us that every species of things is endowed with an occult specific quality by which it acts and produces manifest effects, is to tell us nothing. But to derive two or three general principles of motion from phenomena, and afterwards to tell us how the properties and actions of all corporeal things follow from those manifest principles, would be a very great step in Philosophy, though the causes of those principles were not yet discovered.

NEWTON, *Opticks* (1730)

6
Force, inertia, and motion

THE PRINCIPLE OF INERTIA

THE PRECEDING chapters have treated matter, motion, and force as separate topics. Now we come to the central problem of Newtonian dynamics: How are motions of material objects affected by forces? We shall preface this with a question that is, on the face of it, much simpler: What can we say about the motion of an object that is subjected to *no* forces? It was in the analysis of this problem by Galileo that the science of dynamics really began.[1]

We saw in Chapter 4 how the study of static equilibrium situations leads us to a basic principle: For an object at rest, the net force acting is zero. What could be more natural than to turn this statement around and infer that, if the net force on an object is zero, the object must remain at rest—and, as a kind of corollary to this, to conclude that in order to keep an object moving a net force must be maintained. After all, our experience

---

[1]You will undoubtedly have solved many problems in the use of Newton's laws before reading this chapter. Do not, on that account, assume that the following discussion is superfluous. A wish to get down to business—writing equations and using them—is very sound. The quantitative use of a physical theory is an essential part of the game; physics is not a spectator sport. But to gain real insight and understanding—Where do the equations come from? What do they really say?—one must also examine the basic assumptions and phenomena. And some of the greatest advances in physics have come about in just this way. Einstein arrived at special relativity by thinking deeply about the nature of time. And Newton, when asked once about how he gained his insight into the problems of nature, replied "By constantly thinking unto them."

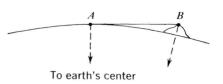

Fig. 6–1 Limitations on inertial motion at the earth's surface. An object starting out horizontally in a truly straight line at A might end up on a hilltop at B. In the process it would be bound to slow down.

shows that moving objects on the earth's surface do come to rest if left to themselves, and a continuing effort does have to be applied to keep an object moving steadily. But, as Galileo was the first to realize, an extrapolation beyond the range of ordinary experience is possible; it is expressed in his principle of inertia. Initially, this simply asserted that an object would, if free of all resistive forces, continue with unchanging speed on a horizontal plane. Galileo himself recognized that this assertion is true only in a limited sense. For a truly flat horizontal plane is tangent to the earth's surface; therefore, if extended far enough, it must be seen as going perceptibly up hill (Fig. 6–1), and objects traveling outward along it must ultimately slow down.[1]

Subsequently Isaac Newton stated the principle of inertia in a generalized form in his "first law" of motion as presented in the *Principia:* "Every body perseveres in its state of rest, or of uniform motion in a right line, unless it is compelled to change that state by forces impressed upon it." It is a familiar statement, which we probably all learn in our first encounter with mechanics, and Fig. 6–2 shows a practical illustration of it. But what does it really say? The first thing we must recognize is that, as discussed in Chapter 2, every statement about the motion of a given object involves a physical frame of reference; we can only measure displacements and velocities with respect to other objects. Thus the principle of inertia is not just a clear-cut statement about the behavior of individual objects; it goes much deeper than that. We can, in fact, turn it around and make a statement that goes roughly as follows:

> **There exist certain frames of reference with respect to which the motion of an object, free of all external forces, is a motion in a straight line at constant velocity (including zero).**

[1] For Galileo's own discussion of these matters, see his *Dialogues Concerning Two New Sciences* (H. Crew and A. de Salvio, translators), Dover Publications, New York.

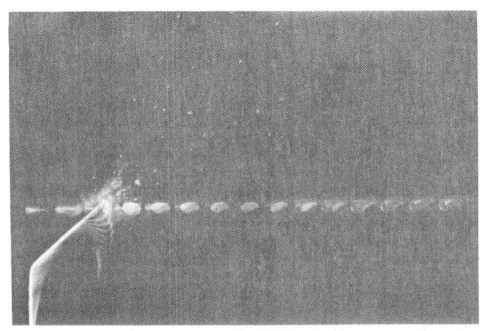

Fig. 6–2   A motion at constant velocity—a bullet traveling at about 1,500 ft/sec, photographed stroboscopically at 30,000 flashes/sec. (Photograph by Prof. Harold E. Edgerton, M.I.T.)

A reference frame in which the law of inertia holds good is called an *inertial frame*, and the question as to whether a given frame of reference is inertial then becomes a matter for observation and experiment. Most observations made within the confines of a laboratory on the earth's surface suggest that a frame of reference attached to that laboratory is suitable. After all, it was on the strength of observations within such a frame that Galileo arrived at the principle of inertia in the first place! A more critical scrutiny shows that this is not quite good enough, and we need to look further afield—but we shall do that later. For the moment we shall limit ourselves to introducing the main principle, which is not affected by the later refinements.

Sir Arthur Eddington, who had a flair for making comments that were both penetrating and witty, offered his own version of the principle of inertia: "Every body continues in its state of rest or uniform motion in a straight line, except insofar as it doesn't."[1] In other words, he regarded it as being, in the last

[1] A. S. Eddington, *The Nature of the Physical World* (Ann Arbor Paperbacks), University of Michigan Press, Ann Arbor, Mich., 1958.

analysis, an expendable proposition. (He was paving the way for a discussion of general relativity and gravitation.) This remark of Eddington's nevertheless draws attention, in a colorful way, to what is the very foundation of Newtonian mechanics: Any deviation from a straight-line path is taken to imply the existence of a force. No deviation, no force—and vice versa. It must be recognized that we cannot "prove" the principle of inertia by an experimental test, because we can never be sure that the object under test is truly free of all external interactions, such as those due to extremely massive objects at very large distances. Moreover, there is the far from trivial question of defining a straight line in a real physical sense: it is certainly not intuitively obvious, nor is it an abstract mathematical question. (How would *you* define a straight line for this purpose?) Nevertheless, it can be claimed that the principle of inertia is a valid generalization from experience; it is a possible interpretation of observed motions, and our belief in its validity grows with the number of phenomena one can correlate successfully with its help.

FORCE AND INERTIAL MASS: NEWTON'S LAW

The law of inertia implies that the "natural" state of motion of an object is a state of constant velocity. Closely linked to this is the recognition that the effect of an interaction between an object and an external physical system is to change the state of motion. For example, we have no doubt that the motion of a tennis ball is affected by the racket, that the motion of a compass needle is affected by a magnet, and that the motion of the earth is affected by the sun. "*Inertial mass*" is the technical phrase for that property which determines how difficult it is for a given applied force to change the state of motion of an object. Let us consider how this description of things can be made quantitative.

"Force" is an abstract term, but we have seen how to associate it with practical operations such as compressing springs, stretching rubber bands, and so on. We can readily study the effect of pushing or pulling on an object by such means, using forces of definite magnitude. The observations become particularly clear-cut if we apply a force to an object that would otherwise move with constant velocity. A very close approximation to this ideal can be obtained by supporting a flat-bottomed object on a cushion of gas—for example by placing the object on

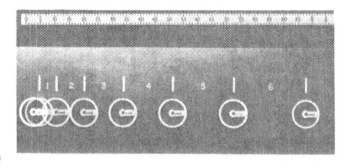

(a)

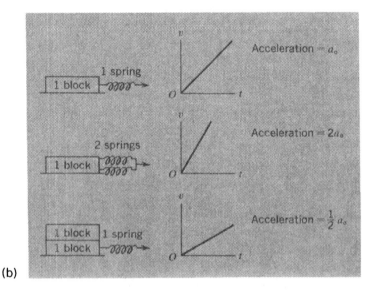

(b)

Fig. 6-3 (a) Stroboscopic photograph of a uniformly accelerated motion. The time interval between light flashes was $\frac{1}{5}$ sec. (From PSSC Physics, D. C. Heath, Lexington, Massachusetts, 1960.) (b) Simple dynamical experiments that can be used as a basis for developing Newton's second law.

a horizontal table pierced with holes through which air is blown from below. It is then possible to pull horizontally on the object and make such observations as the following (see also Fig. 6-3):

1. A spring, stretched by a constant amount, causes the velocity of the object to change linearly with time—the acceleration produced in a given object by a given force is constant.

2. If a second spring, identical with the first and stretch- by the same amount, is used side by side with the first, the acceleration is doubled. That is, if we take a known multiple of a force, according to our criteria for comparing forces in static equilibrium (see Chapter 4), then the acceleration produced in a given object is directly proportional to the total force.

It thus becomes possible to write down simple equations expressing the relation between forces F and accelerations a:

$$a = kF$$

or

$$F = k'a$$

where k and k' describe the inertial properties of the particular object. Which of the above two statements of the results is more convenient? We find the answer in another simple experiment:

3. If we place on the first object a second, identical object, it is observed that all accelerations produced by given arrangements of the springs are reduced to half of what was obtained with one object alone. We can express this most easily by choosing the second of the above equations, so that the inertial property is additive—i.e., the inertial constants k' of two different objects can be simply added together, and the acceleration of the combined system under a given force is immediately given by

$$F = (k'_1 + k'_2)a_{1+2}$$

i.e.,

$$a_{1+2} = \frac{F}{k'_1 + k'_2}$$

It is by such steps that one can be led to the equation that is universally known as "Newton's (second) law":

$$F = ma = m\frac{dv}{dt} \tag{6-1}$$

where the proportionality factor m (identical with k' as defined above) is called the *inertial mass* of the object and F is the net force acting on it. Embodied in this basic statement of Newton's law is the feature that force and acceleration are vector quantities and that the acceleration is always in the same direction as the net force.

An interesting historical fact, often overlooked, is that Newton's own statement of the basic law of mechanics was *not* in the form of Eq. (6-1); the equation $F = ma$ appears nowhere in the *Principia*. Instead, Newton spoke of the change of "motion" (by which he meant momentum) and related this to the value of force × time. In other words, Newton's version of the second law of motion was essentially the following:

$$F\Delta t = m\Delta v \tag{6-2}$$

We shall see in Chapter 9 that Newton's way of formulating the law grew, by inference, out of the particular kind of evidence available to him: the consequences of collision processes. Direct experiments of the kind illustrated in Fig. 6-3 were not possible with the limited techniques of Newton's day. (The only type of uniformly accelerated motion easily accessible to Newton was that of motion under gravity, but this, of course, did not allow any *independent* control over the value of F applied to a given object—unless the use of an inclined plane, giving a driving force $mg\sin\theta$, is regarded as fulfilling this purpose.)

SOME COMMENTS ON NEWTON'S LAW

Simple and familiar as Eq. (6-1) is, it nevertheless contains an enormous wealth of physical concepts—indeed, almost the whole basis of classical dynamics. First comes the assumption that quantitative measurements of displacements and time intervals lead us to a unique value of the acceleration of an object at a given instant. If we remind ourselves that displacements can only be measured with respect to other physical objects, we see that this, like the principle of inertia, cannot be separated from the choice of reference frame. In fact, we tacitly assume that the frame in which the acceleration is measured is an inertial frame. For the kinds of basic experiments illustrated in Fig. 6-3, the earth fills this role.

Next comes the feature, already emphasized, that the acceleration vector is in the direction of the net force vector. This is an important result; it is an expression of the fact that the accelerative effects of several different forces combine in a linear way. Suppose, for example, that an object has two springs attached to it [see Fig. 6-4(a)]. Spring 1 exerts on the object a force $\mathbf{F}_1$ in the x direction. Acting alone it would produce an acceleration F_1/m along x. Spring 2 exerts a force $\mathbf{F}_2$ in the y direction. Acting alone it would produce an acceleration F_2/m along y. It is then a matter of *experiment*—not predictable by pure logic—that the acceleration caused by the two springs acting together is just what one would calculate by adding the vectors $\mathbf{F}_1$ and $\mathbf{F}_2$ to form a resultant force $\mathbf{F}$ [Fig. 6-4(b)] and applying this single force to the mass [Fig. 6-4(c)]. The observed acceleration $\mathbf{a}$ is equal to $\mathbf{F}/m$. This result provides the *dynamical* basis for the "independence of motions" that we discussed as a purely kinematic effect in the trajectory problem of Chapter 3. It tells us that the instantaneous acceleration of an object is the

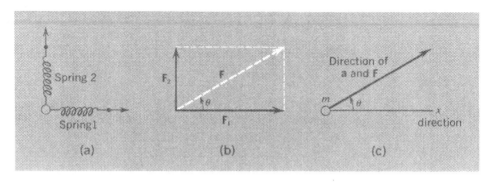

Fig. 6-4 (a) Object pulled by two springs in perpendicular directions. (b) Resultant force calculated according to the laws of vector addition. (c) Observed acceleration of the object agrees with that due to the net force vector as found in (b).

consequence of a *linear superposition* of the applied forces or of the accelerations that they would individually produce. If this result did not hold, the prediction and analysis of motions as produced by forces would become vastly more complicated and difficult.

Let us add a word of explanation and caution here. The linear superposition of *instantaneous* components of acceleration does not mean that we can always automatically proceed to calculate, let us say, the whole course of development of the y component of an object's motion without reference to what is happening in the x direction. To take an example that we shall consider in more detail later, if a charged particle is moving in a magnetic field, the component of force in a given direction depends on the component of velocity perpendicular to that direction. In such a case, we have to keep track of the way in which that perpendicular velocity component changes as time goes on.

In the case of an object subjected to a single force, one may be tempted to think that it is intuitively obvious that the acceleration is in the same direction as the force. It may be worth pointing out, therefore, that this is *not* in general true if high-velocity particles are involved—sufficiently fast to require the modified kinematics and dynamics of special relativity.[1]

Finally comes the assertion that a given force, applied to a particular object, causes the velocity of that object to change at a certain rate **a**, the magnitude of which depends *only* on the

---

[1] See, for example, the volume *Special Relativity* in this series.

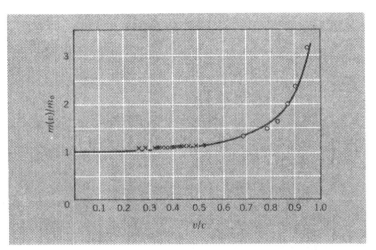

Fig. 6-5 *Increase of inertial mass with speed, as revealed in experiments on high-speed electrons. Based on data of* (open circles) *Kaufmann (1910),* (filled circles) *Bucherer (1909), and* (crosses) *Guye and Lavanchy (1915).* (*After R. S. Shankland,* Atomic and Nuclear Physics, *Macmillan, New York, 1961.*)

magnitude and direction of **F** and on a single scalar quantity characteristic of the object—its inertial mass m. This is a very remarkable result; let us consider it further.

Newton's law asserts that the acceleration produced by any constant force, as for example by a stretched spring, has the same value under all conditions. Thus, according to this statement, it does not matter whether the object is initially stationary or is traveling at high speed. Is this always, and universally true? No! It turns out that for extremely high speeds—speeds that are a significant fraction of the speed of light—the acceleration produced by a given force on a given object *does* depend on v. Under these high-speed conditions Newton's mechanics gives way to Einstein's, as described by the special theory of relativity: The inertia of a given object increases systematically with speed according to the formula

$$m(v) = \frac{m_0}{(1 - v^2/c^2)^{1/2}} \qquad (6\text{-}3)$$

This relation is shown in Fig. 6–5, with experimental data that substantiate it. The quantity m_0, which is called the "rest mass" of the object, represents what we can simply call *the* inertial mass in all situations to which classical mechanics applies, because for any $v \ll c$ the value of m according to Eq. (6-3) is inappreciably different from m_0.

Another implication of Newton's law, as expressed by Eq. (6-1), is that the basic dynamics of an object subjected to a given force does not depend on d^2v/dt^2 or on any of the higher time derivatives of the velocity. The absence of any such complication is in itself a remarkable result, which as far as we know continues to hold good even in the "relativistic" region of very high velocities. It has, however, been pointed out that if one considers physiological effects, not just the basic physics, the existence and magnitude of d^2v/dt^2 ($= da/dt$) can be important. We all know the good feeling of a "smooth acceleration" in a car, and what we mean by that phrase is an acceleration that is close to being constant. A rapid rate of change of acceleration produces great discomfort, and it has even been suggested that a unit of da/dt—to be called a "jerk"—should be introduced as a quantitative measure of such effects!

The conclusion that we can draw from the above discussion is that Newton's law, although ultimately limited in its application, does express with insignificantly small error the relation between the acceleration of an object and the force acting on it for almost everything outside the realm of high-speed atomic particles.

SCALES OF MASS AND FORCE

Granted that a given force produces a unique acceleration of a given object, we can then apply this same force to different objects. Such observations can be used to establish quantitative scales for measuring both inertial masses and forces. In taking this step, we adopt Newton's law as the central feature of mechanics. Although we have hitherto used static situations to compare forces with one another, we now turn to dynamics for defining the absolute magnitudes of forces in terms of the motions they produce. This also means that instead of relying on static measurements to give us prior knowledge of the magnitude of a force, we accept the idea that Newton's law can be used as an analytical tool for deducing the force from the observed acceleration that it produces. Because the measures of force and inertial mass are linked in the single equation $F = ma$, there is danger of circularity in our definitions. But we shall not delve into the subtleties of this problem; we shall simply present a pragmatic method of establishing scales of measures for these quantities.

Our observations permit us to assume that every time a given spring is stretched to the same elongation, it exerts the same magnitude of force on an object attached to it—for we observe a reproducible acceleration. We can then take a number of different objects, labeled 1, 2, 3, ..., pull them one at a time with the same force (i.e., our spring stretched to the same elongation) and measure the individual accelerations: $a_1, a_2, a_3, \ldots$. We can use these experimental results to *define* an inertial mass scale because we can put

$$F = m_1 a_1 = m_2 a_2 = m_3 a_3 \cdots$$

Therefore,

$$\frac{m_2}{m_1} = \frac{a_1}{a_2} \quad \frac{m_3}{m_1} = \frac{a_1}{a_3} \quad \text{etc.} \tag{6-4}$$

One particular object (e.g., m_1) can be chosen (arbitrarily) as a standard unit mass called a "kilogram." A quantitative measure of all other inertial masses can then be obtained in terms of the standard object. Originally, the kilogram was defined to be the mass of 1000 cm^3 of water at its temperature of maximum density (about 4°C), but it is now the mass of a particular cylinder of platinum–iridium alloy, kept at the International Bureau of Weights and Measures in Sèvres, France. (This object was intended to be exactly equivalent to the liter of water, but when a small discrepancy was discovered it was decided to switch to the metal standard as being generally superior in terms of durability, reproducibility, and convenience.)

If the inertial mass is truly a property of the object alone, then the ratios (6-4) must be independent of the particular force used. Repeating the procedure with a spring extended by a different amount and hence with different accelerations, one finds experimentally that the mass ratios are the same as before. The fact that these experimentally determined mass ratios are independent of the magnitude of the force establishes the inertial mass as a characteristic property of the object.

Our quantitative scale for forces likewise stems from Newton's law once the scale of inertial masses has been established, so that in the MKS system, as we mentioned in our first discussion of forces in Chapter 4, the unit force (the *newton*) is defined as the force that imparts to 1 kg an acceleration of 1 m/sec^2:

$$1 \text{ N} = 1 \text{ kg-m/sec}^2$$

Dynamical units of force in other systems of measurement can be defined in an exactly similar way, A newton, as we noted in Chapter 4, is about equal to the gravitational force on an apple, i.e., on a mass of a few ounces.

In describing the kinds of simple experiments on which the formulation of $F = ma$ might be based, we introduced the result that the inertial property represented by mass is additive. You may be tempted to feel that this is obvious and not worthy of an experimental test. As far as ordinary objects are concerned, this commonsense reaction to the problem is sound. But just to recognize that there is, in principle, a legitimate question here, imagine that we have measured the inertial masses of a proton and an electron, separately, and that we then let them come together to form a hydrogen atom. Is the inertial mass of the atom equal to the sum of the masses of the electron and proton? No; it is a shade less. Why? Because in the formation of the atom, with the binding together of the proton and electron, the equivalent of a tiny amount of mass escapes in the form of radiation. Conversely, if an object is made up of various parts held together by cohesive forces, so that effort has to be supplied to separate it into those parts, then the sum of the masses of the individual parts is greater than the mass of the original object. For ordinary objects the difference is immeasurably small, but in atomic and nuclear systems it can become a significant feature of the total mass, and provides the basis of calculations on the energy of nuclear reactions, etc.

Returning now to the observed additivity of inertial masses for macroscopic objects, we can see that this property fully justifies the procedure of making a set of standard masses by constructing blocks of a given material with their volumes in simple numerical ratios—e.g., $1:2:5:10\ldots$. This proportionality of mass to volume was taken by Newton himself as basic, embodying the concept of a constant density for a given material. The very first sentence in the *Principia* is, in fact, a definition of mass as the product of density and volume—a definition that has drawn heavy fire, because some critics regard it as circular. "How," they say, "can one define density except as the quotient of mass and volume?" But Newton had a picture of solid matter as built up of small particles packed together in a uniform manner, and it probably seemed to him more logical to take this inner structure as primary. The calculation of the mass of a lump of matter would then be in essence a matter of counting the number of particles that it contained and multiplying by the mass of a single particle.

THE EFFECT OF A CONTINUING FORCE

Our main concern in this chapter and the next is with the instantaneous effect of a force. Let us, however, take a first look at a question that will be the subject of very extensive analysis later in the book. This question is: What is the effect of a force that is applied to an object and maintained for a while? You are no doubt aware that the answer to this question can be given in more than one way, depending on whether we consider the time or the distance over which the force is applied. To take the simplest possible case, let us suppose that a constant force F is applied to an object of mass m that is at rest at time zero. Then we have $F = ma$, defining a constant acceleration. The resulting motion is thus described by the most elementary versions of the kinematic equations:

$$v = at$$
$$x = \tfrac{1}{2}at^2$$

At time t the object has traveled a total distance x, and we can calculate two possible measures of the total effect of F:

$$Ft = mat = mv$$
$$Fx = (ma)(\tfrac{1}{2}at^2) = \tfrac{1}{2}mv^2$$

Thus we arrive at the two primary dynamical properties that we associate with a moving object: its momentum and its kinetic energy.

The effect of F as measured by the product Ft is called its *impulse;* the effect of F as measured by the product Fx is of course what we call *work*. The quantitative measures of Ft and Fx are newton-seconds and newton-meters. The former, which defines our measure of momentum, does not have a special unit named in its honor. The latter, however, is expressed in terms of the basic unit of work or energy in the MKS system—the joule. Thus we have

$$\text{impulse} \longrightarrow \text{momentum in N-sec} = \text{kg-m/sec}$$
$$\text{work} \longrightarrow \text{kinetic energy in N-m} = \text{joules}$$

THE INVARIANCE OF NEWTON'S LAW; RELATIVITY

We have emphasized how the experimental basis of Newton's law involves the observation of motions with respect to an inertial

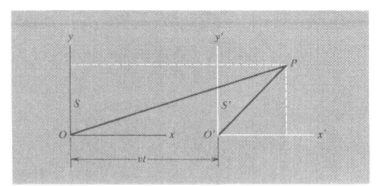

Fig. 6-6 Motion of a particle P referred to two frames that have a relative velocity v.

frame of reference. The actual appearance of a given motion will vary from one such frame to another. It is worth seeing, therefore, how the dynamical conclusions are independent of the particular choice of frame—which means that Newton's mechanics embodies a principle of relativity.

The first point to establish is that, if we have identified any one inertial frame, S (i.e., a frame in which an object under no forces moves uniformly in a straight line), then any other frame, S', having a constant velocity relative to the first is also an inertial frame. This follows directly from the fact that if an object has the instantaneous velocity **u** in S, and if the velocity of S' relative to S is **v**, then the instantaneous velocity of the object relative to S' is given (see Chapter 2) by

$$\mathbf{u}' = \mathbf{u} - \mathbf{v}$$

Thus if **u** and **v** are constant velocities, so also is **u**′, and the object will obey the law of inertia as observed in S'.

To discuss the problem further, let us set up rectangular coordinate systems in both frames, with their x axes along the direction of the velocity **v** (see Fig. 6–6). Let the origins O and O' of the two systems be chosen to coincide at $t = 0$, at which instant, also, the y and z axes of S' coincide with those of S. Let a moving object be at the point P at a later time, t, when the origin O' has moved a distance vt along the x axis of S. Then the coordinates of P in the two systems are related by the following equations. (It is appropriate, in view of Galileo's pioneer work in kinematics and especially of his clear statement of the law of inertia, that they should have become known as the Galilean transformations.)

(*Galilean transformation:*
S' moves relative to S with
a constant speed v in the
$+x$ direction)
$$\begin{cases} x' = x - vt \\ y' = y \\ z' = z \\ t' = t \end{cases} \quad (v = \text{const.}) \tag{6-5}$$

The last of these equations expresses the Newtonian assumption of a universal, absolute flow of time, but it also embodies the specific convention that the zero of time is taken to be the same instant in both frames of reference, so that all the clocks in both frames agree with one another.

We can then proceed to obtain relationships between the components of an instantaneous velocity as measured in the two frames. Thus for the x components we have

$$u'_x = \frac{dx'}{dt'} \quad u_x = \frac{dx}{dt}$$

Putting $x' = x - vt$, and $dt' = dt$, we have

$$u'_x = \frac{d}{dt}(x - vt) = u_x - v$$

The transformations of all three components of velocity are as follows:

$$\begin{aligned} u'_x &= u_x - v \\ u'_y &= u_y \\ u'_z &= u_z \end{aligned} \tag{6-6}$$

Finally, differentiating these velocity components with respect to time, we have (for $v = $ const.) three equalities involving the components of acceleration:

$$\frac{du'_x}{dt'} = \frac{du_x}{dt} \quad \frac{du'_y}{dt'} = \frac{du_y}{dt} \quad \frac{du'_z}{dt'} = \frac{du_z}{dt}$$

Thus the measure of any acceleration is the *same* in both frames:

$$\mathbf{a'} = \mathbf{a} \tag{6-7}$$

Since this identity holds for any two inertial frames, whatever their relative velocity, we say that the acceleration is an *invariant* in classical mechanics. This result is the central feature of relativity in Newtonian dynamics (and it ceases to hold good in the description of motion according to special relativity).

To illustrate the application of these ideas, consider the

Fig. 6–7 Two different views of the trajectory of an object after it has been released from rest with respect to a moving frame, S'.

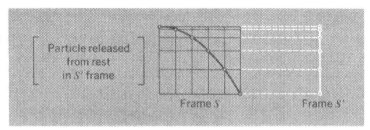

simple and familiar example of a particle falling freely under gravity. Suppose that at $t = 0$, when the axes of the systems S and S' are coincident, an experimenter in S' drops a particle from rest in this frame. The trajectories of the particle, as seen in S and S', are plotted in Fig. 6–7. In each frame the particle is observed to follow the expected trajectory according to the kinematic equations with a vertical acceleration g. In the S frame, the particle has an initial horizontal velocity and therefore it follows a parabolic path, whereas in S' the particle, under the action of gravity, falls straight down. Observers in these two different frames would agree that the equation $\mathbf{F} = m\mathbf{a}$, where they use the same $\mathbf{F}$, accounts properly for the trajectories for any particle launched in any manner in either frame. The frames are thus equivalent as far as dynamical experiments are concerned—either frame may be assumed stationary and the other frame in motion, with the same laws of mechanics providing correct explanations from the observed motions. This is a simple example of the *invariance* of Newton's law itself.

INVARIANCE WITH SPECIFIC FORCE LAWS[1]

In this section we shall consider a little more carefully what is involved in a transformation of Newton's second law of motion. What do we mean by a transformation of this law? Unless we have an explicit *law* of force, $F = ma$ can be regarded as only a prescription for deducing F from the observed motions. So let us consider the force provided by the interaction between two objects. Suppose, for simplicity, that the interaction is provided by something like a stretched rubber band, so that the force is a function only of the distance r between the objects; i.e., we can put $F = f(r)$. For further simplicity, let us assume that the

[1]This section can be omitted without loss of continuity.

motion is confined to the x axis. Then the force exerted on object 2 by object 1, as measured in frame S, can be written:

$$F_{12} = f(x_2 - x_1) \tag{6-8}$$

Newton's law, as applied to object 2, is then stated as follows in terms of measurements in S alone:

$$F_{12} = f(x_2 - x_1) = m_2 a_2 \tag{6-9}$$

We shall now rewrite this equation so that it is expressed entirely in terms of measurements made in the frame S'. From Eq. (6–5) we have

$$x_2 - x_1 = (x_2' + vt) - (x_1' + vt) = x_2' - x_1'$$

Thus

$$F_{12} = f(x_2' - x_1')$$

but according to the assumed law of force, the function $f(x_2' - x_1')$ is precisely the specification of the force F_{12}' in terms of measurements in S'. Hence we can put

$$F_{12} = F_{12}'$$

Turning now to the right-hand side of Eq. (6–9), the Galilean transformations give us $a = a'$; and in Newtonian dynamics the inertial mass is a constant: $m_2 = m_2'$. Thus we are able to write

$$F_{12}' = f(x_2' - x_1') = m_2' a_2' \tag{6-10}$$

We see here in explicit terms how the Newtonian law of motion is invariant with respect to the particular choice of inertial frame, provided that the Galilean transformations correctly describe the transformations of displacements and times between one frame and another. A more complicated force law, as long as it involved only the *relative* positions and velocities of two interacting objects, would possess this same property of invariance. If, however, the force depended on *absolute* positions and velocities—e.g., if the force law were of the form

$$F_{12} = f(x_2^2 - x_1^2)$$

then the form of the equation of motion would cease to be the same in all inertial frames. Nothing in our experience has revealed such a situation, which would make the laws of physics appear different in a laboratory and in a train or plane moving at constant velocity. If the physical laws were different for different observers, this might be a clue to the uniqueness of certain frames of reference. It was, indeed, believed for a long time that a unique reference frame must exist, in the form of a medium, pervading space, in which the waves of light could be carried. (Otherwise, how could light travel from the stars to the earth?) But Einstein showed how the equivalence of all inertial frames and the invariance of all physical laws could be preserved, provided that the kinematics of Galileo and the dynamics of Newton were replaced by new formulations that merged into the old ones in the region of moderate or small velocities.

NEWTON'S LAW AND TIME REVERSAL

The subject of this section might be more dramatically, although less accurately, stated as a question: Is time reversible? Look at the two stroboscopic photographs in Fig. 6-8. The first shows an individual object moving vertically under gravity. Is it falling down or "falling up"? The second shows a collision between two objects. Which were the paths of the objects before collision? In both cases the answer must be "We don't know." A motion picture of these two sequences of events could be run backward and it would be impossible for the viewer to detect any violation of Newton's laws. The reason is that all velocities of a collection of particles can be reversed without violating Newton's law of motion. A *time-reversal* operation (replacing t by $-t$ in the kinematical equations) changes every v to $-v$, but it leaves the acceleration unchanged. (Gravity, for example, remains in the downward direction.) This is because acceleration involves the *second* derivative with respect to time. Thus any conclusions about forces that we reach as a result of watching a dynamical process in reverse sequence are identical with what we would conclude from the process itself. We do not see attractions apparently turn into repulsions, or anything like that.

And yet, when we see an ordinary motion picture in reverse, it quickly becomes apparent from the behavior of inanimate objects—leaving aside the ludicrous effects of reversing human actions, which appear strange for quite different reasons—that most physical actions have a well-defined direction. Imagine, for example, a sequence in which a glass falls from a table and shatters into small fragments on the floor. If we saw a motion

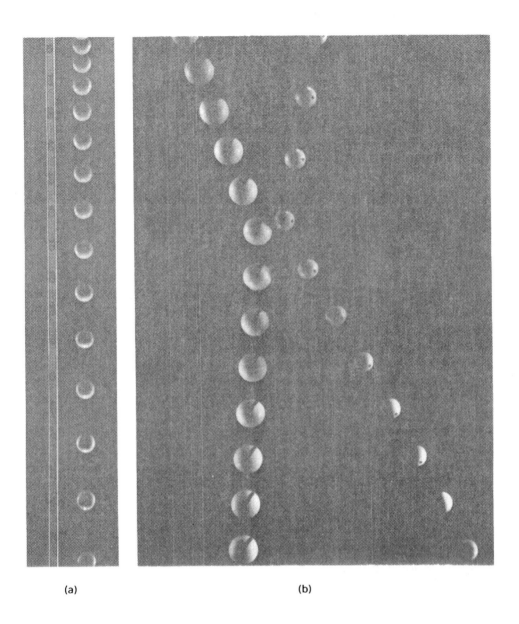

Fig. 6-8 (a) Stroboscopic photograph of an object moving vertically under gravity. Which way is it moving—up, or down? (b) Stroboscopic photograph of an elastic collision. Its time sequence is to all intents and purposes completely reversible. (From PSSC Physics, D. C. Heath, Lexington, Massachusetts, 1965.)

picture in which the fragments gathered themselves together into a whole glass, which then jumped up onto the table, this would clearly be unbelievable—nature doesn't act like that. Yet a "micromovie" of the individual atomic encounters at every stage of the process ought to be perfectly time-reversible.

Thus we are faced with a puzzle: Newton's law implies that the fundamental dynamical behavior of an individual particle is reversible in time, but when one takes a system of very large numbers of particles, apparently the behavior ceases to be time-reversible. The resolution of this mystery is found in the detailed statistical analysis of many-particle systems—the subject known as statistical mechanics. As long as we are dealing, as we shall be here, with systems of only a few particles, the problems associated with time reversal do not arise and we shall not consider them further.

CONCLUDING REMARKS

It will probably have become apparent to you during the course of this chapter that the foundation of classical mechanics, as represented by Newton's second law, is a complex and in many respects subtle matter. The precise content of the law is still a matter for debate, nearly three centuries after Newton stated the first version of it. In a fine discussion entitled "The Origin and Nature of Newton's Laws of Motion," one author (Brian Ellis) says: "But what of Newton's second law of motion? What is the logical status of this law? Is it a definition of force? Of mass? Or is it an empirical proposition relating force, mass, and acceleration?"[1] Ellis argues that it is something of all of these:

> Consider how Newton's second law is actually used. In some fields it is unquestionably true that Newton's second law is used to define a scale of force. How else, for example, can we measure interplanetary gravitational forces? But it is also unquestionably true that Newton's second law is sometimes used to define a scale of mass. Consider, for example, the use of the mass spectrograph. And in yet other fields, where force, mass, and acceleration are all easily and independently measurable, Newton's second law of motion functions as an empirical correlation between these three quantities. Consider, for example,

---

[1] Published in a collection of essays, *Beyond the Edge of Certainty* (Robert G. Colodny, ed.), Prentice-Hall, Englewood Cliffs, N.J., 1965.

the application of Newton's second law in ballistics and rocketry... To suppose that Newton's second law of motion, or *any* law for that matter, must have a unique role that we can describe generally and call the logical status is an unfounded and unjustifiable supposition.

Since force and mass are both abstract concepts and not objective realities, we might conceive of a description of nature in which we dispensed with both of them. But, as one physicist (D. H. Frisch) has remarked, "Whatever we think about ultimate reality it is convenient to follow Newton and split the description of our observations into 'forces,' which are what make masses accelerate, and 'masses,' which are what forces make accelerate. This would be just tautology were it not that the observed phenomena can best be classified as the result of *different forces* acting on the *same* set of masses." Ellis spells out this same idea in more detail:

> Now there are, in fact, many and various procedures by which the magnitudes of the individual forces acting on a given system may be determined—electrostatic forces by charge and distance measurements, elastic forces by measurement of strain, magnetic forces by current and distance determinations, gravitational forces by mass and distance measurements, and so on. And it is an empirical fact that when all such force measurements are made and the magnitude of the resultant force determined, then the rate of change of momentum of the system under consideration is found to be proportional to the magnitude of this resultant force.

And so it is that we obtain an immensely fruitful and accurate description of a very large part of our whole experience of objects in motion, through the simple and compact statement of Newton's second law.[1]

PROBLEMS

6-1 Make a graphical analysis of the data represented by the stroboscopic photograph of Fig. 6-2(a) to test whether this is indeed acceleration under a constant force.

---

[1]For further discussion of these questions, the *American Journal of Physics* is a perennial source. See, for example, the following articles: L. Eisenbud, "On the Classical Laws of Motion," *Am. J. Phys.*, **26**, 144 (1958); N. Austern, "Presentation of Newtonian Mechanics," *Am. J. Phys.*, **29**, 617 (1961); R. Weinstock, "Laws of Classical Motion: What's *F*? What's *m*? What's *a*?", *Am. J. Phys.*, **29**, 698 (1961).

6–2 A cabin cruiser of mass 15 metric tons drifts in toward a dock at a speed of 0.3 m/sec after its engines have been cut. (A metric ton is 10^3 kg.) A man on the dock is able to touch the boat when it is 1 m from the dock, and thereafter he pushes on it with a force of 700 N to try to stop it. Can he bring the boat to rest before it touches the dock?

6–3 (a) A man of mass 80 kg jumps down to a concrete patio from a window ledge only 0.5 m above the ground. He neglects to bend his knees on landing, so that his motion is arrested in a distance of about 2 cm. With what average force does this jar his bone structure?

(b) If the man jumps from a ledge 1.5 m above the ground but bends his knees so that his center of gravity descends an additional distance h after his feet touch the ground, what must h be so that the average force exerted on him by the ground is only three times his normal weight?

6–4 An object of mass 2 kg is acted upon by the following combination of forces in the xy plane: 5 N at $\theta = 0$, 10 N at $\theta = \pi/4$, and 20 N at $\theta = 4\pi/3$. The direction $\theta = 0$ corresponds to the $+x$ direction. At $t = 0$ the object is at the point $x = -6$ m and $y = 3$ m and has velocity components $v_x = 2$ m/sec and $v_y = 4$ m/sec. Find the object's velocity and position at $t = 2$ sec.

6–5 The graphs shown give information regarding the motion in the xy plane of three different particles. In diagrams (a) and (b) the small

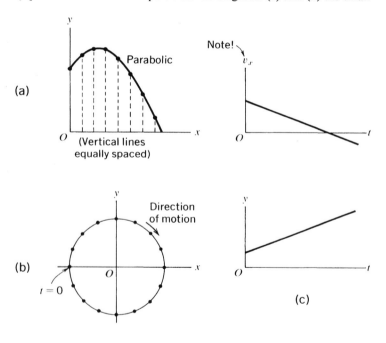

dots indicate the positions at equal intervals of time. For each case, write equations that describe the force components F_x and F_y.

6-6 An observer first measures the velocity of an approaching object to be 10^{-2} m/sec and then, 1 sec later, to be 2×10^{-2} m/sec. No intermediate readings are possible because the observer's instruments take a full second to determine a velocity. If the object has a mass of 5 g, what conclusions can the observer make about
      (a) The size of the force that had been acting?
      (b) The impulse supplied by the force?
      (c) The work done by the force?

6-7 A particle of mass 2 kg oscillates along the x axis according to the equation

$$x = 0.2 \sin\left(5t - \frac{\pi}{6}\right)$$

where x is in meters and t in seconds.
      (a) What is the force acting on the particle at $t = 0$?
      (b) What is the maximum force that acts on the particle?

6-8 A car of mass 10^3 kg is traveling at 28 m/sec (a little over 60 mph) along a horizontal straight road when the driver suddenly sees a fallen tree blocking the road 100 m ahead. The driver applies the brakes as soon as his reaction time (0.75 sec) allows and comes to rest 9 m short of the tree.
      (a) Assuming constant deceleration caused by the brakes, what is the decelerating force? What fraction is it of the weight of the car (take $g = 9.8$ m/sec^2)?
      (b) If the car had been on a downward grade of $\sin^{-1}(\frac{1}{10})$ with the brakes supplying the same decelerating force as before, with what speed would the car have hit the tree?

6-9 A particle of mass m follows a path in the xy plane that is described by the following equations:

$$x = A(\alpha t - \sin \alpha t)$$
$$y = A(1 - \cos \alpha t)$$

      (a) Sketch this path.
      (b) Find the time-dependent force vector that causes this motion. Can you suggest a way of producing such a situation in practice?

6-10 A piece of string of length l, which can support a maximum tension T, is used to whirl a particle of mass m in a circular path. What is the maximum speed with which the particle may be whirled if the circle is (a) horizontal; (b) vertical?

Part II
Classical mechanics at work

Does the engineer ever predict the acceleration of a given body from a knowledge of its mass and of the forces acting upon it? Of course. Does the chemist ever measure the mass of an atom by measuring its acceleration in a given field of force? Yes. Does the physicist ever determine the strength of a field by measuring the acceleration of a known mass in that field? Certainly. Why then, should any one of these roles be singled out as the role of Newton's second law of motion? The fact is that it has a variety of roles.

BRIAN ELLIS, *The Origin and Nature of Newton's Laws of Motion* (1961)

7

Using Newton's law

IT IS WORTH reemphasizing the fact that Newton's law may be used in two primary ways:

1. Given a knowledge of all the forces acting on a body, we can calculate its motion.

2. Given a knowledge of the motion, we can infer what force or forces must be acting.[1]

This may seem like a very obvious and quite trivial separation, but it is not. The first category represents a purely *deductive* activity—using known laws of force and making clearly defined predictions therefrom. The second category includes the *inductive*, exploratory use of mechanics—making use of observed motions to learn about hitherto unknown features of the interactions between objects. Skill in the deductive use of Newton's law is of course basic to successful analytical and design work in physics and engineering and can bring great intellectual satisfaction. But, for the physicist, the real thrill comes from the inductive process of probing the forces of nature through the study of motions. It was in this way that Newton discovered the law of universal gravitation, that Rutherford discovered the atomic nucleus, and that the particle physicists explore the structure of nucleons (although, to be sure, this last field requires analysis in terms of quantum mechanics rather than Newtonian mechanics).

---

[1]If the *law* of force is also known, this can be used to obtain information about an unknown mass. The quotation opposite treats this as a third category.

SOME INTRODUCTORY EXAMPLES

In Chapter 8 we shall have something to say about the way in which Newton arrived at his insight into gravitational forces from the study of planetary motions. But first we shall discuss how one goes about calculating motions from given forces. There will be a certain lack of glamour about some of this initial work—but that is an inescapable aspect of science, as of everything else. And some systematic groundwork at the beginning will pay rich dividends later. We shall restrict ourselves at first to cases in which the forces are constant. Thus we shall be able to use the kinematic equations for constant acceleration [Eq. (3-10)]. For convenience, we quote the equations again here:

$$\left. \begin{array}{l} \text{Kinematic} \\ \text{equations:} \\ \text{(Constant acceleration only)} \end{array} \right\} \begin{array}{l} v = v_0 + at \\ s = s_0 + v_0 t + \tfrac{1}{2} a t^2 \\ v^2 = v_0^2 + 2a(s - s_0) \end{array} \quad (7\text{-}1)$$

In any given problem our procedure will be first to identify all the forces acting on an object and, from $F_{\text{net}} = ma$, to calculate the resultant acceleration. We can then use the kinematic equations to describe the subsequent motion. Generally, this latter step is merely an exercise in mathematics; the real "physics" of the situation lies in the analysis of what forces are present. Do not be misled by the apparently trivial nature of these introductory problems. Taken at face value, they are, indeed, uninteresting and inconsequential. But the *method* of analysis is of salient importance—exactly the same approach is used in far more sophisticated problems. We purposely begin by choosing rather elementary systems to clarify the procedure, so the problems are trite. But keep your eye on the method—it is powerful.

Example 1: Block on a smooth table. By a "smooth" surface we mean one that is incapable of exerting any force tangential to itself. No such surface exists but some come close enough for this to be a useful idealization.

Consider a block of mass m pulled horizontally along a frictionless surface. Question: What is the motion of the block?[1]

We first must determine what forces act *on the block*. To assist the analysis, mentally "isolate the block" with an imag-

---

[1]In this and succeeding examples we make the assumption that an object at rest relative to the earth's surface has *zero* acceleration. This is only approximately true—see Chapter 12 for a full discussion.

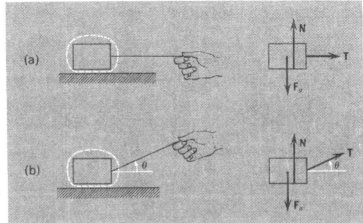

Fig. 7-1 (a) Block pulled horizontally on a perfectly smooth surface. (b) Same block with a string pulling in an arbitrary direction.

inary boundary surface [see Fig. 7-1(a)] and draw a sketch showing all the forces that act from outside through this surface on the block.[1] They are

$\mathbf{F}_g$ = force of gravity

$\mathbf{T}$ = tension in the string (a contact force)

$\mathbf{N}$ = contact force the table exerts on the block; this force is normal to the surface, hence our choice of the letter $\mathbf{N}$

We have introduced vector symbols here because clearly forces acting in more than one direction are involved. The complete statement of Newton's law for this case is thus $\mathbf{T} + \mathbf{N} + \mathbf{F}_g = m\mathbf{a}$. Since we assume that the block has no acceleration in the vertical direction, we conclude that $\mathbf{N}$ and $\mathbf{F}_g$ are equal and opposite vectors; in other words,

$$\mathbf{N} + \mathbf{F}_g = 0$$

Expressing this a different way, we can say that the magnitude F_g of $\mathbf{F}_g$ is equal to the magnitude N of $\mathbf{N}$, because the magnitude of a vector, without regard to its direction, is defined as a positive quantity. The condition of zero net force component along y can then be stated in the equation $N - F_g = 0$. The sum of the forces on the block—the "net" force—is thus (by vector addition) equal to $\mathbf{T}$, and hence we have $T = ma$ (with both $\mathbf{T}$ and $\mathbf{a}$ horizontal).

[1] We shall, in these examples, treat each object as if it were a point particle, with all forces accordingly acting through a single point.

The problem becomes somewhat more substantial if we suppose that the pull of the string is not horizontal [Fig. 7-1(b)]. The initial vector statement of Newton's law is of the same form as before, but the analysis of it into vertical and horizontal components now gives us the following equations:

Vertically: $N + T \sin \theta - F_g = 0$

Horizontally: $T \cos \theta = ma$

The first equation tells us the magnitude of N:

$$N = F_g - T \sin \theta$$

(The vertical component of the tension in the string helps to support the block, and so N becomes less than F_g.) The second equation tells us directly the magnitude of the horizontal acceleration.

We notice that there is a physical limitation on this analysis. Unless the table can pull downward on the block, as well as being able to push upward on it to support its weight, the force **N** is necessarily upward, as shown in the diagram, and the scalar N is necessarily positive. Thus, if $T \sin \theta > F_g$, we cannot satisfy the assumed conditions. What happens then?

Example 2: Block on a rough table. As in Example 1, we shall suppose that the block is pulled by a string with a force **T**, and we shall, for simplicity, take the pull to be horizontal (see Fig. 7-2). Although this is certainly not a difficult problem, we cannot say what will happen unless we have some information about the properties of the frictional force $\mathcal{F}$. If it is "dry friction," we can put $\mathcal{F} \leq \mu N$, where μ is the coefficient of friction. The value of N is equal to F_g in this case, and so, writing F_g as equal to m times the gravitational acceleration g, we have $\mathcal{F} \leq \mu mg$. The inequality expresses the ability of the frictional force to adjust itself, up to a certain limit, to balance the force **T**.

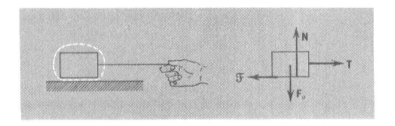

Fig. 7-2 Block pulled horizontally on a rough surface.

Thus the equation $\mathbf{T} + \mathcal{F} = m\mathbf{a}$, which defines the horizontal acceleration, leads to two separate statements:

If $T < \mu m g$: $T - \mathcal{F} = 0$ (and hence $a = 0$)
If $T > \mu m g$: $T - \mu m g = ma$ [and hence $a = (T/m) - \mu g$]

We shall not consider in detail what would happen if **T** were applied at an angle to the horizontal. But one can see that this will modify the value of **N** and hence, in turn, the limiting value of $\mathcal{F}$. You should analyze this case for yourself.

Example 3: Block on a smooth incline. Two forces act on the block (Fig. 7-3): the force **N** normal to the (frictionless) surface and the gravitational force $\mathbf{F}_g$. These are in different directions, they are bound to have a nonzero resultant, and the block must accelerate. We could, if we wished, introduce horizontal and vertical coordinates x and y and write equations for $\mathbf{F} = m\mathbf{a}$ that would define the components a_x and a_y of the vector acceleration **a**:

$$N \cos \theta - F_g = ma_y$$
$$N \sin \theta = ma_x$$

It is clear, however, that **a** is parallel to the slope, and it is simpler in this case to introduce the coordinate s representing distance along the slope in the downward direction. Resolving the forces along and perpendicular to s, we have

$$F_g \sin \theta = ma$$
$$N - F_g \cos \theta = 0$$

If we write the magnitude of the gravitational force as equal to m times the gravitational acceleration g, the first of these equations gives us

$$a = g \sin \theta$$

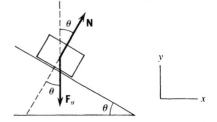

Fig. 7-3 Forces acting on a block on a perfectly smooth incline.

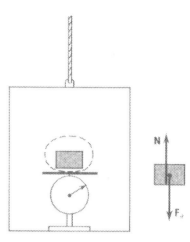

Fig. 7–4 Block at rest with respect to an accelerating elevator; the spring scale records its apparent weight.

Example 4: Block in an elevator. A block sits on a package scale (a spring scale!) in an elevator (Fig. 7–4). Question: What does the scale read as the elevator moves up and down?

Mentally isolating the block, we recognize that only two forces act on it. They are:

$\mathbf{F}_g$ = the force of gravity (downward)
$\mathbf{N}$ = the contact force (upward) exerted by the scale on the block

The package scale records the magnitude of $\mathbf{N}$, because, by the equality of action and reaction for objects in contact, the force exerted on the scale is $-\mathbf{N}$. The value of this reading, under any conditions, is what we shall call the *measured weight* (W) of the block.

If the elevator (and the block and scale with it) has an acceleration $\mathbf{a}$, measured as positive upward, then Newton's law requires

$$N - F_g = ma$$

Thus, if the elevator is stationary *or is moving with constant velocity upward or downward*, the reading of the scale is equal to the gravitational force on the block. But if the elevator has a positive acceleration (upward), then we have

$$N = m(g + a)$$

In this case the weight of the block, as measured by the scale, is greater than the gravitational force on it. One will often notice this effect personally, as an increased force on the soles of the feet, when riding an elevator in two situations: (a) when the elevator is picking up speed, going upward, and (b) when it is slowing down, going downward. Both of these involve upward acceleration. Similarly, if the elevator is slowing down in its upward motion or just beginning its downward motion, **a** is downward, so that the measured weight is *less* than F_g. (Incidentally, the internal discomfort that one sometimes feels in an elevator can be linked directly to Newton's law. If the elevator acquires a positive (upward) acceleration, one's heart and stomach must—literally—sink a little before they experience extra forces from the surrounding tissues to supply the acceleration called for.

Example 5: Two connected masses. Here is a simple example designed to illustrate the important point that one is free to isolate, in one's imagination, *any part* of a complete system, and apply $\mathbf{F} = m\mathbf{a}$ to it alone. Figure 7-5 shows two masses connected by a light (massless) string on a smooth (frictionless) surface. A horizontal force, **P**, pulls at the right-hand mass. What can we deduce about the situation?

First, we can imagine an isolation boundary drawn around both m_1 and m_2 and the string that connects them. The only external horizontal force applied to this system is P, and the total mass is m_1 plus m_2. Hence we have

$$P = (m_1 + m_2)a$$

This at once tells us the acceleration that is common to both masses. Next, we can imagine an isolation boundary surrounding the connecting string alone. In Fig. 7-5 we indicate the forces $\mathbf{T}_1$ and $\mathbf{T}_2$ with which the string pulls on the masses; by the equality of action and reaction, the string has forces equal to $-\mathbf{T}_1$ and $-\mathbf{T}_2$ applied to its ends. The sum of these forces must equal the mass of the string times its acceleration **a**. Assuming

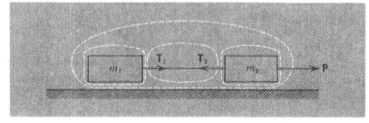

Fig. 7-5 Two connected blocks pulled horizontally on a perfectly smooth surface. Newton's law must apply to any part of the system that one chooses to consider.

the mass of the string to be negligible, this means that T_1 and T_2 would have the same magnitude, T, which we call *the* tension in the string. (This idealized result is, of course, rather obvious; what *is* worth noting is that, in any real situation, there would have to be enough difference of tensions at the ends of the string to supply the requisite accelerative force to the mass of the string itself.)

Finally, we can imagine drawing isolation boundaries around m_1 and m_2 separately, and applying Newton's law to the horizontal motion of each:

$$T = m_1 a$$
$$P - T = m_2 a$$

Adding these equations, we arrive at the equation of motion of the total system once again. But if we take either equation alone, substitution of the already determined value of a will give the value of the tension T in terms of P and the masses.

Once again we shall acknowledge the simple character of these problems. But if they are studied in the spirit in which they are offered—not for their own sake, but for the way in which they exemplify the systematic use of Newton's law—they will be found to suggest a sound approach to almost any problem that involves a direct application of $\mathbf{F} = m\mathbf{a}$.

MOTION IN TWO DIMENSIONS

All the examples in the last section dealt with motion along one dimension only. This is not a serious limitation because, thanks to the "independence of motions," we can always analyze a situation at any given instant in terms of the components of force and acceleration along separate coordinate directions. The simplest case of this occurs when the acceleration **a** is a constant vector. It is then very convenient to choose the direction of **a** itself as one coordinate; along any direction perpendicular to this the acceleration component is, by definition, zero, and the velocity component must be constant. Probably the most familiar example of this procedure is the analysis of motion under gravity in the idealization that air resistance can be ignored. We considered this as a purely kinematic problem in Chapter 3. Another example, with the additional feature that we have control over

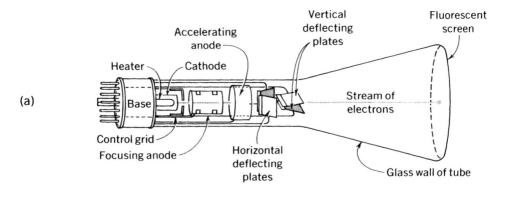

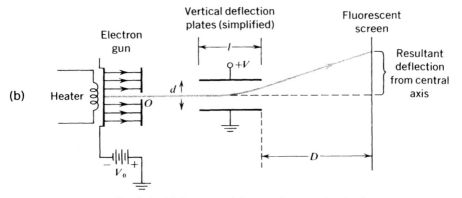

Fig. 7–6 (a) Diagram of the main features of a simple cathode-ray tube. (b) Simplified diagram of electrical connections and electron trajectory.

the applied force, is the motion of an electron beam in a cathode-ray tube. Let us consider this as a dynamical problem.[1]

Figure 7–6(a) shows a sketch of a cathode-ray tube, and Fig. 7–6(b) is a schematic diagram of some principal features. A well-focused beam from an electron gun passes between two pairs of deflection plates which, if appropriately charged, will deflect the beam transversely away from the central axis so as to strike any particular point on the fluorescent screen. We shall call the central axis the z direction, so that the directions of transverse deflection, perpendicular to z and to one another, can be called x and y, just as in a real oscilloscope.

The electron gun accelerates the electrons through a po-

---

[1] We shall make use of certain results concerning the effects of electric forces. If these are not already familiar to you, see, for example, PSSC, *Physics* (2nd ed.), Part IV, Heath, Boston, 1965.

tential difference V_0. This gives to each electron a kinetic energy eV_0 (where e is the elementary charge) so that it acquires a longitudinal velocity component v_z given by

$$\tfrac{1}{2}mv_z^2 = eV_0 \tag{7-2}$$

Therefore,

$$v_z = \left(\frac{2eV_0}{m}\right)^{1/2}$$

After leaving the gun, the electron is not subjected to further acceleration or deceleration along the z direction, so its z coordinate continues to change at the rate v_z.

We shall suppose that a potential difference V is applied between the upper and lower y-deflection plates. This means that if an electron were to travel all the way from one plate to another, it would acquire an energy eV in consequence of the work done on it by the transverse electric force F_y. If we suppose the plates to be parallel and spaced by a distance d that is small compared to their length, the force F_y would have the same value at all points between the plates, so that the gain of energy would also be given by $F_y d$. Hence we have

$$F_y = \frac{eV}{d}$$

Given this value of F_y the electron will, throughout its passage between the plates, have an upward acceleration given by

$$a_y = \frac{F_y}{m} = \frac{eV}{md}$$

How far vertically will the electron be deflected? This will depend on the amount of time it spends between the deflection plates. If the horizontal extent of the plates is l, and the horizontal component of the electron's velocity has the constant value v_z, the time is given simply by

$$t = \frac{l}{v_z}$$

From Eq. (7-1) we can determine the transverse displacement y that occurs during the time this upward force acts:

$$s_y = v_{0y}t + \tfrac{1}{2}a_y t^2$$

$$y = 0 + \frac{1}{2}\frac{eV}{md}\left(\frac{l}{v_z}\right)^2$$

But from Eq. (7-2), mv_z^2/e is just $2V_0$, so we obtain

$$y = \frac{Vl^2}{4V_0d} \tag{7-3}$$

This expression gives the transverse displacement of the electron as it emerges from the deflection plates. The resultant displacement away from the central axis of the spot on the fluorescent screen may be found from trigonometric considerations. The z and y components of the displacement of the electron while between the plates are given by

$$z = v_z t$$
$$y = \tfrac{1}{2}a_y t^2$$

Eliminating t from these equations shows that y is proportional to z^2, and the trajectory is therefore a parabola.

Once the electron leaves the region between the plates, there is no further force on it and it travels in a straight line at an angle equal to arctan (v_y/v_z), where v_y is its transverse velocity component upon leaving the plates. Now we have

$$v_y = a_y t = \frac{eV}{md}\frac{l}{v_z}$$

$$\frac{v_y}{v_z} = \frac{eVl}{mdv_z^2} = \frac{Vl}{2V_0 d}$$

The additional transverse deflection Y as the electron travels the distance D from the deflector plates to the screen is thus given by

$$Y = D\frac{v_y}{v_z} = \frac{VlD}{2V_0 d} \tag{7-4}$$

If $D \gg l$, as is usually the case in practice, most of the total transverse deflection is contained in Y, and we can use Eq. (7-4) to estimate the approximate sensitivity of an actual oscilloscope [sensitivity = (spot deflection)/(deflection voltage).]

We may note in passing a striking feature of Eqs. (7-3) and (7-4). All distinguishing characteristics of the particle as an electron have disappeared. Any negatively charged particle, with any charge or any mass, could in principle pass through the system and end up at the same spot on the screen.

In a real oscilloscope, as Fig. 7-6(a) indicates, the deflector plates are not flat and parallel; thus Eqs. (7-3) and (7-4) do not strictly apply. It remains true, however, that these equations indicate the way in which the deflection depends on the accelerating and deflecting voltages, and also, in a less specific way, on the characteristic dimensions of the tube.

MOTION IN A CIRCLE

The problem of circular motion is often presented as though it were separate from the kinds of applications of Newton's law that we have discussed so far. It may therefore be worth emphasizing that it involves a completely straightforward use of **F** = m**a**. The only special feature is that the radial component of the acceleration is uniquely related to the radius of the path and the instantaneous speed.

Consider first the simplest case, in which we suppose that an object of mass m is traveling around a circle of radius r at a *constant* speed v (Fig. 7–7). Such a motion can be set up, for example, by tethering a puck, by means of a taut string, to a fixed peg on a very smooth horizontal surface (e.g., an air table) and giving the puck an arbitrary velocity at right angles to the string. Then, as we saw in Chapter 3, the acceleration of the object is purely toward the center of the circle and is of magnitude v^2/r. We know that the production of this "centripetal" acceleration necessitates the existence of a corresponding force:

$$F = ma = \frac{mv^2}{r} \tag{7-5}$$

In the hypothetical case that we have described, this force would have to be supplied by the tension T in the tethering string. And if the string is not strong enough to supply a force of the required magnitude, it will break and the object, being now free of forces (at least in the plane of the motion), will fly off along a tangent.

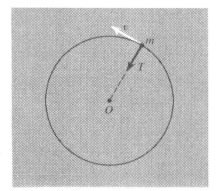

Fig. 7–7 Basic dynamical situation for a particle traveling in a circle at constant speed.

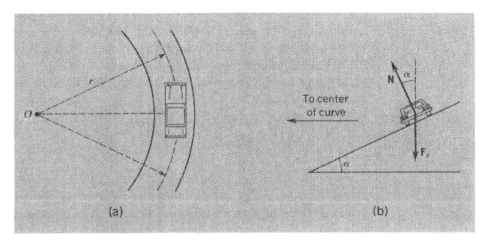

Fig. 7-8 (a) View of a curve in a road, as seen from vertically overhead. (b) Car on the banked curve, as seen from directly behind or in front.

The motion of a car on a banked curve presents an important practical problem of this type. Suppose that a road has a radius of curvature r (measured in a horizontal plane), as indicated in Fig. 7-8(a) and is banked at an angle α as shown in Fig. 7-8(b). A car, traveling into or out of the plane of the latter diagram with speed v, has a centripetal acceleration v^2/r. The purpose of banking is to make it possible for the car, traveling at some reasonable speed, to be held in this curved path by a force exerted on it purely normal to the road surface—i.e., there would be no tangential force, as there would have to be if the road surface were horizontal. (And the inability of the road–tire contact to supply such a tangential force would result in skidding.)

Consider, then, the ideal case as shown in Fig. 7-8(b). Resolving the forces vertically and horizontally, and applying **F** = m**a**, we have

$$N \sin \alpha = \frac{mv^2}{r}$$
$$N \cos \alpha - F_g = 0$$
(7-6)

Replacing F_g by mg, and solving for α, we find that

$$\tan \alpha = \frac{v^2}{gr} \tag{7-7}$$

which defines the correct angle of banking for given values of v

199   Motion in a circle

and r. Alternatively, given r and α, Eq. (7-7) defines the speed at which the curve should be taken. The situations that arise if greater or lesser values of v are used will require the introduction of a frictional force $\mathfrak{F}$ perpendicular to $\mathbf{N}$ (and of limiting magnitude μN) acting inward or outward along the slope of the banked surface (see Problem 7-17).

CURVILINEAR MOTION WITH CHANGING SPEED

If a particle changes its speed as it travels along a circular path, it has, in addition to the centripetal acceleration toward the center of the curvature, a component of acceleration tangent to the path. This tangential acceleration component represents the rate of change of the *magnitude* of the velocity vector. (This is in contrast to the centripetal acceleration component, which depends upon the rate of change of the *direction* of the velocity vector.) We derived the relevant results in Chapter 3 [Eq. (3-18)].

The situation is most easily handled by considering the two components separately:

$$\left.\begin{array}{l}\textit{radial} \text{ component of}\\ \text{acceleration (at right}\\ \text{angles to the path)}^1\end{array}\right\} = a_r = -\frac{v^2}{r} \qquad (r \text{ is the radius of curvature})$$

$$\left.\begin{array}{l}\textit{transverse} \text{ acceleration}\\ \text{(tangent to the path)}\end{array}\right\} = a_\theta = \lim_{\Delta t \to 0} \frac{\Delta v}{\Delta t} = \frac{dv}{dt} \quad \begin{array}{l}\text{[note that this is}\\ \text{the change of}\\ \text{magnitude (only)}\\ \text{of the velocity}\\ \text{vector]}\end{array}$$

(7-8)

With only slight reinterpretation, we may apply these results to the case of motion along any arbitrary curvilinear path. For every point along such a path, there is a center of curvature and an associated radius of curvature (both of which change as one moves along the path). Provided that we interpret the symbols r and v to mean the *instantaneous* values of the radius of curvature and the speed, the above expressions are perfectly general. They give the instantaneous acceleration components—tangent to the

---

[1] In our formal discussion of coordinate systems in Chapter 2 we defined the outward radial direction as positive. We are introducing here the symbol a_r to denote the acceleration component along this direction. The minus sign indicates that this acceleration is in fact toward the center (i.e., centripetal), as we have discussed. See also the discussion on pp. 106-108.

Fig. 7–9 *Acceleration components at a point on an arbitrarily curved path.*

path, and normal to the path—for the general curvilinear motion. In this case Fig. 7–9 indicates a more appropriate notation for these acceleration components.

A particle of dust that rides, without slipping, on a phonograph turntable as it starts up provides a simple and familiar example of a particle possessing both radial and tangential acceleration components. Its total acceleration **a** (Fig. 7–10) is the combination of a_r and a_θ at right angles: $a = (a_r^2 + a_\theta^2)^{1/2}$. The net horizontal force applied to the particle by the turntable must be in the direction of **a** and of magnitude ma. If the contact between the turntable and the particle cannot supply a force of this magnitude, such circular motion is not possible and the particle will slip relative to the surface.

If we analyze the process of whirling an object at the end of a string, so as to bring it from rest up to some high speed of circular motion (as, for example, in the athletic event "throwing the hammer") we see that the string must perform two functions: (1) supply the tangential force to increase the speed of the object, and (2) supply the force of the appropriate magnitude mv^2/r

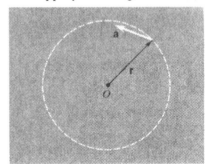

Fig. 7–10 *Net acceleration vector of a particle attached to a disk if the angular velocity of the disk is changing.*

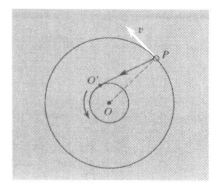

Fig. 7-11 Particle travels in a circular path about O. To increase its speed, the string PO' must provide a force component tangential to the circle.

toward the center of the circle. To fill this dual role, the pull supplied by the string must "lead" the object, so that the tension supplied by it has a tangential component, as shown in Fig. 7-11. This can be maintained if the inner end, O', of the string is continually moved around in a circular path, as indicated. This is the kind of thing we do more or less instinctively in practice.

CIRCULAR PATHS OF CHARGED PARTICLES IN UNIFORM MAGNETIC FIELDS

One of the most important examples of circular motion is the behavior of electrically charged particles in magnetic fields. The separate section on pp. 205–206 summarizes the properties of the magnetic force and describes how this force is in general given by the following equation:

$$\mathbf{F}_{mag} = q\mathbf{v} \times \mathbf{B} \tag{7-9}$$

where q is the electric charge of the particle (positive or negative).

Let us now imagine a charged particle of (positive) charge q, moving in the plane of this page, in a region in which the magnetic field **B** points perpendicularly down into the paper. Then from Eq. (7-9) we have

$$F = qvB \tag{7-10}$$

This is a pure deflecting force, always at right angles to the particle's motion. Hence the magnitude of the velocity **v** cannot change, but its direction changes uniformly with time. Thus, although there is no center of force in the usual sense, the particle describes a circular path of radius r with center O [Fig.

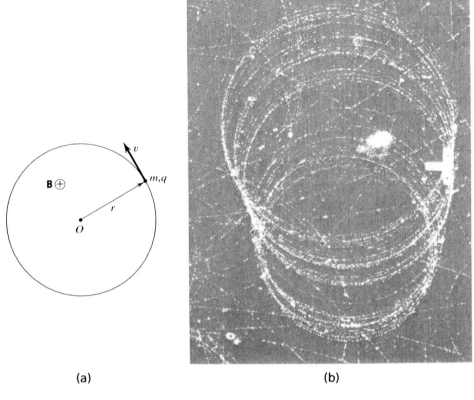

(a) (b)

Fig. 7–12   (a) Charged particle following a circular path in a uniform magnetic field. (b) Magnetically curved tracks of electrons and positrons in a cloud chamber. The main feature is the path of an electron of initial energy 30 MeV going through a succession of almost circular orbits of decreasing radius as it loses energy. (Courtesy of the University of California Lawrence Radiation Laboratory, Berkeley.)

7–12(a)]. The centripetal acceleration is v^2/r, and so by Newton's law we have

$$qvB = \frac{mv^2}{r}$$

whence

$$r = \frac{mv}{qB} \tag{7-11}$$

Thus the radius of the circle is a measure of the momentum mv

of a particle of given charge in a given magnetic field. This fact underlies the nuclear physicist's method for determining the momenta of charged particles in a cloud or bubble chamber [see Fig. 7-12(b)]. From the radius of the circular track which a charged particle generates in a cloud chamber placed between the poles of a magnet the momentum can be readily found if the charge of the particle is known. To get the period T (or the angular velocity), we write

$$qvB = m\omega v$$

or

$$\omega = \frac{2\pi}{T} = \frac{q}{m} B \qquad (7\text{-}12)$$

which is independent of the particle's speed. The angular frequency given by Eq. (7-12) is called the "cyclotron" frequency and depends, for a fixed magnetic field, only on the ratio of charge to mass of the particle. It was the recognition of this by E. O. Lawrence that led him in 1929 to design the first cyclotron, in which protons could be raised to high energies by the application of an alternating electric field of constant frequency. The holding of charged particles in an orbit of a given radius by means of a magnetic field is an essential feature of most high-energy nuclear accelerators.

Having the velocity in a plane perpendicular to **B** is, of course, a very special case. But if this condition is not satisfied, we can imagine **v** to be resolved into one component perpendicular to **B** and a second component parallel to **B**. The latter, by Eq. (7-9), has no magnetic force associated with it; the former

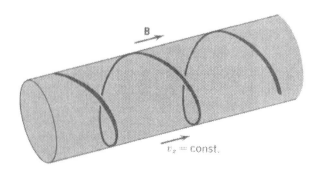

Fig. 7-13 Helical path of charged particle having a velocity component parallel to a magnetic field.

changes with time in precisely the way described above. Thus the resultant motion is a helix, whose projection on the plane perpendicular to **B** is a fixed circle, as shown in Fig. 7-13.

CHARGED PARTICLE IN A MAGNETIC FIELD

It is a matter of experimental fact that an electrically charged particle may, at a given point in space, experience a force when it is moving which is absent if it is at the same point but stationary. The existence of such a force depends on the presence somewhere in the neighborhood (although perhaps quite far away) of magnets or electric currents. Detailed observations reveal the following features [cf. Fig. 7-14]:

1. The force is always exerted at right angles to the direction of the velocity **v** of the particle. The force is reversed if the direction of the velocity is reversed.

2. The force is proportional to the amount of charge, q, carried by the particle. The force reverses if the sign of the charge is reversed.

3. For a given value of q and a given direction of **v**, the size of the force is proportional to the magnitude of **v**.

4. For motion parallel to one certain direction at a given point, the force is zero. This direction coincides with the direction in which a compass needle placed there would align itself. It is called the direction of the magnetic *field* at that point.

5. For any other direction of motion, the direction of the

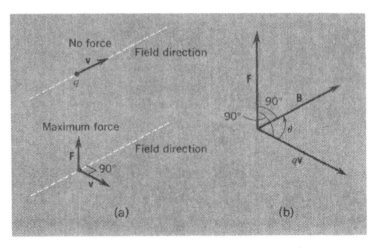

Fig. 7–14 (a) Situations of zero force and maximum force for a charged particle moving in a magnetic field. (b) General vector relationship of velocity, magnetic field, and magnetic force.

force is perpendicular to the plane formed by **v** and the field direction.

6. The magnitude of the force is proportional, for given values of q and **v** and for a fixed magnetic field arrangement, to the sine of the angle between **v** and the field direction.

All the above results can be summarized in a very compact mathematical statement. We are going to introduce a quantitative measure of the magnetic field strength and denote it by the vector symbol **B**, in such a way that Fig. 7-14 shows the relation of the directions of q**v**, **B**, and **F** for a charged particle. (Note that q may be of either sign and that **F** is normal to the plane containing **v** and **B**.) Then the value of **F**, in both magnitude and direction, is given by the following equation, involving the cross product of vectors (see p. 127):

$$\mathbf{F} = q\mathbf{v} \times \mathbf{B}$$

We can then use this equation to define the quantitative measure of the magnetic field strength, such that a field of unit strength, applied at right angles (sin $\theta = 1$) to a charge of 1 C moving with a speed of 1 m/sec, would produce a force of 1 N.

MASS SPECTROGRAPHS

The characteristic curvature of a particle of given q/m in a magnetic field provides the basis of nearly all methods of obtaining precise relative values of atomic and isotopic masses. Such devices often make use of the magnetic force in two ways, first as a velocity selector and second in the manner described in the last section. The velocity selection takes advantage of the fact that the magnetic force on a charged particle is proportional to v, whereas the electric force is not. If a beam of charged particles travels between parallel plates a distance d apart, with voltage difference V, the electric force is given by

$$F_{el} = \frac{qV}{d} \tag{7-13}$$

But the magnetic force is given by Eq. (7-10),

$$F_{mag} = qvB$$

These forces can be arranged to be in opposite directions by

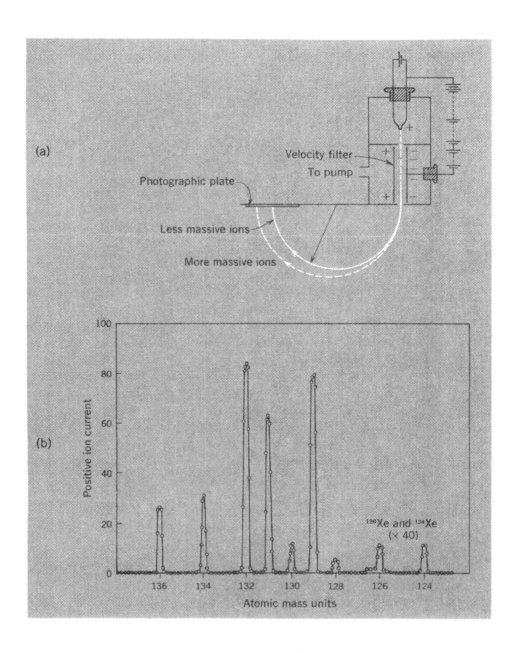

Fig. 7-15 (a) Schematic design of a simple form of mass spectrometer. (b) Example of isotopic separation and mass analysis for isotopes of xenon with a spectrometer like that in (a). [After A. O. Nier, Phys. Rev., **52**, 933 (1937).]

applying the magnetic field at right angles to the electric field. There is then only one speed at which particles can travel through these "crossed fields" undeflected:

$$v = \frac{V}{Bd} \qquad (7\text{-}14)$$

Figure 7-15(a) is a diagram of a simple mass spectrograph that uses a velocity filter of this type, followed by a region in which the charged particles (ions) travel in a semicircular path under the influence of the magnetic field alone. Figure 7-15(b) shows an example of the separation of isotopes by such a device.

THE FRACTURE OF RAPIDLY ROTATING OBJECTS

The question of the stresses set up in a rotating object, and the possibility of fracture if they become excessive, provides another good example of Newton's law applied to uniform circular motion. Whenever an object such as a wheel is rotating, every portion of it has an acceleration toward the axis of rotation, and a corresponding accelerative force is required. Suppose, for example, that a thin wheel or hoop of radius r is rotating about its axis at n revolutions per second (rps) [see Fig. 7-16(a)]. Then any small section of the hoop, such as the one shown shaded, must be supplied with a force equal to its mass, m, multiplied by its centripetal acceleration v^2/r. In this case the magnitude of the angular velocity ω is defined by the equation

$$\omega = 2\pi n$$

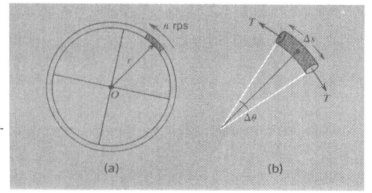

Fig. 7-16 (a) Rotating hoop. (b) Forces acting on a small element of the hoop.

Thus the instantaneous speed is given by

$$v = \omega r = 2\pi r n \tag{7-15}$$

If we make an isolation diagram [Fig. 7-16(b)] for the small portion of material shown shaded in Fig. 7-16(a), it is clear that the forces acting on it must be supplied via its contact with adjacent material of the rim.[1] These forces must, by symmetry, be tangential to the rim at each point. (For example, if the force exerted on Δm at one end had a component radially outward, then by the equality of action and reaction the portion of material with which it was in contact would be subjected to a force with a component radially inward. But in a uniform hoop, all portions such as Δm are equivalent; there is no basis for any asymmetries of this kind.) Thus we can picture the small portion of the rim being acted on by a force of magnitude T at each end. If the length of arc represented by this portion is Δs, it subtends an angle $\Delta \theta$, equal to $\Delta s/r$, at the center O, and each force has a component equal to $T \sin(\Delta\theta/2)$ along the bisector of $\Delta\theta$. Thus by Newton's law we have

$$2T\sin(\Delta\theta/2) = \Delta m \frac{v^2}{r}$$

Putting $\sin(\Delta\theta/2) \approx \Delta\theta/2$, this then gives us

$$T\Delta\theta \approx \Delta m \frac{v^2}{r}$$

i.e.,

$$T\frac{\Delta s}{r} \approx \Delta m \frac{v^2}{r}$$

or

$$T\Delta s = v^2 \Delta m \tag{7-16}$$

Let us now express the mass Δm in terms of the density, ρ, of the material and the volume of the piece of the rim. If the cross-sectional area of the rim is A, we have

$$\Delta m = \rho A \Delta s$$

[1] We shall assume that the spokes of the wheel serve primarily to give just a geometrical connection between the rim and the axis and have almost negligible strength.

Substituting this into Eq. (7–16) and substituting $v = 2\pi r n$ from Eq. (7–15), we obtain the following result:

$$T = 4\pi^2 n^2 r^2 \rho A \tag{7-17}$$

Now, it is an experimental fact that a bar or rod of a given material will fracture under tension if the ratio of the applied force to the cross-sectional area exceeds a certain critical value—the ultimate tensile strength, S. Thus we can infer from Eq. (7–17) that a hoop of the kind we have been discussing has a critical rate of rotation, above which it will burst. We have, in fact,

$$n_{\max} = \frac{1}{2\pi r}\left(\frac{S}{\rho}\right)^{1/2} \tag{7-18}$$

Suppose, for example, that we have a steel hoop of radius 1 ft (i.e., about 0.3 m). The density of steel is about 7600 kg/m^3, and its ultimate strength is about 10^9 N/m^2. These values lead to a value of $n_{\max}$ of about 500 rps, or 30,000 rpm—much faster than any such wheel would normally be driven. However, the rotors of ultracentrifuges are, in fact, driven at speeds of this order, up to a significant fraction of the bursting speed for their particular radius (see p. 513).

MOTION AGAINST RESISTIVE FORCES

We shall now consider an important class of problems in which an object is subjected to a constant driving force, $\mathbf{F}_0$, but has its motion opposed by a resistive force, $\mathbf{R}$, that always acts in a direction opposite to the instantaneous velocity. Typical of such forces are the frictional resistance as an object is pulled along a solid surface, or the air resistance to falling raindrops, moving cars, and so on. In general this resistive force is a function of speed, so that the statement of Newton's law must be written as follows:

$$F_0 - R(v) = m\frac{dv}{dt} \tag{7-19}$$

As we saw in Chapter 5, the resistive force of dry friction is in fact nearly independent of v, as indicated in Fig. 7–17(a), so that we can put

$$R(v) = \mathcal{F} \approx \text{const.}$$

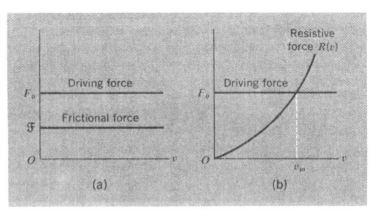

Fig. 7-17 (a) Driving and resistive forces for an object resisted by dry friction. (b) Driving and resistive forces for an object resisted by fluid friction.

Equation (7-19) then reduces to the simple case of acceleration under a constant net force, as described by the kinematics of Eqs. (7-1). The situation is very different in the case of fluid resistance, for which $R(v)$ increases monotonically with v, as indicated in Fig. 7-17(b) and as described [Eq. (5-4)] by the relation

$$R(v) = Av + Bv^2 \qquad (7\text{-}20)$$

In this case, if we consider an object starting out from rest under the force F_0, the initial acceleration is F_0/m, but the net driving force is immediately reduced to a value below F_0, because as soon as the object has any appreciable velocity it is exposed, in its own frame of reference, to a wind or flow of fluid past it at the speed v. The statement of Newton's law as it applies to this problem must now be written

$$m\frac{dv}{dt} = F_0 - Av - Bv^2 \qquad (7\text{-}21)$$

(In this equation F_0, A, B, and v are all taken to be positive. One must be careful to consider what is the appropriate statement of Newton's law for this system if the direction of **v** is reversed.)

The solving of Eq. (7-21) is not at all such a simple matter as our familiar problems involving constant forces or forces perpendicular to the velocity. We are now faced with finding the solution to an awkward differential equation. We do not intend to plunge into all the formal mathematics of this problem. Instead, this is a suitable moment to point to the value of approximate numerical methods—in other words, the method of the digital computer, using finitely small intervals. We outlined the principle of the method in Chapter 3 [Eq. (3-11)]; here we have a good case for using it. First, however, let us consider some individual features of the solution:

1. For some range of small values of v, the acceleration will be almost constant and v will start off as a linear function of t with slope F_0/m.

2. As v increases, a decreases monotonically, giving a steadily decreasing slope in the graph of v versus t (Fig. 7-18).

3. There is a *limiting speed*, v_m, under any given applied force. It is the speed at which the graph of $R(v)$ versus v is intersected by a horizontal line at the ordinate equal to F_0 [see Fig. 7-17(b)]. Algebraically, it is the positive root of the quadratic equation

$$Bv^2 + Av - F_0 = 0$$

(What is the status of the negative root, and why is it to be discarded?)

Notice the contrast between the sharply defined value of v_m, in the graph of $R(v)$ versus v, and the gradual manner in which this velocity is approached (and in principle never quite reached) if one considers v as a function of time (Fig. 7-18).

It is well worth taking a moment to consider the dynamical situation represented by $v = v_m$. It is a motion with zero acceleration under zero net force, but it seems a far cry from the unaccelerated motion of objects moving under no force at all; and it is certainly not an application of what we understand by the principle of inertia. But let us emphasize that, like static equilibrium, it *is* a case of $\sum F = 0$. Every time we see a car

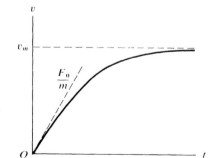

Fig. 7-18 Asymptotic approach to terminal speed for object in a fluid resistive medium.

hurtling along a straight road at a steady 80 mph, or a jet plane racing through the air at a constant 600 mph, we are seeing objects traveling under zero *net* force. This basic dynamical fact tends to be obscured because what matters in practical terms is the large value of the driving force F_0 needed to maintain the steady motion once it has been established.

DETAILED ANALYSIS OF RESISTED MOTION

In order to see how Newton's law for resisted motion [Eq. (7-21)] works in practice we need to introduce one additional feature already described in Chapter 5—that the two terms in the resistive force differ not only in their dependence on v but also in their dependence on the linear dimensions of the object (see Fig. 7-19). Specifically, for a sphere of radius r we have

$$A = C_1 r$$
$$B = C_2 r^2$$

and thus

$$R(v) = C_1 rv + C_2 r^2 v^2 \tag{7-22}$$

The two terms in the resistance thus become equal at a critical speed, v_c, defined by the formula

$$v_c = \frac{C_1}{C_2 r} \tag{7-23}$$

We know that the term proportional to v will dominate the picture if v is sufficiently small (since $Bv^2/Av \sim v$), but we know

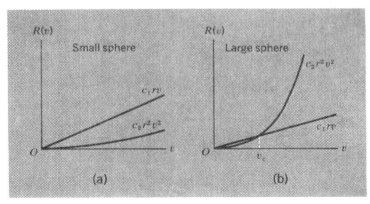

Fig. 7-19 (a) Linear and quadratic terms in fluid resistance for a small object, with viscous resistance predominant. (b) Similar graph for a large object, for which the v^2 term predominates at all except low speeds.

equally that for speeds much in excess of v_c (say $v \gtrsim 10v_c$) the quadratic term takes over. If the resistive medium is air, the coefficients c_1 and c_2 have the following approximate values:

$$c_1 = 3.1 \times 10^{-4} \text{ kg m}^{-1} \text{ sec}^{-1}$$
$$c_2 = 0.87 \text{ kg/m}^3$$

Thus, if r is expressed in meters, we have

$$v_c \text{ (m/sec)} = \frac{3.6 \times 10^{-4}}{r} \tag{7-24}$$

This means that for an object such as a small pebble, with $r \approx 1$ cm, the value of v_c is only a few centimeters per second; a speed equal to 10 times this (say 0.5 m/sec) would be acquired in free fall under gravity within a time of about 0.05 sec and a distance of about 0.5 in. (Check these numbers!) Thus for most problems of practical interest (we shall consider an exception in the next section) the motion under a constant applied force in a resisting medium can be extremely well described with the help of the following simplified version of Eq. (7-21):

$$m \frac{dv}{dt} \approx F_0 - Bv^2 \tag{7-25}$$

The resistive term, Bv^2, is quite important in the motion of ordinary objects falling through the air under gravity. This becomes very apparent if we calculate the *terminal speed*, v_t, by putting $dv/dt = 0$ in Eq. (7-25), with F_0 set equal to the gravitational force, mg. Take, for example, the case of our pebble of radius 1 cm. The density of stone or rock is about 2.5 times that of water, i.e., about 2500 kg/m^3, so we have

$$m = \frac{4\pi}{3} \rho r^3 \approx 4(2.5 \times 10^3) \times (10^{-2})^3 \approx 10^{-2} \text{ kg}$$

The value of F_0 is thus about 0.1 N. The value of the coefficient B ($= c_2 r^2$) for an object of this size is about 10^{-4} kg/m. Substituting these values in the equation

$$F_0 - Bv_t^2 = 0$$

we find

$$v_t \approx 30 \text{ m/sec}$$

Under the assumptions of genuinely free fall, this speed would be attained in a time of about 3 sec and a vertical distance of about 150 ft. We can be sure, then, that the effects of the resistance become quite significant within times and distances appreciably less than these. This means that many of the idealized problems of free fall under gravity, of the sort that everyone meets in his first encounter with mechanics, do not correspond very well with reality.

Let us now consider the computer procedure for solving these problems. Given a computer, it is almost as little trouble to handle the full equation [Eq. (7–21)] as it is to use the approximation represented by Eq. (7–25). But for the purposes of discussing the method, we shall use the simpler, approximate form.

We choose some convenient small increment of time, Δt, and will measure time from $t = 0$ in terms of integral multiples (n) of Δt. In the simplest approach, we assume that the acceleration during each small interval remains constant at the value calculated for the *beginning* of that interval. Thus, for an initial velocity equal to zero, the acceleration during the first Δt is set equal to a_0 ($= F_0/m$). The velocity v_1 at the end of this interval is thus given by

$$v_1 = 0 + a_0 \Delta t$$

Using this velocity, we calculate the acceleration a_1 at $t = \Delta t$:

$$a_1 = a_0 - \frac{B}{m} v_1^2$$

Applying this acceleration to the next interval, the velocity at $t = 2 \Delta t$ is given by

$$v_2 = v_1 + a_1 \Delta t$$

We know that the first step in the calculation, as performed in this way, leads to an overestimate of v_1, but we see that in this particular problem the error is in some measure compensated by leading in turn to an *under*estimate of a_1. The calculated values of a in successive time intervals are as indicated in Fig. 7–20(a). Applying an exactly similar treatment to the changes of position (x_n) we would take our set of values of velocities, as given by the above calculation and represented graphically in Fig. 7–21(a), and would use the following formula:

$$x_{n+1} = x_n + v_n \Delta t$$

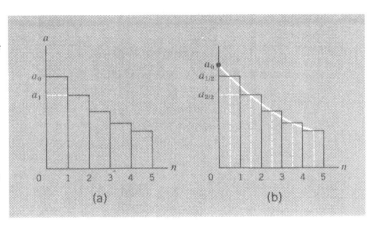

Fig. 7-20 (a) Basis of simple calculation of acceleration versus time for object starting from rest in a resistive medium. The acceleration is calculated from the speed at the beginning of each time interval. (b) Improved approach to the same problem, using acceleration calculated from speed at midpoint of each time interval.

Thus we should have

$$x_1 = v_0 \Delta t = 0 \quad \text{(clearly an underestimate)}$$
$$x_2 = x_1 + v_1 \Delta t = v_1 \Delta t$$
$$x_3 = x_2 + v_2 \Delta t$$
$$\vdots$$

Our graph of x versus t would be the result of summing the areas of the rectangles in Fig. 7-21(a) up to successively greater values of n.

A more sophisticated analysis takes account of the fact that a better average value of the acceleration or velocity during a given time interval is provided by the instantaneous value at the *midpoint* of that interval. Thus the acceleration between $n \Delta t$ and $(n + 1) \Delta t$ is set equal to the instantaneous value at $(n + \frac{1}{2}) \Delta t$. This leads to the following formulas:

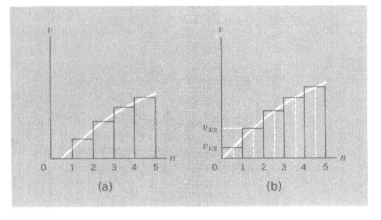

Fig. 7-21 (a) Velocity-time graph based on velocities at the beginnings of the successive time intervals. (b) Improved graph based on velocities evaluated at midpoints of the time intervals.

$$v_{n+1} = v_n + a_{n+1/2}\Delta t$$
$$x_{n+1} = x_n + v_{n+1/2}\Delta t \tag{7-26}$$

This looks fine, but we now run into a slight snag when we try to get the calculation started. To find v_1, the velocity at the end of the first interval, we now need $v_0\ (= 0)$ and $a_{1/2}$. The latter, however, by Eq. (7-25), depends on a knowledge of $v_{1/2}$, which we do not yet know. (Notice that in the first, crude method, we were able to start out directly from the initial conditions v_0 and a_0.) We are thus forced to compromise a little, although we

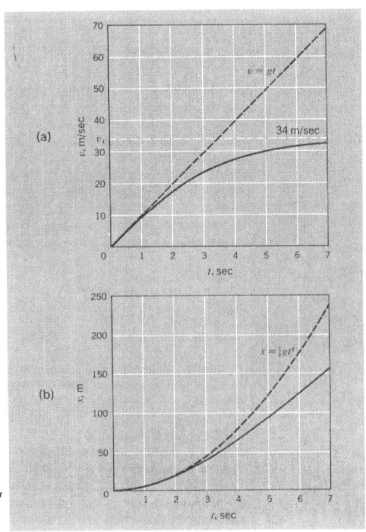

Fig. 7-22 (a) Comparison of idealized (resistanceless) and actual dependence of speed on time for a falling pebble of radius 1 cm. (b) Idealized and actual distances fallen by such a pebble.

are still left with a better treatment than before. What we do is to calculate an approximate value of $v_{1/2}$ from the equation

$$v_{1/2} \approx a_0 \frac{\Delta t}{2} \tag{7-27}$$

and then we are under way. Figures 7–20(b) and 7–21(b) show what this method means in terms of graphs of a and v against time.

Figure 7–22 shows graphs of the actual calculated variation of speed and distance with time for our 1-cm-radius pebble falling in air, under gravity; the idealized free-fall curves are given for comparison.

MOTION GOVERNED BY VISCOSITY

If we are dealing with microscopic or near-microscopic objects, such as particles of dust, then, in contrast to the situations discussed above, the resistance is due almost entirely to the viscous term, Av, up to quite high values of v. If, for example, we consider a tiny particle of radius $1\,\mu$ ($= 10^{-6}$ m), then Eq. (7–24) tells us that the critical speed, v_c, at which the contributions Av and Bv^2 become equal, is 360 m/sec. This implies a wide range of lower speeds for which the motion is controlled by viscous resistance alone, and the statement of Newton's law can be written, without any appreciable error, in the following form:

$$m\frac{dv}{dt} = F_0 - Av \tag{7-28}$$

with

$$A = c_1 r$$

Motion under these conditions played the central role in R. A. Millikan's celebrated "oil-drop" experiment to determine the elementary electric charge. The basic idea was to measure the electric force exerted on a small charged object by finding the terminal speed of the object in air. If the radius of the particle is known, the resistive force is completely determined, and the driving force must be equal and opposite to it at $v = v_t$.

In Millikan's original experiments the charged particles were tiny droplets of oil in the mist of vapor from an "atomizer." Such droplets have a high probability of carrying a net electric

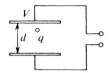

Fig. 7-23 *Basic parallel-plate arrangement in Millikan experiment.*

charge of either sign when they are produced. In order to apply electric forces to them, Millikan used the arrangement shown schematically in Fig. 7-23. Two parallel metal plates, spaced by a small fraction of their diameter, are connected to the terminals of a battery.[1] The force on a particle of charge q anywhere between these plates is given by

$$F_{el} = q\frac{V}{d} \qquad (7\text{-}29)$$

where V is the voltage difference applied to the plates and d is their separation in meters. If q is measured in coulombs, F is given in newtons by this equation. (See also our discussion of the cathode-ray oscilloscope, p. 195.) Thus if a dynamic balance were set up between this electric driving force and the resistive force Av, we should have

$$q\frac{V}{d} = Av_t = c_1 r v_t$$

The droplets randomly produced in a mist of oil vapor are of various sizes. The ones that Millikan found most suitable for his experiments were the smallest (partially because they had the lowest terminal speed under their own weight). But these droplets were so tiny that even through a medium-power microscope they appeared against a dark background merely as points of light; no direct measurement could be made of their size. Millikan, however, used the clever trick of exploiting the law of viscous resistance a second time by applying it to the fall of a droplet under the gravitational force alone, with no voltage between the plates. Under these conditions we have

$$F_0 = mg = \frac{4\pi}{3}\rho r^3 g$$

where ρ is the density of oil (about 800 kg/m^3). The terminal speed of fall under gravity is then given by

---

[1]Millikan himself used plates about 20 cm across and 1 cm apart and several thousand volts. Most modern versions of the apparatus use smaller values of all three quantities.

$$\frac{4\pi}{3} \rho r^3 g = c_1 r v_g$$

(Gravitational) $v_g = \dfrac{4\pi}{3} \dfrac{\rho g}{c_1} r^2$ (7-30)

Putting in the approximate numerical values we find

(Gravitational) $v_g \approx 10^8 r^2$ (v_t in m/sec, r in m)

Putting $r \approx 1\mu = 10^{-6}$ m, we have

$v \approx 10^{-4}$ m/sec $= 0.1$ mm/sec

Such a droplet would take over 1 min to fall 1 cm in air under its own weight, thus allowing precision measurements of its speed. (It is clear, incidentally, that for such motions as this the resistive term Bv^2 is utterly negligible.)

It is worth noting the dynamical stability of this system, and indeed of any situation involving a constant driving force and a resistive force that increases monotonically with speed. If by chance the droplet should slow down a little, there is a net force that will speed it up. Conversely, if it should speed up, it is subjected to a net retarding force. If one could observe the motion of a falling droplet in sufficient detail, this behavior would doubtless be found, because the air, being made up of individual molecules, does not behave as a perfectly homogeneous fluid. In other words, the speed of the droplet would fluctuate about some average value, although the fluctuations would be exceedingly small.

In the Millikan experiment proper, the vertical motion of the charged droplets is studied with the electric force either aiding or opposing the gravitational force. Thus if we measure velocities as positive downward, the terminal velocity in both magnitude and sign will be defined by the equation

$$mg + \frac{qV}{d} = c_1 r v_t$$ (7-31)

where V is the voltage of the upper plate relative to the lower and q is the net charge on the drop (positive or negative). Although in principle the terminal velocity is approached but never quite reached (see Fig. 7-18), the small droplets under the conditions of the Millikan experiment do, in effect, reach this speed within a very short time—much less than 1 msec in most cases (see the next section).

Millikan was able to follow the motion of a given droplet

for many hours on end, using its electric charge as a handle by which to pull it up or down at will. In the course of such protracted observations the charge on the drop would often change spontaneously, and several different values of v_t would be obtained. The crucial observation was that in any such experiment, with a given value of the voltage V, the speed v_t was limited to a set of sharp and distinct values, implying that the electric charge itself comes in discrete units. But Millikan went further and obtained the first precise value of the absolute magnitude of the elementary charge.[1]

You might wonder why Millikan used such a roundabout method to measure the electric force exerted on a charged particle. After all, it would in principle be possible to hang such an electrified particle on a balance and measure the force in a static arrangement. However, in practice, when only a few elementary charges are involved, the forces are extremely weak and such a method is not feasible. For example, the force on a particle with a net charge of 10 elementary units, between plates 1 cm apart with 500 V between them, is only about 10^{-13} N, equal to the gravitational force on only 10 $\mu\mu$g!

GROWTH AND DECAY OF RESISTED MOTION

We have seen how the velocity under a constant force in a resistive fluid medium rises asymptotically toward the terminal value. What happens if the driving force is suddenly removed? We can guess that the velocity will decay away toward zero in a similar asymptotic way, as indicated in Fig. 7–24. If the initial speed is small enough, the whole decay depends on viscous resistance alone and is governed by a special, simplified form of Eq. (7–28) in which F_0 is set equal to zero:

$$m \frac{dv}{dt} = -Av$$

or (7-32)

$$\frac{dv}{dt} = -\alpha v$$

where $\alpha = A/m$.

[1] For his own full and interesting account (and much other good physics), see R. A. Millikan, *The Electron* (J. W. M. Du Mond, ed.), University of Chicago Press (Phoenix Series), Chicago, 1963.

Fig. 7-24 Growth and decay of the velocity of a particle controlled by a viscous resistive force proportional to v.

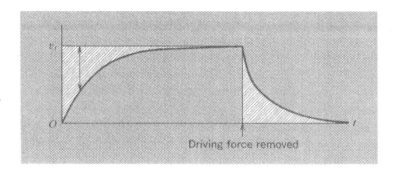

You may recognize Eq. (7-32) as the basic differential equation of all forms of exponential decay. Whether you do or not, you may like to see how this equation can be solved by what is in effect just arithmetic, using the approach of numerical analysis and digital-computer techniques. We divide up the time into a large number of equal intervals Δt, and interpret Eq. (7-32) as telling us that the change of v between $t = n\Delta t$ and $t = (n+1)\Delta t$ is proportional to the mean velocity during that interval:

$$\Delta v = v_{n+1} - v_n \approx -\alpha \left(\frac{v_{n+1} + v_n}{2} \right) \Delta t$$

Solving this, we have

$$\frac{v_{n+1}}{v_n} \approx \frac{1 - \alpha \Delta t/2}{1 + \alpha \Delta t/2} = f$$

where f is a constant ratio, less than unity. The velocities at equal intervals of time thus decrease in geometric proportion. If the initial velocity is v_0, the velocity at time $t (= k\Delta t)$ is given by

$$v(t) \approx v_0 f^k$$

Substituting the expression for f from the preceding equation and putting $k = t/\Delta t$, we thus get

$$v(t) \approx v_0 \left(\frac{1 - \alpha \Delta t/2}{1 + \alpha \Delta t/2} \right)^{t/\Delta t}$$

We shall now consider what happens to this expression as we imagine the intervals Δt to be made shorter and shorter, and their number correspondingly greater. To simplify the discussion, we shall put

$$\alpha \Delta t = \frac{1}{z}$$

The quantity z is then a large number that we shall allow to approach infinity.

Besides substituting $\alpha \Delta t = 1/z$ in the above equation for $v(t)$, we shall also replace the exponent $t/\Delta t$ by its equivalent quantity, $\alpha z t$. Thus we have

$$v(t) \approx v_0 \left(\frac{1 - 1/2z}{1 + 1/2z}\right)^{\alpha z t}$$

$$= v_0 \left[\left(\frac{1 - 1/2z}{1 + 1/2z}\right)^z\right]^{\alpha t}$$

i.e.,

$$v(t) \approx v_0 p^{\alpha t}$$

where

$$p(z) = \left(\frac{1 - 1/2z}{1 + 1/2z}\right)^z \tag{7-33}$$

Let us look at the behavior of $p(z)$ as z is made larger and larger (Table 7-1). As z increases, the number p is clearly ap-

TABLE 7-1

z	$p(z)$	*Decimal value of p*
1	$(1/3)^1$	0.3333 ...
2	$(3/5)^2$	0.3600
3	$(5/7)^3$	0.3644
4	$(7/9)^4$	0.3659
5	$(9/11)^5$	0.3667
⋮	⋮	⋮
10	$(19/21)^{10}$	0.3676

proaching a limiting value; this value is 0.367879 ... and is the reciprocal of the famous number e ($= 2.71828 \ldots$), which forms the base of natural or Napierian logarithms. Thus in Eq. (7-32) we can put

$$\lim_{z \to \infty} p(z) = 0.367879\ldots = e^{-1}$$

and hence we have the following expression, now exact, for the value of $v(t)$:

$$v(t) = v_0 e^{-\alpha t} \tag{7-34}$$

The reciprocal of α in Eq. (7-34) is of the dimension of time and represents a characteristic *time constant*, τ, for the exponential decay process. In the particular case of small spheres moving through the air, τ is defined by the equation

$$\tau = \frac{m}{A} = \frac{m}{c_1 r}$$

We can express this in much more vivid terms by introducing the terminal velocity of fall under gravity, v_g. For from Eq. (7-30) we have

$$v_g = \frac{mg}{c_1 r}$$

It follows that we have

$$\tau = \frac{v_g}{g} \tag{7-35}$$

Thus τ is equal to the time that a particle would take to reach a velocity equal to the terminal velocity under conditions of free fall. For an oil drop of radius 1 μ, with $v_g \approx 0.1$ mm/sec, this time would be only about 10^{-5} sec ($= 10$ μsec).

In a time equal to any substantial multiple of τ, the value of $v(t)$ as given by Eq. (7-34) falls to a very small fraction of v_0. For example, if we take the basic equation

$$v(t) = v_0 e^{-t/\tau} \tag{7-36}$$

and put $t = 10\tau$, then the value of v becomes less than 10^{-4} of v_0, which for many purposes can be taken to be effectively zero.

It is more or less intuitively clear that the *growth* of velocity toward its limiting value, after a particle starts from rest under the action of a suddenly applied driving force, must be a kind of upside-down version of the decay curve (see Fig. 7-24). In fact, if the terminal velocity is v_t, the approach to it, under these conditions of viscous resistance, is described by

$$v(t) = v_t(1 - e^{-t/\tau}) \tag{7-37}$$

If one chooses to write this in the form

$$v_t - v(t) = v_t e^{-t/\tau}$$

one can see explicitly how closely the growth and decay curves are related. Indeed, one could almost deduce Eq. (7-37) from Eq. (7-36) plus the recognition of this symmetry.

AIR RESISTANCE AND "INDEPENDENCE OF MOTIONS"

When Galileo put forward the proposition that the motion of a projectile could be analyzed into separate horizontal and vertical parts, with a constant velocity horizontally and a constant acceleration under gravity vertically, he made a great contribution to the conceptual basis of mechanics. It may be worth pointing out, however, that this "independence of motions" breaks down if one takes into account the resistive force exerted on objects of ordinary size. We have seen how for such objects the magnitude of the resistive force is proportional to the square of the speed.

Consider an object moving in a vertical plane (Fig. 7–25). At a given instant let its velocity **v** be directed at an angle θ above the horizontal, as shown. The object is then subjected to the gravitational force, $\mathbf{F}_g$, and a resistive force $\mathbf{R}(v)$, of magnitude Bv^2, in a direction opposite to **v**. Newton's law applied to the x and y components of the motion at any instant, thus gives us

$$m\frac{dv_x}{dt} = -R_x = -Bv^2 \cos\theta$$

$$m\frac{dv_y}{dt} = -mg - Bv^2 \sin\theta$$

Since $v\cos\theta = v_x$, and $v\sin\theta = v_y$, we can rewrite these equations as follows:

$$m\frac{dv_x}{dt} = -(Bv)v_x$$

$$m\frac{dv_y}{dt} = -mg - (Bv)v_y$$

Thus the equation governing each separate component of the velocity involves a knowledge of the magnitude of the *total* velocity and hence of what is happening at each instant along the other coordinate direction. The larger the magnitude of **v**, the more important does this cross connection between the different components of the motion become. Thus we really cannot calculate the vertical motion of a falling body without reference

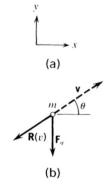
(a)

(b)

Fig. 7–25 *Resistive and gravitational force vectors for a particle moving in a vertical plane.*

to the horizontal component. Recognition of this fact may be salutary for those who are accustomed to taking for granted the idealized equations of falling objects. In the folklore of physics there is the story of the impecunious student who would make money off his nonscientific acquaintances by a bet. He asked them to consider two identical objects. One is dropped from rest at a certain height above the ground; the other is fired off horizontally at the same instant. Which one will reach the ground first? The victim (poor ignoramus!) would often have some vague idea that the fast horizontal motion of the second object must somehow keep it in the air longer. But of course that is demonstrably false, isn't it?—look at the analysis of motion under gravity in Chapter 3. And even a direct test would appear to prove the point if the velocities were kept small. But the complete equations, as written above, show that a large initial horizontal velocity *would* increase the time taken to descend a given vertical distance. Moral: Beware of facile idealizations.

Notice that if the resistance is purely viscous, varying as the first power of v, then the x and y components of the motion *can* be handled entirely separately, even though the problem is not that of idealized free fall. Thus, although one can always take a statement of Newton's law as the starting point of any problem of a particle exposed to forces, the way of proceeding from there to the analysis of the complete motion may depend critically on the precise nature of the individual problem.

SIMPLE HARMONIC MOTION

One of the most important of all dynamical problems is that of a mass attracted toward a given point by a force proportional to its distance from that point. If the motion is assumed to be confined to the x axis we have

$$F(x) = -kx \tag{7-38}$$

A good approximation to this situation is provided by an object on a very smooth horizontal table (e.g., with air suspension) and a horizontal coiled spring, as shown in Fig. 7-26(a). The object would normally rest at a position in which the spring is neither compressed nor extended. The force brought into play by a slight displacement of the mass in either direction is then well described by Eq. (7-38). The constant k is called the *spring*

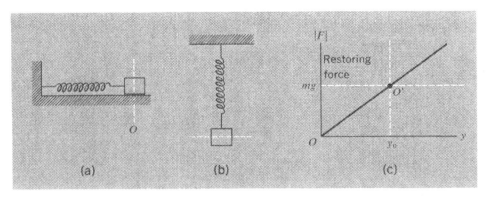

Fig. 7-26 (a) Mass–spring system on a frictionless horizontal surface. (b) Mass hanging from a vertical spring. (c) Graph of force versus displacement with a linear force law; the equilibrium situations O and O' correspond to (a) and (b), respectively.

constant and is measured in newtons per meter. This linear force law for a spring was discovered by Robert Hooke in 1676 and is named after him.[1] An even simpler arrangement in practice, although slightly more complicated theoretically, is to suspend a mass at the bottom end of a vertical spring, as in Fig. 7-26(b). In this case the normal resting situation already involves an extension of the spring, sufficient to support the weight of the object. A further displacement up or down from this equilibrium position leads, however, to a net restoring force exactly of the form of Eq. (7-38). This is indicated in Fig. 7-26(c), which shows the magnitude of the force exerted by the spring as a function of its extension y. If one takes as a new origin of this graph the point O', then one has a net restoring force $(F - mg)$ proportional to the extra displacement $(y - y_0)$.

The great importance of this dynamical problem of a mass on a spring is that the behavior of very many physical systems under small displacements from equilibrium obeys the same basic equation as Eq. (7-38). We shall discuss this in more detail in Chapter 10; for the present we shall just concern ourselves with solving the problem as such.

[1]He first announced it in a famous anagram—ceiiinosssttuv—which 2 years later he revealed as a Latin sentence, "ut tensio, sic vis" ("as the extension, so the force"). By this device (popular in his day) Hooke could claim prior publication for his discovery without actually telling his competitors what it was!

This makes a good case in which to begin with the computer method of solution rather than with formal mathematics. The basic equation of motion, expressed in the form $ma = F$, is as follows:

$$m \frac{d^2x}{dt^2} = -kx \tag{7-39}$$

Rewriting this as

$$\frac{d^2x}{dt^2} = -\frac{k}{m} x$$

we recognize that k/m is a constant, of dimension (time)$^{-2}$. Denoting this by ω^2, our basic equation thus becomes

$$\frac{d^2x}{dt^2} = -\omega^2 x \tag{7-40}$$

We can read this as a direct statement of the way in which dx/dt is changing with time and can proceed to calculate the approximate change of dx/dt in a small interval of time Δt:

$$\frac{d}{dt}\left(\frac{dx}{dt}\right) = -\omega^2 x$$

and so

$$\Delta\left(\frac{dx}{dt}\right) \approx -\omega^2 x \, \Delta t \tag{7-41}$$

This is the reverse of the process by which d^2x/dt^2 was originally defined (cf. Chapter 3).

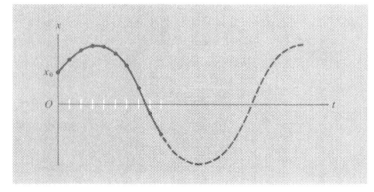

Fig. 7–27 Displacement versus time in simple harmonic motion.

Suppose, to be specific, that we start out at $t = 0$ with $x = x_0$ and v ($= dx/dt) = v_0$ both positive, as shown in Fig. 7-27. Then at time $t = \Delta t$ we have

$$x \approx x_0 + v_0 \Delta t$$
$$\frac{dx}{dt} \approx v_0 - \omega^2 x_0 \Delta t$$

The displacement is a little bigger, the slope a little less. Using these new values, we take another step of Δt, and so on. Several features can be read from Eq. (7-41):

1. As long as x is positive, the slope dx/dt decreases when we go from t to $t + \Delta t$. That is, if dx/dt is positive it gets smaller; if dx/dt is negative it becomes more negative.

2. The rate of change of dx/dt is proportional to x. The graph has its greatest curvature at the largest x, and as $x \to 0$ it becomes almost a straight line.

3. As soon as x becomes negative, dx/dt becomes less negative or more positive with each time increment Δt.

Using these considerations, we can construct the picture of a curve that is always curving toward the line $x = 0$ (i.e., the t axis), necessarily forming a repetitive wavy pattern.

Now anyone who has ever drawn graphs of trigonometric functions will recognize that Fig. 7-27 looks remarkably like a sine or cosine curve. More specifically, it suggests a comparison with the following analytic expression for the distance x as a function of time:

$$x = A \sin(\alpha t + \varphi_0)$$

where A is the maximum value attained by x during the motion, α a constant with the dimension (time)$^{-1}$, and φ_0 an adjustable angle that allows us to fit the value of x at $t = 0$.

Testing this trial function against the original differential equation of motion [Eq. (7-40)] requires differentiating x twice with respect to t:

$$v = \frac{dx}{dt} = \alpha A \cos(\alpha t + \varphi_0)$$
$$a = \frac{d^2 x}{dt^2} = -\alpha^2 A \sin(\alpha t + \varphi_0) = -\alpha^2 x$$

We see that the solution does indeed fit, provided that we put $\alpha = \omega$. This then brings us to the following final result:

229 Simple harmonic motion

$$x(t) = A \sin(\omega t + \varphi_0) \quad \text{where } \omega = \left(\frac{k}{m}\right)^{1/2} \tag{7-42a}$$

Equation (7-42a) is the characteristic equation of what is called *simple harmonic motion* (SHM), and any system that obeys this equation of motion is called a harmonic oscillator. The constant A is the *amplitude* of the motion, and φ_0 (Gk: phi) is what is called the initial phase (at $t = 0$). The complete argument, $\omega t + \varphi_0$, of the sine function is called just "the phase" of the motion at any given instant. The result represented by Eq. (7-42a) could be equally well expressed by writing x as a cosine function, rather than a sine function, of t:

$$x(t) = A \cos(\omega t + \varphi_0') \tag{7-42b}$$

with an appropriate value of the constant φ_0'. This form of the solution is found more convenient for some purposes.

The harmonic motion is characterized by its *period*, T, which defines successive equal intervals of time at the end of which the state of the motion reproduces itself exactly in both displacement and velocity. The value of T is readily obtained from Eqs. (7-42) by noting that each time the phase angle ($\omega t + \varphi_0$) changes by 2π, both x and v have passed through a complete cycle of variation. Thus we can put

$$\varphi_1 = \omega t_1 + \varphi_0$$
$$\varphi_1 + 2\pi = \omega(t_1 + T) + \varphi_0$$

Therefore, by subtraction,

$$2\pi = \omega T$$

or

$$T = \frac{2\pi}{\omega} = 2\pi \left(\frac{m}{k}\right)^{1/2} \tag{7-43}$$

The form of this result corresponds to one's commonsense knowledge that if m gets bigger the oscillation goes more slowly, and if the spring is made stiffer (larger k) the oscillation becomes more rapid.

Example. A spring of negligible mass hangs vertically from a fixed point. When a mass of 2 kg is hung from the bottom end of the spring, the spring is stretched by 3 cm. What is the period of simple harmonic vibration of this system?

First we calculate the force constant k. The gravitational force on the mass is 2×9.8 N $= 19.6$ N. This force causes an extension of 0.03 m; therefore, $k = 19.6/0.03 = 653$ N/m. Using Eq. (7–43) we then have

$$T = 2\pi \left(\frac{2}{653}\right)^{1/2} = 0.35 \text{ sec}$$

(You might like to note that we need not have specified the mass. Any spring that extends 3 cm when a certain mass is hung on it will have a vibration period of 0.35 sec with this same mass. Why?)

MORE ABOUT SIMPLE HARMONIC MOTION

Fitting the initial conditions

We have come to recognize, both as a general principle and through various examples, that the complete solution of any problem in the use of Newton's law requires not only a knowledge of the force law but also the specification of two independent quantities that correspond to the constants of integration introduced as we go from $a \ (= d^2x/dt^2)$ to x. Most commonly we have talked of giving the initial position, x_0, and the initial velocity, v_0. Here, in our analysis of the motion of a harmonic oscillator, we also need initial conditions or their equivalent. Actually they appear as the two constants A and φ_0 in Eq. (7–42). We have already identified A as the amplitude of the motion and φ_0 as the initial phase. If we are given the values of x_0 and v_0 we can readily solve for A and φ_0 as follows:

From Eq. (7–42a),

$$x = A \sin(\omega t + \varphi_0)$$
$$v = \frac{dx}{dt} = \omega A \cos(\omega t + \varphi_0) \tag{7–44}$$

Therefore,

$$x_0 = A \sin \varphi_0$$
$$v_0 = \omega A \cos \varphi_0$$

It follows that

$$A = \left[x_0^2 + \left(\frac{v_0}{\omega}\right)^2\right]^{1/2}$$
$$\tan \varphi_0 = \frac{\omega x_0}{v_0} \tag{7–45}$$

A geometrical representation of SHM

There is a very basic connection between simple harmonic motion and uniform motion in a circular path. This fact leads to a particularly simple way of displaying and visualizing SHM. Imagine that a horizontal disk, of radius A, rotates with constant angular speed ω about a vertical axis through its center. Suppose that a peg P is mounted on the rim of the disk as shown in Fig. 7-28(a). Then, if the disk is viewed edge on (horizontally) the peg will seem to move back and forth along a horizontal straight line [Fig. 7-28(b)]. Its motion along this line will correspond exactly to Eq. (7-42a) if the angular position of the peg at $t = 0$ is correctly chosen so that the angle SOP in Fig. 7-28(c) is equal to $\omega t + \varphi_0$. In this representation the quantity ω is seen as the actual angular velocity of P as it travels around the circle. The peg has a velocity $\mathbf{v}$ that changes direction but always has the magnitude ωA. Resolving $\mathbf{v}$ parallel to Ox at once gives Eq. (7-44). Thus the motion of the point B in Fig. 7-28(c) corre-

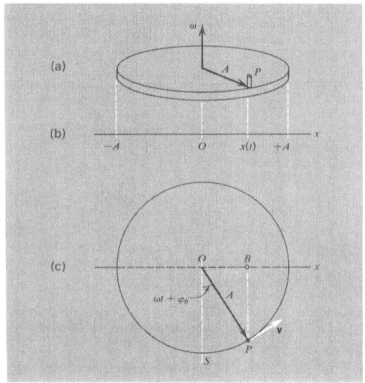

Fig. 7-28 (a) Peg on a uniformly rotating disk. (b) Displacement of the peg as viewed in a plane containing the disk. (c) Detailed relation of circular motion to simple harmonic motion.

sponds in every respect to that of a particle performing SHM along the x axis.

Dynamical relation between SHM and circular motion

Although the analysis just described is a purely geometrical one, it suggests a close dynamical connection, also, between the actual linear motion of a harmonic oscillator and the projection of a uniform circular motion. The acceptance of $\mathbf{F} = m\mathbf{a}$ means that the same motion implies the same force causing it, whatever the particular origin of the force may be. This equivalence can be understood with the help of Fig. 7–29. A particle of mass m is kept moving in a circle of radius A by means of a string attached to a fixed support at the center of the circle, O. If the particle has a constant speed, v, the tension, T, in the string must be given by

$$T = \frac{mv^2}{A}$$

Here v^2/A is the magnitude of the instantaneous acceleration, a_r, of m toward the center of the circle.

Now the total force and the total acceleration at any instant can be resolved into their x and y components in a rectangular coordinate system. (Normally we would not be interested in doing this, because T and a_r have well-defined constant mag-

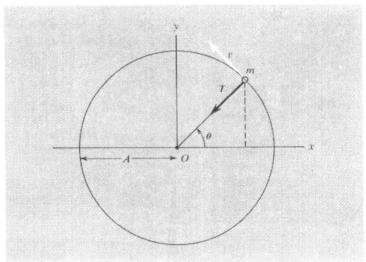

Fig. 7-29 Dynamical relationship between circular motion and linear harmonic motion.

nitudes.) Taking the x components alone, we have

$$F_x = -T\cos\theta = -\frac{mv^2}{A}\cos\theta$$

$$a_x = -\frac{v^2}{A}\cos\theta$$

Thus the x component of the complete vector equation, $\mathbf{F} = m\mathbf{a}$, is $F_x = ma_x$, with the values of F_x and a_x stated above.

In order to display the dynamical identity of this component motion with SHM, we can take the expressions for F_x and a_x separately, introducing the angular velocity ω and putting $v = \omega A$. We then have

$$F_x = -m\omega^2 A \cos\theta = -m\omega^2 x$$

$$a_x = -\omega^2 A \cos\theta = -\omega^2 x$$

The first of these equations defines a restoring force proportional to displacement, exactly in accord with our initial statement of Hooke's law [Eq. (7-38)]. The second corresponds exactly to the equation [Eq. (7-40)] that was our starting point for the kinematic analysis of the problem:

$$\frac{d^2x}{dt^2} = -\omega^2 x$$

Thus we see that the dynamical correspondence is complete in every respect. It tells us, moreover, that we could, if we wished, go the other way and treat a uniform circular motion as a superposition of two simple harmonic motions at right angles. This is, in fact, an extremely important and useful procedure in some contexts, although we shall not take time to follow it up here and now.

PROBLEMS

7-1 Two identical gliders, each of mass m, are being towed through the air in tandem, as shown. Initially they are traveling at a constant speed and the tension in the tow rope A is T_0. The tow plane then begins to accelerate with an acceleration a. What are the tensions in A and B immediately after this acceleration begins?

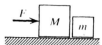

7-2 Two blocks, of masses $M = 3$ kg and $m = 2$ kg, are in contact on a horizontal table. A constant horizontal force $F = 5$ N is applied to block M as shown. There is a *constant* frictional force of 2 N between the table and the block m but *no* frictional force between the table and the first block M.

(a) Calculate the acceleration of the two blocks.

(b) Calculate the force of contact between the blocks.

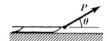

7-3 A sled of mass m is pulled by a force of magnitude P at angle θ to the horizontal (see the figure). The sled slides over a horizontal surface of snow. It experiences a tangential resistive force equal to μ times the perpendicular force N exerted on the sled by the snow.

(a) Draw an isolation diagram showing all the forces exerted on the sled.

(b) Write the equations corresponding to $F = ma$ for the horizontal and vertical components of the motion.

(c) Obtain an expression for the horizontal acceleration in terms of P, θ, m, μ, and g.

(d) For a given magnitude of P, find what value of θ gives the biggest acceleration.

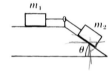

7-4 A block of mass m_1 rests on a frictionless horizontal surface; it is connected by a massless string, passing through a frictionless eyelet, to a second block of mass m_2 that rests on a frictionless incline (see the figure).

(a) Draw isolation diagrams for the masses and write the equation of motion for each one separately.

(b) Find the tension in the string and the acceleration of m_2.

(c) Verify that, for $\theta = \pi/2$, your answers reduce to the expected results.

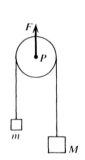

7-5 In the figure, P is a pulley of negligible mass. An external force F acts on it as indicated.

(a) Find the relation between the tensions on the right-hand and left-hand sides of the pulley. Find also the relation between F and the tensions.

(b) What relation among the motions of m, M, and P is provided by the presence of the string?

(c) Use the above results and Newton's law as applied to each block to find the accelerations of m, M, and P in terms of m, M, g, and F. Check that the results make sense for various specialized or simplified situations.

7-6 A man is raising himself and the platform on which he stands with a uniform acceleration of 5 m/sec^2 by means of the rope-and-pulley arrangement shown. The man has mass 100 kg and the platform is 50 kg. Assume that the pulley and rope are massless and move without friction, and neglect any tilting effects of the platform.

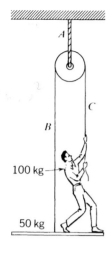

Assume g = 10 m/sec².

(a) What are the tensions in the ropes A, B, and C?

(b) Draw isolation diagrams for the man and the platform and draw a separate force diagram for each, showing all the forces acting on them. Label each force and clearly indicate its direction.

(c) What is the force of contact exerted on the man by the platform?

7-7 In an equal-arm arrangement, a mass $5m_0$ is balanced by the masses $3m_0$ and $2m_0$, which are connected by a string over a pulley of negligible mass and prevented from moving by the string A (see the figure). Analyze what happens if the string A is suddenly severed, e.g., by means of a lighted match.

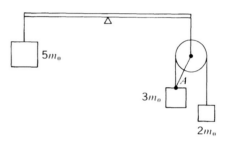

7-8 A prisoner in jail decides to escape by sliding to freedom down a rope provided by an accomplice. He attaches the top end of the rope to a hook outside his window; the bottom end of the rope hangs clear of the ground. The rope has a mass of 10 kg, and the prisoner has a mass of 70 kg. The hook can stand a pull of 600 N without giving way. If the prisoner's window is 15 m above the ground, what is the least velocity with which he can reach the ground, starting from rest at the top end of the rope?

7-9 (a) Suppose that a uniform rope of length L, resting on a frictionless horizontal surface, is accelerated along the direction of its length by means of a force F pulling it at one end. Derive an expression for the tension T in the rope as a function of position along its length. How is the expression for T changed if the rope is accelerated vertically in a constant gravitational field?

(b) A mass M is accelerated by the rope in part (a). Assuming the mass of the rope to be m, calculate the tension for the horizontal and vertical cases.

7-10 In 1931 F. Kirchner performed an experiment to determine the charge-to-mass ratio, e/m, for electrons. An electron gun (see the figure) produced a beam of electrons that passed through two "gates," each gate consisting of a pair of parallel plates with the upper plates connected to an alternating voltage source. Electrons could pass

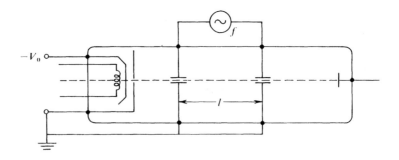

straight through a gate only if the voltage on the upper plate were momentarily zero. With the gates separated by a distance l equal to 50.35 cm, and with a gate voltage varying sinusoidally at a frequency f equal to 2.449×10^7 Hz (1 Hz = 1 cycle/sec), Kirchner found that electrons could pass completely undeflected through both gates when the initial accelerating voltage (V_0) was set at 1735 V. Under these conditions the flight-time between the gates corresponded to one half-cycle of the alternating voltage.

(a) What was the electron speed, deduced directly from l and f?
(b) What value of e/m is implied by the data?
(c) Were corrections due to special relativity significant?

[For Kirchner's original paper, see *Ann. Physik*, **8**, 975 (1931).]

7-11 A certain loaded car is known to have its center of gravity halfway between the front and rear axles. It is found that the drive wheels (at the rear) start slipping when the car is driven up a 20° incline. How far back must the load (weighing a quarter the weight of the empty car) be shifted for the car to get up a 25° slope? (The distance between the axles is 10 ft.)

7-12 A child sleds down a snowy hillside, starting from rest. The hill has a 15° slope, with a long stretch of level field at the foot. The child starts 50 ft up the slope and continues for 100 ft on the level field before coming to a complete stop. Find the coefficient of friction between the sled and the snow, assuming that it is constant throughout the ride. Neglect air resistance.

7-13 A beam of electrons from an electron gun passes between two parallel plates, 3 mm apart and 2 cm long. After leaving the plates the electrons travel to form a spot on a fluorescent screen 25 cm farther on. It is desired to get the spot to deflect vertically through 3 cm when 100 V are applied to the deflector plates. What must be the accelerating voltage V_0 on the electron gun? (Show first, in general, that if the linear displacement caused by the deflector plates can be neglected, the required voltage is given by $V_0 = V(lD/2Yd)$, where Y is the linear displacement of the spot on the screen. The notation is that used on p. 197.)

7–14 A ball of mass m is attached to one end of a string of length l. It is known that the string will break if pulled with a force equal to nine times the weight of the ball. The ball, supported by a frictionless table, is made to travel a horizontal circular path, the other end of the string being attached to a fixed point O. What is the largest number of revolutions per unit time that the mass can make without breaking the string?

7–15 A mass of 100 g is attached to one end of a very light rigid rod 20 cm long. The other end of the rod is attached to the shaft of a motor so that the rod and the mass are caused to rotate in a vertical circle with a constant angular velocity of 7 rad/sec.

(a) Draw a force diagram showing all the forces acting on the mass for an arbitrary angle θ of the rod to the downward vertical.

(b) What are the magnitude and the direction of the force exerted by the rod on the mass when the rod points in a horizontal direction, i.e., at $\theta = 90°$?

7–16 You are flying along in your Sopwith Camel at 60 mph and 2000-ft altitude in the vicinity of Saint Michel when suddenly you notice that the Red Baron is just 300 ft behind you flying at 90 mph. Recalling from captured medical data that the Red Baron can withstand only 4 g's of acceleration before blacking out, whereas you can withstand 5 g's, you decide on the following plan. You dive straight down at full power, then level out by flying in a circular arc that comes out horizontally just above the ground. Assume that your speed is constant after you start to pull out and that the acceleration you experience in the arc is 5 g's. Since you know that the Red Baron will follow you, you are assured he will black out and crash. Assuming that both planes dive with 2 g's acceleration from the same initial point (but with initial speeds given above), to what altitude must you descend so that the Red Baron, in trying to follow your subsequent arc, must either crash or black out? Assuming that the Red Baron is a poor shot and must get within 100 ft of your plane to shoot you down, will your plan succeed? After starting down you recall reading that the wings of your plane will fall off if you exceed 300 mph. Is your plan sound in view of this limitation on your plane?

7–17 A curve of 300 m radius on a level road is banked for a speed of 25 m/sec ($\approx$ 55 mph) so that the force exerted on a car by the road is normal to the surface at this speed.

(a) What is the angle of bank?

(b) The friction between tires and road can provide a maximum tangential force equal to 0.4 of the force normal to the road surface. What is the highest speed at which the car can take this curve without skidding?

7–18 A large mass M hangs (stationary) at the end of a string that

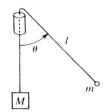

passes through a smooth tube to a small mass m that whirls around in a circular path of radius $l\sin\theta$, where l is the length of the string from m to the top end of the tube (see the figure). Write down the dynamical equations that apply to each mass and show that m must complete one orbit in a time of $2\pi(lm/gM)^{1/2}$. Consider whether there is any restriction on the value of the angle θ in this motion.

7-19 A model rocket rests on a frictionless horizontal surface and is joined by a string of length l to a fixed point so that the rocket moves in a horizontal circular path of radius l. The string will break if its tension exceeds a value T. The rocket engine provides a thrust F of constant magnitude along the rocket's direction of motion. The rocket has a mass m that does not decrease appreciably with time.

(a) Starting from rest at $t = 0$, at what later time t_1 is the rocket traveling so fast that the string breaks? Ignore air resistance.

(b) What was the magnitude of the rocket's instantaneous net acceleration at time $t_1/2$? Obtain the answer in terms of F, T, and m.

(c) What distance does the rocket travel between the time t_1 when the string breaks and the time $2t_1$? The rocket engine continues to operate after the string breaks.

7-20 It has been suggested that the biggest nuclear accelerator we are likely to make will be an evacuated pipe running around the earth's equator. The strength of the earth's magnetic field at the equator is about 0.3 G or 3×10^{-5} MKS units (N-sec/C-m). With what speed would an atom of lead (at. wt. 207), singly ionized (i.e., carrying one elementary charge), have to move around such an orbit so that the magnetic force provided the correct centripetal acceleration? ($e = 1.6 \times 10^{-19}$ C.) Through what voltage would a singly ionized lead atom have to be accelerated to give it this correct orbital speed?

7-21 A trick cyclist rides his bike around a "wall of death" in the form of a vertical cylinder (see the figure). The maximum frictional force parallel to the surface of the cylinder is equal to a fraction μ of the normal force exerted on the bike by the wall.

(a) At what speed must the cyclist go to avoid slipping down?

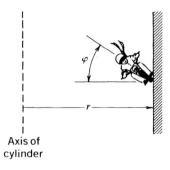

Axis of cylinder

(b) At what angle (φ) to the horizontal must he be inclined?

(c) If $\mu \approx 0.6$ (typical of rubber tires on dry roads) and the radius of the cylinder is 5 m, at what minimum speed must the cyclist ride, and what angle does he make with the horizontal?

7-22 The following expression gives the resistive force exerted on a sphere of radius r moving at speed v through air. It is valid over a very wide range of speeds.

$$R(v) = 3.1 \times 10^{-4} rv + 0.87 r^2 v^2$$

where R is in N, r in m, and v in m/sec. Consider water drops falling under their own weight and reaching a terminal speed.

(a) For what range of values of small r is the terminal speed determined within 1% by the first term alone in the expression for $R(v)$?

(b) For what range of values of larger r is the terminal speed determined within 1% by the second term alone?

(c) Calculate the terminal speed of a raindrop of radius 2 mm. If there were no air resistance, from what height would it fall from rest before reaching this speed?

7-23 An experiment is performed with the Millikan oil-drop apparatus. The plates are 8 mm apart. The experiment is done with oil droplets of density 896 kg/m^3. The droplets are timed between two horizontal lines that are 2.58 mm apart. With the plates uncharged, a droplet is observed to take 23.6 sec to fall from one line to the other. When the upper plate is made 1100 V positive with respect to the lower, the droplet rises and takes 22.0 sec to cover the same distance. Assume that the resistive force is $3.1 \times 10^{-4} rv$ (MKS units).

(a) What is the radius of the droplet?

(b) What is its net charge, measured as a number of elementary charges? ($e = 1.6 \times 10^{-19}$ C.)

(c) What voltage would hold the droplet stationary? [Use the precise value of the charge deduced in part (b)].

7-24 Two solid plastic spheres of the same material but of different radii, R and $2R$, are used in a Millikan experiment. The spheres carry equal charges q. The *larger* sphere is observed to reach terminal speeds as follows: (1) plates uncharged: terminal speed $= v_0$ (downward), and (2) plates charged: terminal speed $= v_1$ (upward). Assuming that the resistive force on a sphere of radius r at speed v is $c_1 rv$, find, in terms of v_0 and v_1, the corresponding terminal speeds for the smaller sphere.

7-25 Analyze in retrospect the legendary Galilean experiment that took place at the leaning tower of Pisa. Imagine such an experiment done with two iron spheres, of radii 2 and 10 cm, respectively, dropped simultaneously from a height of about 15 m. Make calculations to determine, approximately, the difference in the times at which they hit the ground. Do you think this could be detected without special measuring devices? (Density of iron ≈ 7500 kg/m³.)

7-26 Estimate the terminal speed of fall (in air) of an air-filled toy balloon, with a diameter of 30 cm and a mass (not counting the air inside) of about 0.5 g. About how long would it take for the balloon to come to within a few percent of this terminal speed? Try making some real observations of balloons inflated to different sizes.

7-27 A spring that obeys Hooke's law in both extension and compression is extended by 10 cm when a mass of 2 kg is hung from it.

(a) What is the spring constant k?

(b) The spring and the 2-kg mass are placed on a smooth table. The mass is pulled so as to extend the spring by 5 cm and is then released at $t = 0$. What is the equation of the ensuing motion?

(c) If, instead of being released from rest, the mass were started off at $x = 5$ cm with a speed of 1 m/sec in the direction of increasing x, what would be the equation of motion?

7-28 When the mass is doubled in diagram (a), the end of the spring descends an additional distance h. What is the frequency of oscillation for the arrangement in diagram (b)? All individual springs shown are identical.

7-29 Any object, partially or wholly submerged in a fluid, is buoyed up by a force equal to the weight of the displaced fluid. A uniform cylinder of density ρ and length l is floating with its axis vertical, in a fluid of density ρ_0. What is the frequency of small-amplitude vertical oscillations of the cylinder?

7-30 (a) A small bead of mass m is attached to the midpoint of a string (itself of negligible mass). The string is of length L and is under constant tension T. Find the frequency of the SHM that the mass describes when given a slight transverse displacement.

(b) Find the frequency in the case where the mass is attached at a distance D from one end instead of the midpoint.

7-31 A block rests on a tabletop that is undergoing simple harmonic motion of amplitude A and period T.

(a) If the oscillation is vertical, what is the maximum value of A that will allow the block to remain always in contact with the table?

(b) If the oscillation is horizontal, and the coefficient of friction between block and tabletop is μ, what is the maximum value of A that will allow the block to remain on the surface without slipping?

7-32 The springs of a car of mass 1200 kg give it a period when empty of 0.5 sec for small vertical oscillations.

(a) How far does the car sink down when a driver and three passengers, each of mass 75 kg, get into the car?

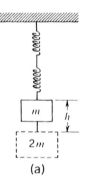

(a)

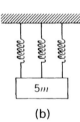

(b)

(b) The car with its passengers is traveling along a horizontal road when it suddenly runs onto a piece of new road surface, raised 2 in. above the old surface. Assume that this suddenly raises the wheels and the bottom ends of the springs through 2 in. before the body of the car begins to move upward. In the ensuing rebound, are the passengers thrown clear of their seats? Consider the maximum acceleration of the resulting simple harmonic motion.

If it universally appears, by experiments and astronomical observations, that all bodies about the earth gravitate towards the earth . . . in proportion to the quantity of matter that they severally contain; that the moon likewise . . . gravitates towards the earth . . . and all the planets one towards another; and the comets in like manner towards the sun; we must, in consequence of this rule, universally allow that all bodies whatsoever are endowed with a principle of mutual gravitation.

NEWTON, *Principia* (1686)

8
Universal gravitation

THE DISCOVERY OF UNIVERSAL GRAVITATION

IN CHAPTERS 6 and 7 we have built up the kind of foundation in dynamics that Newton himself was the first to establish. In a nutshell, it is the quantitative identification of force as the cause of acceleration, coupled with the purely kinematic problem of relating accelerations to velocities and displacements. We shall now consider, as a topic in its own right, the first and most splendid example of how a *law of force* was deduced from the study of motions.

It is convenient, and historically not unreasonable, to consider separately three aspects of this great discovery:

1. The analysis of the data concerning the orbits of the planets around the sun, to the approximation that these orbits are circular with the sun at the center. Several people besides Newton were closely associated with this problem.

2. The proof that gravitation is universal, in the sense that the law of force that governs the motion of objects near the earth's surface is also the law that controls the motion of celestial bodies. It seems clear that Newton was the true discoverer of this result, through his analysis of the motion of the moon.

3. The proof that the true planetary orbits, which are ellipses rather than circles, are explained by an inverse-square law of force. This achievement, certainly, was the product of Newton's genius alone.

In the present chapter we shall be able to discuss the first of these questions quite fully, using only our basic results in the kinematics and dynamics of particles. The second question requires us to learn (as Newton himself originally had to) how to analyze the gravitating properties of a body, like the earth, which is so obviously not a geometrical point when viewed from close to its surface. We shall present one approach to the problem here and complete the story in Chapter 11, where this special feature of the gravitational problem is discussed. The third question, concerning the exact mathematical description of the orbits, is something that we shall not go into at all at this stage; such orbit problems will be the exclusive concern of Chapter 13.

THE ORBITS OF THE PLANETS

We have described in Chapter 2 how the knowledge of the motions of the classical planets—Mercury, Venus, Mars, Jupiter, and Saturn—was already exceedingly well developed by the time of the astronomer Ptolemy around 150 A.D. By this we mean that the angular positions of these planets as a function of time had been catalogued with remarkable accuracy and over a long enough span for their periodic returns to the same position in the sky to be extremely well known. We have pointed out previously, however, that the interpretation of such results depends on the model of the solar system that one uses. Let us now look more carefully at the original observational data and the conclusions that can be drawn from them.

The first thing to recognize is that, whether or not one accepts the earth as the real center of the universe, it *is* the center as far as all primary observations are concerned. From this vantage point, the motion of each planet can be described, to a first approximation, as a small circle (the epicycle) whose center moves around a larger circle (the deferent). Now there are some facts about the motions of two particular planets—Mercury and Venus—that point the way to some far-reaching conclusions. These are

1. That for these two planets, the time for the center C of the epicycle [Fig. 8-1(a)] to travel once around the deferent is exactly 1 solar year—i.e., the same time that it takes the sun to complete one circuit around the ecliptic.

2. The planets Mercury and Venus never get far from the sun. They are always found within a limited angular range from the line joining the earth to the sun (about $\pm 22\frac{1}{2}°$ for Mercury,

Fig. 8–1 (a) Motions of the sun and Venus as seen from the earth. Venus always lies within the angular range ±θ_m of the sun's direction. (b) Heliocentric picture of the same situation.

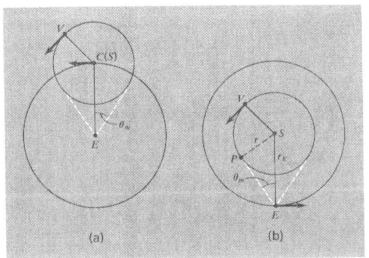

±46° for Venus). Both of these facts are beautifully accounted for if we go over to the heliocentric, Copernican system [Fig. 8–1(b)]. We see that the larger circle of Fig. 8–1(a) corresponds in this case to the earth's own orbit around the sun, of radius r_E, and the smaller circle—the epicycle—represents the orbit of the other planet (Venus or Mercury, as the case may be). Given this interpretation, we can proceed to make quantitative inferences about the radii of the planetary orbits themselves. This is a crucial advance of the Copernican scheme over the Ptolemaic. Although Ptolemy had excellent data, they were for him just the source of purely geometric parameters, but with Copernicus we arrive at the basis of a truly physical model. Thus in Fig. 8–1(b) the maximum angular deviation, θ_m, of the planet P from the earth–sun line ES defines the planet's orbit radius r by the equation

$$\frac{r}{r_E} = \sin \theta_m \qquad (r_E > r) \tag{8-1a}$$

The radius of the earth's orbit is clearly a natural unit for measuring other astronomical distances, and has long been used for this purpose:

1 astronomical unit (AU) = mean distance from earth to sun
(1.496 × 10¹¹ m)

In terms of this unit, we then have

For Mercury: $r \approx \sin 22\frac{1}{2}° \approx 0.38$ AU
For Venus: $r \approx \sin 46° \approx 0.72$ AU

When it comes to the other planets (Mars, Jupiter, and Saturn) the tables are turned. These planets are not closely linked to the sun's position; they progress through the full 360° with respect to the earth–sun line. This can be readily explained if we interchange the roles of the two component circular motions, so that the large primary circle (the deferent) is taken to be the orbit of the planet, now larger than that of the earth, and the epicycle is seen as the expression of the earth's orbit around the sun. In the case of Jupiter, for example, the Ptolemaic picture is represented by Fig. 8-2(a) and the Copernican picture by Fig. 8-2(b). Thus the periodic angular swing, $\pm\theta_m$, of the epicycle is now related to the ratio of orbital radii through the equation

$$\frac{r_E}{r} = \sin \theta_m \qquad (r_E < r) \tag{8-1b}$$

in which the roles of r and r_E are reversed with respect to Eq. (8-1a). Ptolemy's recorded values of θ_m for Mars, Jupiter, and Saturn were about 41°, 11°, and 6°, respectively. These would then lead to the following results:

For Mars: $r \approx \csc 41° \approx 1.5$ AU
For Jupiter: $r \approx \csc 11° \approx 5.2$ AU
For Saturn: $r \approx \csc 6° \approx 9.5$ AU

Thus with the Copernican scheme (and this was its great triumph) it became possible to use the long-established data to construct

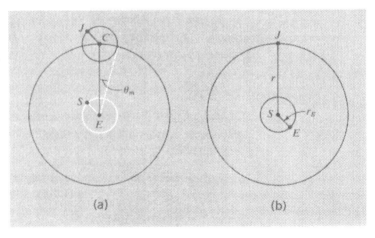

Fig. 8-2 (a) Motions of the sun and Jupiter as seen from the earth. The angle θ_m here characterizes the magnitude of the retrograde (epicyclic) motion. (b) Heliocentric picture of the same situation.

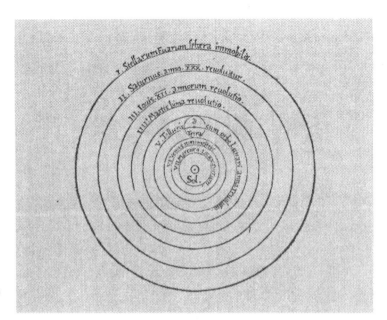

Fig. 8–3 Universe according to Copernicus. (Reproduced from his historic work, De Revolutionibus.)

a picture of the planets in their orbits in order of their increasing distance from the sun. Figure 8–3 is a reproduction of the historic diagram by which Copernicus displayed the results in his book (*De Revolutionibus*) in 1543.

The data with which Copernicus worked (and Ptolemy, too, 1400 years before him) were actually far too good to permit a simple picture of the planets describing circular paths at constant speed around a common center. Thus Copernicus carried out a detailed analysis to find out how far the center of the orbit of each planet was offset from the sun. But even with this adjustment, the detailed change with time of the angular positions of the planets could not be fitted unless the motion around the orbit was made nonuniform. Copernicus, like Ptolemy before him, introduced auxiliary circular motions to deal with the problem, but this, as we now know, was not the answer and we shall not discuss its complexities. For the moment we shall use the basic idealization of uniform circular orbits and set aside until later the refinements that were first mastered by Kepler when he recognized the planetary paths as being ellipses.

PLANETARY PERIODS

The problem of determining the periodicities of the planets, like that of finding the shapes of their orbits, must begin with what

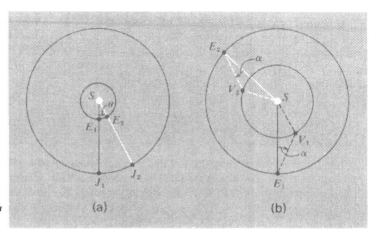

Fig. 8-4 (a) Relative positions of the sun, the earth, and Jupiter at the beginning (SE_1J_1) and end (SE_2J_2) of one synodic period. (b) Comparable diagram for the sun, the earth, and Venus, allowing for the fact that Venus must be offset from the line between sun and earth if it is to be visible.

can be observed from the moving platform that is our earth. The recurring situation that can be most easily recognized is the one in which the sun, the earth, and another planet return, after some characteristic time, to the same positions relative to one another. The length of this recurrence time is known as the *synodic period* of the planet in question. In terms of a heliocentric model of the solar system, this is easily related to the true (sidereal) period of one complete orbit of the planet around the sun.

Consider first the case of one of the outer planets, say Jupiter. Figure 8-4(a) shows a situation that can be observed from time to time. The positions of the sun, the earth, and Jupiter lie in a straight line. Observationally this could be established by finding the date on which Jupiter passes across the celestial meridian at midnight, thus placing it 180° away from the sun.[1]

Now if one such alignment is represented by the positions E_1 and J_1 of the earth and Jupiter, the next one will occur rather more than 1 year later, when the earth has gained one whole revolution on Jupiter. This is shown by the positions E_2 and J_2. Jupiter has traveled through the angular distance θ while the earth goes through $2\pi + \theta$. Both Ptolemy and Copernicus knew

---

[1]The celestial meridian is the projection, on the celestial sphere, of a plane containing the earth's axis and the point on the earth's surface where the observer is located. It is thus a great circle on the celestial sphere, running from north to south through the observer's zenith point, vertically above him. Noon is the instant at which the sun crosses this celestial meridian in its daily journey from east to west.

that the length of the synodic period separating these two configurations is close to 399 days. Let us denote the synodic period in general by the symbol τ. Then if the earth makes n_E complete revolutions per unit time and Jupiter makes n_J revolutions per unit time, we have

$$n_E \tau = n_J \tau + 1$$

But n_E and n_J are the reciprocals of the periods of revolution T_E and T_J of the two planets. Thus we have

$$\frac{\tau}{T_E} = \frac{\tau}{T_J} + 1$$

and solving this for T_J we have

$$T_J = \frac{T_E}{1 - T_E/\tau} \qquad (8\text{-}2a)$$

Putting $T_E/\tau \approx 365/399 \approx 0.915$, we thus find that

$$T_J \approx \frac{T_E}{0.085} \approx 11.8 \text{ yr}$$

The same type of observation and calculation can be applied to Mars and Saturn and the other outer planets that we now know. When we come to Venus and Mercury, however, the situation, as with the determination of orbital radii, is a little different. First is the practical difficulty that we cannot, at least with the naked eye, see these planets when they are in line with the sun, because it would require looking directly toward the sun to do so. We can easily get around this by considering any other situation [see Fig. 8-4(b)] in which the angle between the directions ES and EV is measured. This particular diagram shows Venus as a morning star, appearing above the horizon 1 hr or so before the sun as the earth rotates from west to east. The same value of the angle α will recur after one synodic period. This takes over $1\frac{1}{2}$ yr—about 583 days, to be more precise. In this case, however, it is Venus that has gained one revolution on the earth. Thus instead of the form of the equation that applies to the outer planets, we now have

$$n_V \tau = n_E \tau + 1$$

leading to the result

$$T_V = \frac{T_E}{1 + T_E/\tau} \qquad (8\text{-}2b)$$

Putting $T_E/\tau \approx 365/583 \approx 0.627$, we find that

$$T_V \approx \frac{T_E}{1.627} \approx 224 \text{ days}$$

It is a curious fact that Copernicus, in the introductory general account of his model of the solar system, quotes values of the planetary periods which are so rough that some of them could even be called wrong. These values are marked on his diagram (Fig. 8–3) and are repeated in his text: Saturn, 30 yr; Jupiter, 12 yr; Mars, 2 yr; Venus, 9 months; Mercury, 80 days. The worst cases are Mars (2 yr instead of about $1\frac{1}{2}$) and Venus (9 months instead of about $7\frac{1}{2}$). This seems to have led some people to think that Copernicus had only a crude knowledge of the facts, which was certainly not the case. Perhaps he was careless about quoting the periods because his real interest was in the geometrical details of the planetary orbits and distances. The truth of the matter, in any event, is that his quantitative knowledge of both periods and radii, as spelled out in detail later in his book, was so good that the best modern values do not, for the most part, differ significantly from the ones he quoted. This is shown in Table 8–1, which lists both the Copernican and the modern data on the classical planets. (Incidentally, the values to be extracted from Ptolemy's data are almost iden-

TABLE 8-1: DATA ON PLANETARY ORBITS

Planet	Orbital radius, AU		Synodic period, days	Sidereal period	
	Copernicus	Modern	Copernicus	Copernicus	Modern
Mercury	0.376	0.3871	115.88	87.97 days	87.97 days
Venus	0.719	0.7233	538.92	224.70 days	224.70 days
Earth	1.000	1.0000	—	365.26 days	365.26 days
Mars	1.520	1.5237	779.04	1.882 yr	1.881 yr
Jupiter	5.219	5.2028	398.96	11.87 yr	11.862 yr
Saturn	9.174	9.5389	378.09	29.44 yr	29.457 yr

tical with those of Copernicus, an astonishing tribute to those astronomers whose measurements, from about 750 B.C. up to the time of Ptolemy's own observations around 130 A.D., provided the basis of his analysis.)

KEPLER'S THIRD LAW

The data of Table 8–1 point clearly to a systematic relationship between the planetary periods and distances. This is displayed

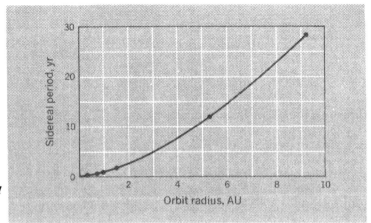

Fig. 8-5 Smooth curve relating the periods and the orbital radii of the planets.

graphically in Fig. 8-5. The precise form of the relationship was first discovered by Johann Kepler in 1618 and published by him the following year in his book *The Harmonies of the World*. In it he triumphantly wrote: "I first believed I was dreaming... But it is absolutely certain and exact that the ratio which exists between the periodic times of any two planets is precisely the ratio of the $\frac{3}{2}$th powers of the mean distances...." Table 8-2

TABLE 8-2: KEPLER'S THIRD LAW

Planet	Radius r of orbit of planet, AU	Period T, days	r^3/T^2, $(AU)^3/(day)^2 \times 10^{-6}$
Mercury	0.389	87.77	7.64
Venus	0.724	224.70	7.52
Earth	1.000	365.25	7.50
Mars	1.524	686.98	7.50
Jupiter	5.200	4,332.62	7.49
Saturn	9.510	10,759.20	7.43

shows the data used by Kepler and a test of the near constancy of the ratio r^3/T^2. Figure 8-6 is a different presentation of the planetary data (actually in this case the data of Copernicus from Table 8-1) plotted in modern fashion on log-log graph paper so as to show this relationship:

$$T \sim r^{3/2} \tag{8-3}$$

This is known as Kepler's third law, having been preceded, 10 years earlier, by the statement of his two great discoveries (quoted

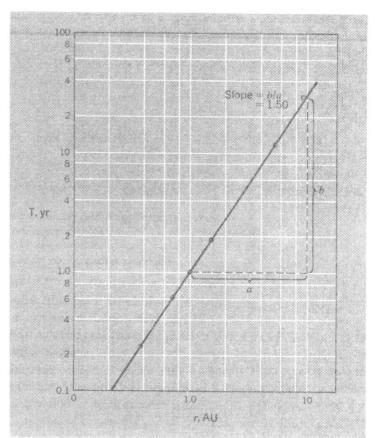

Fig. 8-6 Log-log plot of planetary period T versus orbit radius r, using data quoted by Copernicus. The graph shows that T is proportional to $r^{3/2}$ (Kepler's third law).

in the Prologue) concerning the elliptical paths of the individual planets.

The dynamical explanation of Kepler's third law had to await Newton's discussion of such problems in the *Principia*. A very simple analysis of it is possible if we again use the simplified picture of the planetary orbits as circles with the sun at the center. It then becomes apparent that Eq. (8-3) implies that an inverse-square law of force is at work. For in a circular orbit of radius r we have

$$a_r = -\frac{v^2}{r}$$

Expressing v in terms of the known quantities, r and T, we have

$$v = \frac{2\pi r}{T}$$

$$a_r = -\frac{4\pi^2 r}{T^2} \quad \text{(toward the center)} \tag{8-4}$$

From Newton's law, then, we infer that the force on a mass in a circular orbit must be given by

$$F_r = ma_r = -\frac{4\pi^2 mr}{T^2} \tag{8-5}$$

From Kepler's third law, however, we have the relation

$$\frac{r^3}{T^2} = K \tag{8-6}$$

where K might be called Kepler's constant—the same value of it applies to all the planets traveling around the sun. From Eq. (8-6) we thus have $1/T^2 = K/r^3$, and substituting this in Eq. (8-5) gives us

$$F_r = -\frac{4\pi^2 Km}{r^2} \tag{8-7}$$

The implication of Kepler's third law, therefore, when analyzed in terms of Newton's dynamics, is that the force on a planet is proportional to its inertial mass m and inversely proportional to the square of its distance from the sun. Newton's contemporaries Halley, Hooke, and Huygens all appear to have arrived at some kind of formulation of an inverse-square law in the planetary problem, although Newton's, in terms of his definite concept of forces acting on masses, seems to have been the most clear-cut. The general idea of an influence falling off as $1/r^2$ was probably not a great novelty, for it is the most natural-seeming of all conceivable effects—something spreading out and having to cover spheres of larger and larger area, in proportion to r^2, so that the intensity (as with light from a source) gets weaker according to an inverse-square relationship.

The proportionality of the force to the attracted mass, as required by Eq. (8-7), was a feature of which only Newton appreciated the full significance. With his grasp of the concept of interactions exerted mutually between pairs of objects, Newton saw that the reciprocity in the gravitational interaction must mean that the force is proportional to the mass of the attracting object just as it is to the mass of the attracted. Each object is the attracting agent as far as the other one is concerned. Hence the magnitude of the force exerted on either one of a mutually gravitating pair of particles must be expressed in the famous mathematical statement of universal gravitation:

$$F = -\frac{Gm_1 m_2}{r^2} \tag{8-8}$$

where G is a constant to be found by experiment, and m_1 and m_2 are the inertial masses of the particles. We shall return to the matter of determining G in practice, but first we shall discuss the famous problem that led Newton toward some of his greatest discoveries concerning gravitation.

THE MOON AND THE APPLE

It is an old story, but still an enthralling one, of how Newton, as a young man of 23, came to think about the motion of the moon in a way that nobody had ever done before. The path of the moon through space, as referred to the "fixed" stars, is a line of varying curvature (always, however, bending toward the sun), which crosses and recrosses the earth's orbit. But of course there is a much more striking way of looking at it—the familiar earth-centered view, which shows the moon describing an approximately circular orbit around the earth. To this extent it is quite like the planetary-orbit problem that we have just been discussing. But Newton, with his extraordinary insight, constructed an intellectual bridge between this motion and the behavior of falling objects—the latter being such a commonplace phenomenon that it needed a genius to recognize its relevance. He saw the moon as being just an object falling toward the earth like any other—as, for example, an apple dropping off a tree in his garden. A very special case, to be sure, because the moon was so much farther away than any other falling object in our experience. But perhaps it was all part of the same pattern.

As Newton himself described it,[1] he began in 1665 or 1666 to think of the earth's gravity as extending out to the moon's orbit, with an inverse-square relationship already suggested by Kepler's third law. We could of course just restate the centripetal acceleration formula and apply it to the moon, but it is illuminating to trace the course of Newton's own way of discussing the problem. In effect he said this: Imagine the moon at any point A in its orbit (Fig. 8-7). If freed of all forces, it

[1] See the Prologue of this book.

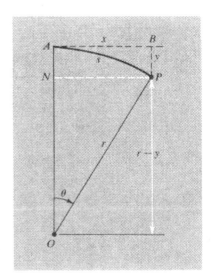

Fig. 8-7 Geometry of a small portion of a circular orbit, showing the deviation y from the tangential straight-line displacement AB (=x) that would be followed in the absence of gravity.

would travel along a straight line AB, tangent to the orbit at A. Instead, it follows the arc AP. If O is the center of the earth, the moon has in effect "fallen" the distance BP toward O, even though its radial distance r is unchanged. Let us calculate how far the moon falls, in this sense, in 1 sec, and compare it with the distance of about 16 ft that an object projected horizontally near the earth's surface would fall in that same time.

First, a bit of analytic geometry. If we denote the distance AB as x, and the distance BP as y, it will be an exceedingly good approximation to put

$$y \approx \frac{x^2}{2r} \tag{8-9}$$

One way of obtaining this result is to consider the right triangle ONP, in which we have

$$ON = r - y \qquad NP = x \qquad OP = r$$

Hence, by Pythagoras' theorem,

$$(r - y)^2 + x^2 = r^2$$
$$x^2 = 2ry - y^2$$

Since $y \ll r$ for any small value of the angle θ, Eq. (8-9) follows as a good approximation. Furthermore, since (again for small θ) the arc length AP ($= s$) is almost equal to the distance AB, we can equally well put

$$y \approx \frac{s^2}{2r} \qquad (8\text{-}10)$$

In order to put numbers into this formula we need to know both the radius and the period of the moon's orbital motion. The distance to the moon, as known to Newton, depended on the two-step process devised by the ancient astronomers—finding the earth's radius and finding the moon's distance as a multiple of the earth's radius. A reminder of these classic measurements is given in Figs. 8-8 and 8-9 and the accompanying discussion (pp. 259–261). The final result, familiar to everyone, is that the moon's orbit radius r is about 240,000 miles $\approx 3.8 \times 10^8$ m. Its period T is 27.3 days $\approx 2.4 \times 10^6$ sec. Therefore, in 1 sec it travels a distance along its orbit given by

$$(\text{in 1 sec}) \qquad s = \frac{2\pi \times 3.8 \times 10^8}{2.4 \times 10^6} \approx 1000 \text{ m}$$

During this same time it falls a vertical distance, which we will denote y_1 to identify it, given [via Eq. (8-10)] by

$$(\text{in 1 sec}) \qquad y_1 \approx \frac{10^6}{7.6 \times 10^8} = 1.3 \times 10^{-3} \text{ m}$$

In other words, in 1 sec, while traveling "horizontally" through a distance of 1 km, the moon falls vertically through just over 1 mm, or about $\frac{1}{20}$ in.; its deviation from a straight-line path is indeed slight. On the other hand, for an object near the earth's surface, projected horizontally, the vertical displacement in 1 sec is given by

$$y_2 = \tfrac{1}{2}gt^2 = 4.9 \text{ m}$$

Thus

$$\frac{y_1}{y_2} = 2.7 \times 10^{-4} \approx \frac{1}{3700}$$

Newton knew that the radius of the moon's orbit was about 60 times the radius of the earth itself, as the ancient Greeks had first shown. And with an inverse-square law, if it applied equally well at all radial distances from the earth's center, we would expect y_1/y_2 to be about 1/3600. It *must* be right! And yet, what an astounding result. Even granted an inverse-square law of attraction between objects separated by many times their diameters, one still has the task of proving that an object a few feet above the earth's surface is attracted as though the whole mass of the earth were concentrated at a point 4000 miles below the ground. Newton did not prove this result until 1685, nearly 20 years after his first great insight into the problem. He published nothing, either, until it all came out, perfect and complete, in the *Principia* in 1687. One way of solving the problem follows on p. 262 (after the special section below).

FINDING THE DISTANCE TO THE MOON

The earth's radius

About 225 B.C. Eratosthenes, who lived and worked at Alexandria near the mouth of the Nile, reported on measurements made on the shadows cast by the sun at noon on midsummer day. At Alexandria (marked A in Fig. 8–8) the sun's rays made an angle of 7.2° to the local vertical, whereas corresponding measurements made 500 miles farther south at Syene (now the site of the Aswan Dam) showed the sun to be exactly overhead at noon. (In other words, Syene lay almost exactly on the Tropic of Cancer.) It follows at once from these figures that the arc AS, of length 500 miles, subtends an angle of 7.2° or $\frac{1}{8}$ rad at the center of the earth. Hence

$$\frac{500}{R_E} = \frac{1}{8}$$

or

$$R_E \approx 4000 \text{ miles}$$

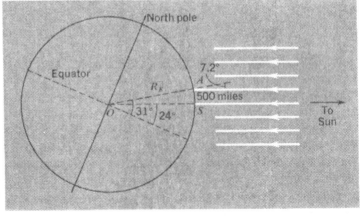

Fig. 8–8 Basis of the method used by Eratosthenes to find the earth's radius. When the midday sun was exactly overhead at Syene (S) its rays fell at 7.2° to the vertical at Alexandria (A).

The moon's distance measured in earth radii

Hipparchus, a Greek astronomer who lived mostly on the island of Rhodes, made observations in about 130 B.C. from which he obtained a remarkably accurate estimate of the moon's distance. His method was one suggested by another great astronomer, Aristarchus, about 150 years earlier.

The method involves a clear understanding of the positional relationships of sun, earth, and moon. We know that sun and moon subtend almost exactly the same angle α at the earth. Hipparchus measured this angle to be $0.553°$ ($\approx 1/103.5$ rad); he also knew what Aristarchus before him had found—that the sun is far more distant than the moon. Hipparchus used this knowledge in an analysis of an eclipse of the moon by the earth (Fig. 8-9). The shaded region indicates the area that is in complete shadow; its boundary lines PA and QB make an angle α with one another, because this is the angle between rays coming from the extreme edges of the sun. The moon passes through the shadowed region, and from the measured time that this passage took, Hipparchus deduced that the angle subtended at the earth by the arc BA was 2.5 times that subtended by the moon itself. Thus $\angle AOB \approx 2.5\alpha$.

Let us now do some geometry. If the distance from the earth's center to the moon is D, the length of the arc BA is very nearly equal to the earth's diameter PQ diminished by the amount αD:

$$AB \approx 2R_E - \alpha D$$

Fig. 8-9 Basis of the method used by Hipparchus to find the moon's distance. The method depended on observing the duration (and hence the angular width) of the moon's total eclipse in the shadow of the earth, as represented by the arc AB.

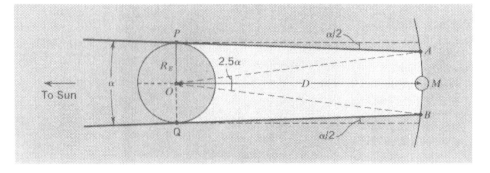

But we also have

$$\frac{AB}{D} \approx 2.5\alpha$$

or

$$AB \approx 2.5\alpha D$$

Substituting this in the first equation we have

$$3.5\alpha D \approx 2R_E$$

or

$$\frac{D}{R_E} \approx \frac{2}{3.5\alpha}$$

Since $\alpha \approx 1/103.5$ rad, this gives

$$\frac{D}{R_E} \approx \frac{207}{3.5} \approx 59$$

Combining this with the value of R_E itself, we have

$$D \approx 236{,}000 \text{ miles}$$

Modern methods

Refined triangulation techniques give a mean value of 3,422.6″, or 0.951°, for the angle subtended at the moon by the earth's radius. Using the modern value of the earth's radius

$$(R_E = 6378 \text{ km} = 3986 \text{ miles})$$

one obtains almost exactly 240,000 miles for the moon's mean distance. Such traditional methods, however, are far surpassed by the technique of making a precision measurement of the time for a radar echo or laser reflection to return to earth. The flight time of such signals (only about 2.5 sec for the roundtrip) can be measured to a fraction of a microsecond, giving range determinations that are not only of unprecedented accuracy but are also effectively instantaneous.

THE GRAVITATIONAL ATTRACTION OF A LARGE SPHERE

It has long been suggested that Newton's failure to publicize his discovery about an inverse-square law of the earth's gravity extending to the moon was due primarily to an actual numerical discrepancy, resulting from his use of an erroneous value for the earth's radius. This would, via Eq. (8-10), falsify the value of the moon's distance of fall, since r (the radius of the moon's orbit) was calculated, according to the method discovered by Hipparchus, in terms of the earth's radius. When Newton first did the calculation he was home in the countryside, out of reach of reference books, and it is reliably recorded that he calculated the earth's radius by assuming that 1° of latitude is 60 miles, instead of the correct figure of nearly 70 miles. Be this as it may, it remains almost certain that Newton, with his outstandingly thorough and critical approach to problems, would never have regarded the theory as complete until he had solved the problem of gravitation by large objects. Let us now consider a way of analyzing this problem. (In Chapter 11 we shall tackle it again in a more sophisticated way.)

Suppose we have a large solid sphere, of radius R_0, as shown in Fig. 8-10(a), and wish to calculate the force with which it attracts a small object of mass m at an arbitrary point P. We shall assume that the density of the material of the sphere may vary with distance from the center (as is the case for the earth, to a very marked degree) but that the density is the same at all points equidistant from the center. We can then consider the solid sphere to be built up of a very large number of thin uniform

Fig. 8-10 (a) A solid sphere can be regarded as built up of a set of thin concentric spherical shells. (b) The gravitational effect of an individual shell can be found by treating it as an assemblage of circular zones.

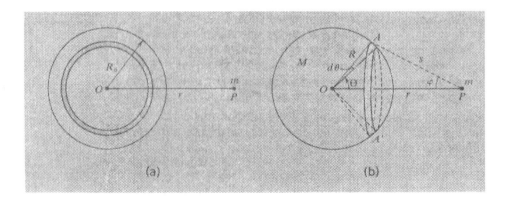

spherical shells, like the successive layers of an onion. The total gravitational effect of the sphere can be calculated as the superposition of the effects of all these individual shells. Thus the basic problem becomes that of calculating the force exerted by a thin spherical shell of arbitrary radius, assuming that the fundamental law of force is that of the inverse square between point masses.

In Fig. 8-10(b) we show a shell of mass M, radius R, of negligible thickness, with a particle of mass m at a distance r from the center of the shell. If we consider a small piece of the shell, near point A, the force that it exerts on m is along the line AP. It is clear from the symmetry of the system, however, that the resultant force due to the whole shell must be along the line OP; any component of force transverse to OP due to material near A will be canceled by an equal and opposite contribution from material near A'. Thus if we have an element of mass dM near A, we need only consider its contribution to the force along OP, i.e., the radial direction from the center of the shell to m. Hence we have

$$dF_r = -\frac{Gm\,dM}{s^2}\cos\varphi \tag{8-11}$$

Let us now consider a complete belt or zone of the shell, shown shaded in the diagram. It represents the portion of the shell that is contained between the directions θ and $\theta + d\theta$ to the axis OP, and the same mean values of s and φ apply to every part of it. Thus, if we calculate its mass, we can substitute this value as dM in Eq. (8-11) to obtain the contribution of the whole belt to the resultant gravitational force along OP. Now the width of the belt is $R\,d\theta$ and its circumference is $2\pi R \sin\theta$; thus its area is $2\pi R^2 \sin\theta\,d\theta$. The area of the whole shell is $4\pi R^2$; hence the mass of the belt is given by

$$dM = \frac{2\pi R^2 \sin\theta\,d\theta}{4\pi R^2} M = \frac{M}{2}\sin\theta\,d\theta$$

Thus Eq. (8-11) gives us

$$dF_r = -\frac{GMm}{2}\frac{\cos\varphi \sin\theta\,d\theta}{s^2} \tag{8-12}$$

Our task now is to sum the contributions such as dF_r over the whole of the shell, i.e., over the whole range of values of s, φ, and θ. This looks like a formidable task, but with the help of a little calculus (another of Newton's inventions!) the solution turns out to be surprisingly straightforward.

From the geometry of the situation [Fig. 8–10(b)], it is possible to express both of the angles θ and φ in terms of two fixed distances, r and R, and the variable distance s. By two separate applications of the cosine rule we have

$$\cos\theta = \frac{r^2 + R^2 - s^2}{2rR} \qquad \cos\varphi = \frac{r^2 + s^2 - R^2}{2rs}$$

From the first of these, by differentiation, we have

$$\sin\theta\, d\theta = \frac{s\, ds}{rR}$$

Hence, substituting the values of $\cos\varphi$ and $\sin\theta\, d\theta$ in Eq. (8–12), we obtain

$$dF_r = -\frac{GMm}{4r^2R}\frac{(r^2 + s^2 - R^2)\, ds}{s^2}$$

The total force is found by integrating this expression from the minimum value of s ($= r - R$) to its maximum value ($r + R$). Thus we have

$$F_r = -\frac{GMm}{4r^2R}\int_{r-R}^{r+R}\frac{r^2 + s^2 - R^2}{s^2}\, ds \qquad (8\text{–}13)$$

The integral is just the sum of two elementary forms; we have

$$\int\frac{r^2 + s^2 - R^2}{s^2}\, ds = \int ds + (r^2 - R^2)\int\frac{ds}{s^2}$$

$$= s - \frac{r^2 - R^2}{s}$$

Inserting the limits, we then find that

$$\int_{r-R}^{r+R}\frac{r^2 + s^2 - R^2}{s^2}\, ds = [(r + R) - (r - R)]$$

$$- \left(\frac{r^2 - R^2}{r + R} - \frac{r^2 - R^2}{r - R}\right)$$

$$= 2R - (r - R) + (r + R)$$

$$= 4R$$

Substituting this value of the definite integral in Eq. (8–13) we have

$$F_r = -\frac{GMm}{r^2} \tag{8-14}$$

What a wonderful result! It is of extraordinary simplicity, and the radius R of the shell does not appear at all. It is uniquely a consequence of an inverse-square law of force between particles; no other force law would yield such a simple result for the net effect of an extended spherical object.

Once we have Eq. (8–14), the total effect of a solid sphere follows at once. Regardless of the particular way in which the density varies between the center and the surface (provided that it depends only on R) the complete sphere does indeed act as though its total mass were concentrated at its center. It does not matter how close the attracted particle P is to the surface of the sphere, as long as it is in fact outside. Take a moment to consider what a truly remarkable result this is. Ask yourself: Is it obvious that an object a few feet above the apparently flat ground should be attracted as though the whole mass of the earth (all 6,000,000,000,000,000,000,000 tons of it!) were concentrated at a point (the earth's center) 4000 miles down? It is about as far from obvious as could be, and there can be little doubt that Newton had to convince himself of this result before he could establish, to his own satisfaction, the grand connection between terrestrial gravity and the motion of the moon and other celestial objects.

OTHER SATELLITES OF THE EARTH

Newton's thinking quite explicitly embraced the possibility—at least theoretically—of having other satellites of the earth. Figure 8–11 is an illustration from Newton's book, *The System of the World* (which is incorporated in the *Principia*); it shows the transition from the effectively parabolic trajectories of short-range projectiles (although the apparent parabolas are really small parts of ellipses) to a perfectly circular orbit and then to other elliptic orbits of arbitrary dimensions.

Let us derive the formulas for the required velocity v and the period T of a satellite launched horizontally in a circular orbit at a distance r from the center of the earth. The necessary

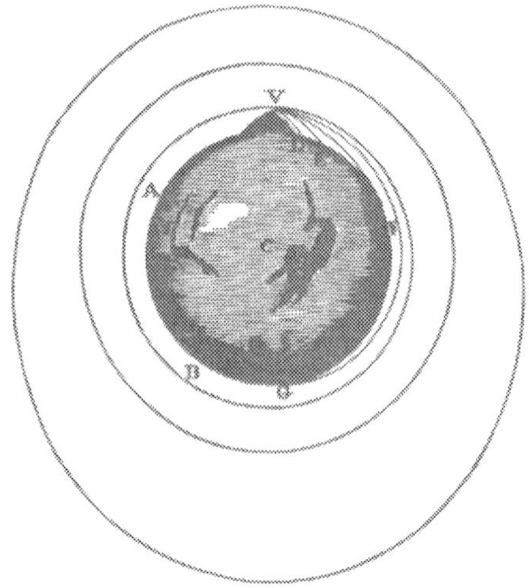

Fig. 8-11 *Newton's diagram showing the transition from normal parabolic trajectories to complete orbits encircling the earth. (From* The System of the World.*)*

force to maintain circular motion is provided by gravitational attraction:

$$\frac{mv^2}{r} = G\frac{M_E m}{r^2}$$

where M_E is the mass of the earth, m the mass of the satellite, and G the universal gravitational constant. Solving for v,

$$v = \left(\frac{GM_E}{r}\right)^{1/2} \tag{8-15}$$

It is often convenient to express this result in terms of more familiar quantities. We can do this by noticing that, for an object of mass m at the earth's surface, the gravitational force on it, by Eq. (8-8), is

$$F_g = \frac{GM_E m}{R_E^2}$$

But this is the force that can be set equal to mg for the mass in question. Hence we have

$$mg = \frac{GM_E m}{R_E^2}$$

or

$$GM_E = gR_E^2$$

Substituting this in Eq. (8-15), we get

$$v = \left(\frac{gR_E^2}{r}\right)^{1/2}$$

The period, T, of the satellite is then given by

$$T = \frac{2\pi r}{v} = \frac{2\pi r^{3/2}}{g^{1/2}R_E} \tag{8-16}$$

Putting $g = 9.8$ m/sec^2, $R_E = 6.4 \times 10^6$ m, we have a numerical formula for the period of any satellite in a circular orbit of radius r around the earth:

(Earth satellites) $\quad T \approx 3.14 \times 10^{-7} r^{3/2}$ \hfill (8-17)

where T is in seconds and r in meters.

For example, a satellite at minimum practicable altitude (about 200 km, say), has $r \approx 6.6 \times 10^6$ m, and hence

$$T \approx 5.3 \times 10^3 \text{ sec} \approx 90 \text{ min}$$

The first man-made satellite, Sputnik I (October 1957) had an orbit as shown in Fig. 8-12(a). Its maximum and minimum distances from the earth's surface were initially 228 and 947 km, respectively, giving a mean value of r equal to about 6950 km.

Fig. 8-12 (a) Orbit of Sputnik I, the first man-made satellite (October 1957). (b) Synchronous satellite communication system. Orbital diameter in relation to earth's diameter is approximately to scale.

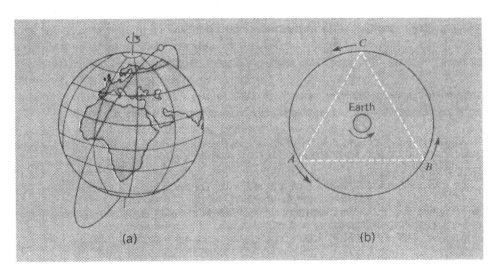

With this value of r, Eq. (8-17) gives an orbital period of about 96 min, which agrees closely with the observed figure.

Particular interest attaches to synchronous satellites that have an orbital period equal to the period of the earth's rotation on its axis. If placed in orbit in the earth's equatorial plane, such satellites will remain above the same spot on the earth's surface, and a set of three of them, ideally in a regular triangular array as shown in Fig. 8-12(b), can provide the basis of a worldwide communications system with no blind spots. Putting $T = 1$ day in Eq. (8-17), one finds $r \approx 42{,}000$ km or 26,000 miles. Thus such satellites must be about 22,000 miles above the earth's surface, i.e., about $5\frac{1}{2}$ earth radii overhead. The first such satellite to be successfully launched was Syncom II in July 1963.

Equation (8-16), on which the above calculations are based, has a very noteworthy feature. A satellite traveling in a circular orbit of a given radius has a period independent of the mass of the satellite. Thus a massive spaceship of many tons will, for the same value of r, have precisely the same orbital period as a flimsy object such as one of the Echo balloons, with a mass of only about 100 kg—or, for that matter, a small piece of interplanetary debris with a mass of only a few kilograms. This result is a direct consequence of the fact that the gravitational force on any object is strictly proportional to its own mass.

THE VALUE OF G, AND THE MASS OF THE EARTH

Although the result expressed by Eq. (8-14) was obtained by considering a large sphere and a small particle, one can quickly convince oneself that it is also the correct statement of the force between any two spherical objects whose centers are a distance r apart. For suppose that we have two such spheres, as shown in Fig. 8-13(a). The calculation that we have carried out shows that one sphere (say the one on the left) attracts every particle of the other as if the left-hand mass were a point [Fig. 8-13(b)]. This therefore reduces the problem to the mutual gravitational attraction between a sphere (the right-hand sphere, of mass m) and a point particle of mass M. But now we can apply the result of the last section a second time. Thus we arrive at Fig. 8-13(c), with two point masses separated by a distance r, as a rigorously correct basis for calculating the force of attraction between the two extended masses shown in Fig. 8-13(a).

The above result is important in the analysis of the experiment, already described in Chapter 5, for finding the universal

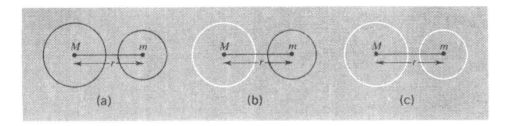

Fig. 8–13 (a) Two gravitating spheres at small separation. (b) Effect of one sphere (M) can be calculated by treating it as a point mass. (c) The argument can be repeated, so that the attraction between the spheres can be calculated as though both were point masses.

gravitation constant, G, from the measured force between two spheres of known masses. In order to get the biggest possible effect with an interaction that is so extremely weak, it is usual to arrange things so that the centers of the spheres are separated by only a little more than the sum of the radii. It is then a great convenience to be able to calculate the force, even under these conditions, on the basis of Eq. (8–14). Notice, however, that the result holds only for spheres. Some of the measurements to determine G have made use of cylindrical masses, because of the greater ease of machining them to high precision. In such cases it becomes necessary to calculate the net force by an explicit integration over the spatial distribution of material.

The presently accepted value of G, as obtained from laboratory measurements of the force exerted between two known masses, is (as already quoted in Chapter 5):

$$G = 6.670 \times 10^{-11} \text{ m}^3/\text{kg-sec}^2 \qquad (8\text{–}18)$$

Newton himself did not know the value of G, although he made a celebrated guess at the mean density of the earth, from which he could have obtained a conjectural figure. In Book III of the *Principia*, he remarks at one point as follows: "Since ... the common matter of our earth on the surface thereof is about twice as heavy as water, and a little lower, in mines, is found about three, or four, or even five times heavier, it is probable that the quantity of the whole matter of the earth may be five or six times greater than if it consisted all of water...."

If we denote the mean density of the earth as ρ and its radius as R, the gravitational force exerted on a particle of mass m just at the earth's surface is given by

$$F = \frac{GMm}{R^2} \tag{8-19}$$

where

$$M = \frac{4\pi}{3} \rho R^3$$

Hence

$$F = \frac{4\pi}{3} (G\rho R) m$$

Since, however, this is just the force that gives the particle an acceleration g in free fall, we also have

$$F = mg$$

It follows, then, that

$$g = \frac{4\pi}{3} G\rho R \tag{8-20}$$

If in this equation we put $g \approx 9.8$ m/sec^2, $R \approx 6.37 \times 10^6$ m, and (using Newton's estimate) $\rho \approx 5000$ to 6000 kg/m^3, we find that

$$G \approx (6.7 \pm 0.6) \times 10^{-11} \text{ m}^3/\text{kg-sec}^2$$

Thus Newton's estimate was almost exactly on target. In practice, of course, the calculation is done the other way around. Given the directly determined value of G [Eq. (8–18)] we substitute in Eqs. (8–19) and (8–20) to find the mass and the mean density of the earth. The result of these substitutions (with $R = 6.37 \times 10^6$ m) is

$$M = 5.97 \times 10^{24} \text{ kg}$$
$$\rho = 5.52 \times 10^3 \text{ kg/m}^3$$

LOCAL VARIATIONS OF g

If we take the idealization of a perfectly spherical, symmetrical earth, then the gravitational force on an object of mass m at a distance h above the surface is given by

$$F = \frac{GMm}{(R+h)^2}$$

If we identify F with m times the value of g at the point in question, we have

$$g(h) = \frac{GM}{(R+h)^2} \qquad (8\text{-}21)$$

For $h \ll R$, this would imply an almost exactly linear decrease of g with height. Using the binomial theorem, we can rewrite Eq. (8-21) as follows:

$$\begin{aligned} g(h) &= \frac{GM}{R^2}\left(1 + \frac{h}{r}\right)^{-2} \\ &= \frac{GM}{R^2}\left(1 - \frac{2h}{R} + \frac{3h^2}{R^2}\cdots\right) \end{aligned}$$

Hence, for small h, we have

$$g(h) \approx g_0\left(1 - \frac{2h}{R}\right) \qquad (8\text{-}22)$$

where $g_0 = GM/R^2$, the value of g at points extremely close to the earth's surface. [Alternatively, we can use a calculus method that can be extremely useful whenever we want to consider the *fractional* variation of a quantity. It is based on the fact that the differentiation of the natural logarithm of a quantity leads at once to the fractional variation. In the present case we have

$$g(r) = \frac{GM}{r^2}$$

Therefore,

$$\ln g = \text{const.} - 2\ln r$$

Differentiating,

$$\frac{\Delta g}{g} \approx -2\frac{\Delta r}{r}$$

Hence, putting $r = R$, $g = g_0$, and $\Delta r = h$, we have

$$\Delta g \approx -2g_0\frac{h}{r}$$

which leads us back to Eq. (8-22). Notice how this method frees us of the need to concern ourselves with the values of any multiplicative constants that appear in the original equation— e.g., the value of GM in the equation for g. A recognition of this fact can enable one to avoid a lot of unnecessary arithmetic in the computation of small changes of one quantity that depends upon another according to some well-defined functional relationship.]

Newton's contemporary, Robert Hooke, made several efforts to detect a variation of the gravitational attraction with height. He did this by looking for any changes in the measured weights of objects at the tops of church towers and the bottoms of deep wells. Not surprisingly, he was unable to find any difference. By Eq. (8–22) one would have to ascend to a point about 1000 ft above ground (e.g., the top of the Empire State Building) before the decrease of g was even as great as 1 part in 10,000. As we shall see in a moment, however, such variations are detected with the greatest of ease by modern techniques.

Superimposed on the systematic variations of the gravitational force with height are the variations produced by inhomogeneities in the material of the earth's crust. For example, if one is standing above a subterranean deposit of salt or sand, much lower in density than ordinary rocks, one would expect the value of g to be reduced. Such changes, although extremely small, can be measured with remarkable accuracy by modern instruments and have become a very valuable tool in geophysical prospecting.

Almost all modern gravity meters are static instruments, in which a mass is in equilibrium under the combined action of gravity and an elastic restoring force supplied by a spring. In other words, it is just a very sensitive spring balance. A change in g as the instrument is moved from one point to another leads to a minute change in the equilibrium position, and this is detected by sensitive optical methods or electrically by, for example, making the suspended mass part of a tuned circuit whose capacitance, and hence frequency, is changed by the slight displacement. To be useful, such instruments must be capable of detecting fractional changes of g of 10^{-7} or less. Figure 8–14(a) shows the basic construction of one such device. With it one can trace out contours of constant g over a region of interest. Figure 8–14(b) shows the results of such a survey, after allowance has been made for effects due to varying altitude, surface features, and so on. Such contours can give good indications of ore concentrations. The primary unit of measurement in these gravity surveys is known as the *gal* (after Galileo):

$$1 \text{ gal} = 1 \text{ cm/sec}^2 \approx 10^{-3} g$$

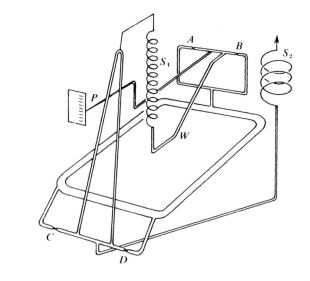

(a)

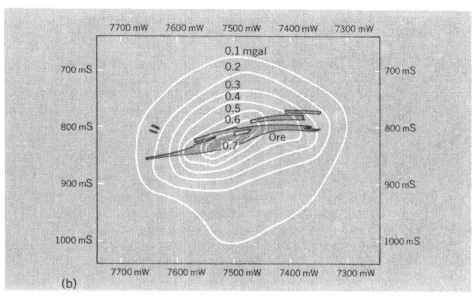

(b)

Fig. 8–14 (a) Sketch of basic features of a sensitive gravimeter, made of fused quartz. The arm marked W acts as the main weight. It is pivoted at A and B and carries a pointer P. The restoring force is provided by a control spring S_1 and a null reading can be obtained with the help of the calibrated spring S_2. (b) Example of a gravity survey over an area about 400 by 500 m, with contours of constant g indicating an ore deposit. (After a survey made by the Boliden Mining Co., Sweden, and reproduced in D. S. Parasnis, Mining Geophysics, *Elsevier, Amsterdam, 1960.)*

This is far too large for convenience, so most surveys, like that of Fig. 8–14(b), show contours labeled in terms of milligals (1 mgal = 10^{-3} cm sec$^2 \approx 10^{-6} g$). Under the best conditions, relative measurements accurate to 0.01 mgal may be achievable. One can appreciate how impressive this is by noting that a change of g by 0.01 mgal (1 part in 10^8) corresponds to a change in elevation of only about 3 cm at the earth's surface!

THE MASS OF THE SUN

Let us return to the simple picture of the solar system in which each planet describes a circular orbit about a fixed central sun (Fig. 8–15). We have seen, in the discussion of Kepler's third law, how the use of Newton's law of motion implies that the force on the planet is given, in terms of its mass, orbital radius, and period, by the following equation [Eq. (8–5)]:

$$F_r = -\frac{4\pi^2 mr}{T^2}$$

According to the basic law of the force, however, as expressed by Newton's law of universal gravitation [Eq. (8–8)], the value of F_r is given by the equation

$$F_r = -\frac{GMm}{r^2}$$

where M is the mass of the sun. From the equality of these two expressions, we obtain the following result:

$$T^2 = \frac{4\pi^2 r^3}{GM} \tag{8-23}$$

We may again note the feature, already commented on in con-

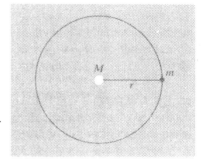

Fig. 8–15 Planetary orbit approximated by a circle with the sun at the center.

nection with earth satellites, that the period is independent of the mass of the orbiting object, in this case the earth itself or some other planet. What *does* matter is the mass of the sun, and if we turn Eq. (8–23) around, we have an equation that tells us the value of this mass, M, in terms of observable quantities:

$$M = \frac{4\pi^2}{G} \frac{r^3}{T^2} \tag{8-24}$$

Kepler's third law expresses the fact that the value of r^3/T^2 is indeed the same for all the planets. The statement of this result does not, however, require the use of absolute values of r—or, for that matter, of T either. It is sufficient to know the values of r and T for the various planets as multiples or fractions of the earth's orbital radius and period. In order to deduce the mass of the sun from Eq. (8–24), however, the use of absolute values is essential. We have seen, earlier in this chapter, how the length of the earth's year has been known with great accuracy since the days of antiquity. A knowledge of the distance from the earth to the sun is, however, rather recent. The development of this knowledge makes an interesting story, which is summarized in the special section following. The final result, expressed as a mean distance in meters, can be substituted as the value of r in Eq. (8–24), along with the other necessary quantities as follows:

$r_E \approx 1.50 \times 10^{11}$ m
$T_E \approx 3.17 \times 10^7$ sec
$G = 6.67 \times 10^{-11}$ m^3/kg-sec^2

We then find that

$M_{\text{sun}} \approx 2.0 \times 10^{30}$ kg

FINDING THE DISTANCE TO THE SUN

The first attempt to estimate the distance of the sun was made by the great Greek astronomer, Aristarchus, in the third century B.C., and he arrived at a result which, although quite erroneous, held the field for many centuries. His method, sound in principle but made ineffectual by the great remoteness of the sun, is indicated in Fig. 8–16(a). He knew that one half of the moon was illuminated by the sun and that the phases of the moon were the result of viewing this illuminated hemisphere from the earth.

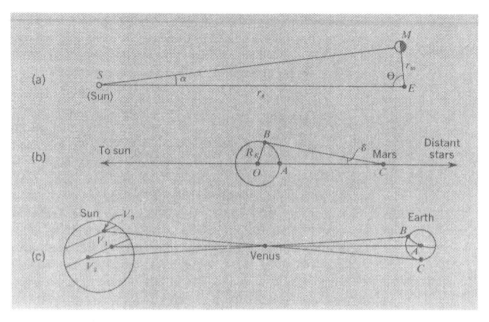

Fig. 8-16 (a) Method attempted by Aristarchus to find the sun's distance by measuring the angle SEM at half-moon. (b) Triangulation method of establishing the scale of the solar system by finding the distance of Mars, using the earth's radius as a base line. (c) Direct determination of the sun's distance by observing the transit of Venus from different points on the earth.

When the moon is exactly half full, the angle SME is $90°$. If, in this situation, an exact measurement can be made of the angle θ, the difference in directions of the sun and moon as seen from a point on the earth, we can deduce the angle α ($= 90° - \theta$) subtended at the sun by the earth–moon distance r_M. Aristarchus judged θ to be about $87°$, which gives $\alpha \approx 3° \approx \frac{1}{20}$ rad and hence $r_S \approx 20 r_M$. Since, however, the measured angle is θ and not α, the error in the final result may be (and is) very great. Our present knowledge tells us that the value of θ in the situation represented by Fig. 8-16(a) is actually about $89.8°$ instead of $87°$; this relatively small change in θ raises the ratio r_S/r_M to several hundred instead of 20.

A completely different attack on the problem was initiated by Kepler, although its full exploitation was not possible until much later. Even so it at once became clear that the sun is more distant than Aristachus had concluded. The basis of the method is indicated in Fig. 8-16(b). It involves observations on the planet Mars. When Mars is closest to the earth, it lies on a line joining both planets to the sun. Under these conditions the distance between them is the difference between their orbital radii. Now if Mars is viewed from two different points on the earth, it should appear in slightly different directions with respect to the vastly more distant background of "fixed" stars. The particular angular difference, δ, for observers placed at A and B is called the *parallax;* it is the angle subtended by the earth's radius at the position of Mars. To measure this angle one does not need to have observers at two different points on the earth; the rotation of the earth itself would carry an observer from A to B in about 6 hr during a given night. Now Kepler was able to deduce from the very careful observations of his master, Tycho Brahe, that the value of δ must be less than 3 minutes of arc, which is about $1/1200$ rad; he could conclude that the distance to Mars in this configuration must be greater than 1200 earth radii or about 5 million miles. Then, using the known *relative* values of orbital radii from the Copernican scheme (Table 8–1), it follows that the distance of the sun from the earth is more than 2400 earth radii, i.e., more than 10 million miles.

John Flamsteed, a contemporary of Newton to whose observations Newton owed a great deal (he was Astronomer Royal from 1675 to 1720), reduced the upper limit on the parallax of Mars to about 25 seconds of arc, and concluded that the sun's distance was at least 80 million miles. An Italian astronomer, Cassini, arrived at a specific value of about 87 million miles at about the same time, using observations made by himself in Europe and by a French astronomer, Richer, at Cayenne in South America. Another contemporary of Newton's—Edmund Halley[1]—proposed a method that finally led, 100 years later, to the first precision measurements of the sun's distance. The method involved what is known as a *transit of Venus*, i.e., a passage of Venus across the sun's disk, as seen from the earth. Figure 8–16(c) illustrates the basis of the method. As it passes across the sun, Venus looks like a small black dot. Its apparent path, and also the times at which the transit begins or ends, depend on the position of the observer on earth. Since the motion of Venus

---

[1] Edmund Halley, best known for the comet named after him, succeeded Flamsteed in 1720 as Astronomer Royal. Long before this, however, he had been very active in physics and astronomy. He became a devoted friend and admirer of Newton, and it was largely through his help and persuasion that the *Principia* was published.

is accurately known, the timing of the transit can be used to yield accurate measures of the differences in angular positions of Venus as seen by observers at different positions. From such observations the parallax of Venus can be inferred, after which one can use an analysis just like that for Mars. These transits are fairly rare, because the orbits of the earth and Venus are not in the same plane, but Halley pointed out that a pair of them would occur in 1761 and 1769, and again in 1874 and 1882, and then in 2004 and 2012. From the first two of these (both occurring long after Halley's own death) the solar parallax was found to be definitely between about 8.5 and 9.0 seconds of arc, corresponding to a distance of between about 92 and 97 million miles. Thus the currently accepted result was approached. (The best measurements of this type have been made on the asteroid Eros at its closest approach to the earth.)

Further refinements came with the observations made in the late nineteenth century. One of the most notable of these was the use of an accurately known value of the speed of light to deduce the diameter of the earth's orbit from the accumulated time lag, over a period of 6 months, in the observed eclipses of the moons of Jupiter. The situation is indicated in Fig. 8-17. While the earth moves from E_1 to E_2, Jupiter moves only from J_1 to J_2. This introduces an extra transit time of about 16 min for the light that tells us that one of Jupiter's moons has, for example, just appeared from behind the planet. Knowing that the speed of light is 186,000 miles/sec in empty space, we can deduce that the earth's orbital radius is equal to this speed times about 480 sec, or about 90 million miles. (The calculation was originally done just the other way around, by the Danish astronomer Roemer in 1675. Using an approximate value of the distance from earth to sun, he made the first quantitative estimate of the speed of light.)

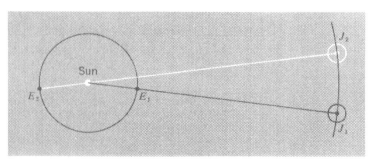

Fig. 8-17 Measurement of the diameter of the earth's orbit by observing the eclipses of Jupiter's moons and the apparent delays due to the travel time of light through space.

Although the modern measurements of the sun's distance are of great accuracy, we must still reckon with the fact that this distance varies during the course of a year. If we ignore this relatively small variation, however, we can make use of the average value, already quoted near the beginning of this chapter:

$$r_E = 1 \text{ AU} = 1.496 \times 10^{11} \text{ m}$$

MASS AND WEIGHT

Perhaps the most profound contribution that Newton made to science was the fundamental connection that he recognized between the inertial mass of an object and the earth's gravitational force on it—a force roughly equal to the measured weight of the object. (Remember, we have defined *weight* as being equal and opposite to the force, as measured for example on a spring balance, that holds the object at rest relative to the earth's surface.)

It had been known since Galileo's time that all objects near the earth fall with about the same acceleration, **g**. Until Newton it was just a kinematic fact. But in terms of Newton's law it took on a much deeper significance. If an object is observed to have this acceleration, there must be a force $\mathbf{F}_g$ acting on it given by $\mathbf{F} = m\mathbf{a}$, i.e.,

$$\mathbf{F}_g = m\mathbf{g} \tag{8-25}$$

It then becomes a vitally significant *dynamical* fact that, since the acceleration **g** is the same for all objects, the force $\mathbf{F}_g$ causing it is strictly proportional to the inertial mass. To appreciate how remarkable this result is, imagine starting from scratch to investigate the force of attraction between two objects in a purely static experiment. One measures the force by balancing it with a springlike device—a torsion fiber. One finds a quantity, which might be called (by analogy with electrical interactions) a gravitational charge, q_g. This "charge" is characteristic of any object and has, as far as these experiments are concerned, nothing at all to do with the inertial mass, which is defined solely in terms of acceleration (under the action of forces produced, for example, by stretched springs). One experiments with objects of all sorts of materials, in different states of aggregation, and so on. It then turns out that, in each and every instance, the gravitational charge is strictly proportional to an independently established quantity, the inertial mass. Is this just a remarkable coincidence, or does it point to something very fundamental? For a long time this apparent coincidence was regarded as one of the unexplained mysteries of nature. It took the sagacity of an Einstein to suspect that gravitation may, in a sense, be *equivalent* to acceleration. Einstein's "postulate of equivalence," that the gravitational charge q_g and the inertial mass m are measures of the same quantity, provided the basis of his own theory of gravitation as embodied in the general theory of relativity. We shall come back to this in Chapter 12, when we discuss noninertial frames of reference.

We are quite accustomed to exploiting the proportionality of F_g to m in our use of the equal-arm balance [Fig. 8–18(a)]. What we are doing is balancing the torques of two forces, but what we are actually interested in is the equality of the amounts of material. We make use of the fact that, to very high accuracy, the value of g is the same at the positions of both masses, and we do not need to bother about what its particular value is. Thanks to the proportionality of gravitational force to mass, we could, with an equal-arm balance and a set of standard weights, measure out a required quantity of any substance equally well on the earth, the moon, or Mars. The spring balance [Fig. 8–18(b)], on the other hand, has a calibration that depends directly on a particular value of g. Its readings are in effect readings of force, even though we use them as a basis for measuring out required amounts of mass. We might find it convenient to use a spring balance for this purpose on the moon, but if its scale were marked in kilograms, we should have to mask this out and attach a fresh calibration with the help of standard masses.

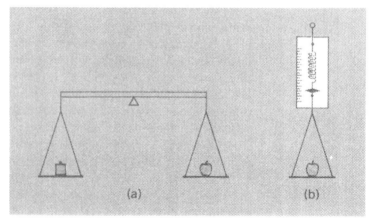

Fig. 8–18 (a) Weighing with an equal-arm balance—in effect a direct comparison of masses, valid whatever the value of g.
(b) Weighing with a spring balance—a measurement of the gravitational force, directly dependent on the value of g.

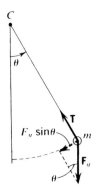

Fig. 8-19 *Forces acting on the bob of a simple pendulum.*

Newton himself recognized that the strict proportionality of the gravitational force to the inertial mass, as evidenced by the identical local acceleration of falling objects, was a key feature in his own statement of universal gravitation as expressed in Eq. (8–8). He therefore made a series of very careful pendulum experiments to test whether a pendulum of a given length, but with a variety of objects used as the bob, always had the same period. To see how this works, consider an object of inertial mass m hung on a string (Fig. 8–19). The two forces on it (ignoring air resistance) are the tension $\mathbf{T}$ and the gravitational force $\mathbf{F}_g$. The tension $\mathbf{T}$ is at every instant perpendicular to the path of the pendulum bob and so has no effect on the tangential acceleration a_θ. The tangential acceleration is due to the tangential component of $\mathbf{F}_g$. From Fig. 8–19,

$$F_\theta = ma_\theta = -F_g \sin \theta$$

from which

$$a_\theta = -(F_g/m) \sin \theta \qquad (8\text{–}26)$$

At every angle θ the *acceleration* a_θ depends on the ratio F_g/m. Therefore—for given initial conditions—the *velocity* of the bob at every angle θ will be determined by this ratio. So also the *period* for one complete round trip will depend upon the ratio (F_g/m). Newton observed the periods of pendulums with different bobs but with equal lengths. From his observation that the periods of all such pendulums were equal within his experimental error, Newton concluded that F_g was proportional to m to better than 1 part in 1000.

More recent experiments (beginning with Baron Eötvös in Budapest in the nineteenth century) have made use of a very clever idea that permits a *static* measurement. It depends on recognizing that an object hanging *at rest relative to the earth* in fact has an acceleration toward the earth's axis because, by virtue of the earth's rotation, it is traveling in a circular path of radius $r = R \cos \lambda$, where λ is the latitude (see Fig. 8–20). This means that a net force of magnitude $m\omega^2 r$ must be acting on it, where m is the inertial mass. How is this force provided? The answer is that, when a body hangs on a string near the earth's surface, the string, exerting a force $\mathbf{T}$, is not in quite the same

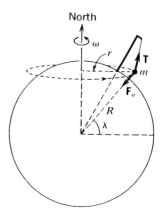

Fig. 8-20 Basis of the Eötvös method for comparing the inertial mass and the gravitational mass of an object that is at rest relative to the earth and hence is being accelerated toward the earth's axis.

Fig. 8-21 Principle of the Eötvös torsion-balance measurement: (a) Two approximately equal masses hang from a torsion bar. (b) If the objects do not have identical ratios of inertial to gravitational mass, the tensions in the suspending strings must be in slightly different directions. In equilibrium, the direction of the main supporting fiber must be intermediate between the directions of T_1 and T_2. (c) This implies the possibility of a net torque that twists the torsion bar about a vertical axis.

direction as the gravitational force F_g. And if F_g is not strictly proportional to m, the angle between T and F_g will be different for different objects. To search for any such variations, a very sensitive torsion balance is used, carrying dissimilar objects at the two ends of a horizontal bar [Fig. 8-21(a)]. If the directions of the tensions T_1 and T_2 are different [Fig. 8-21(b)], there will be small horizontal components [Fig. 8-21(c)] that act in opposite directions with respect to the horizontal bar but that give torques in the same sense. On the other hand, if the directions of T_1 and T_2 are identical, even if their magnitudes are not quite the same, there is no net torque tending to twist the torsion fiber. To test for the existence of any such torque, Eötvös placed the whole apparatus in a case that could be rotated. The hori-

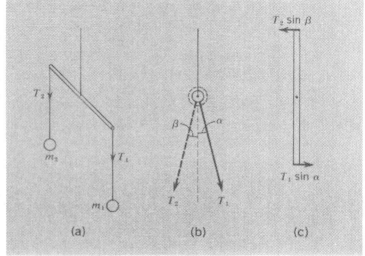

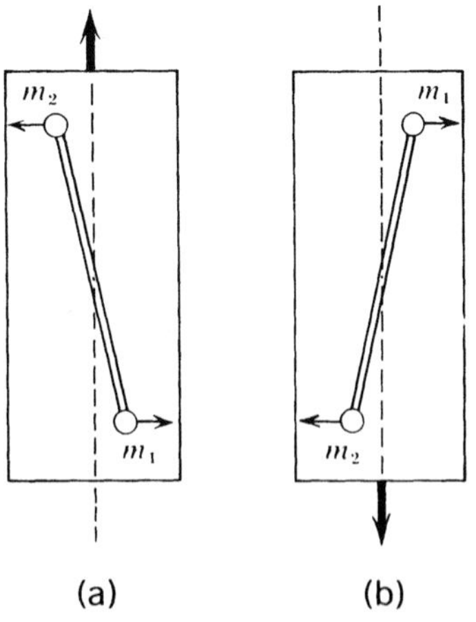

Fig. 8-22 To see whether a net torque exists in the Eötvös experiment, the whole apparatus is turned through 180°. This would reverse the sense of the torque.

zontal beam carrying the two masses was aligned in an east–west direction [Fig. 8–22(a)] and a reading was taken of its position with respect to the case. The whole system was then rotated through 180°, as in Fig. 8–22(b). If you analyze both situations on the basis of Fig. 8–21, you will find that, with respect to the center line of the case, the angle of twist would be reversed by this operation; hence, if any net torque existed, its existence would be revealed.

More recently, some elegant modernized experiments of this type have been performed by R. H. Dicke and his collaborators.[1] By such experiments it has been shown that the strict proportionality of F_g to m holds to 1 part in 10^{10} or better.

The description of the above experiments points to a closely related phenomenon—a systematic variation with latitude of the measured weight of an object. If we take the idealization of a perfectly spherical earth (Fig. 8–23), the equilibrium of the object is maintained by applying a force of magnitude W at an angle α to the radius such that the following conditions are satisfied:

$$W \sin \alpha = m\omega^2 r \sin \lambda$$
$$F_g - W \cos \alpha = m\omega^2 r \cos \lambda$$

where $r = R \cos \lambda$. Since α is certainly a very small angle, it is justifiable to put $\cos \alpha \approx 1$ in the second equation, thus giving the result

$$W \approx F_g - m\omega^2 R \cos^2 \lambda$$

[1] See R. H. Dicke, *Sci. Am.*, **205**, 84 (Dec. 1961).

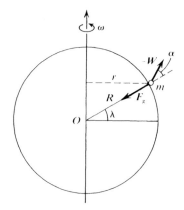

Fig. 8-23 *The force needed to balance the weight of an object is different in both direction and magnitude from the force of gravity.*

It follows that

$$W(\lambda) \approx W_0 + m\omega^2 R \sin^2 \lambda$$

where W_0 is the measured weight on the equator. Putting $W_0 = mg_0$, we can also obtain a corresponding expression for the latitude dependence of g:

$$g(\lambda) = g_0 + \omega^2 R \sin^2 \lambda$$

If in this expression we substitute $\omega = 2\pi/86{,}400 \text{ sec}^{-1}$ and $R = 6.4 \times 10^6$ m, we find $\omega^2 R \approx 3.4 \times 10^{-2}$ m/sec^2, which with $g_0 \approx 9.8$ m/sec^2 gives us

$$g(\lambda) \approx 9.8(1 + 0.0035 \sin^2 \lambda) \quad \text{m/sec}^2$$

This formula is more successful than it deserves to be, for we have no right to ignore the significant flattening of the earth, due again to the rotation, which makes the equatorial radius of the earth about 1 part in 300 greater than the polar radius. This ellipticity has two consequences: It puts a point on the equator farther away from the earth's center than it otherwise would be, but it also in effect adds an extra belt of gravitating material around the equatorial region. The resultant value of g at sea level, taking these effects into account, is quite well described by the following formula:

$$g(\lambda) = 9.7805(1 + 0.00529 \sin^2 \lambda) \tag{8-27}$$

Thus our simple calculation has the correct form, but its value for the numerical coefficient of the latitude-dependent correction is only about two thirds of the true figure.

WEIGHTLESSNESS

It is appropriate, after the detailed discussion of the relations among mass, gravitational force, and weight, to devote a few words to the property that is called weightlessness. The very explicit distinction that we have drawn between the gravitational force on an object and its measured weight makes use of what is called an *operational* definition of the latter quantity. The weight, as we have defined it, is equal and opposite to the force that will hold an object at rest relative to the earth. Our definition of weightlessness derives very naturally from this: *An object is weightless whenever it is in a state of completely free fall.* In this state each part of the object undergoes the same acceleration, of whatever value corresponds to the strength of gravity at its location. (In saying this we assume that g does not change appreciably over the linear extent of the object.) An object that is prevented from falling, by being restrained or supported, inevitably has internal stresses and deformations in its equilibrium state. This may become very obvious, as when a drop of mercury flattens somewhat when it rests on a horizontal surface. All such stresses and deformations are removed in the weightless state of free fall. The mercury drop, for example, is free to take on a perfectly spherical shape.

The above definition of weightlessness can be applied in any gravitational environment, and this is the way it should be. The bizarre dynamical phenomena of life in a space capsule do not depend on getting into regions far from the earth, where the gravitational forces are much reduced, but simply on the fact that the capsule, and everything in it, is falling freely with the same acceleration, which in consequence goes undetected. For example, if a spacecraft is in orbit around the earth, 200 km above the earth's surface, the gravitational force on the spacecraft, and on everything in it, is still about 95% of what one would measure at sea level, but the phenomena associated with what we call weightlessness are just as pronounced under these conditions as they are in another spacecraft 200,000 miles from earth, where the earth's gravitational attraction is down to $\frac{1}{2500}$ of that at the earth's surface. In both situations an object released inside the spacecraft will remain poised in midair. The same would be true in a spacecraft that was simply falling radially toward the earth's center rather than pursuing a circular or elliptical orbit around the earth. When viewed in these terms the phenomena of weightlessness are not in the least mysterious, although they are still startling because they conflict so strongly with our normal experience.

LEARNING ABOUT OTHER PLANETS

The recognition of the universality of gravitation gives us a powerful tool for obtaining information about planets other than the earth, and indeed about celestial objects in general. In particular, if a planet has satellites of its own, we can find its mass by an analysis exactly similar to the one we used for deducing the mass of the sun from the motions of the planets themselves. This provides the simplest way of finding the mass of any planet that has satellites. Such satellites, if a planet has more than one of them, also provide a further test of Kepler's third law, taking the planet itself as the central gravitating body.

Newton himself applied an analysis of this kind to Jupiter, using data for its four most prominent satellites. These were the satellites (or "moons") that made history when Galileo discovered them with his new astronomical telescope in 1610 (see pp. 287–290). Figure 8–24(a) shows their changing positions as seen through a modern telescope, and Fig. 8–24(b) reproduces a few of the sketches that Galileo himself made, night after night, over a period of many months. Figure 8–24(c) is a graph constructed from Galileo's quantitative records, using the readings that can be unambiguously associated with the outermost of the four satellites. A period of about 16 days can be inferred. Galileo had no hesitation in interpreting his observations in terms of the four satellites following circular orbits that were seen edgewise—giving, as we would describe it, the appearance of simple harmonic motion at right angles to our line of sight. On the basis of further measurements Galileo arrived at rather accurate values of the orbital periods of all four satellites and at moderately good values of the orbital radii expressed as multiples of the radius of Jupiter itself.

Newton, in the *Principia*, used similar data of greater precision, obtained by his contemporary John Flamsteed. Table 8–3 (p. 290) lists these data, and in Fig. 8–25 they are plotted logarithmically (cf. Fig. 8–6) in such a way as to show how they give another demonstration of the correctness of Kepler's third law. The slope is accurately $\frac{3}{2}$.

The use of Jupiter's radius as a unit for measuring the orbital radii was not merely a convenience. As we have already noted in our discussion of the mass of the sun, the absolute scale of the solar system was not known with any great certainty in Newton's day. It is interesting, however, that using the data as presented in Table 8-3, without absolute values of the radii, one can deduce the mean density, ρ_J, of Jupiter. By analogy with our analysis of earth satellites [p. 266, and in particular Eq. (8-15)], we have

$$v = \frac{2\pi r}{T} = \left(\frac{GM_J}{r}\right)^{1/2}$$

whence

$$M_J = \frac{4\pi^2 r^3}{GT^2}$$

Putting

$$M_J = \frac{4\pi}{3} \rho_J R_J^3$$

we get

$$\rho_J = \frac{3\pi}{G} \frac{n^3}{T^2}$$

Substituting $n^3/T^2 \approx 7.5 \times 10^{-9}$ sec^{-2}, and $G = 6.67 \times 10^{-11}$ m^3/kg-sec^2, we find

$$\rho_J \approx 1050 \text{ kg/m}^3$$

i.e., about the same density as water.

If a planet does not have satellites of its own, the magnitude of its mass may be inferable from a detailed analysis of the tiny disturbing effects—called perturbations—that planets exert on one another's orbits. This technique has been used for Mercury and Venus. The unraveling of these mutual interactions is a complicated and difficult matter, however, and in at least one case it posed a problem that was not adequately answered for a long time. This was the interaction between the two most massive planets, Jupiter and Saturn, which caused irregularities of a very puzzling kind in the orbits of both. It was even considered possible that the basic law of gravitation would need to be modified slightly away from a precise inverse-square relationship. The solution to the mystery was finally achieved, almost a century after the publication of the *Principia*, by the French mathematician Laplace, building on work by his great fellow

The moons of Jupiter

In 1608 Hans Lippershey, in Holland, patented what may have been the first successful telescope. Galileo learned of this, and soon made telescopes of his own design. His third instrument, of more than 30 diameters' magnification, led him to a dramatic discovery, as recounted by him in his book, The Starry Messenger:

> "On the seventh day of January in this present year 1610, at the first hour of the night, when I was viewing the heavenly bodies with a telescope, Jupiter presented itself to me, and . . . I perceived that beside the planet there were three small starlets, small indeed, but very bright. Though I believed them to be among the host of fixed stars, they aroused my curiosity somewhat by appearing to lie in an exact straight line parallel to the ecliptic . . . I paid no attention to the distances between them and Jupiter, for at the outset I thought them to be fixed stars, as I have said. But returning to the same investigation on January eighth — led by what, I do not know — I found a very different arrangement. . . ."

A few more nights of observation were enough to convince Galileo what he was seeing: "I had now [by January 11] decided beyond all question that there existed in the heavens three stars wandering about Jupiter as do Venus and Mercury about the sun. . . . Nor were there just three such stars; four wanderers complete their revolutions about Jupiter. . . . Also I measured the distances between them by means of the telescope. . . . Moreover I recorded the times of the observations . . . for the revolutions of these planets are so speedily completed that it is usually possible to take even their hourly variations."

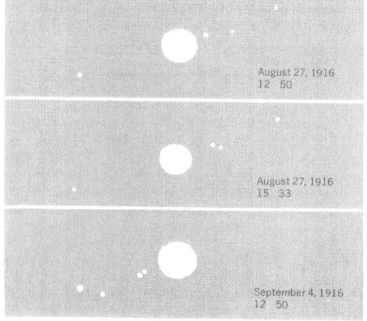

Fig. 8–24(a) Jupiter and its four most prominent satellites as seen through a modern telescope. The first and second photographs illustrate Galileo's observation that noticeable changes occur within a single night. (Yerkes Observatory photographs)

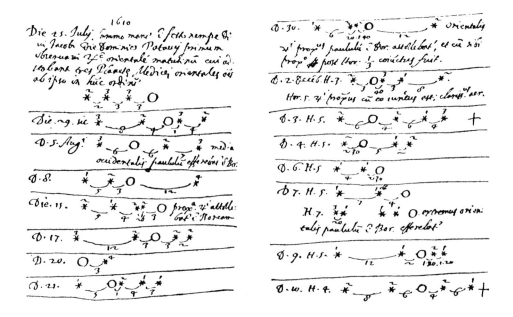

Fig. 8-24(b) Facsimile of a page of Galileo's own handwritten records of his observations during the later months (July–October) of 1610.

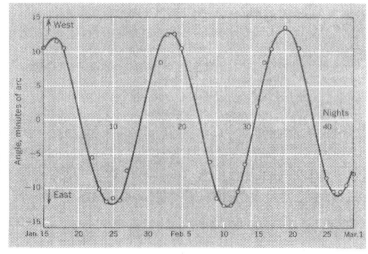

Fig. 8-24(c) A graph constructed from Galileo's own records, showing the periodic motion of Callisto, the outermost of the four satellites visible to him. The period of about 16 3/4 days is clearly exhibited.

countryman Lagrange. It turned out that a curious kind of resonance effect was at work, resulting from the fact that the periods of Jupiter and Saturn are almost in a simple arithmetic

TABLE 8-3: DATA ON SATELLITES OF JUPITER[a]

Satellite	$n = r/R_J$	Period (T)	n^3/T^2, sec^{-2}
Io	5.578	1.7699 days ≈ 1.53 × 10^5 sec	7.4 × 10^{-9}
Europa	8.876	3.5541 days ≈ 3.07 × 10^5 sec	7.5 × 10^{-9}
Ganymede	14.159	7.1650 days ≈ 6.19 × 10^5 sec	7.5 × 10^{-9}
Callisto	24.903	16.7536 days ≈ 1.45 × 10^6 sec	7.4 × 10^{-9}

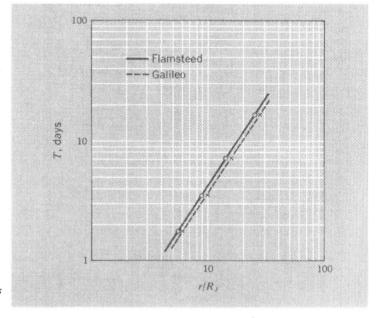

Fig. 8-25 A log-log graph displaying the applicability of Kepler's third law to the Galilean satellites of Jupiter. It may be seen that Galileo's results are little different from those obtained by John Flamsteed nearly 100 years later.

[a]These same data have been presented in a striking way in Eric Rogers, *Physics for the Inquiring Mind*, Princeton University Press, Princeton, N.J., 1960:

Satellite	r^3, (miles)^3	T^2, (hours)^2
Io	1.803 × 10^{16}	1.803 × 10^3
Europa	7.261 × 10^{16}	7.264 × 10^3
Ganymede	29.473 × 10^{16}	29.484 × 10^3
Callisto	160.440 × 10^{16}	160.430 × 10^3

How would you convince your friends that this close *numerical* coincidence is not evidence of a new fundamental law?

relationship ($5T_J \approx 2T_S$). This made large an otherwise negligible term in the perturbation, with a repetition period so long (~900 yr) that it seemed to be increasing without limit. When the mystery was finally resolved the belief in Newton's theory was, of course, strengthened still further.

THE DISCOVERY OF NEPTUNE

Probably the most vivid illustration of the power of the gravitational theory has been the prediction and discovery of planets whose very existence had not previously been suspected. It is noteworthy that the number of known planets remained unchanged from the days of antiquity until long after Newton. Then, in 1781, William Herschel noticed the object that we now know as Uranus. He was engaged in a systematic survey of the stars, and his only clue to start with was that the object seemed slightly less pointlike than the neighboring stars. Then, having a telescope with various degrees of magnification, he confirmed that the size of the image increased with magnification, which is not true for the stars—they remain below the limit of resolution of even the biggest telescopes, and always produce images indistinguishable from those due to ideal point sources.

Once his attention had been drawn to the object, Herschel returned to it night after night and confirmed that it was moving with respect to the other stars. Also, as has happened in various other cases, it was found that the existence of the object had in fact been recorded in earlier star maps (first by John Flamsteed in 1690). These old data suddenly became extremely valuable, because they were a ready-made record of the object's movements dating back through nearly a century. When combined with new measurements carried out over many months, they showed that the object (finally to be called Uranus) was indeed a member of our solar system, following an almost circular orbit with a mean radius of 19.2 AU and a period of 84 years.

This is where our main story begins. Once it was discovered, Uranus and its motions became the subject of a continuing study, and evidence began to accumulate that there were some extremely small irregularities in its motion that could not be ascribed to perturbing effects from any known source. Figure 8-26(a), a tribute to the wonderful precision of astronomical observation, shows the anomaly as a function of time since 1690. The suspicion began to grow that perhaps there was yet another

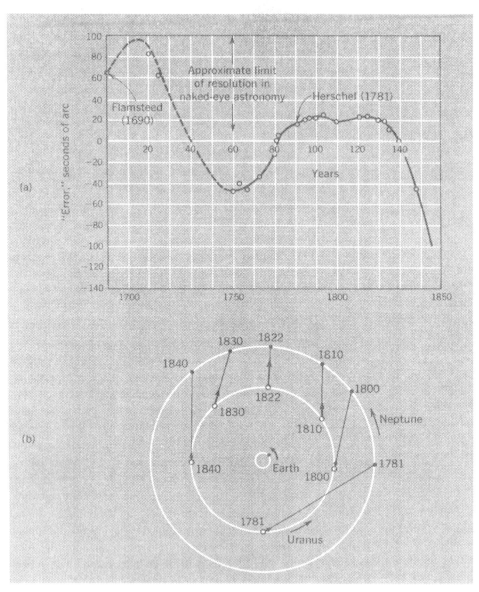

Fig. 8-26 (a) Unexplained residual deviations in the observed positions of Uranus between 1690 and 1840. (b) Basis of ascribing the deviations to the influence of an extra planet. The arrows indicate the relative magnitude of the perturbing force at different times.

planet beyond Uranus, unknown in mass, period, or distance. Two men—J. C. Adams in England and U. J. LeVerrier in France—independently worked on the problem. Both men used as a starting point the assumption that the orbital radius of the unknown planet was almost exactly twice that of Uranus. The basis of this was a curious empirical relation, known as Bode's law (actually discovered by J. D. Titius in 1772, but publicized by J. Bode), which expresses the fact that the orbital radii of the known planets can be roughly fitted by the following formula:

$$R_n \text{ (AU)} = 0.4 + (0.3)(2^n)$$

where n is an appropriate integer for each planet. Putting $n = 0$, 1, 2, we get the approximate radii for Venus, earth, and Mars (Mercury requires $n = -\infty$, which is hard to defend). Using $n = 4$, 5, and 6 one gets quite good values for Jupiter, Saturn, and Uranus. (The missing integer, $n = 3$, corresponds to the asteroid belt.) Figure 8-27 shows this relation of orbital radii with the help of a semilog plot; it is clear that a simple exponential relation (linear on this graph) works almost as well, but if one accepts Bode's law, then $n = 7$ gives $r = 38.8$ AU, and this is what Adams and LeVerrier used.

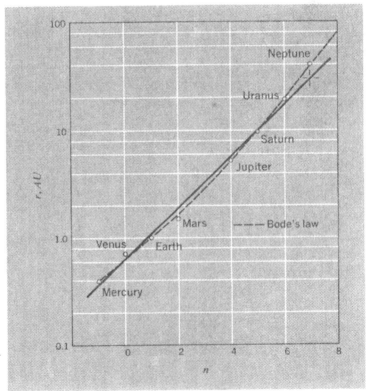

Fig. 8-27 Graph for predicting the orbital radius of the new planet with the help of Bode's law.

Given the radius, the period is automatically defined by Kepler's third law, and then it becomes possible to construct a definite picture, as shown in Fig. 8-26(b), of the way in which the new planet could alternately accelerate and retard Uranus in its orbital motion, depending on their relative positions. With the help of laborious analysis, one can then deduce where in its orbit the new planet should be on a particular date. Adams supplied such information in October 1845 to the British Astronomer Royal, G. B. Airy, who acknowledged Adams' letter, raised a question of detail, but otherwise did nothing. LeVerrier did not complete his own calculations until August 1846, but the astronomer to whom he wrote (J. G. Galle, in Germany) made an immediate search and identified the new planet (Neptune) on his very first night of observation. It was only about a degree from the predicted position (see Fig. 8-28). The next night it had visibly shifted, thereby confirming its planetary status.

Although the discovery of Neptune is in some respects a great success story, it is also a story of luck, both good and bad, and of human frailty. Adams was really first in the field, but he received no support from his seniors (he was fresh from his bachelor's degree when he began his calculations). Airy missed

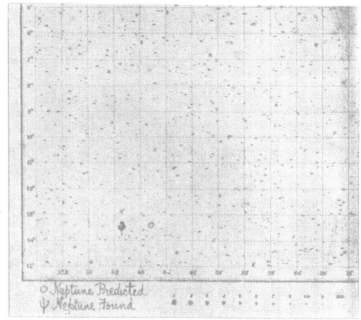

Fig. 8-28 Star map showing the discovery of Neptune, September 23, 1846. (From Herbert Hall Turner, Astronomical Discovery, Edward Arnold, London, 1904.)

the credit, which he might readily have won, of being the man who first identified Neptune. But the locations that both Adams and LeVerrier predicted might well have been hopelessly misleading, for in their reliance on Bode's law they used an orbital radius (and hence a period) that was far from correct. The true value is about 30 AU instead of nearly 40 as they had assumed, which means that they overestimated the period by nearly 50%.[1] It was therefore largely a lucky accident that the planet was so near to its predicted position on the particular date that Galle sought and found it. But let this not be taken as disparagement. A great discovery was made, with the help of the laws of motion and the gravitational force law, and it remains as the most triumphant confirmation of the dynamical model of the universe that Newton invented.[2] The discovery of Pluto by C. Tombaugh in 1930, on the basis of a detailed record of the irregularities of Neptune's own motion, provided an echo of the original achievement.

GRAVITATION OUTSIDE THE SOLAR SYSTEM

When Newton wrote his *System of the World*, nothing was known about the distances or possible motions of the stars. They simply provided a seemingly fixed background against which the dynamics of the solar system proceeded. There were exceptions. A few prominent stars—e.g., Sirius—known since antiquity by naked-eye astronomy, were found to have shifted position within historic time. But the serious and systematic investigation of stellar motions was begun by William Herschel. His observations, continued and refined by his son John Herschel and by other astronomers, revealed two classes of results. The first was the continuing apparent displacement of individual stars in a way that suggested that the solar system is itself involved in a general movement of the stars in our neighborhood, at a speed of the order of 10 miles/sec (comparable to the earth's own orbital speed around the sun). This, as it stood, was just an empirical fact. But the second type of result pointed directly

---

[1] This also means that they overestimated the mass necessary to produce the observed perturbations of Uranus. LeVerrier gave a figure of about 35 times the mass of the earth; the currently accepted value is about half this.

[2] For a detailed account of the whole matter, see H. H. Turner, *Astronomical Discovery*, Edward Arnold, London, 1904. A shorter but more readily accessible account may be found in an essay entitled "John Couch Adams and the Discovery of Neptune," by Sir H. Spencer Jones, in *The World of Mathematics* (J. R. Newman, ed.), Simon and Schuster, New York, 1956.

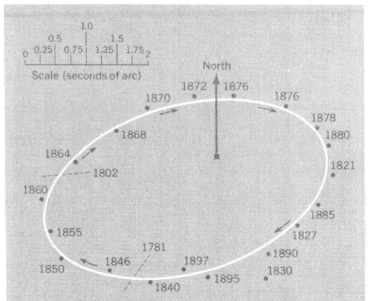

Fig. 8-29 Variation with time of the relative position vector of the members of a double-star system. (After Arthur Berry, A Short History of Astronomy, 1898; reprinted by Dover Publications, New York, 1961.)

to the operation of Newton's dynamics. For the Herschels discovered numerous pairs of stars that were evidently orbiting around one another as *binary systems*. Figure 8-29 shows one of the best documented early examples, and the first to be subjected to a detailed analysis in terms of Kepler's laws. (It is ξ-Ursae, in one of the hind paws of the constellation known as the Great Bear.)

The period of a binary star depends on the *total* mass of the system, not on the individual masses. This is easily proved in the case in which the orbits are assumed to be circles around the common center of mass [see Fig. 8-30(a)].[1] The individual stars are always at opposite ends of a straight line passing through the center, C. If we write the statement of $F = ma$ for one of the stars, say m_1, we have

$$G\frac{m_2 m_1}{(r_1 + r_2)^2} = \frac{m_1 v_1^2}{r_1} = m_1 \omega^2 r_1$$

where $\omega \ (= 2\pi/T)$ is the angular velocity common to both stars. Hence

$$\omega^2 = \frac{Gm_2}{r_1(r_1 + r_2)^2}$$

[1] For a full discussion of the concept of center of mass, see Ch. 9, p. 337.

(a)

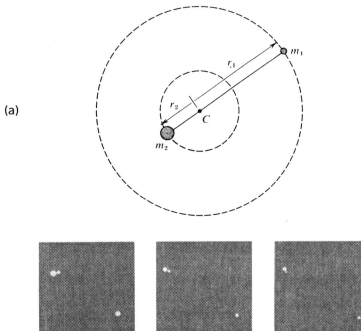

(b)

Fig. 8–30 (a) Motion of the members of a binary star system with respect to the center of mass, C, for the case of circular orbits. (b) Direct visual evidence of the motion of a binary system—Krueger 60, photographed by E. E. Barnard. (Yerkes Observatory photograph.)

However, by the definition of the center of mass, we have

$$r_1 = \frac{m_2}{m_1 + m_2} r$$

where

$$r = r_1 + r_2$$

It follows at once that

$$\omega^2 = \frac{G(m_1 + m_2)}{r^3}$$

Thus if the distance r between the stars can be obtained by direct astronomical observation (e.g., starting from a knowledge of their angular separation), the sum of the masses is at once determined. Finding the individual masses entails the somewhat harder job of measuring the motion of each star in absolute terms against the background of the "fixed" stars. Figure 8–30(b) shows convincing direct evidence of the orbital motion of an actual binary system.

Fig. 8-31 Rotating galaxy (spiral galaxy NGC 5194 in the constellation Canes Venatici). (Photograph from the Hale Observatories.)

With the development of modern astronomy, the systematic motions of our sun and its neighbors came to be seen as part of a greater scheme of movements controlled by gravity. All around us throughout the universe were the immense stellar systems—the galaxies—most of them vividly suggesting a state of general rotation, as in Fig. 8-31, for example. The most difficult structure to elucidate was the one in which we ourselves are embedded, i.e., the Milky Way galaxy. It finally became clear, however, that its basic structure is very much like that of Fig. 8-31, and that in it our sun is describing some kind of orbit around the center, with a radius of about 3×10^{20} m ($\approx$ 30,000 light-years) and an estimated period of about 250 million years ($\approx 8 \times 10^{15}$ sec). Using these figures we can infer the approximate gravitating mass, inside the orbit, that would define this motion. From Eq. (8–24) we have

$$M = \frac{4\pi^2}{G} \frac{r^3}{T^2}$$

With $G \approx 7 \times 10^{-11}$ m³/kg-sec², we find that

$$M \approx \frac{40}{7 \times 10^{-11}} \times \frac{3 \times 10^{61}}{6 \times 10^{31}} \approx 3 \times 10^{41} \text{ kg}$$

Since the mass of the sun (a typical star) is about 2×10^{30} kg, we see that a core of about 10^{11} stars is implied. This is not really a figure that can be independently checked. It is a kind of ultimate tribute to our belief in the universality of the gravitational law that it is confidently used to draw conclusions like those above concerning masses of galactic systems.

EINSTEIN'S THEORY OF GRAVITATION

We have described earlier how Newton recognized that the proportionality of weight to inertial mass is a fact of fundamental significance; it played a central role in leading him to the conclusion that his law of gravitation must be a general law of nature. For Newton this was a strictly dynamical result, expressing the basic properties of the force law. But Albert Einstein, in 1915, looked at the situation through new eyes. For him the fact that all objects fall toward the earth with the same acceleration g, whatever their size or physical state or composition, implied that this must be in some truly profound way a kinematic or geometrical result, not a dynamical one. He regarded it as being on a par with Galileo's law of inertia, which expressed the tendency of objects to persist in straight-line motion.

Building on these ideas, Einstein developed the theory that a planet (for example) follows its characteristic path around the sun because in so doing it is traveling along what is called a *geodesic* line—that is to say, the most economical way of getting from one point to another. His proposition was that although in the absence of massive objects the geodesic path is a straight line in the Euclidean sense, the presence of an extremely massive object such as the sun modifies the geometry locally so that the geodesics become curved lines. The state of affairs in the vicinity of a massive object is, in this view, to be interpreted not in terms of a gravitational field of force but in terms of a "curvature of space"—a facile phrase that covers an abstract and mathematically complex description of non-Euclidean geometries.

For the most part the Einstein theory of gravitation gives results indistinguishable from Newton's; the grounds for preferring it might seem to be conceptual rather than practical. But in one celebrated instance of planetary motions there is a real discrepancy that favors Einstein's theory. This is in the so-called "precession of the perihelion" of Mercury. The phenomenon is that the orbit of Mercury, which is distinctly elliptical in shape, very gradually rotates or precesses in its own plane, so that the major axis is along a slightly different direction at the end of each complete revolution. Most of this precession (amounting to about 10 minutes of arc per century) can be understood in terms of the disturbing effects of the other planets according to Newton's law of gravitation.[1] But there remains a tiny, obstinate residual rotation equal to 43 seconds of arc per century. The attempts to explain it on Newtonian theory—for example by postulating an unobserved planet inside Mercury's own orbit—all came to grief by conflicting with other facts of observation concerning the solar system. Einstein's theory, on the other hand, without the use of any adjustable parameters, led to a calculated precession rate that agreed exactly with observation. It corresponded, in effect, to the existence of a very small force with a different dependence on distance than the dominant $1/r^2$ force of Newton's theory. The way in which a disturbing effect of this kind causes the orbit to precess is discussed in Chapter 13. Other empirical modifications of the basic law of gravitation—small departures from the inverse-square law—had been tried before Einstein developed his theory, but apart from their arbitrary character they also led to false predictions for the other planets. In Einstein's theory, however, it emerged automatically that the size of the disturbing term was proportional to the square of the angular velocity of the planet and hence was much more important for Mercury, with its short period, than for any of the other planets.

---

[1]The apparent amount of precession as viewed from the earth is actually about 1.5° per century, but most of this is due to the continuous change in the direction of the earth's own axis (the precession of the equinoxes — see Chapter 14).

PROBLEMS

8-1 Given a knowledge of Kepler's third law as it applies to the solar system, together with the knowledge that the disk of the sun subtends an angle of about $\frac{1}{2}°$ at the earth, deduce the period of a hypothetical planet in a circular orbit that skims the surface of the sun.

8-2 It is well known that the gap between the four inner planets and the five outer planets is occupied by the asteroid belt instead of by a tenth planet. This asteroid belt extends over a range of orbital radii from about 2.5 to 3.0 AU. Calculate the corresponding range of periods, expressed as multiples of the earth's year.

8-3 It is proposed to put up an earth satellite in a circular orbit with a period of 2 hr.
  (a) How high above the earth's surface would it have to be?
  (b) If its orbit were in the plane of the earth's equator and in the same direction as the earth's rotation, for how long would it be continuously visible from a given place on the equator at sea level?

8-4 A satellite is to be placed in synchronous circular orbit around the planet Jupiter to study the famous "red spot" in Jupiter's lower atmosphere. How high above the surface of Jupiter will the satellite be? The rotation period of Jupiter is 9.9 hr, its mass M_J is about 320 times the earth's mass, and its radius R_J is about 11 times that of the earth. You may find it convenient to calculate first the gravitational acceleration g_J at Jupiter's surface as a multiple of g, using the above values of M_J and R_J, and then use a relationship analogous to that developed in the text for earth satellites [Eq. (8–16) or (8–17)].

8-5 A satellite is to be placed in a circular orbit 10 km above the surface of the moon. What must be its orbital speed and what is the period of revolution?

8-6 A satellite is to be placed in synchronous circular earth orbit. The satellite's power supply is expected to last 10 years. If the maximum acceptable eastward or westward drift in the longitude of the satellite during its lifetime is 10°, what is the margin of error in the radius of its orbit?

8-7 The springs found in retractable ballpoint pens have a relaxed length of about 3 cm and a spring constant of perhaps 0.05 N/mm. Imagine that two lead spheres, each of 10,000 kg, are placed on a frictionless surface so that one of these springs just fits, in its uncompressed state, between their nearest points.
  (a) How much would the spring be compressed by the mutual gravitational attraction of the two spheres? The density of lead is about 11,000 kg/m³.

(b) Let the system be rotated in the horizontal plane. At what frequency of rotation would the presence of the spring become irrelevant to the separation of the masses?

8-8 During the eighteenth century, an ingenious attempt to find the mass of the earth was made by the British Astronomer Royal, Nevil Maskelyne. He observed the extent to which a plumbline was pulled out of true by the gravitational attraction of a mountain. The figure illustrates the principle of the method. The change of direction of the plumbline was measured between the two sides of the mountain.

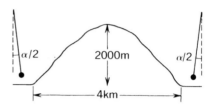

(This was done by sighting on stars.) After allowance had been made for the change in direction of the local vertical because of the curvature of the earth, the residual angular difference α was given by $2F_M/F_E$, where $\pm F_M$ is the horizontal force on the plumb-bob due to the mountain, and $F_E = GM_E m/R_E^2$. (m is the mass of the plumb-bob.)

The value of α is about 10 seconds of arc for measurements on opposite sides of the base of a mountain about 2000 m high. Suppose that the mountain can be approximated by a cone of rock (of density 2.5 times that of water) whose radius at the base is equal to the height and whose mass can be considered to be concentrated at the center of the base. Deduce an approximate value of the earth's mass from these figures. (The true answer is about 6×10^{24} kg.) Compare the gravitational deflection α to the change of direction associated with the earth's curvature in this experiment.

8-9 Imagine that in a certain region of the ocean floor there is a roughly cone-shaped mound of granite 250 m high and 1 km in diameter. The surrounding floor is relatively flat for tens of kilometers in all directions. The ocean depth in the region is 5 km and the density of the granite is 3000 kg/m³. Could the mound's presence be detected by a surface vessel equipped with a gravity meter that can detect a change in g of 0.1 mgal?

(*Hint:* Assume that the field produced by the mound at the surface can be approximated by the field of a mass point of the same total mass located at the level of the surrounding floor. Note that in calculating the change in *g* you must keep in mind that the mound has displaced its own volume of water. The density of water, even at such depths, can be taken as about equal to its surface value of about 1000 kg/m³.

8-10 Show that the period of a particle that moves in a circular orbit close to the surface of a sphere depends only on G and the mean density of the sphere. Deduce what this period would be for any sphere having a mean density equal to the density of water. (Jupiter almost corresponds to this case.)

8-11 Calculate the mean density of the sun, given a knowledge of G, the length of the earth's year, and the fact that the sun's diameter subtends an angle of about $0.55°$ at the earth.

8-12 An astronaut who can lift 50 kg on earth is exploring a planetoid (roughly spherical) of 10 km diameter and density 3500 kg/m^3.

    (a) How large a rock can he pick up from the planetoid's surface, assuming that he finds a well-placed handle?

    (b) The astronaut observes a rock falling from a cliff. The rock's radius is only 1 m and as it approaches the surface its velocity is 1 m/sec. Should he try to catch it? (This is obviously a fanciful problem. One would not expect a planetoid to have cliffs or loose rocks, even if an astronaut were to get there in the first place.)

8-13 It is pointed out in the text that a person can properly be termed "weightless" when he is in a satellite circling the earth. The moon is a satellite, yet it is noted in many discussions that we would weigh $\frac{1}{6}$ of our normal weight there. Is there a contradiction here?

8-14 A dedicated scientist performs the following experiment. After installing a huge spring at the bottom of a 20-story-high elevator shaft, he takes the elevator to the top, positions himself on a bathroom scale inside the airtight car with a stopwatch and with pad and pencil to record the scale reading, and directs an assistant to cut the car's support cable at $t = 0$. Presuming that the scientist survives the first encounter with the spring, sketch a graph of his measured weight versus time from $t = 0$ up to the beginning of the second bounce. (*Note:* Twenty stories is ample distance for the elevator to acquire terminal velocity.)

8-15 A planet of mass M and a single satellite of mass $M/10$ revolve in circular orbits about their stationary center of mass, being held together by their gravitational attraction. The distance between their centers is D.

    (a) What is the period of this orbital motion?

    (b) What fraction of the total kinetic energy resides in the satellite?

(Ignore any spin of planet and satellite about their own axes.)

8-16 We have considered the problem of the moon's orbit around the earth as if the earth's center represented a fixed point about which the motion takes place. In fact, however, the earth and the moon revolve about their common center of mass.

(a) Calculate the position of the center of mass, given that the earth's mass is 81 times that of the moon and that the distance between their centers is 60 earth radii.

(b) How much longer would the month be if the moon were of negligible mass compared to the earth?

8–17 The sun appears to be moving at a speed of about 250 km/sec in a circular orbit of radius about 25,000 light-years around the center of our galaxy. (One light-year $\simeq 10^{16}$ m.) The earth takes 1 year to describe an almost circular orbit of radius about 1.5×10^{11} m around the sun. What do these facts imply about the total mass responsible for keeping the sun in its orbit? Obtain this mass as a multiple of the sun's mass M. (Note that you do not need to introduce the numerical value of G to obtain the answer.)

8–18 (A good problem for discussion.) In 1747 Georges Louis Lesage explained the inverse-square law of gravitation by postulating that vast numbers of invisible particles were flying through space in all directions at high speeds. Objects like the sun and planets block these particles, leading to a shadowing effect that has the same quantitative result as a gravitational attraction. Consider the arguments for and against this theory.

(*Suggestion:* First consider a theory in which opaque objects block the particles completely. This proposal is fairly easy to refute. Next consider a theory in which the attenuation of the particles by objects is incomplete or even very small. This theory is much harder to dismiss.)

8–19 The continuous output of energy by the sun corresponds (through Einstein's relation $E = Mc^2$) to a steady decrease in its mass M, at the rate of about 4×10^6 tons/sec. This implies a progressive increase in the orbital periods of the planets, because for an orbit of a given radius we have $T \sim M^{-1/2}$ [cf Eq. (8–23)].

A precise analysis of the effect must take into account the fact that as M decreases the orbital radius itself increases—the planets gradually spiral away from the sun. However, one can get an order-of-magnitude estimate of the size of the effect, albeit a little bit on the low side, by assuming that r remains constant. (See Problem 13–21 for a more rigorous treatment.)

Using the simplifying assumption of constant r, estimate the approximate increase in the length of the year resulting from the sun's decrease in mass over the time span of accurate astronomical observations—about 2500 years.

8–20 It is mentioned at the end of the chapter how Einstein's theory of gravitation leads to a small correction term on top of the basic Newtonian force of gravitation. For a planet of mass m, traveling at speed v in a circular orbit of radius r, the gravitational force becomes in effect the following:

$$F = \frac{GMm}{r^2}\left(1 + 6\frac{v^2}{c^2}\right)$$

where c is the speed of light. (Correction terms of the order of v^2/c^2 are typical of relativistic effects.)

(a) Show that, if the period under a pure Newtonian force GMm/r^2 is denoted by T_0, the modified period T is given approximately by

$$T \simeq T_0\left(1 - \frac{12\pi^2 r^2}{c^2 T_0^2}\right)$$

(Treat the relativistic correction as representing, in effect, a small fractional increase in the value of G, and use the value of v corresponding to the Newtonian orbit.)

(b) Hence show that, in each revolution, a planet in a circular orbit would travel through an angle greater by about $24\pi^3 r^2/c^2 T_0^2$ than under the pure Newtonian force, and that this is also expressible as $6\pi GM/c^2 r$, where M is the mass of the sun.

(c) Apply these results to the planet Mercury and verify that the accumulated advance in angle amounts to about 43 seconds of arc per century. This corresponds to what is called the precession of its orbit.

God . . . created matter with motion and rest in its parts, and . . . now conserves in the universe, by His ordinary operations, as much of motion and of rest as He originally created.

                    RENÉ DESCARTES, *Principia Philosophiae* (1644)

9
Collisions and conservation laws

IN THIS CHAPTER we shall be discussing some concepts and results that are at the very heart of mechanics. They are rooted in the fact that it takes at least two particles to make a dynamical system. Until now we have rather glossed over that fact, by talking in terms of individual particles subjected to forces of various kinds. Thus, for example, the motion of a planet around the sun, or of an electron between the deflection plates of a cathode-ray tube, was discussed as the problem of a single particle exposed to the force supplied by some completely immovable body or structure. But this is a very special way of looking at things and is not in general justified. The sun pulls on a planet and gives it an acceleration, to be sure, but the law of gravitation doesn't play favorites—it has a completely symmetrical form—and the planet pulls back on the sun with an equal and opposite force. So the sun must accelerate, too, and will follow some kind of path under the combined action of all the planets. It happens that the sun is hundreds of times more massive than the rest of the solar system put together, so that to a first approximation we can ignore its wanderings. But this is only an accident of disparity. The basic dynamical system is made up of two interacting particles, the motions of both of which must be considered. The experimental study of such systems, via collision processes, was in fact the true starting point of dynamics. The study of collisions has lost none of its importance to physics in the 300 years since it first became a subject of exact investigation; let us then consider it with care.

THE LAWS OF IMPACT

In 1668 the Royal Society of London[1] put out a request for the experimental clarification of collision phenomena. Contributions were submitted shortly afterward by John Wallis (mathematician), Sir Christopher Wren (architect), and Christian Huygens (physicist, and Newton's great Dutch contemporary). The results embody what we are familiar with today as the rules governing exchanges of momentum and energy in the collision between two objects. Most importantly, they involve at one stroke the concept of inertial mass and the principle of conservation of linear momentum.

These experiments revealed that the decisive quantity in a collision is what Newton called "the quantity of motion," defined as the product of the velocity of a body and the quantity of matter in it—this latter being what we have called the inertial mass. We have already (in Chapter 6) discussed Newton's straightforward concept of mass: how he was quite ready to assume that the total mass of two objects together is just the sum of their separate masses, and how he took it for granted that the masses of differently sized portions of a homogeneous material are proportional to their volumes. You will recall from our earlier discussion that these assumptions, however reasonable they may seem, are not always strictly justified. On the other hand, their use does not lead to detectable inconsistencies in the dynamics of ordinary objects. And what Newton and his contemporaries found was that, if these commonsense ideas were accepted as a quantitative basis for relating masses, then a very simple description of all collisions could be made, which expressed mathematically is as follows:

$$m_1 u_1 + m_2 u_2 = m_1 v_1 + m_2 v_2 \tag{9-1}$$

where u_1 and u_2 are the velocities before impact, and v_1 and v_2 are the velocities after impact (a one-dimensional motion is assumed). This is a very powerful generalization, because it

---

[1] The Society was formally chartered in 1662. (Newton was its President from 1703 until his death in 1727.) Several other great European scientific academies came into existence at about the same period.

applies to quite different sorts of collisions—those with almost perfect rebound, such as occur between two glass spheres, down to those with no rebound at all, as between two lumps of putty. Qualitatively, collisions can be described in terms of their degree of elasticity—i.e., bounciness. A quantitative but purely empirical measure of this, used by Newton and others, is the ratio of $v_2 - v_1$ (the relative velocity after impact) to $u_1 - u_2$ (the relative velocity before impact). If this ratio is zero, the collision is called completely inelastic. If the ratio is unity, the collision is called elastic (sometimes, more vividly, "perfectly elastic"). A hardened steel ball is much more elastic than a rubber ball in this sense. In the early investigations Wallis confined his studies to completely inelastic collisions, Wren and Huygens to almost perfectly elastic objects, and Newton, some time later, added experiments on objects of intermediate elasticity.

THE CONSERVATION OF LINEAR MOMENTUM

The physicist is always on the lookout for quantities that remain *conserved* (i.e., unchanged) in physical processes. Once he has discovered such quantities, they become powerful tools in his analysis of phenomena. They start out by being just an aid in codifying *past* experience. But as they are found to be applicable to more and more new instances, their value grows and one begins to make confident *predictions* with their help. The statement about conservation of a particular quantity is promoted to the status of a *conservation law*. If in some new instance the conservation law appears to break down, one's faith in the law may be so great that one hunts around for the missing piece. Should it be found, the conservation law is strengthened still further. Thus, for example, the law of conservation of mass in chemical reactions is accepted as a guide to all possible measurements on the masses of the reacting materials. When the chemical balance was first applied to the study of chemical reactions, it appeared that in some processes mass was gained, in some it was lost, and in some it remained unchanged. But when Lavoisier proved, through numerous examples, that mass was merely *transferred*, and that in a *closed* system it was *conserved*, the whole scheme became clear. Chemists could then exploit the conservation law. For example, they could confidently infer the mass of a gaseous product (escaping from an unclosed system) from easily made measurements on solid and liquid reactants.

Some of the most powerful aspects of our physical description of the world are embodied in statements about conserved quantities. In mechanics the law of conservation of linear momentum is such a statement—one might even claim that it is the most important single principle in dynamics. It is based directly on the results of the experiments on impact as summarized in Eq. (9-1). If, for a given particle, we introduce the single word *momentum* to describe the product mv, then we have a compact statement:

> **The total momentum of the system of two colliding particles remains unchanged by the collision, i.e., the total linear momentum is a conserved quantity.**

Underlying this generalization concerning two-particle collisions is the tacit assumption that the system is effectively *isolated*—the particles interact with each other but not with anything else. In the experiments of Newton and others this was achieved by hanging the colliding objects on long strings, so that they swung as pendulums and collided at the lowest point of their swing. In the brief duration of the impact, therefore, the objects were essentially free of all *horizontal* forces except those provided by their mutual interaction.

Since the momentum carried by an object is given, in Newton's words, by "the velocity and the quantity of matter conjointly,"—i.e., by the value of the product mv and not by the values of m and v separately[1]—it is convenient to introduce a single symbol, p, to represent the momentum of a mass m moving along a given line with velocity v. The relation

$$p_1 + p_2 = \text{const.} \tag{9-2}$$

then describes the conservation of momentum in two-body collisions of the type studied by Newton and his contemporaries.

MOMENTUM AS A VECTOR QUANTITY

We have based our discussion so far on one-dimensional collisions. It is important to appreciate, however, that—as is apparent from its definition—the momentum of a particle is a vector quantity having the same direction as the velocity of the

---

[1]So that, for example, a body of mass $\frac{1}{2}m$ traveling with velocity $2v$ has the same momentum as m traveling with v.

particle. Thus our statement of the conservation of linear momentum in a collision between two bodies should really be written as follows:

$$\mathbf{p}_{1i} + \mathbf{p}_{2i} = \mathbf{p}_{1f} + \mathbf{p}_{2f} \tag{9-3}$$

where the subscripts i and f are used to denote initial and final values, respectively (i.e., precollision and postcollision).

This single vector equation defines the magnitude and the direction of any one of the momentum vectors if the other three are known. It will very often be convenient to separate Eq. (9-3) into three equations in terms of the resolved parts of the vectors along three mutually orthogonal axes (x, y, z). Each of these component equations must then be separately satisfied. Thus, for example, if two bodies, of masses m_1 and m_2, have initial and final velocities $\mathbf{u}_1, \mathbf{u}_2$ and $\mathbf{v}_1, \mathbf{v}_2$, Eq. (9-3) becomes

$$m_1 \mathbf{u}_1 + m_2 \mathbf{u}_2 = m_1 \mathbf{v}_1 + m_2 \mathbf{v}_2 \tag{9-4}$$

which contains the following three *independent* statements:

$$\begin{aligned} m_1 u_{1x} + m_2 u_{2x} &= m_1 v_{1x} + m_2 v_{2x} \\ m_1 u_{1y} + m_2 u_{2y} &= m_1 v_{1y} + m_2 v_{2y} \\ m_1 u_{1z} + m_2 u_{2z} &= m_1 v_{1z} + m_2 v_{2z} \end{aligned} \tag{9-5}$$

In carrying out actual numerical calculations this resolution of the vectors will often be necessary. But manipulations involving unspecified masses and velocities are best made in terms of the unresolved equations (9-3) or (9-4), without reference to any particular coordinate system. This is, indeed, one of the main strengths (and economies) of using vector notation.

Example. An object of mass 5 kg, traveling horizontally on a frictionless surface at 16 m/sec, strikes a stationary object of mass 3 kg. After the collision, the 5-kg object is observed to have a velocity of magnitude 12 m/sec at 30° to its original direction of motion, as shown in Fig. 9-1. What is the velocity of the 3-kg object? We can choose an xy plane that contains all the velocity vectors. Let the $+x$ axis lie along the original direction of motion of the 5-kg object (particle 1). Let us denote the initial momentum of the 5-kg object as $\mathbf{p}_{1i}$, and the final momenta as $\mathbf{p}_{1f}$ and $\mathbf{p}_{2f}$. Then we have

$$\mathbf{p}_{1i} = \mathbf{p}_{1f} + \mathbf{p}_{2f} \quad (\text{since } \mathbf{p}_{2i} = 0)$$
$$\mathbf{p}_{2f} = \mathbf{p}_{1i} - \mathbf{p}_{1f} = \mathbf{p}_{1i} + (-\mathbf{p}_{1f})$$

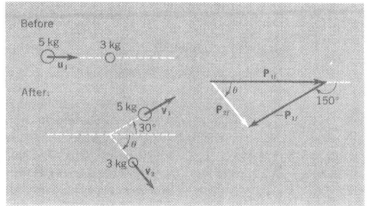

Fig. 9-1 Conservation of the total vector momentum in a simple collision.

Figure 9-1 shows the vector construction by which $\mathbf{p}_{2f}$ can be found. The length of $\mathbf{p}_{1i}$ represents, on some appropriate scale, the initial momentum of 80 kg-m/sec. The length of $-\mathbf{p}_{1f}$ is 60 kg-m/sec, and its direction is as shown. The length and direction of $\mathbf{p}_{2f}$ can be found directly from the vector triangle, and $\mathbf{v}_{2f}$ is found as $\mathbf{p}_{2f}/m_2$.

Alternatively, we can write down the momentum conservation in component form. First, let us list the known quantities:

$m_1 = 5$ kg $m_2 = 3$ kg
$u_{1x} = 16$ m/sec $u_{2x} = 0$
$u_{1y} = 0$ $u_{2y} = 0$
$v_{1x} = 6\sqrt{3}$ m/sec
$v_{1y} = 6$ m/sec

Thus we have [using Eq. (9-5)]

Along x: $80 = 30\sqrt{3} + 3v_{2x}$
Along y: $0 = 30 + 3v_{2y}$

Hence

$v_{2x} = (80 - 30\sqrt{3})/3 \approx 9.3$ m/sec
$v_{2y} = -10.0$ m/sec
$v_2 = [(9.3)^2 + (10.0)^2]^{1/2} \approx 13.6$ m/sec

The direction of $\mathbf{v}_2$ is at an angle θ to the x axis such that

$$\tan \theta = \frac{v_{2y}}{v_{2x}} = -\frac{10.0}{9.3}$$

$$\theta \approx -47°$$

Notice that this result has been obtained through momentum conservation alone; it requires no knowledge of the detailed interaction between the objects. Relating the momentum changes of the individual objects to the forces acting on them during the collision will, however, be our next concern.

ACTION, REACTION, AND IMPULSE

We shall begin this discussion with an analysis along the lines that Newton himself used. Later we shall draw attention to a somewhat different approach, which can be more readily adapted to relativistic dynamics, although it is in harmony with Newton's analysis within the confines of classical mechanics.

Newton interpreted the collision experiments from the standpoint that the concept of inertial mass of an individual object is already established by experiments and arguments like those we made use of in Chapter 6. In that case, Eq. (9-4) is a summary of the actual experimental observations; the "quantity of motion" is conserved. One can then, by using $\mathbf{F} = m\mathbf{a}$, draw conclusions about the forces acting during the collision. A collision is a process involving two objects, each of which exerts a force on the other. Object 1 exerts a force $\mathbf{F}_{12}$ on object 2; object 2 exerts a force $\mathbf{F}_{21}$ on object 1 (Fig. 9-2). We make no assumptions about the relationship between the two forces, *except* that they act for equal times. This last assumption is certainly reasonable, because we recognize that the forces come into being as a result of the collision, and surely the duration of the collision must be the same for both objects.[1] We can then take the statement of $\mathbf{F} = m\mathbf{a}$ as it applies to each object separately:

$$\mathbf{F}_{21} = m_1\mathbf{a}_1 \qquad \mathbf{F}_{12} = m_2\mathbf{a}_2$$

i.e.,

$$\mathbf{a}_1 = \frac{\mathbf{F}_{21}}{m_1} \qquad \mathbf{a}_2 = \frac{\mathbf{F}_{12}}{m_2} \qquad (9\text{-}6)$$

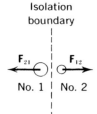

Fig. 9-2 Inferring the equality of action and reaction forces from the fact of momentum conservation.

[1] Yet this is another of those intuitively "obvious" conditions that is not binding and indeed has to be qualified in some of the collision problems that require the use of special relativity.

Suppose, to simplify this present argument, that each force remains constant throughout a collision that lasts for a time Δt. Then we have

$$\mathbf{v}_1 = \mathbf{u}_1 + \frac{\mathbf{F}_{21}}{m_1}\Delta t \qquad \mathbf{v}_2 = \mathbf{u}_2 + \frac{\mathbf{F}_{12}}{m_2}\Delta t \qquad (9\text{-}7)$$

where $\mathbf{u}_1$ and $\mathbf{u}_2$ are the initial velocities of the two objects and $\mathbf{v}_1$ and $\mathbf{v}_2$ are their final velocities. From these equations we therefore have

$$m_1\mathbf{v}_1 = m_1\mathbf{u}_1 + \mathbf{F}_{21}\Delta t$$
$$m_2\mathbf{v}_2 = m_2\mathbf{u}_2 + \mathbf{F}_{12}\Delta t$$

Adding these two, we thus get

$$m_1\mathbf{v}_1 + m_2\mathbf{v}_2 = m_1\mathbf{u}_1 + m_2\mathbf{u}_2 + (\mathbf{F}_{21} + \mathbf{F}_{12})\Delta t \qquad (9\text{-}8)$$

Experimentally, however, we have Eq. (9-4). We *deduce*, therefore, that

$$\mathbf{F}_{21} + \mathbf{F}_{12} = 0$$

i.e.,

$$\mathbf{F}_{21} = -\mathbf{F}_{12} \qquad (9\text{-}9)$$

This is, of course, the famous statement known usually as Newton's third law, that "action and reaction are equal and opposite."

We have already made extensive use of this result, tacitly or explicitly, earlier in this book, and in Chapter 4 we indicated the kind of experimental support that one can supply for it in static equilibrium situations. Newton, in developing the result for dynamic situations, with forces changing from instant to instant, made it quite clear that he regarded Eq. (9-9) as an inference from observations. In the *Principia* he describes an experiment in which he floated a magnet and a piece of iron on water, and released them from rest. He noted that there was no motion of the combined mass after the magnetic attraction had pulled them together, and took this as a demonstration of what he himself called the third law of motion. His description of his pendulum collision experiments, also in the *Principia*, is couched in the same terms.

Having introduced these forces of interaction, we can now relate them to the changes of momentum of the individual objects in a collision. Thus, if Δt is the duration of the collision and $\mathbf{F}_{21}$ is a constant force exerted by particle 2 on particle 1, the change of momentum $\Delta \mathbf{p}_1$ of particle 1 in the collision is given by

$$\mathbf{F}_{21} \Delta t = \Delta \mathbf{p}_1$$

More generally, if any constant force $\mathbf{F}$ acts on a particle for some short time interval Δt, the change of momentum that it generates is given by

$$\mathbf{F} \Delta t = \Delta \mathbf{p}$$

As we noted in Chapter 6, Newton's own approach to dynamics was rooted in this equation rather than in $\mathbf{F} = m\mathbf{a}$. We also introduced the terminology by which the product $\mathbf{F} \Delta t$ is called the *impulse* of the force in Δt. If $\mathbf{F}$ is varying in magnitude and/or direction during the time span Δt, one can proceed to the limit of vanishingly small time and so obtain the following equation:

$$(\text{Newton's law reformulated}) \qquad \mathbf{F} = \frac{d\mathbf{p}}{dt} \qquad (9\text{-}10)$$

This equation, written on the assumption that $\mathbf{F}$ is the net force acting on a particle, then becomes the basic statement of Newtonian dynamics. It is, in a sense, broader in scope than $\mathbf{F} = m\mathbf{a}$, or at least a more efficient statement of it. For example, a given force applied in turn to a number of different masses causes the same rate of change of momentum in each but not the same acceleration. In short, we come to recognize that momentum is a valuable *single* quantity to be accepted in its own right and most importantly for its property of being *conserved* in an isolated system of interacting particles.

In our earlier discussion of the collision problem, we took the forces of interaction as being constant in time. In most cases that would be quite unrealistic. It is easy to see, however, that through the integration of Eq. (9-10) we obtain the net momentum change caused by a varying force. Thus, for example, if a force has some arbitrary variation between $t = 0$ and $t = \Delta t$, the momentum change that it generates is given by

$$\Delta \mathbf{p} = \int_0^{\Delta t} \mathbf{F} \, dt \qquad (9\text{-}11)$$

In a two-body collision, the duration of which is Δt, all that we actually observe is that the total momentum after the collision is equal to the total momentum before the collision. In terms of the impulses $\Delta \mathbf{p}_1$ and $\Delta \mathbf{p}_2$ given to the separate objects this result is expressed by the condition

$$\Delta \mathbf{p}_1 + \Delta \mathbf{p}_2 = 0$$

From this we infer that

$$\int_0^{\Delta t} \mathbf{F}_{21}\, dt = -\int_0^{\Delta t} \mathbf{F}_{12}\, dt$$

In principle, $\mathbf{F}_{12}$ and $\mathbf{F}_{21}$ could have quite unrelated values at any particular instant, as long as the above integrals are equal. However, failing any evidence to the contrary we assume that they are equal and opposite at each and every instant. Thus in a one-dimensional collision the graphs of these forces as a function of time, whatever their exact form, are taken to be mirror images of each other as shown in Fig. 9-3. It is important to realize, however, that this *is* a postulate. And it is not always true! There is no difficulty as far as "contact" collisions between ordinary objects are concerned. But in situations in which objects influence one another at a distance, as for example through the long-range forces of electricity or gravitation, Newton's third law *may* cease to apply. For no interaction is transmitted instantaneously, and if the propagation time cannot be ignored in comparison with the time scale of the motion, the concept of instantaneous action and reaction can no longer be used. A simple mechanical model of such a delayed interaction is sug-

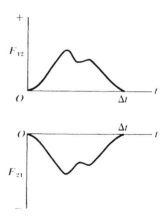

Fig. 9-3 Corresponding variations of action and reaction forces during the course of a collision.

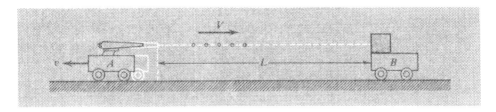

Fig. 9-4 Interaction with a time delay, mediated by particles traversing the gap between two separated objects.

gested in Fig. 9-4. A cart, A, carries a gun that fires off a stream of bullets with speed V. At a distance L away there is a second cart, B, carrying a block in which the bullets are caught. Suppose that the bullets carry plenty of momentum, thanks to a large value of V, but are so light that they represent a negligible transfer of mass from the first cart to the second. Suppose further that they are invisible to an observer standing some distance away. (We might make them extremely small, or perhaps paint them black and set up a dark background.) Then if a brief burst of bullets is fired off from A, we see this cart begin to recoil, apparently spontaneously, while B remains at rest. Not until a time L/V later does B begin to recoil in the opposite direction. There is, in effect, a breakdown of the equality between action and reaction in such a case. By the time the whole interaction has been completed, with all the bullets reabsorbed in B, we recognize that momentum has ultimately been conserved, but if one restricts attention to the carts it does not appear to be conserved, instant by instant, during the interaction.

The above example may seem artificial because, after all, we could save the action–reaction principle by looking more closely and observing the bullets at the instant of leaving A or striking B. Nevertheless, it offers an interesting parallel to what are probably the most important delayed interactions—those of electromagnetism. We know that the interaction between two separated charges takes place via the electromagnetic field, and the propagation of such a field takes place at a speed which, although extremely large, is still finite—the speed of light. The transfer of momentum from one charge to another (resulting, let us say, from a sudden movement of the first charge) involves a time lag equal to the distance between them divided by c. If we looked at the charged particles alone, we would see a sudden change in the momentum of the first charge without an equal and opposite change in the momentum of the other at the same time.[1] There would thus appear to be a failure of momentum conservation, instant by instant, unless we associate some momentum with the electromagnetic field that carries the interaction. And that is precisely what electromagnetic theory suggests. The picture becomes even more vivid when we introduce the quantization of the electromagnetic field and recognize that radiation is carried in the form of photons, or light quanta. By the time we have associated a momentum and an energy with each photon individually, we are remarkably close to our mechanical model. There is even a well-developed theoretical description of the static interaction between electric charges in terms of a continual exchange of so-called "virtual photons." In this case, however, since the forces are constant in time, the equality of action and reaction holds good at every instant, and the existence of a finite time for propagating the interaction ceases to be apparent.

EXTENDING THE PRINCIPLE OF MOMENTUM CONSERVATION[2]

It is perhaps worth amplifying the remarks of the last section a little. We have seen how, in Newton's view, the law of conservation of momentum is closely tied in with the action–reaction idea. There is, however, an alternative approach to simple collisions which loosens this connection and eases the transition to non-Newtonian mechanics.

This approach can be defined in terms of a question: What do we actually *observe* in a collision experiment? The answer is that our observations are purely kinematic ones—measurements of the velocities of the two objects before and after impact. Suppose that two objects, A and B, with initial velocities $\mathbf{u}_1$, and $\mathbf{u}_2$, respectively, collide with one another and afterward have velocities $\mathbf{v}_1$ and $\mathbf{v}_2$. In any individual collision of this type, it is always possible to find a set of four scalar multiples (α) that permit one to write an equation of the following form:

$$\alpha_1 \mathbf{u}_1 + \alpha_2 \mathbf{u}_2 = \alpha_3 \mathbf{v}_1 + \alpha_4 \mathbf{v}_2 \tag{9-12}$$

[1] In circumstances where such effects are important, the very concept of "the same time" — i.e., simultaneity — comes into question. It is just here that Newtonian mechanics itself ceases to be adequate and the revised formulation of dynamics according to special relativity becomes essential.

[2] This section may be omitted without destroying the continuity, but it is recommended if you want to see how the bases of classical mechanics can be approached in more than one way.

This as it stands is a quite uninteresting statement. But experiments for all sorts of values of $\mathbf{u}_1$ and $\mathbf{u}_2$ reveal the remarkable result that in every such collision, for two given objects, we can obtain a vector identity by putting $\alpha_1 = \alpha_3 = \alpha_A$ (a scalar property of A) and $\alpha_2 = \alpha_4 = \alpha_B$ (a corresponding scalar property of B). In other words, the purely kinematic observations on a collision process permit us to introduce a unique dynamical property of each object. Notice that this simple situation ceases to hold if any of the velocities involved become comparable to that of light. In that case it is still possible to construct a vector balance equation in the form of Eq. (9-12), but only if the parameters α are made explicit functions of speed. In fact, we arrive at the relativistic formula for the variation of mass with speed [Eq. (6-3)].[1]

Let us return now to the results of experiments on collisions at low velocities. The basic statement of these results is that, in the impact of any two given objects, the velocity change of one always bears a constant, negative ratio to the velocity change of the other:

$$\mathbf{v}_2 - \mathbf{u}_2 = -\text{const}(\mathbf{v}_1 - \mathbf{u}_1) \qquad (9\text{-}13)$$

It is precisely because this ratio of velocity changes is found to come out to the same value, whatever the type, strength, or duration of the interaction between the objects, that we can infer that it provides a measure of some intrinsic property of the objects themselves. We *define* this property as the inertial mass.

One can set up an inertial mass scale for a number of objects 1, 2, 3, ... by finding the velocity changes pairwise in such interaction processes and defining the inertial mass ratio, let us say of objects 1 and 2, by

$$\frac{m_2}{m_1} = \frac{|\Delta \mathbf{v}|_1}{|\Delta \mathbf{v}|_2}$$

Similarly for objects 1 and 3,

$$\frac{m_3}{m_1} = \frac{|\Delta \mathbf{v}'|_1}{|\Delta \mathbf{v}'|_3}$$

and so on. If m_1 is the standard kilogram, we then have an operation to determine the inertial masses of any other objects.

---

[1]For further discussion, see the volume *Special Relativity* in this series.

We must however, do more. If we let objects 2 and 3 interact, then the ratio

$$\frac{m_3}{m_2} = \frac{|\Delta \mathbf{v}''|_2}{|\Delta \mathbf{v}''|_3}$$

must be consistent with the same ratio obtained from the first two measurements. In fact it is, and this internal experimental consistency then allows us to use the values $m_2, m_3, \ldots$ as measures of the inertial masses of the respective objects.

Having set up a consistent measure of inertial mass in this way, we can then rewrite Eq. (9-13) in the form

$$\frac{\mathbf{v}_1 - \mathbf{u}_1}{\mathbf{v}_2 - \mathbf{u}_2} = -\frac{m_2}{m_1} \tag{9-14}$$

which when rearranged gives us

$$m_1 \mathbf{u}_1 + m_2 \mathbf{u}_2 = m_1 \mathbf{v}_1 + m_2 \mathbf{v}_2 \tag{9-15}$$

Thus an equation identical in appearance with the usual momentum-conservation relation emerges, but notice how this contrasts with the Newtonian analysis. We have used the collision processes themselves to define mass ratios through Eq. (9-13). Once this has been done, the terms in Eq. (9-15) *automatically* add up to the same total before and after the collision.

What we have done here, in effect, is to give primacy of place to momentum conservation. The question as to whether or not action equals reaction does not arise. And this can be very valuable. For when one is confronted with non-Newtonian interactions (i.e., those for which action and reaction are *not* equal opposite forces at each instant) one faces the problem of how to incorporate them into physics—whether to abandon the law of momentum conservation in its limited form or to extend the idea of momentum and retain the conservation law. The momentum-conservation principle has proved so extremely powerful that the latter course has been chosen, and conservation of linear momentum is a central feature of relativistic dynamics.

This way of analyzing the basic results of collision processes exposes the very intimate relation that exists between kinematics and dynamics. If we change our description of space, time, and motion, then we can expect that our dynamics must be changed also. This is, in fact, precisely the situation as we make the transition from the kinematics of Galileo and Newton to the kinematics of Einstein's special theory of relativity.

THE FORCE EXERTED BY A STREAM OF PARTICLES

The impact of a stream of particles or fluid on a solid surface provides an instructive application of the laws of collision and the conservation of linear momentum. Suppose that a stream of particles, each having mass m and speed v, strikes a block of mass M [Fig. 9-5(a)] and that the particles all lodge in the block. Let the rate (i.e., number per second) at which particles pass through an imaginary fixed plane at right angles to the stream be denoted by R.

We know that momentum must be conserved, although if the block were extremely massive one might, in watching the process, receive the impression that the momentum brought in by the particles was simply destroyed, because the velocity acquired is unnoticeably small. Suppose, for example, that the block is stationary when the first particle hits it. At the end of a short time Δt the number of particles that has arrived is $R \Delta t$, each carrying momentum mv. If the stream were cut off at this instant, the block and the particles together would continue to move with some constant velocity u (we are assuming that the block is not restrained in any way as far as its horizontal motion is concerned). By conservation of linear momentum and mass we then have

$$Rmv \, \Delta t = (M + \Delta M)u$$

where

$$\Delta M = Rm \, \Delta t$$

Fig. 9-5 (a) Stream of individual particles striking a massive object. (b) Initial phase of building up to a constant average force produced by the particle stream.

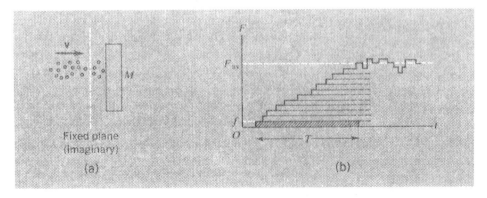

Letting Δt approach zero, we arrive at two equations that describe the acceleration and the rate of increase of mass of the block:

$$\frac{du}{dt} = \frac{Rmv}{M} \tag{9-16}$$

$$\frac{dM}{dt} = Rm = \mu \tag{9-17}$$

where we introduce the single symbol μ to denote the mean rate of transport of mass in the beam. Clearly, if M is sufficiently large, the block will appear to remain at rest; its displacement and its increase of velocity can remain negligible for some time.[1] But in any event, if the block is stationary at some instant, the force F exerted on it at that instant must be equal to $M\,du/dt$. Hence from Eq. (9–16) we have

$$F = M\frac{du}{dt} = Rmv = \mu v \tag{9-18}$$

Thus a stream of particles striking a stationary surface exerts an average force on it given by Eq. (9–18). The word "average" should be emphasized. Our calculation treats the stream as truly continuous, in the sense that we let $\Delta t \rightarrow 0$ in calculating F. But this is a purely mathematical step that does not correspond to the physical reality. The force is produced by the impact of individual discrete particles, and on a sufficiently short time scale we might be able to detect this. Figure 9–5(b) is an attempt to portray the hypothetical results of such observations on the basis of the following very simple model. Suppose that each individual particle, upon striking the surface, undergoes a constant deceleration that brings it to rest in a time T. During this time it must be subjected to a force, f, given by

$$f = \frac{mv}{T}$$

If the force exerted on the block as a result of the arrival of one particle were plotted as a function of time, it would then be a rectangle of height f and width T, as indicated by the small shaded area in Fig. 9–5(b). The force would suddenly come into existence at a certain instant and, a time T later, it would suddenly fall to zero. Suppose now that we consider what happens as a function of time after the beam of particles is first turned on (e.g., by

---

[1] The correctness of Eqs. (9–16) and (9–17) depends, in fact, on this condition. If the block has a speed u, the rate at which the particles strike it is reduced from R to $R(1 - u/v)$, and the values of du/dt and dM/dt are reduced accordingly — see Problem 9–13.

opening a shutter that was previously preventing them from reaching the block). Then as successive particles arrive, each adds its contribution f to the total force, which thus rises in an irregular stepwise fashion. However, at a time T after arrival of the first particle, the latter's contribution to the total force would vanish. Thereafter there would be no net increase in the total force, because on the average the earlier particles drop out of the picture as fast as the new ones appear. The total force thus levels off at some value F but will exhibit statistical fluctuations about that mean value. The force will appear to be almost constant if the effects of the individual particles overlap considerably in time, as shown in the figure. This corresponds to making the average time interval between successive particles very short compared to the average time that it takes for an individual particle to be brought to rest.

We can use the above microscopic picture to recalculate the total force F. It is equal to the force per particle, f, multiplied by the number of particles that are effective at any one instant. This number is equal to the total number of particles that arrive within one deceleration period, T. [Figure 9-5(b) should make this clear.] Since the rate of arrival is R, this number is just RT. Thus we can put

$$F_{av} = (RT)f = RT\frac{mv}{T}$$

i.e.,

$$F = Rmv$$

as before.

It is noteworthy that the average force the beam exerts against the plate (and hence also the average force the plate exerts against the beam of atoms) is quite independent of the actual magnitude of the deceleration that each atom undergoes. The average force is simply equal to the rate of change of momentum of the beam.

This force exerted by a stream of particles has been exploited by W. Paul and G. Wessel to measure the average speed of a beam of silver atoms.[1] A beam of atoms, evaporated from an "oven," was directed downward onto the pan of a very delicate balance. The force exerted on the balance pan was thus made up of two parts: a steady force, as given by Eq. (9-18), and a

[1]W. Paul and G. Wessel, Z. Phys. **124**, 691 (1948).

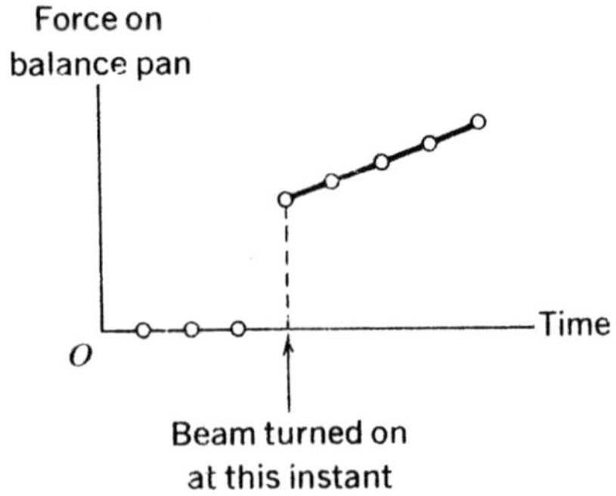

Fig. 9-6 *Time dependence of a total force due to the momentum transfer and the weight of a stream of particles striking an object.*

force increasing linearly with time, due to the increasing mass as given by Eq. (9-17) (see Fig. 9-6):

$$F_{total} = \mu v + \mu g t$$

The forces involved were equivalent to the weight of a small number of micrograms. The experimental values for an oven temperature of 1363°K (= 1090°C) were approximately as follows:

$$\mu v = 3.4 \times 10^{-7} \text{ N}$$
$$\mu = 5.6 \times 10^{-10} \text{ kg/sec}$$

leading to a value for the average speed, v, of silver atoms at 1363°K:

$$v \approx 600 \text{ m/sec}$$

REACTION FROM A FLUID JET

Just as the impact of a stream of particles on a surface produces a push, so the production of such a stream in the first place must cause a force of reaction on the system that gives the stream its momentum. In a normal jet of liquid or gas, the basic granularity of the stream is too fine to be noticed, and any device that sends out such a jet will experience a steady force of reaction, as given by Eq. (9-18). Thus if we imagine a bench test of a rocket engine, for example, with the engine clamped to a rigid structure, then the burnt fuel is thrown out backward with speed v_0, as shown in Fig. 9-7(a), and the forward push, P, exerted on the rocket is given by

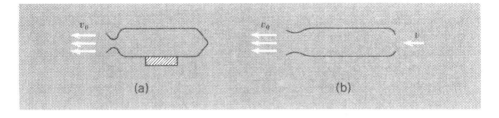

Fig. 9–7 Schematic diagrams of a test-bed arrangement of (a) a rocket engine and (b) a jet engine with air intake at the front.

$$P = \mu v_0$$

One sees this same principle at work on a smaller scale in garden sprinklers, fire hoses, and so on.

A jet engine of an aircraft presents a somewhat more complicated application of these dynamical results. In this case the air that enters at the front of the engine, and leaves as part of the exhaust gases at the rear, plays an important role in the over-all process of momentum transfer. The main function of the fuel that is carried with the plane is to give the ejected gases a high speed with respect to the plane, and most of the moving mass is supplied by the air. It is very convenient to analyze the dynamics of this system from the standpoint of a reference frame in which the engine is instantaneously at rest. If the plane is traveling forward at a velocity v, the air is seen as entering the engine at the front with an equal and opposite velocity, as shown in Fig. 9–7(b). Then, at the rear, all the ejected material has a backward velocity v_0 in this frame. Thus, if air is being carried through the engine at the rate μ_{air} (kg/sec) and fuel is being burnt at the rate μ_{fuel}, the total rate of change of momentum defines a total forward force on the engine according to the following equation:

$$P = \mu_{fuel} v_0 + \mu_{air}(v_0 - v) \tag{9-19}$$

Example. A jet aircraft is traveling at a speed of 250 m/sec. Each of its engines takes in 100 m^3 of air per second, corresponding to a mass of 50 kg of air per second at the plane's flying altitude. The air is used to burn 3 kg of fuel per second, and all the gases coming from the combustion chamber are ejected with a speed of 500 m/sec relative to the aircraft. What is the thrust of each engine? Substituting directly in Eq. (9–19) we have

$$P = 3 \times 500 + 50(500 - 250)$$
$$= 1.4 \times 10^4 \text{ N}$$

Four engines of this type would thus give a total driving force of about 12,000 lb, a more or less realistic figure.

The most spectacular manifestation of these reaction forces, at least in man-made systems, is of course the initial thrust from the rocket engines in a launching at Cape Kennedy (Fig. 9-8).

Fig. 9-8 Launching of a Saturn V rocket. (N.A.S.A. photograph.)

The following approximate figures are based on published data[1] on the first stage of the Saturn V system:

Rate of burning	15 tons/sec	$= 1.4 \times 10^4$ kg/sec
Total thrust	7.6×10^6 lb	$= 3.4 \times 10^7$ N
Total mass at liftoff	3100 tons	$= 2.8 \times 10^6$ kg
Burn time	2.5 min	$= 150$ sec

One can infer from these figures that the speed of the ejected gases is about 2500 m/sec and that the initial acceleration is about 2 m/sec^2. (Remember that in a vertical liftoff one must first overcome the downward gravitational pull on the rocket!) By the end of the first-stage burn, about 2250 tons of fuel have been consumed, and the total mass is down to only about a quarter of the initial value. The analysis of the accelerative process itself, as mass is continually lost, is the subject of the next section.

ROCKET PROPULSION

This has become such a very large subject in recent years that it is clear that we shall do no more than touch upon the underlying dynamical principles. For anything like a substantial discussion you should look elsewhere.[2] The fact remains, however, that the operation of any rocket does depend on the basic laws of conservation of momentum, as applied to a system made up of the rocket and its ejected fuel. For simplicity, let us consider the motion of a rocket out in a region of space where the effects of gravity are sufficiently small to be ignored in the first approximation. Under this assumption, the only force acting on the body of the rocket is the thrust from the ejected fuel. Suppose that the burnt fuel has a speed v_0 relative to the rocket. Between time t and time $t + \Delta t$, a mass Δm of fuel is burnt and becomes separated from the rocket. The situations before and after ejection are shown in Fig. 9-9, where m is the total mass of the rocket plus its remaining fuel at $t + \Delta t$, and v is its velocity at time t.

[1]NASA Facts: Space Launch Vehicles.
[2]See, for example, M. Barrère et al., *Rocket Propulsion*, Elsevier, Amsterdam, 1960; S. L. Bragg, *Rocket Engines*, Geo. Newnes, London, 1962; G. P. Sutton, *Rocket Propulsion Elements*, Wiley, New York, 1963.

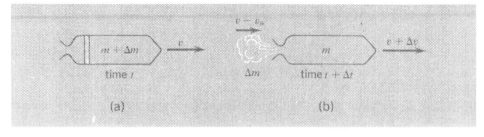

Fig. 9-9 *Situations just before and just after the ejection of an element Δm of mass by a rocket.*

The ejection of the fuel is a kind of inelastic collision in reverse, since initially the masses m and Δm have the same velocity. By conservation of linear momentum we have

$$(m + \Delta m)v = m(v + \Delta v) + \Delta m(v - v_0)$$

Therefore,

$$mv + v\,\Delta m = mv + m\,\Delta v + v\,\Delta m - v_0\,\Delta m$$

Therefore,

$$m\,\Delta v = v_0\,\Delta m$$

or

$$\Delta v = v_0 \frac{\Delta m}{m} \qquad (9\text{--}20)$$

This equation is not quite exact. (Why?) But as we let Δt approach zero, the error approaches zero. As long as $\Delta v/v_0$ is much less than unity, Eq. (9-20) is an excellent approximation. Given the initial total mass m_i of the rocket plus fuel, and the final mass m_f of the rocket at burnout, the velocity gained by the rocket can be evaluated. If the initial and final velocities of the rocket are v_i and v_f, Eq. (9-20) tells us that

$$v_f - v_i = v_0 \sum \frac{1}{m}\,\Delta m$$

The answer could be obtained numerically, by drawing a graph (Fig. 9-10) of $1/m$ against m, and finding the area (shaded) between the limits m_f and m_i. This is a pure number, which when multiplied by v_0 gives the increase of velocity. Analytically, if m is used to denote the mass of the rocket (plus its remaining fuel) at any instant, it is more satisfactory to interpret dm as the

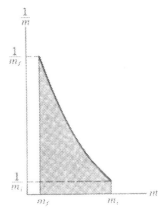

Fig. 9–10 *Graph of $1/m$ versus m. The shaded area gives a relative measure of the gain of velocity resulting from a given mass change.*

change of mass *of the rocket* in a time dt. Defined in this way, dm is actually a negative quantity, which when integrated from the beginning to the end of the burning process gives the value of $m_f - m_i$ (< 0). In these terms the change of velocity can be written as the following integral in closed form:

$$v_f - v_i = -v_0 \int_{m_i}^{m_f} \frac{dm}{m} = v_0 \ln\left(\frac{m_i}{m_f}\right) \tag{9-21}$$

You should satisfy yourself that this is indeed equivalent to the final result of the numerical-graphical method described above.

Notice that the time does not enter into the calculation at all, although of course it would do so if we wanted to consider the *rate* of increase of v or the magnitude of the thrust.

Notice also that we would be entitled, at each and every instant, to look at the situation from the frame of reference of the rocket itself. In this frame the velocity of the ejected fuel is always just $-v_0$, and Eq. (9–20) follows immediately.

It is worth examining some of the implications of Eq. (9–21). The first thing to notice is that the gain of velocity is directly proportional to the speed of the ejected gases. Thus it pays to make v_0 as great as possible. The highest values of v_0 attainable through chemical burning processes are of the order of 5000 m/sec, and in practice, because of incomplete burning and other losses, it is hard to do better than about 50% of the ideal theoretical value for a given fuel (cf. the figure of 2500 m/sec for the LOX–kerosene mixture of Saturn V, first stage). These velocities are, of course, very high in ordinary terms, but they are small compared to the velocities that can be given to charged particles by electrical acceleration. Hence the interest in developing ion-gun engines, or even using the highest available speed (that of light) by making an exhaust jet of radiation. The trouble with both of these, however, is the very small rate of ejection of mass, which makes the attainable thrust very small.

The other main feature of Eq. (9-21) is the way in which the increase of velocity varies logarithmically with the mass ratio. This places rapidly increasing demands on the amount of fuel needed to confer larger and larger final velocities on a given payload. Suppose, for example, that we wanted to attain a velocity equal to the exhaust velocity v_0, starting from rest. Then, by Eq. (9-21) we have

$$v_f - v_i = v_0 = v_0 \ln\left(\frac{m_i}{m_f}\right)$$

Therefore,

$$\frac{m_i}{m_f} = e = 2.718\ldots$$

But to attain twice this velocity, we need to have

$$\ln\left(\frac{m_i}{m_f}\right) = 2$$

i.e.,

$$\frac{m_i}{m_f} = e^2 \approx 7.4$$

Table 9-1 presents the results of such calculations in more convenient form. The last column represents the extra mass needed,

TABLE 9-1

$v_f - v_i$	m_i/m_f	$(m_i - m_f)/m_f$
v_0	2.7	1.7
$2v_0$	7.4	6.4
$3v_0$	20.1	19.1
$4v_0$	54.5	53.5

as a multiple of the payload. The practical problems of producing very large mass ratios are prohibitive, but the use of multistage rockets (which also have other advantages) avoids this difficulty (see Problem 9-12, p. 359).

A result that may seem surprising at first sight is that there is nothing, in principle, to prevent us from giving a rocket a forward velocity that is considerably greater than the speed v_0 of the exhaust gases. Thus at a late stage in the motion one would see both the rocket and the ejected material moving forward with respect to the frame in which the rocket started out from rest. No violation of dynamical principles would be involved, and if one made a detailed accounting of the motion of all the material that was in the rocket initially, one would find that the total momentum of the system had remained at zero (as long as the effects of external forces, including gravity, could be ignored). It should of course be emphasized that our whole analysis would hold good, as it stands, only in the absence of gravity and of resistive forces due to the air.

COLLISIONS AND FRAMES OF REFERENCE

The seventeenth-century investigations of collision processes, by which momentum conservation was established, were chiefly experimental. But one of the men involved—Christian Huygens—applied the spirit of twentieth-century physics in a brilliant analysis of the particular case of two objects with perfect rebound. He based his argument on symmetry and on the equivalence of frames related by a constant velocity. The analysis is simple but of great intrinsic interest.

Figure 9-11[1] provides a suitably seventeenth-century background to the situation. But though the picture may appear quaint and archaic, the thinking is sharp and fresh. Huygens imagines two equal elastic masses, colliding with equal and opposite speeds $\pm v$. He assumes that they rebound so that each velocity is exactly reversed; this is a symmetry argument. Next, Huygens imagines a precisely similar collision to take place on a boat that is itself moving with speed v relative to the shore. This collision, viewed by a man on terra firma, appears as a stationary mass being struck by a mass with velocity $2v$. After the impact,

[1] From Ernst Mach, *The Science of Mechanics*, trans. T. J. McCormack, Open Court Publishing Co., La Salle, Ill., 1960. The original figure comes from Huygens' treatise on impact (ca. 1700). Mach's book (which is extremely readable and uses a minimum of mathematics) is itself a landmark in the discussion of the fundamental principles of physics. His speculations on the origin of inertia were a significant part of the background to Einstein's thinking about general relativity and cosmology (see Chapter 12).

Fig. 9-11 Huygens' visualization of an elastic collision between two equal masses, as shown in his book, De Motu Corporum ex Percussione *(1703). (Reprinted in Vol. 16 of C. Huygens,* Oeuvres Complètes, *Martinus Nijhoff, The Hague, Netherlands, 1940.)*

the first mass has acquired the velocity $2v$, and the second mass is stationary.

More generally, if the boat has a speed u, different from v, the velocities exchange as follows:

	Body 1	Body 2
Before impact:	$u + v$	$u - v$
After impact:	$u - v$	$u + v$

Thus Huygens predicts, on theoretical grounds, the results of all possible one-dimensional experiments on the perfectly elastic collision of two identical masses. The feature common to all of them is that the magnitude of the *relative* velocity has the same value after the collision as it had before the collision.

Huygens went even further. Again using a kind of symmetry argument, he deduced a general property of perfectly elastic collisions between *unequal* masses. If a moving object A strikes an unequal stationary object B, then after the collision they will be moving with velocities v and w as indicated in Fig. 9-12(a). Imagine that this is viewed from a boat moving to the right with velocity $w/2$. Then before collision the object B is seen moving to the left, with velocity $-w/2$, as shown in Fig. 9-12(b). Huygens argues that the exact reversal of the motion of B, as seen from this second frame, must also imply the exact reversal of the motion of A in this perfectly elastic process. Thus the final velocity of A must be $-(u - w/2)$ in this frame. But this velocity is also equal to $v - w/2$, because the final velocity of A as seen from

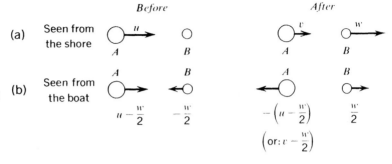

Fig. 9–12 *Elastic collision between two unequal objects as seen (a) from the laboratory frame, and (b) from a frame in which both velocities are simply reversed.*

the shore is v. Hence we have

$$-(u - w/2) = v - w/2$$
$$w - v = u \tag{9-22}$$

Thus Huygens concludes that in the elastic collision of any two objects whatsoever, the magnitude of the *relative* velocity remains unaltered. Notice, however, that something beyond the conservation of linear momentum is involved here. The extra something is what we have learned to call kinetic energy, and its conservation defines a very special class of collisions. Let us consider this further.

KINETIC ENERGY IN COLLISIONS

Suppose that a one-dimensional collision takes place as shown in Fig. 9–13(a). If this system is effectively free of external influences we have conservation of linear momentum:

$$m_1 u_1 + m_2 u_2 = m_1 v_1 + m_2 v_2$$

Suppose further that the collision is perfectly elastic, in the sense defined by Eq. (9-22). In the present problem this implies the following relationship:

$$u_1 - u_2 = v_2 - v_1$$

We can solve these two equations for the final velocities, v_1 and v_2. The results are

$$v_1 = \frac{m_1 - m_2}{m_1 + m_2} u_1 + \frac{2m_2}{m_1 + m_2} u_2$$

$$v_2 = \frac{2m_1}{m_1 + m_2} u_1 - \frac{m_1 - m_2}{m_1 + m_2} u_2$$

Using these values of v_1 and v_2, it is a straightforward piece of algebra to arrive at the following result:

$$m_1v_1^2 + m_2v_2^2 = m_1u_1^2 + m_2u_2^2 \tag{9-23}$$

You will recognize that this equation, apart from the absence of factors of $\frac{1}{2}$ throughout, corresponds to a statement that the total kinetic energy after the collision is equal to the kinetic energy initially. We have arrived at this result on the basis of symmetry considerations and without any explicit mention of forces or work. We shall bring these latter concepts into the picture in Chapter 10, but for the moment we do not need them. Some of the early workers in mechanics recognized the conservation property represented by Eq. (9-23) for perfectly elastic collisions, and referred to the quantity mv^2 as the "living force" (Latin, *vis viva*) of a moving object.

Having, in effect, introduced kinetic energy into the description of collision processes, we shall now develop an important result that holds good whether or not the total kinetic energy is conserved. Suppose that Fig. 9-13(a) represents an arbitrary one-dimensional collision as observed in a reference frame S. Let us consider this same collision from the standpoint of another frame S' that has the velocity v with respect to S. (In other words, S and S' are related by the Galilean transformations.) We then have

$$u_1' = u_1 - v \qquad u_2' = u_2 - v$$
$$v_1' = v_1 - v \qquad v_2' = v_2 - v$$

We shall now show that the *change* of kinetic energy between initial and final states is an invariant—i.e., it has the same value in both frames.

In S:

$$K_{\text{initial}} = \tfrac{1}{2}m_1u_1^2 + \tfrac{1}{2}m_2u_2^2$$
$$K_{\text{final}} = \tfrac{1}{2}m_1v_1^2 + \tfrac{1}{2}m_2v_2^2$$
$$\Delta K = K_f - K_i = (\tfrac{1}{2}m_1v_1^2 + \tfrac{1}{2}m_2v_2^2)$$
$$\qquad - (\tfrac{1}{2}m_1u_1^2 + \tfrac{1}{2}m_2u_2^2)$$

In S':

$$K'_{\text{initial}} = \tfrac{1}{2}m_1(u_1 - v)^2 + \tfrac{1}{2}m_2(u_2 - v)^2$$
$$K'_{\text{final}} = \tfrac{1}{2}m_1(v_1 - v)^2 + \tfrac{1}{2}m_2(v_2 - v)^2$$
$$\Delta K' = (\tfrac{1}{2}m_1v_1^2 + \tfrac{1}{2}m_2v_2^2) - (\tfrac{1}{2}m_1u_1^2 + \tfrac{1}{2}m_2u_2^2)$$
$$\qquad - v[(m_1v_1 + m_2v_2) - (m_1u_1 + m_2u_2)]$$

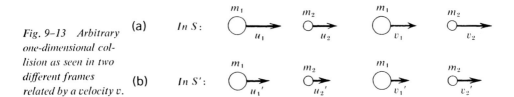

Fig. 9–13 Arbitrary one-dimensional collision as seen in two different frames related by a velocity v.

By momentum conservation, for *any* kind of collision, the combination of terms in square brackets in $\Delta K'$ is zero. It follows that

$$\Delta K' = \Delta K$$

Thus, if the change of kinetic energy for a given collision process is specified in one frame, it has the same value in all frames. Given a knowledge of this quantity, and the fact of momentum conservation, we can predict the values of v_1 and v_2 in any one-dimensional collision process for which the initial velocities u_1 and u_2 are specified. The value of ΔK may be positive, negative, or zero. The first of these corresponds to an explosive process, in which extra kinetic energy is given to the separating particles as a result of the interaction. The second and third possibilities correspond to inelastic and elastic collisions, respectively, in the sense in which we have already been using these terms.

THE ZERO-MOMENTUM FRAME

The momentum of a particle or of a system of particles is not an invariant; it depends on the frame of reference in which one observes the motion. If, however, one compares the descriptions of the motion in two different inertial frames, related by a constant velocity, the difference between the measured values of the momentum of a particle is always a constant vector of magnitude $m\mathbf{v}$, where m is the mass of the particle and $\mathbf{v}$ is the velocity of one frame relative to the other. One can always find a reference frame in which the total momentum of any particle vanishes; it is evidently a frame in which the particle is at rest at the instant of time one determines its momentum. One can likewise find a reference frame in which the total momentum of any system of particles is zero, and this zero-momentum frame of reference is of great importance, not only as a convenience for looking at collisions and interactions in general but also, as we shall show

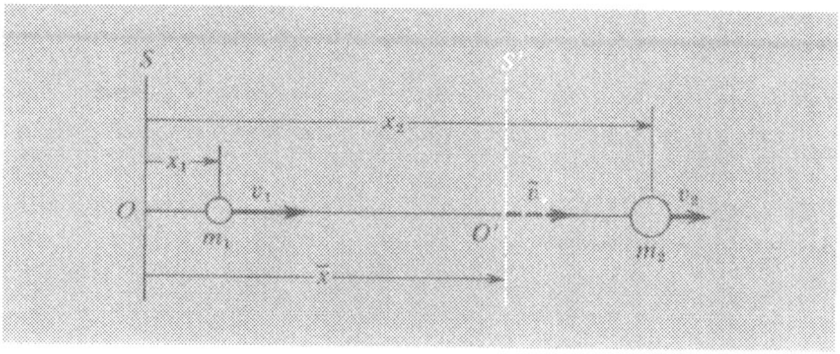

Fig. 9-14 Basis of defining the zero-momentum (center-of-mass) reference frame.

shortly, for its *dynamical* implications. To see how we identify this zero-momentum frame, let us start with a simple example; two particles m_1 and m_2 move along the x axis with constant velocities v_1 and v_2 in a frame of reference S (Fig. 9-14). Our task is to find the velocity $\bar{v}$ of a reference frame S' relative to S such that in S' the total momentum $m_1 v'_1 + m_2 v'_2$ is equal to zero.

Let O' be the origin of S' moving with velocity $\bar{v}$ relative to O. From Fig. 9-14 we see that the velocities of m_1 and m_2 relative to O are given by

$$v'_1 = v_1 - \bar{v}$$
$$v'_2 = v_2 - \bar{v}$$

Hence the momentum of the two particles in S' is

$$m_1 v'_1 + m_2 v'_2 = m_1 v_1 - m_1 \bar{v} + m_2 v_2 - m_2 \bar{v}$$

and if this is to be zero, we must have

$$(m_1 + m_2)\bar{v} = m_1 v_1 + m_2 v_2 \qquad (9\text{-}24)$$

Equation (9-24) fixes the velocity of the reference frame S' relative to S but leaves the choice of the position ($\bar{x}$) of O' arbitrary. We use this freedom to choose the location of O' relative to the positions of m_1 and m_2 as simply as possible. If we rewrite Eq. (9-24) in the equivalent form

$$(m_1 + m_2)\frac{d\bar{x}}{dt} = m_1 \frac{dx_1}{dt} + m_2 \frac{dx_2}{dt}$$

or better as

$$\frac{d}{dt}[(m_1 + m_2)\bar{x}] = \frac{d}{dt}[m_1 x_1 + m_2 x_2]$$

evidently the simplest way of satisfying this is to equate the two

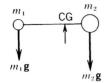

Fig. 9-15 *Basis of defining the center of gravity, not necessarily identical with the center of mass.*

sets of square brackets. The difference between them must be a constant and we choose the constant equal to zero. We then have for the position of O' in S,

$$\bar{x} = \frac{m_1 x_1 + m_2 x_2}{m_1 + m_2} \qquad (9\text{-}25)$$

The origin of a zero-momentum reference frame chosen in this way is called the *center of mass* of the two-particle system made up of m_1 and m_2. You will recognize that Eq. (9-25) also defines what we are accustomed to calling the center of gravity of two objects (cf. Fig. 9-15). In principle, however, these need not be identical points. The center of gravity is literally the point through which a single force, equal to the sum of the gravitational forces on the separate objects, effectively acts. If the values of g at the positions of the two objects were not quite identical, then the forces F_{g1} and F_{g2} would not be strictly proportional to m_1 and m_2, in which case the center of gravity and the center of mass would not quite coincide. The difference is negligible for ordinary purposes, but it should be recognized that the center of mass, as given by Eq. (9-25), is defined without reference to the uniformity or even the existence of gravitational forces. Perhaps you recognized that we have already used the zero-momentum frame in our earlier section on collisions and frames of reference. When Huygens wanted to apply symmetry arguments to collisions, this was the frame he started with. But now we have identified it in a much more explicit fashion.

The introduction of the zero-momentum or center-of-mass (CM) frame leads to a very important and immensely useful way of analyzing the dynamics of a system of particles. To develop the essential idea in the simplest possible way we shall again consider a one-dimensional system of two particles. In frame S (let us suppose that this is the frame defined by our laboratory) the particles have velocities v_1 and v_2. In the zero-momentum frame (S') they have velocities v_1' and v_2' given by

$$v_1' = v_1 - \bar{v}$$
$$v_2' = v_2 - \bar{v}$$

where $\bar{v}$ is defined by Eq. (9-24). Putting $m_1 + m_2 = M$, we can put

$$\bar{v} = \frac{m_1v_1 + m_2v_2}{M} = \frac{P}{M} \tag{9-26}$$

where P is the total linear momentum, which remains unchanged in any collision process:

$$P = m_1v_1 + m_2v_2 = \text{const.} \tag{9-27}$$

We now proceed to express the total kinetic energy K, as measured in the laboratory frame, in terms of the center-of-mass velocity $\bar{v}$ and the velocities v_1' and v_2' of m_1 and m_2 relative to the CM. We have

$$\begin{aligned} K &= \tfrac{1}{2}m_1(v_1' + \bar{v})^2 + \tfrac{1}{2}m_2(v_2' + \bar{v})^2 \\ &= (\tfrac{1}{2}m_1v_1'^2 + \tfrac{1}{2}m_2v_2'^2) + (m_1v_1' + m_2v_2')\bar{v} \\ &\quad + \tfrac{1}{2}(m_1 + m_2)\bar{v}^2 \end{aligned} \tag{9-28}$$

Now, by the very definition of the zero-momentum frame, we have

$$m_1v_1' + m_2v_2' = 0$$

Hence the middle term on the right-hand side of Eq. (9-28) is automatically zero. The first term we recognize as the total kinetic energy K' of the two individual particles *relative to the CM*; the last term is equal to the kinetic energy of a single particle of mass M moving with the velocity $\bar{v}$ of the CM. Thus we can write

$$K = K' + \tfrac{1}{2}M\bar{v}^2 \tag{9-29}$$

This is such an important result that we shall express it in words also:

> **The kinetic energy of a system of two particles is equal to the kinetic energy of motion relative to the center of mass of the system (the internal kinetic energy), plus the kinetic energy of a single particle of mass equal to the total mass of the system moving with the center of mass.**

The great importance of this separation of the kinetic energy into two parts—and note well that it works *only* if the new reference frame is the zero-momentum frame—is that it opens the way to a very powerful and simplifying procedure. *We have the possibility of analyzing the internal motion of a system (relative to the CM) without reference to the bodily motion of the system as a whole*. (We shall show shortly that the result holds good not only for one-dimensional motions, but in general.)

One of the implications of Eq. (9–29) is that a certain amount of kinetic energy is locked up, as it were, in the motion of the center of mass. In the absence of external forces the velocity $\bar{v}$ remains constant throughout the course of a collision process, and the kinetic energy $\frac{1}{2}M\bar{v}^2$ must likewise remain unchanged. This means that in the collision of two objects, only a certain fraction of their total kinetic energy, as measured in the laboratory, is available for conversion to other purposes. The amount of this available kinetic energy is, in fact, identical with K'. We can calculate it with the help of Eq. (9–29):

$$K' = K - \tfrac{1}{2}M\bar{v}^2$$
$$= (\tfrac{1}{2}m_1 v_1^2 + \tfrac{1}{2}m_2 v_2^2) - \tfrac{1}{2}(m_1 + m_2)\bar{v}^2$$

Substituting for $\bar{v}$ from Eq. (9–20), this leads by simple algebra to the following result:

$$K' = \frac{1}{2}\frac{m_1 m_2}{m_1 + m_2}(v_2 - v_1)^2 \tag{9-30a}$$

The value of K', as expressed by this equation, is thus the kinetic energy of what can be regarded as being effectively a single mass, of magnitude $m_1 m_2/(m_1 + m_2)$, moving at a velocity equal to the *relative* velocity of the colliding particles. The effective mass is called the *reduced mass* of the system and is given the symbol μ. Thus we may write Eq. (9–30a) in the more compact form

$$K' = \tfrac{1}{2}\mu v_{\text{rel}}^2 \tag{9-30b}$$

where

$$\mu = \frac{m_1 m_2}{m_1 + m_2}$$
$$v_{\text{rel}} = v_2 - v_1 = v_2' - v_1'$$

For example, if a moving object of mass 2 units strikes a stationary object of mass 1 unit, two thirds of the initial kinetic energy is locked up in center-of-mass motion, and only one third is available for the purpose of producing deformations, and so on, when the objects collide.

COLLISION PROCESSES IN TWO DIMENSIONS

In general the velocity vectors of two colliding objects define a plane, and it is therefore important to extend our analysis of collision processes into two-dimensional space. Actually one

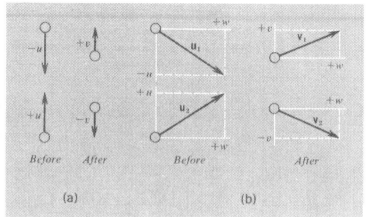

Fig. 9-16 (a) Collision seen as occurring in one dimension. (b) As seen in another reference frame, the collision appears two-dimensional.

must stand ready to go all the way into three dimensions, because the plane defined by the velocity vectors of two particles after a collision may not be the same as that defined by the initial velocities. Many collision processes are, however, purely two-dimensional, and for simplicity we shall limit ourselves to such cases in discussing specific examples, even though the theory applies equally well to three-dimensional problems.

One thing that is worth recognizing at the outset is that the analysis of purely one-dimensional problems, of the type we have discussed so far, can in fact lead to predictions about certain types of two-dimensional collisions, because we can turn a one-dimensional collision into a two-dimensional one by simply changing our point of view. Consider, for example, an imperfectly elastic collision between two identical spheres that approach one another along the y axis with equal and opposite velocities, $\pm u$ [see Fig. 9-16(a)]. Suppose that as a result of this collision they recoil back along the y axis with reduced velocities $\pm v$ (symmetry requires that these final velocities be equal and opposite). Figure 9-16(b) then shows how the same collision would appear to an observer who had a velocity $-w$ parallel to the x axis. This defines a whole class of oblique collision problems, all of which can be solved with reference to the simple head-on collision of Fig. 9-16(a). One can go even further yet, by viewing the collision of Fig. 9-16(b) from a reference frame that has some velocity parallel to the y direction. This removes all the apparent symmetry, yet it is basically the same collision. *All the collisions for which, at the moment of impact, the two spheres have a relative velocity of given magnitude along the line joining their centers are dynamically equivalent*

This suggests a very important way of simplifying the analysis of collision problems. Imagine yourself in the frame in which the collision is most simply described; this will be the center-of-mass frame, in which by definition the colliding particles approach one another along a straight line with equal and opposite momenta. Solve the problem in this CM frame, and transfer the result to the laboratory frame by means of a Galilean transformation.

The great simplicity of a collision process as viewed in the zero-momentum frame is illustrated in Fig. 9–17. Since the total momentum in this frame is zero, the particles separate, as well as approach, with equal and opposite momenta, as shown in Fig. 9–17(a). Also the magnitudes of the final momentum vectors are independent of their direction in the CM frame. Thus, as shown in Fig. 9–17(b), the vectors $\mathbf{p}'_{1f}$ and $\mathbf{p}'_{2f}$ can be represented with their tips lying at opposite ends of the diameter of a circle.

The relative directions of $\mathbf{p}'_{1i}$ and $\mathbf{p}'_{1f}$ can be anything, depending on the details of the interaction. The relative lengths of these vectors also depend, of course, on the detailed mechanism. In a perfectly elastic collision we have $p'_f = p'_i$. In a completely inelastic collision we have $p'_f = 0$. (In an explosive process, for example rocket propulsion, we have the converse situation, where $p'_i = 0$, $p'_f > 0$.) And in atomic, chemical or nuclear reaction processes we may find p'_f less than, equal to, or greater than p'_i. To begin with, let us look at one or two examples of perfectly elastic collisions so as to appreciate the beauties of a view from the center-of-mass frame. We are defining an *elastic* collision

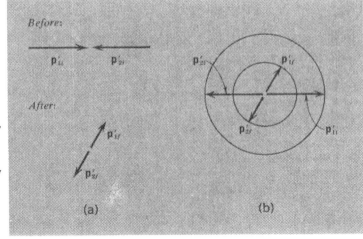

Fig. 9–17 (a) An arbitrary collision as described by equal and opposite momentum vectors in the CM frame. (b) In the CM frame, the end points of the momentum vectors lie on circles.

here as one in which the magnitude of the relative velocity is unchanged by the collision. As we have seen, this is equivalent to saying that the total kinetic energy, as well as the momentum, is conserved. The definition in terms of kinetic energy conservation will later become the dominant one.

ELASTIC NUCLEAR COLLISIONS

The nuclear physicist lives in two worlds. One is his laboratory, the other is the center-of-mass frame of the particles whose collisions and interactions he is studying. By learning to skip nimbly from one frame to the other he gets the best of both worlds. Let us see how.

Proton–proton collisions

Figure 9–18 shows a collision between two protons, as recorded in a photographic emulsion. One of the protons belonged to a hydrogen atom in the emulsion and was effectively stationary before the collision took place; the other entered the emulsion with a kinetic energy of about 5 MeV.

The most notable feature of the collision is that the paths of the two protons after collision make an angle of 90° with each

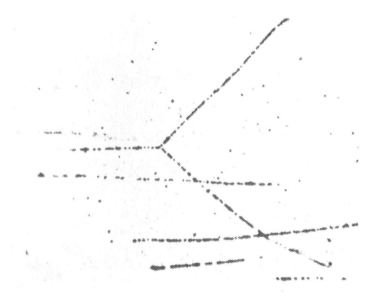

Fig. 9–18 Elastic collision between an incident proton and an initially stationary proton in a photographic emulsion. (From C. F. Powell and G.P.S. Occhialini, Nuclear Physics in Photographs, *Oxford University Press, New York, 1947.)*

other. This is true for all such proton-proton collisions, until we get up to energies so high that Newtonian mechanics is no longer adequate to describe the situation. By first looking into the center-of-mass frame we can readily understand this. Let the velocity of the incident proton as observed in the laboratory be v. Then the zero-momentum frame has velocity $v/2$, and in this frame the protons approach and recede with equal and opposite velocities as shown in Fig. 9-19(a). Suppose one proton emerges from the collision in the direction θ', so that the other is at $\pi - \theta'$ on the other side of the line of approach. To get back to the laboratory frame we add the velocity $v/2$ to each proton, parallel to the original line of motion, as shown in Fig. 9-19(b). But the triangles ABC and AEF are both isosceles, so the directions θ_1 and θ_2 of the protons as observed in the laboratory are given by

$$\theta_1 = \theta'/2$$
$$\theta_2 = (\pi - \theta')/2$$

Therefore,

$$\theta_1 + \theta_2 = \pi/2$$

Moreover, we can easily find the laboratory velocities of the two protons after the collision, for we have

$$v_1 = 2(v/2) \cos \theta_1 = v \cos \theta_1$$
$$v_2 = 2(v/2) \cos \theta_2 = v \sin \theta_1$$

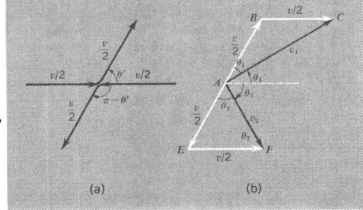

Fig. 9-19 (a) Elastic collision between two equal masses, as seen in the CM frame. (b) Transformation to the laboratory frame, showing a 90° angle between the final velocities.

We see that in any such collision the total kinetic energy is conserved, because

Initial KE $= \frac{1}{2}mv^2$
Final KE $= \frac{1}{2}m(v\cos\theta_1)^2 + \frac{1}{2}m(v\cos\theta_2)^2$
$= \frac{1}{2}mv^2(\cos^2\theta_1 + \sin^2\theta_1)$
$= \frac{1}{2}mv^2$

Figure 9-20 shows a similar collision between equal macroscopic objects (billiard balls); it is not perfectly elastic, but it is impressively close.

Neutron–nucleus collisions

In nuclear fission processes, neutrons are ejected with a variety of energies, but the average energy is of the order of 1 MeV.

Fig. 9-20 Stroboscopic photograph of an almost perfectly elastic collision between equal masses. (From PSSC Physics, D. C. Heath, Lexington, Massachusetts, 1965.)

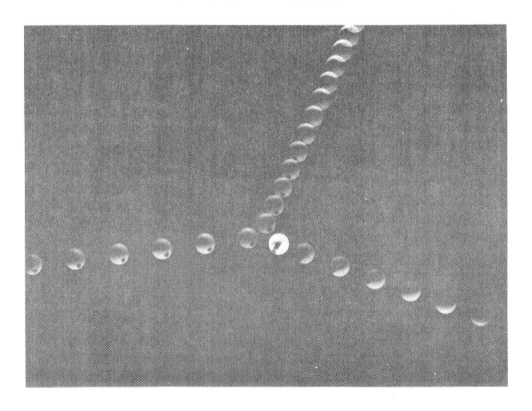

These neutrons are however, most effective in causing further fissions if they are reduced to energies of the order of 10^{-2} or 10^{-1} eV (thermal energies). Thus an essential feature of every slow-neutron reactor is a means of slowing down the neutrons. And elastic collisions of neutrons with other nuclei (those composing the moderator material of the reactor) do most of the job.

Suppose that a neutron of mass m makes an elastic collision with a nucleus of mass M. Let the initial velocity of the neutron in the laboratory be $\mathbf{v}_0$; in this frame the struck nucleus will be assumed stationary. Figure 9–21(a) shows the collision as seen in the center-of-mass frame, which has a velocity $\bar{\mathbf{v}}$ relative to the laboratory frame given by Eq. (9–24):

$$\bar{\mathbf{v}} = \frac{m}{M + m} \mathbf{v}_0 \qquad (9\text{–}31)$$

If the collision turns the neutron through an angle θ' in the zero-momentum frame, its final velocity in the laboratory frame is the vector $\mathbf{v}$ shown in Fig. 9–21(b). The magnitudes of the final velocities, as measured in the laboratory, for any given value of θ', are readily calculated. (*Exercise:* Do this calculation for yourself, and verify that in this case, as with two equal masses, the total kinetic energy, as measured in the laboratory, remains unchanged as a result of the collision.)

The biggest energy loss for a neutron as seen in the laboratory frame occurs if it is scattered straight backward ($\theta' = \pi$). In this case we have

$$v(\pi) = \bar{v} - (v_0 - \bar{v}) = -(v_0 - 2\bar{v})$$

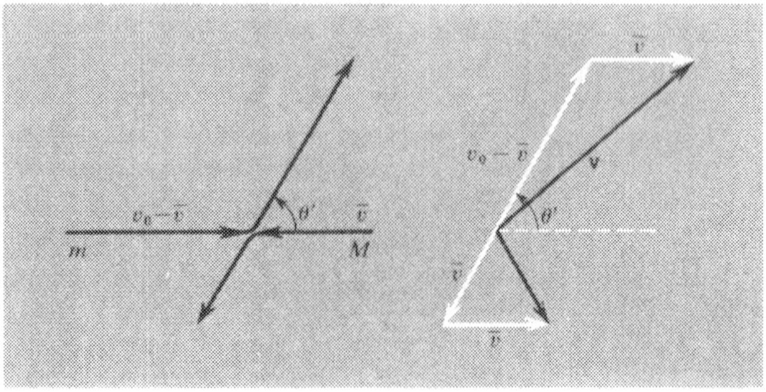

Fig. 9–21 (a) Elastic collision between two unequal masses, as seen in the CM frame. (b) Transformation to the laboratory frame; the angle between the velocities is different from 90°.

$$v(\pi) = -v_0\left(1 - \frac{2m}{M+m}\right)$$
$$= -\frac{M-m}{M+m}v_0$$

For $\theta' = 0$ the neutron loses no energy at all. (What sort of a collision is this?) Thus the kinetic energy of the neutron after the collision lies between the following limits:

$$K_{max} = \tfrac{1}{2}mv_0^2$$
$$K_{min} = \tfrac{1}{2}m\left(\frac{M-m}{M+m}\right)^2 v_0^2 \tag{9-32}$$

Since K_{max} is independent of the mass M of a moderator nucleus, it is the expression for K_{min} that tells us what value of M is most likely to lead to the greatest reduction of the average neutron energy. And we see that $M = m$ makes K_{min} equal to zero; we cannot do as well as this for any other value of M, whether it be bigger or smaller than m. Thus if no other considerations were involved, ordinary hydrogen would make the best moderator, since in this case M (the proton mass) is equal to m within about 1 part in 10^3. Protons, however, also capture slow neutrons rather effectively, thereby making them unavailable for causing further fissions, and it turns out that certain other light nuclei (e.g., deuterium, beryllium, and carbon) offer a better compromise between moderating and trapping of the fission neutrons.

INELASTIC AND EXPLOSIVE PROCESSES

We shall turn now to processes in which there may be a net loss or gain of kinetic energy as a result of the collision. In analyzing such processes, it is important to confirm that the change of total kinetic energy is indeed an invariant quantity, and that the result we obtained on the basis of a one-dimensional collision does hold good in general. In order to show this, we first redevelop Eq. (9–29) for two particles moving in arbitrary directions. Let the particles have masses m_1 and m_2 and velocities $\mathbf{v}_1$ and $\mathbf{v}_2$ in the laboratory frame (S). Then the velocity, $\bar{\mathbf{v}}$, of the CM frame (S') is defined by the equation

$$\bar{\mathbf{v}} = \frac{m_1\mathbf{v}_1 + m_2\mathbf{v}_2}{m_1 + m_2} \tag{9-33}$$

If the velocities of the particles as measured in S' are $\mathbf{v}_1'$ and $\mathbf{v}_2'$, we then have

$$v_1 = v_1' + \bar{v}$$
$$v_2 = v_2' + \bar{v}$$

We now write down the total kinetic energy in S, using the fact that the kinetic energy $\frac{1}{2}mv^2$ of a particle can also be expressed as $\frac{1}{2}m(v \cdot v)$, i.e., in terms of the scalar product of the vector v with itself. Thus we have

$$K = \tfrac{1}{2}m_1(v_1 \cdot v_1) + \tfrac{1}{2}m_2(v_2 \cdot v_2)$$
$$= \tfrac{1}{2}m_1(v_1' + \bar{v}) \cdot (v_1' + \bar{v}) + \tfrac{1}{2}m_2(v_2' + \bar{v}) \cdot (v_2' + \bar{v})$$

Now consider one of these scalar products, using the distributive and commutative laws that apply to the dot products of vectors:

$$(v_1' + \bar{v}) \cdot (v_1' + \bar{v}) = v_1' \cdot v_1' + 2v_1' \cdot \bar{v} + \bar{v} \cdot \bar{v}$$
$$= v_1'^2 + 2v_1' \cdot \bar{v} + \bar{v}^2$$

Using this result and its counterpart in the second term of the above expression for K, we have

$$K = (\tfrac{1}{2}m_1 v_1'^2 + \tfrac{1}{2}m_2 v_2'^2) + (m_1 v_1' + m_2 v_2') \cdot \bar{v} + \tfrac{1}{2}(m_1 + m_2)\bar{v}^2$$

But again we note that, by the definition of the zero-momentum frame, the second term on the right is zero, so that we come back to the simple result of Eq. (9–29):

$$K = K' + \tfrac{1}{2}M\bar{v}^2$$

Applying this to the states of a two-particle system before and after collision, we have

$$K_i = K_i' + \tfrac{1}{2}M\bar{v}^2$$
$$K_f = K_f' + \tfrac{1}{2}M\bar{v}^2$$

We assume that the total mass remains unchanged and, in the absence of external forces, so does $\bar{v}$. Thus we again arrive at the result

$$K_f - K_i = K_f' - K_i' = Q$$

where Q is the amount by which the final kinetic energy exceeds (algebraically) the initial kinetic energy. The actual value of Q may, of course, be negative.

The results of the above analysis can be extended, with virtually no modification, to such processes as nuclear reactions, in which the actual identity of the particles in the final state may be quite different from what one has at the beginning. Suppose, for example, that there is a collision between two nuclei, of masses m_1 and m_2, which react to produce two different nuclei, of masses m_3 and m_4. Then we can write the following statements of conservation:

Mass: $m_1 + m_2 = m_3 + m_4 = M$

Momentum: $m_1 v_1 + m_2 v_2 = m_3 v_3 + m_4 v_4 = M\bar{v}$
$m_1 v'_1 + m_2 v'_2 = m_3 v'_3 + m_4 v'_4 = 0$

Thus the initial and final kinetic energies can be written as follows:

$$K_i = \tfrac{1}{2} m_1 (v'_1)^2 + \tfrac{1}{2} m_2 (v'_2)^2 + \tfrac{1}{2} M\bar{v}^2$$
$$K_f = \tfrac{1}{2} m_3 (v'_3)^2 + \tfrac{1}{2} m_4 (v'_4)^2 + \tfrac{1}{2} M\bar{v}^2$$

with $K_f - K_i = Q$.

Example: The D—D reaction. One of the most famous nuclear reactions (and an important one for the process of energy generation by nuclear fusion) is the reaction of two nuclei of deuterium (hydrogen 2) to form a helium 3 nucleus and a neutron[1]:

$${}^2_1\text{H} + {}^2_1\text{H} \rightarrow {}^3_2\text{He} + {}^1_0 n + 3.27 \text{ MeV}$$

The 3.27 MeV represents the extra amount of kinetic energy, Q, made available because the masses of the product particles (their rest masses, to be precise) add up to a little less than the masses of the initial particles; the total energy, including mass equivalents, remains constant, of course.

Suppose now that a deuteron with a kinetic energy of 1 MeV strikes a stationary deuteron. What is the final state of affairs as viewed in the CM frame? (See Fig. 9–22.) First, let us calculate the velocity $\bar{v}$. The mass of a deuteron is about 2 amu, or about 3.34×10^{-27} kg. Now, 1 MeV = 1.6×10^{-13} J. Therefore,

$$v_1 = \left(\frac{2K_1}{m_1}\right)^{1/2} \approx 1.0 \times 10^7 \text{ m/sec}$$

---

[1] In this equation, the subscript before the letter for a given nucleus denotes the number of protons and the superscript shows the total number of nucleons. D and ${}^2_1$H both stand for a deuteron.

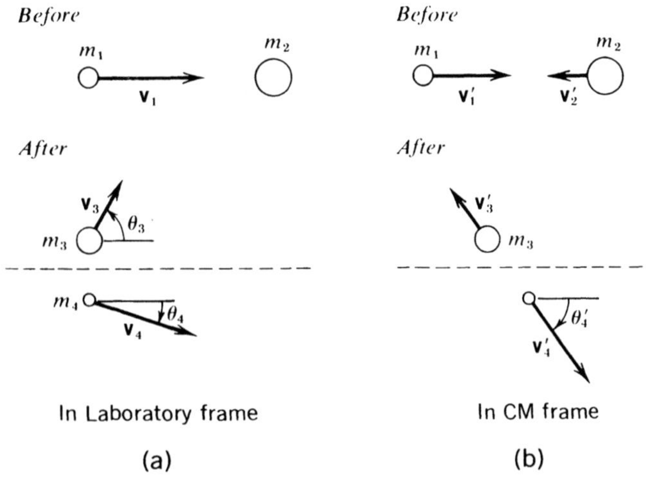

Fig. 9-22 (a) Reaction process, in which the collision of the particles m_1 and m_2 leads to the formation of two different particles, m_3 and m_4. (b) Same process as seen in the CM frame.

Since $m_2 = m_1$ and m_2 is initially at rest, we have

$$\bar{v} = v_1/2 = 0.5 \times 10^7 \text{ m/sec}$$

In the CM frame the deuterons have equal and opposite velocities of magnitude equal to $\bar{v}$. Hence we have

$$K'_{\text{initial}} = 2 \times \tfrac{1}{2}m_1\bar{v}^2 = \tfrac{1}{4}m_1 v_1^2$$

i.e.,

$$K'_{\text{initial}} = \tfrac{1}{2}K_1 = 0.5 \text{ MeV}$$

[We see here a particular application of Eq. (9–30). If a moving object collides with a stationary one of equal mass, only half the initial kinetic energy is available for their relative motion in the CM frame.]

Now consider the result of the nuclear reaction. The final total kinetic energy in the CM frame is given by

$$K'_{\text{final}} = K'_{\text{initial}} + 3.27 \text{ MeV}$$
$$= 3.77 \text{ MeV} = 6.03 \times 10^{-13} \text{ J}$$

This is partitioned between the ^{3}He and the neutron in such a way that the momenta are numerically equal. Denoting the masses and velocities of the ^{3}He and the neutron as m_3 and $\mathbf{v}'_3$ and m_4 and $\mathbf{v}'_4$, respectively, we have

$$m_3 \mathbf{v}'_3 + m_4 \mathbf{v}'_4 = 0$$

$$v'_4 = \frac{m_3}{m_4} v'_3$$

Then

$$K'_{\text{final}} = \tfrac{1}{2}m_3(v'_3)^2 + \tfrac{1}{2}m_4(v'_4)^2$$
$$= \tfrac{1}{2}m_3(v'_3)^2 + \tfrac{1}{2}m_4\left(\frac{m_3}{m_4}\right)^2(v'_3)^2$$
$$= \frac{1}{2}\frac{m_3(m_3+m_4)}{m_4}(v'_3)^2$$

Putting $m_3 \approx 3$ amu and $m_4 \approx 1$ amu, this gives us

$$v'_3 = \left(\frac{2 \times 6.03 \times 10^{-13}}{12 \times 1.67 \times 10^{-27}}\right)^{1/2} \approx 0.77 \times 10^7 \text{ m/sec}$$

and so $v'_4 (= 3v'_3) \approx 2.3 \times 10^7$ m/sec.

We thus have a full picture of the final situation as viewed in the CM frame for any specified direction θ'_4 of the outgoing neutron. To go back to the laboratory frame we have simply to add the CM velocity $\bar{\mathbf{v}}$ to each of the vectors $\mathbf{v}'_3$ and $\mathbf{v}'_4$.

The great advantage of using the CM frame in this way is that, regardless of the final directions as specified by θ'_4, the *magnitudes* of $\mathbf{v}'_3$ and $\mathbf{v}'_4$ always have the same values, whereas in the laboratory frame $\mathbf{v}_3$ (and also $\mathbf{v}_4$) has a different magnitude for each direction. This does not mean, however, that it is *always* desirable or necessary to go into the CM frame. For example, one may wish to answer the question: What is the speed of a neutron emitted at some given direction θ_4 in the laboratory with respect to the initial direction of a deuteron beam? In such a case it is easiest to work directly from the equations for energies and momenta as measured in the laboratory:

$$K_1 + Q = K_3 + K_4$$
$$\mathbf{p}_1 = \mathbf{p}_3 + \mathbf{p}_4 \tag{9-34}$$

(In the first equation Q represents the amount by which K_{final} differs from K_{initial}, so that in the example we have just considered we have $Q = +3.27$ MeV. Q may be positive, negative, or zero; the last of these represents an elastic collision.)

Note that Eq. (9-34) represents three independent equations (one for kinetic energy, and two for momentum—treating this as a two-dimensional problem). In the final state there are four unknowns: a magnitude and a direction for each of the vectors $\mathbf{v}_3$ and $\mathbf{v}_4$. The situation is indeterminate unless we put in one more piece of information, as for example the direction of one of the particles. Q is here taken to be a known quantity (as are K_1

and $\mathbf{p}_1$), but one could *deduce* it from a complete set of measurements of $\mathbf{v}_3$ and $\mathbf{v}_4$.

WHAT IS A COLLISION?

We are so used to associating the word "collision" with some abrupt, violent event that it may be well to point out that the results we have developed can be applied to other, quite gentle interactions. The only *essential* features are these:

1. That the interaction is confined, for all practical purposes, within some limited interval of time, so that it can be said to have a beginning and an end.

2. That over the duration of the collision, the effect of any external forces can be ignored, so that the system behaves as though it were isolated.

Figure 9–23 shows stroboscopic photographs of a collision between two frictionless pucks carrying permanent magnets, mounted vertically so as to repel each other. There is no contact in the usual sense, but this is certainly a collision within the

Fig. 9–23 (a) A "soft" collision, with no contact in the ordinary sense, occurring between two objects (2:1 mass ratio) carrying permanent magnets. (b) The identical collision photographed by a camera moving with the center of mass of the two objects. (From the PSSC film, "Moving with the Center of Mass", by Herman R. Branson, Education Development Center Film Sudio, Newton, Mass., 1965.)

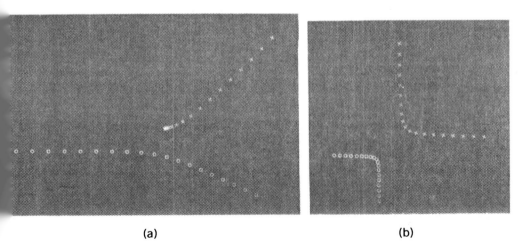

(a)   (b)

physicist's meaning of the term. One can see that the velocities are effectively constant in magnitude and direction except over the limited region of close approach of the objects.

INTERACTING PARTICLES SUBJECT TO EXTERNAL FORCES

Having discussed the conservation of linear momentum, and the general relation of force to rate of change of momentum, we can now consider the motion of a system of interacting particles that are *not* free of external influence. This is, of course, a very important extension of our ideas, because in practice a system is never completely isolated from its surroundings.

It will suffice to look at a two-particle system. The extension to any number of particles is quite simple but will be deferred to Chapter 14. Let m_1 and m_2 be the masses of the two particles, $\mathbf{F}_1$ and $\mathbf{F}_2$ the *external* forces acting on them, and $\mathbf{f}_{21}$ and $\mathbf{f}_{12}$ the internal interaction forces—$\mathbf{f}_{21}$ the force exerted on particle 1 by particle 2, and $\mathbf{f}_{12}$ the force exerted on particle 2 by particle 1.

Newton's law of motion applied to the particles individually states that

$$\mathbf{F}_1 + \mathbf{f}_{21} = m_1 \frac{d\mathbf{v}_1}{dt}$$

$$\mathbf{F}_2 + \mathbf{f}_{12} = m_2 \frac{d\mathbf{v}_2}{dt}$$

(9-35)

If the interactions are Newtonian, and we shall consider this the case, then the third law of Newton requires that

$$\mathbf{f}_{12} + \mathbf{f}_{21} = 0$$

Adding the two equations (vectorially), we then get

$$\mathbf{F}_1 + \mathbf{F}_2 = m_1 \frac{d\mathbf{v}_1}{dt} + m_2 \frac{d\mathbf{v}_2}{dt}$$

or

$$\mathbf{F}_1 + \mathbf{F}_2 = \frac{d}{dt}(m_1 \mathbf{v}_1 + m_2 \mathbf{v}_2)$$

(9-36)

in which the internal forces $\mathbf{f}_{12}$ and $\mathbf{f}_{21}$ have disappeared. This equation states that the resultant of all the external forces acting on the system equals the rate of change of the *total* vector momentum of the system.

We can express this result in another, very compact way by introducing the concept of the center of mass of the system. If the individual positions of the two particles are given by the vectors $\mathbf{r}_1$ and $\mathbf{r}_2$, drawn from some origin, the position of the center of mass is given by

$$\bar{\mathbf{r}} = \frac{m_1\mathbf{r}_1 + m_2\mathbf{r}_2}{m_1 + m_2} \tag{9-37}$$

This is the three-dimensional analogue of (9–25) and corresponds to the vector velocity of the center of mass as already defined by Eq. (9–33), so that

$$m_1\mathbf{v}_1 + m_2\mathbf{v}_2 = M\bar{\mathbf{v}}$$

where $M\ (= m_1 + m_2)$ is the total mass of the system. Accordingly, we have

$$\mathbf{F}_1 + \mathbf{F}_2 = \mathbf{F} = M\frac{d\bar{\mathbf{v}}}{dt} \tag{9-38}$$

and this proves the result we are after:

The motion of the center of mass of a system of two particles is the same as the motion of a single particle of mass equal to the total mass of the system acted on by the resultant of all the external forces which act on the individual particles.

The implications of this result are significant. First, it suggests that a fundamental method for treating the motion of a system of particles is to analyze its motion as the combination of (1) the motion of its center of mass and (2) the motions of the particles relative to their center of mass. The latter motion, the internal motion of the system, is one of zero momentum, as we saw earlier. Furthermore, Eq. (9–38) allows us to treat some aspects of the motions of extended objects by the laws of dynamics for a simple particle. In particular, when an extended object moves in translation, i.e., when there is no motion of any particle in the object relative to the center of mass, Eq. (9–38) tells the whole story. We shall return to such questions in Chapter 14.

Incidentally, Eqs. (9–36) and (9–38) also provide a basis for a criterion as to whether or not a system of colliding particles is effectively isolated. The conservation of momentum in a collision process holds good only to the extent that the effect of any external forces can be ignored. If external forces are indeed present, the duration Δt of the collision must be so short that the product **F** Δt is negligible. A different way of stating this same condition is that the forces of interaction between the colliding particles must be much greater than any external forces which may be acting.

THE PRESSURE OF A GAS

As our last example of collision processes and momentum transfers we shall take the calculation of the pressure of a gas, on the basis that this is due to the perfectly elastic collisions of the molecules of the gas with the atoms that comprise the wall of the container.

We shall assume that the gas is made up of n particles, each of mass m_0, per unit volume. The most naive possible calculation makes the following assumptions:

1. All the particles have the same speed, v.
2. The particles can be regarded as though one third of them were, at any instant, traveling parallel to a given direction in space, with the other two thirds traveling parallel to two other directions, perpendicular to each other and to the first direction.
3. The gas is in a rectangular container with perfectly flat, hard walls.

None of these assumptions is in the least realistic as it stands. The first two are certainly false, the choice of a rectangular container is very special, and on an atomic scale the walls are knobbly and sticky, not flat and hard. Nevertheless, the calculation we shall make, using these assumptions, comes very close to yielding the correct result. The reason is that, *on the average*, with the huge number of particles present in a sample of gas, there is an essential symmetry in the aggregate motion and an effectively exact conservation of the total kinetic energy. An individual molecule may strike the wall, sit there a short while, and jump off again at a different angle with a different speed—perhaps faster, perhaps slower. But on the average, we may treat the collisions as perfectly elastic because the kinetic energy of the gas as a whole neither increases nor decreases with time.

On this basis, we make the following very simple calculation. Consider an element of area ΔA in one wall of the containing vessel (Fig. 9-24). Resolve the motions of the molecules so that the normal to ΔA is one of the three mutually perpendicular directions along which the molecules are assumed to move. Any

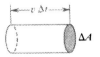

Fig. 9-24 Simplest possible approach to calculating the pressure of a gas. During a time Δt the element of wall ΔA can be reached only by molecules that lie initially within the cylinder of length $v \Delta t$.

one molecule approaches the wall with momentum $m_0 v$ and recoils with momentum $-m_0 v$; it thus provides an impulse $2m_0 v$. How many molecules strike ΔA in a time Δt? If we assume that all of them have the same speed v, molecules farther than $v \Delta t$ from the wall cannot arrive until after the time Δt has elapsed. Thus our attention is limited to molecules within a cylinder of cross section ΔA and length $v \Delta t$. There are n molecules per unit volume. Hence

number of molecules within cylinder = $nv \Delta t \Delta A$

But of these, only *one sixth* are moving, not just along a line perpendicular to ΔA, but specifically approaching ΔA instead of receding from it. This, then, gives us

number of impacts with ΔA in $\Delta t = \frac{1}{6} nv \Delta t \Delta A$
total impulse communicated to ΔA in $\Delta t = 2m_0 v \times \frac{1}{6} nv \Delta t \Delta A$
$= \frac{1}{3} nm_0 v^2 \Delta t \Delta A$

But the average force ΔF exerted on ΔA is the impulse divided by Δt. Hence

$\Delta F = \frac{1}{3} nm_0 v^2 \Delta A$

The mean force per unit area, exerted normal to the containing wall, is what we call the pressure, p.[1] Thus we arrive at the result

$$p = \frac{\Delta F}{\Delta A} = \frac{1}{3} nm_0 v^2 \qquad (9\text{-}39)$$

Since nm_0 is the total mass of gas per unit volume, which is its density ρ, we can alternatively put

$$p = \frac{1}{3} \rho v^2$$

or

$$v = \left(\frac{3p}{\rho}\right)^{1/2} \qquad (9\text{-}40)$$

[1] In this section, p always refers to pressure, not momentum.

Thus, if our calculation is justified, we can infer the speed of the invisible molecules from perfectly straightforward measurements of the bulk properties of the gas. Taking nitrogen at ordinary atmospheric pressure and room temperature, for example, we have

$$p \approx 10^5 \text{ N/m}^2 = 10 \text{ kg/m-sec}^2$$
$$\rho \approx 1.15 \text{ kg/m}^3$$

Therefore,

$$v \approx 500 \text{ m/sec}$$

How defensible is Eq. (9-39)? A more careful calculation is clearly called for. We have seen experimental evidence (Chapter 3) that the molecules of a gas at a given temperature have a wide spread of speeds. So the v^2 in our formula should be replaced by some kind of an average squared speed, v_m^2. And certainly we could not, on the strength of our own calculation, place much faith in the numerical factor $\frac{1}{3}$—although the rigorous theory confirms it, in fact. An acceptable treatment of the problem must investigate the consequences of having molecules approaching the wall in quite arbitrary directions; it is only an accident that this does not change the result in detail. (It would, if we considered the *numbers* of particles striking ΔA, instead of the force they exert.) Nevertheless, the simple analysis that we have presented gives us a remarkably useful beginning for the understanding of bulk properties in microscopic terms. And, since the detailed kinetic theory of gases is not our present concern, we shall rest content with that.

THE NEUTRINO

At the beginning of this chapter we pointed out that any conservation law or conservation principle in physics is provisional, but that if, in the face of apparent failure, it is finally vindicated, then its status may be greatly strengthened. The most dramatic success of the conservation laws of dynamics took place in connection with the neutrino—that elusive, neutral particle emitted in the process of radioactive beta decay. The prediction of its existence stemmed from an apparent nonconservation of energy and angular momentum, but perhaps the most beautiful and direct dynamical evidence for it is furnished by the apparent nonconservation of linear momentum.

Fig. 9–25 Evidence for the neutrino. The visible tracks of the electron and the recoiling lithium 6 nucleus in the beta decay of helium 6 in a cloud-chamber are not collinear. [From J. Csikai and A. Szalay, Soviet Physics JETP, 8, 749 (1959).]

The situation can be simply stated as follows: It is known that the process of beta decay involves the ejection of an electron from a nucleus, as a result of which the nuclear charge goes up by one unit (if the electron is an ordinary negative electron). If no other particles were involved, the process could be written

$$A \rightarrow B + e^-$$

where A is the initial nucleus and B the final nucleus. If A were effectively isolated, and initially stationary, our belief in linear momentum conservation would lead us to predict that, whatever the direction (or energy) of the ejected electron, the nucleus B would inevitably recoil in the opposite direction. Any departure from this, regardless of all other details, would demand the involvement of another particle.

Figure 9-25 shows a cloud-chamber photograph of the beta decay of helium 6. The decay takes place at the position of the sharp knee near the top of the picture. The short stubby track pointing in a "northwesterly" direction is the recoiling nucleus of lithium 6; the other track is the electron. There *must* be another particle—the neutrino—if the final momentum vectors are to add up to have the same resultant—i.e., zero—as the initially stationary ^{6}He nucleus. It fails to reveal itself because its lack of charge, or of almost any other interaction, allows it to escape unnoticed—so readily, in fact, that the chance would be only about 1 in 10^{12} of its interacting with any matter in passing right through the earth.

PROBLEMS

9-1 A particle of mass m, traveling with velocity v_0, makes a completely inelastic collision with an initially stationary particle of mass

M. Make a graph of the final velocity v as a function of the ratio m/M from $m/M = 0$ to $m/M = 10$.

9-2 Consider how conservation of linear momentum applies to a ball bouncing off a wall.

9-3 A mouse is put into a small closed box that is placed upon a table. Could a clever mouse control the movements of the box over the table? Just what maneuvers could the mouse make the box perform? If you were such a mouse, and your object were to elude pursuers, would you prefer that the table have a large, small, or negligible coefficient of friction?

9-4 In the *Principia*, Newton mentions that in one set of collision experiments he found that the relative velocity of separation of two objects of a certain kind of material was five ninths of their relative velocity of approach. Suppose that an initially stationary object, of mass $3m_0$, of this material was struck by a similar object of mass $2m_0$, traveling with an initial velocity v_0. Find the final velocities of both objects.

9-5 A particle of mass m_0, traveling at speed v_0, strikes a stationary particle of mass $2m_0$. As a result, the particle of mass m_0 is deflected through 45° and has a final speed of $v_0/2$. Find the speed and direction of the particle of mass $2m_0$ after this collision. Was kinetic energy conserved?

9-6 Two skaters (*A* and *B*), both of mass 70 kg, are approaching one another, each with a speed of 1 m/sec. *A* carries a bowling ball with a mass of 10 kg. Both skaters can toss the ball at 5 m/sec relative to themselves. To avoid collision they start tossing the ball back and forth when they are 10 m apart. Is one toss enough? How about two tosses, i.e., *A* gets the ball back? If the ball weighs half as much but they can throw twice as fast, how many tosses do they need? Plot the entire incident on a time versus displacement graph, in which the positions of the skaters are marked along the abscissa, and the advance of time is represented by the increasing value of the ordinate. (Mark the initial positions of the skaters at $x = \pm 5$ m, and include the space-time record of the ball's motion in the diagram.) This situation serves as a simple model of the present view of interactions (repulsive, in the above example) between elementary particles. An attractive interaction can be simulated by supposing that the skaters exchange a boomerang instead of a ball. [These theoretical models were presented by F. Reines and J. P. F. Sellschop in an article entitled "Neutrinos from the Atmosphere and Beyond," *Sci. Am.*, **214**, 40 (Feb. 1966).]

9-7 Find the average recoil force on a machine gun firing 240 rounds (shots) per minute, if the mass of each bullet is 10 g and the muzzle velocity is 900 m/sec.

9–8 Water emerges in a vertical jet from a nozzle mounted on one end of a long horizontal metal tube, clamped at its other end and thin enough to be rather flexible. The jet rises to a height of 2.5 m above the nozzle, and the rate of water flow is 2 liter/min. It has been previously found by static experiments that the nozzle is depressed vertically by an amount proportional to the applied force, and that a mass of 10 g, hung upon it, causes a depression of 1 cm. How far is the nozzle depressed by the reaction force from the water jet? [This problem is based on a demonstration experiment described by E. F. Schrader, *Am. J. Phys.*, **33**, 784 (1965).]

9–9 A "standard fire stream" employed by a city fire department delivers 250 gallons of water per minute and can attain a height of 70 ft on a building whose base is 63 ft from the nozzle. Neglecting air resistance:

(a) What is the nozzle velocity of the stream?

(b) If directed horizontally against a vertical wall, what force would the stream exert? (Assume that the water spreads out over the surface of the wall without any rebound, so that the collision is effectively inelastic.)

9–10 A helicopter has a total mass M. Its main rotor blade sweeps out a circle of radius R, and air over this whole circular area is pulled in from above the rotor and driven vertically downward with a speed v_0. The density of air is ρ.

(a) If the helicopter hovers at some fixed height, what must be the value of v_0?

(b) One of the largest helicopters of the type described above weighs about 10 tons and has $R \simeq 10$ m. What is v_0 for hovering in this case? Take $\rho = 1.3$ kg/m^3.

9–11 A rocket of initial mass M_0 ejects its burnt fuel at a constant rate $|dM/dt| = \mu$ and at a speed v_0 relative to the rocket.

(a) Calculate the initial acceleration of the rocket if it starts vertically upward from its launch pad.

(b) If $v_0 = 2000$ m/sec, how many kilograms of fuel must be ejected per second to give such a rocket, of mass 1000 tons, an initial upward acceleration equal to 0.5 g?

9–12 This rather complicated problem is designed to illustrate the advantage that can be obtained by the use of multiple-stage instead of single-stage rockets as launching vehicles. Suppose that the payload (e.g., a space capsule) has mass m and is mounted on a two-stage rocket (see the figure). The *total* mass—both rockets fully fueled, plus the payload—is Nm. The mass of the second-stage rocket plus the payload, after first-stage burnout and separation, is nm. In each stage the ratio of burnout mass (casing) to initial mass (casing plus fuel) is r, and the exhaust speed is v_0.

(a) Show that the velocity v_1 gained from first-stage burn,

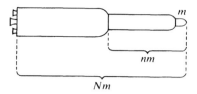

starting from rest (and ignoring gravity), is given by

$$v_1 = v_0 \ln\left[\frac{N}{rN + n(1-r)}\right]$$

(b) Obtain a corresponding expression for the additional velocity, v_2, gained from the second-stage burn.

(c) Adding v_1 and v_2, you have the payload velocity v in terms of N, n, and r. Taking N and r as constants, find the value of n for which v is a maximum.

(d) Show that the condition for v to be a maximum corresponds to having equal gains of velocity in the two stages. Find the maximum value of v, and verify that it makes sense for the limiting cases described by $r = 0$ and $r = 1$.

(e) Find an expression for the payload velocity of a single-stage rocket with the same values of N, r, and v_0.

(f) Suppose that it is desired to obtain a payload velocity of 10 km/sec, using rockets for which $v_0 = 2.5$ km/sec and $r = 0.1$. Show that the job can be done with a two-stage rocket but is impossible, however large the value of N, with a single-stage rocket.

(g) If you are ambitious, try extending the analysis to an arbitrary number of stages. It is possible to show that once again the greatest payload velocity for a given total initial mass is obtained if the stages are so designed that the velocity increment contributed by each stage is the same.

9-13 A block of mass m, initially at rest on a frictionless surface, is bombarded by a succession of particles each of mass δm ($\ll m$) and of initial speed v_0 in the positive x direction. The collisions are perfectly elastic and each particle bounces back in the negative x direction. Show that the speed acquired by the block after the nth particle has struck it is given very nearly by $v = v_0(1 - e^{-\alpha n})$, where $\alpha = 2\delta m/m$. Consider the validity of this result for $\alpha n \ll 1$ as well as for $\alpha n \to \infty$.

9-14 Newton calculated the resistive force for an object traveling through a fluid by supposing that the particles of the fluid (supposedly initially stationary) rebounded elastically when struck by the object.

(a) On this model, the resistive force would vary as some power, n, of the speed v of the object. What is the value of n?

(b) Suppose that a flat-ended object of cross-sectional area A is moving at speed v through a fluid of density ρ. By picturing the fluid

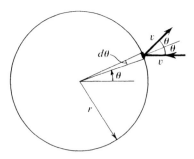

as composed of n particles, each of mass m, per unit volume (such that $nm = \rho$), obtain an explicit expression for the resistive force if each particle that is struck by the object recoils elastically from it.

(c) If the object, instead of being flat-ended, were a massive sphere of radius r, traveling at speed v through a medium of density ρ, what would the magnitude of the resistive force be? The whole calculation can be carried out from the standpoint of a frame attached to the sphere, so that the fluid particles approach it with the velocity $-v$. Assume that in this frame the fluid particles are reflected as by a mirror—angle of reflection equals angle of incidence (see the figure). You must consider the surface of the sphere as divided up into circular zones corresponding to small angular increments $d\theta$ at the various possible values of θ.

9–15 A particle of mass m_1 and initial velocity u_1 strikes a stationary particle of mass m_2. The collision is perfectly elastic. It is observed that after the collision the particles have equal and opposite velocities. Find

(a) The ratio m_2/m_1.
(b) The velocity of the center of mass.
(c) The total kinetic energy of the two particles in the center of mass frame expressed as a fraction of $\frac{1}{2}m_1u_1^2$.
(d) The final kinetic energy of m_1 in the lab frame.

9–16 A mass m_1 collides with a mass m_2. Define relative velocity as the velocity of m_1 observed in the rest frame of m_2. Show the equivalence of the following two statements:
(1) Total kinetic energy is conserved.
(2) The magnitude of the relative velocity is unchanged.
(It is suggested that you solve the problem for a one-dimensional collision, at least in the first instance.)

9–17 A collision apparatus is made of a set of n graded masses suspended so that they are in a horizontal line and not quite in contact with one another (see the figure). The first mass is fm_0, the second is f^2m_0, the third f^3m_0, and so on, so that the last mass is $f^n m_0$. The first mass is struck by a particle of mass m_0 traveling at a speed v_0. This produces a succession of collisions along the line of masses.

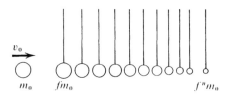

(a) Assuming that all the collisions are perfectly elastic, show that the last mass flies off with a speed v_n given by

$$v_n = \left(\frac{2}{1+f}\right)^n v_0$$

(b) Hence show that, if f is close to unity, so that it can be written as $1 \pm \epsilon$ (with $\epsilon \ll 1$), this system can be used to transfer virtually all the kinetic energy of the incident mass to the last one, even for large n.

(c) For $f = 0.9$, $n = 20$, calculate the mass, speed, and kinetic energy of the last mass in the line in terms of the mass, speed, and kinetic energy of the incident particle. Compare this with the result of a direct collision between the incident mass and the last mass in the line.

9-18 A 2-kg and an 8-kg mass collide elastically, compressing a spring bumper on one of them; the bumper returns to its original length as the masses separate. Assume that the collision takes place along a single line and that you can cause the collision to occur in different ways, each having the same initial energy:

Case A: The 8-kg mass has 16 J of kinetic energy and hits the stationary 2-kg mass.

Case B: The 2-kg mass has 16 J of kinetic energy and hits the stationary 8-kg mass.

(a) Which way of causing the collision to occur will result in the greater compression of the spring? Arrive at your choice without actually solving for the compression of the spring.

(b) Keeping the condition of a total initial kinetic energy of 16 J, how should this energy be divided between the two masses to obtain the greatest possible compression of the spring?

9-19 In a certain road accident (this is based on an actual case) a car of mass 2000 kg, traveling south, collided in the middle of an intersection with a truck of mass 6000 kg, traveling west. The vehicles locked and skidded off the road along a line pointing almost exactly southwest. A witness claimed that the truck had entered the intersection at 50 mph.

(a) Do you believe the witness?

(b) Whether or not you believe him, what fraction of the total initial kinetic energy was converted into other forms of energy by the collision?

9-20 A nucleus A of mass $2m$, traveling with a velocity u, collides with a stationary nucleus of mass $10m$. The collision results in a change of the total kinetic energy. After collision the nucleus A is observed to be traveling with speed v_1 at 90° to its original direction of motion, and B is traveling with speed v_2 at angle θ ($\sin \theta = 3/5$) to the original direction of motion of A.

(a) What are the magnitudes of v_1 and v_2?

(b) What fraction of the initial kinetic energy is gained or lost as a result of the interaction?

9-21 A particle of mass m and initial velocity u collides *elastically* with a particle of mass M initially at rest. As a result of the collision the particle of mass m is deflected through 90° and its speed is reduced to $u/\sqrt{3}$. The particle of mass M recoils with speed v at an angle θ to the original direction of m. (All speeds and angles are those observed in the laboratory system.)

(a) Find M in terms of m, and v in terms of u. Find also the angle θ.

(b) At what angles are the particles deflected in the center-of-mass system?

9-22 Make measurements on the stroboscopic photographs of a collision of two magnetized pucks (Fig. 9-23) to test the conservation of linear momentum and total kinetic energy between the initial state (first three time units) and the final state (last three time units).

9-23 A particle of mass $2m$ and of velocity u strikes a second particle of mass $2m$ initially at rest. As a result of the collision, a particle of mass m is produced which moves off at 45° with respect to the initial direction of the incident particle. The other product of this rearrangement collision is a particle of mass $3m$. Assuming that this collision involves no significant change of total kinetic energy, calculate the speed and direction of the particle of mass $3m$ in the Lab and in the CM frame.

9-24 In a historic piece of research, James Chadwick in 1932 obtained a value for the mass of the neutron by studying elastic collisions of fast neutrons with nuclei of hydrogen and nitrogen. He found that the maximum recoil velocity of hydrogen nuclei (initially stationary) was 3.3×10^7 m/sec, and that the maximum recoil velocity of nitrogen 14 nuclei was 4.7×10^6 m/sec with an uncertainty of $\pm 10\%$. What does this tell you about

(a) The mass of a neutron?

(b) The initial velocity of the neutrons used?

(Take the uncertainty of the nitrogen measurement into account. Take the mass of an H nucleus as 1 amu and the mass of a nitrogen 14 nucleus as 14 amu.)

9–25 A cloud-chamber photograph showed an alpha particle of mass 4 amu with an initial velocity of 1.90×10^7 m/sec colliding with a nucleus in the gas of the chamber. The collision changed the direction of motion of the alpha particle by 12° and reduced its speed to 1.18×10^7 m/sec. The other particle, initially stationary, acquired a velocity of 2.98×10^7 m/sec at 18° with respect to the initial forward direction of the alpha particle. What was this second particle? Was the collision elastic? (In interpreting your results, take account of the fact that these cloud-chamber measurements of speeds and angles are subject to errors of up to a few percent.)

9–26 A nuclear reactor has a moderator of graphite. The carbon nuclei in the atoms of this crystal lattice can be regarded as effectively free to recoil if struck by fast neutrons, although they cannot be knocked out of place by thermal neutrons. A fast neutron, of kinetic energy 1 MeV, collides elastically with a stationary carbon 12 nucleus.

(a) What is the initial speed of each particle in the center of mass frame?

(b) As measured in the center-of-mass frame, the velocity of the carbon nucleus is turned through 135° by the collision. What are the final speed and direction of the neutron as measured in the lab frame?

(c) About how many elastic collisions, involving random changes of direction, must a neutron make with carbon nuclei if its kinetic energy is to be reduced from 1 MeV to 1 keV? Assume that the mean energy loss is midway between maximum and minimum values.

9–27 (a) A moving particle of mass M collides perfectly elastically with a stationary particle of mass $m < M$. Show that the maximum possible angle through which the incident particle can be deflected is $\sin^{-1}(m/M)$. (Use of the vector diagrams of the collision in Lab and CM systems will be found helpful.)

(b) A particle of mass m collides perfectly elastically with a stationary particle of mass $M > m$. The incident particle is deflected through 90°. At what angle θ with the original direction of m does the more massive particle recoil?

9–28 The text (p. 348) gives an example of the analysis of the dynamics of a nuclear reaction between two deuterons. Another possible reaction is the following:

$$^2_1H + {}^2_1H \rightarrow {}^1_1H + {}^3_1H + 4.0 \text{ MeV}$$

In this case the reaction products are a proton and a triton (the latter being the nucleus of the unstable isotope hydrogen 3 or tritium). Suppose that a stationary deuteron is struck by an incident deuteron of kinetic energy 5 MeV.

(a) What are the maximum and minimum possible values of the kinetic energy of the proton produced in this reaction?

(b) What is the maximum angle (as observed in the laboratory) that the direction of the triton can make with the direction of the incident deuteron? What is its kinetic energy as measured in the laboratory when it is emitted in this direction? (This problem can be conveniently handled with the use of amu and MeV as units throughout.)

9-29 A boat of mass M and length L is floating in the water, stationary; a man of mass m is sitting at the bow. The man stands up, walks to the stern of the boat, and sits down again.

(a) If the water is assumed to offer no resistance at all to motion of the boat, how far does the boat move as a result of the man's trip from bow to stern?

(b) More realistically, assume that the water offers a viscous resistance given by $-kv$, where k is a constant and v is the velocity of the boat. Show that in this case one has the remarkable result that the boat should eventually return to its initial position.

(c) Consider the paradox presented by the fact that, according to (b), any nonzero value of k, however small, implies that the boat ends up at its starting point, but a strictly zero value implies that it ends up somewhere else. How do you explain this discontinuous jump in the final position when the variation of k can be imagined as continuous, down to zero? For enlightenment see the short and clear analysis by D. Tilley, *Am. J. Phys.*, **35**, 546 (1967).

Conservation of Energy as a principle . . . has played a fundamental role, the most fundamental as some would say, in the last half-century of classical physics, and since then in quantum mechanics. . . . Yet it is a curiosity that this, which is the first law of thermodynamics, should have so fruitfully watered the stonier, the more austere terrain of mechanics.

C. G. GILLISPIE, *The Edge of Objectivity* (1960)

It may be thought strange that we have brought our discussion of the foundations of mechanics so far with hardly a mention of energy. . . . We have indeed had no desire to slight the significance of this concept, but have wished to emphasize that the logical foundation of mechanics is quite possible without it. . . . The question at once arises: why then should it have been introduced at all? This is what we wish to discuss.

R. B. LINDSAY AND H. MARGENAU,
Foundations of Physics (1936)

10

Energy conservation in dynamics; vibrational motions

INTRODUCTION

OF ALL THE physical concepts, that of energy is perhaps the most far-reaching. Everyone, whether a scientist or not, has an awareness of energy and what it means. Energy is what we have to pay for in order to get things done. The word itself may remain in the background, but we recognize that each gallon of gasoline, each Btu of heating gas, each kilowatt-hour of electricity, each car battery, each calorie of food value, represents, in one way or another, the wherewithal for doing what we call *work*. We do not think in terms of paying for force, or acceleration, or momentum. *Energy* is the universal currency that exists in apparently countless denominations; and physical processes represent a conversion from one denomination to another.

The above remarks do not really *define* energy. No matter. It is worth recalling once more the opinion that H. A. Kramers expressed: "The most important and most fruitful concepts are those to which it is impossible to attach a well-defined meaning."[1] The clue to the immense value of energy as a concept lies in its transformation. It is *conserved*—that is the point. Although we may not be able to define energy in general, that does not mean

[1] See p. 62, where we quoted Kramers' remark in connection with our discussion of the concept of time.

that it is only a vague, qualitative idea. We have set up quantitative measures of various specific *kinds* of energy: gravitational, electrical, magnetic, elastic, kinetic, and so on. And whenever a situation has arisen in which it seemed that energy had disappeared, it has always been possible to recognize and define a new form of energy that permits us to save the conservation law. And conservation laws, as we remarked at the beginning of Chapter 9, represent one of the physicist's most powerful tools for organizing his description of nature.

In this book we shall be dealing only with the two main categories of energy that are relevant to classical mechanics—the kinetic energy associated with the bodily motion of objects, and the potential energy associated with elastic deformations, gravitational attractions, electrical interactions, and the like. If energy should be transferred from one or another of these forms into chemical energy, radiation, or the random molecular and atomic motion we call heat, then from the standpoint of mechanics it is lost. This is a very important feature, because it means that, if we restrict our attention to the purely mechanical aspects, the conservation of energy is not binding; it must not be blindly assumed. Nevertheless, as we shall see, there are many physical situations for which the conservation of the total mechanical energy holds good, and in such contexts it is of enormous value in the analysis of physical problems.

It is an interesting historical sidelight that in pursuing the subject of energy we are temporarily parting company with Newton, although not with what we may properly call Newtonian mechanics. In the whole of the *Principia*, with its awe-inspiring elucidation of the dynamics of the universe, the concept of energy is never once used or even referred to![1] For Newton, $F = ma$ was enough. But we shall see how the energy concept, although rooted in $F = ma$, has its own special contributions to make. We shall begin with the quantitative connection between work and kinetic energy.

INTEGRALS OF MOTION

In Chapter 6 we briefly presented the basic notion of the work

---

[1] As we remarked in Chapter 9, however, some of Newton's contemporaries, particularly Huygens and Leibnitz, did recognize the importance of an energy-like quantity—the "vis viva" mv^2—in certain contexts, such as collisions and pendulum motions.

done by a force acting on an object, producing a corresponding increase in the kinetic energy of the object. We shall now return to this topic and develop it considerably further.

Let us again take, for a start, the simple and familiar case of an object of mass m acted on by a constant force F (we shall begin by assuming a straight-line motion for which vector symbols are unnecessary). Then by Newton's law we have

$$F = \frac{dp}{dt} = m\frac{dv}{dt} = ma$$

Let the force act for a time t during which the velocity changes from v_1 to v_2. Then we have

$$Ft = mat = m(v_2 - v_1) \qquad (10\text{-}1a)$$

This expresses the fact that the impulse of F is equal to the change of linear momentum. But suppose we multiply the force by the distance over which it acts, instead of by the time. In this case we obtain

$$Fx = max = ma\frac{v_1 + v_2}{2}t$$
$$= \tfrac{1}{2}m(at)(v_1 + v_2)$$
$$= \tfrac{1}{2}m(v_2 - v_1)(v_1 + v_2)$$

Therefore,

$$Fx = \tfrac{1}{2}mv_2^2 - \tfrac{1}{2}mv_1^2 \qquad (10\text{-}2a)$$

This expresses the fact that the work done by F is equal to the change of kinetic energy. We have no reason to declare a preference between Eqs. (10-1a) and (10-2a) as statements of the effect of the force. In fact, they will in general tell us different things. But before amplifying that last remark, let us note that the restriction to constant force and constant acceleration is unnecessary. For we have

$$F = m\frac{dv}{dt}$$

Multiplying by dt and integrating gives us

$$\int_{t_1}^{t_2} F\,dt = m\int_{v_1}^{v_2} dv = m(v_2 - v_1) \qquad (10\text{-}1b)$$

Multiplying by dx and integrating gives us

$$\int_{x_1}^{x_2} F\,dx = m\int_{x_1}^{x_2} \frac{dv}{dt}\,dx$$

But

$$\frac{dv}{dt}\,dx = \frac{dx}{dt}\,dv = v\,dv$$

Thus

$$\int_{x_1}^{x_2} F\,dx = m\int_{v_1}^{v_2} v\,dv = \tfrac{1}{2}m(v_2{}^2 - v_1{}^2) \qquad (10\text{--}2b)$$

The left-hand side of Eq. (10–1b) is the total impulse of the force, and the left-hand side of Eq. (10–2b) is of course defined as the total *work* done by the force. Each of these integrals can be represented as the area under the appropriate graph of F against t or x between certain limits (see Fig. 10–1).

The general similarity of Eqs. (10–1b) and (10–2b) is, however, deceptive, because force, displacement, and velocity are *vector* quantities. The result of applying a force of a given magnitude depends very much on its direction relative to the direction of motion of the object on which it acts. Thus in circular motion, as we have seen, a net force is continually applied to an object, the momentum (also a vector) changes continuously, but the magnitude of the velocity does not change at all.

The situation becomes clear if we return to the proper vector statement of Newton's law:

$$\mathbf{F} = m\frac{d\mathbf{v}}{dt}$$

It is still possible to integrate directly with respect to t:

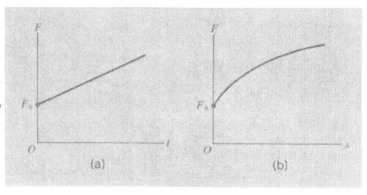

Fig. 10-1 (a) Force that varies linearly with time. The area under the curve measures the total impulse. (b) The same force plotted as a function of displacement. The area under the curve measures the total work.

$$\int_{t_1}^{t_2} \mathbf{F}\, dt = m(\mathbf{v}_2 - \mathbf{v}_1)$$

The left-hand side of the expression is an instruction that, if the direction of the force changes with time, we must form the appropriate vector sum of all the small impulses $\mathbf{F}\,\Delta t$ applied in successive time intervals Δt between t_1 and t_2. That is exactly how Newton himself conceived the action of the varying force to which a planet, for example, finds itself exposed as it moves along its orbit around the sun (cf. Chapter 13). But what about integrating F over the elements of distance along a path, where these elements ($\Delta \mathbf{r}$) are themselves vectors? This is essentially a physical question, not a mathematical one. We are asking: What is the effect of applying a force at some arbitrary direction to the motion of an object? Figure 10–2 helps to supply the answer. If a force $\mathbf{F}_1$ is applied *at right angles* to $\mathbf{v}$ for some very short time Δt, it changes the direction of the velocity without appreciably changing its magnitude. If a force $\mathbf{F}_2$ is applied *along* the direction of $\mathbf{v}$, it changes the magnitude of $\mathbf{v}$ without changing the direction. Thus if we want to fix our attention on changes in the magnitude of $\mathbf{v}$, we should restrict ourselves to forces or force components along the direction of $\mathbf{v}$. If the net force $\mathbf{F}$ acts at an angle θ to $\mathbf{v}$, we have its component along $\mathbf{v}$ given by

$$|\mathbf{F}|\cos\theta = m\,\frac{|\Delta\mathbf{v}|\cos\theta}{\Delta t}$$

where $\Delta\mathbf{v}$ is the vector change in $\mathbf{v}$, equal to $\mathbf{F}\,\Delta t/m$ [cf. Fig. 10–2(c)]. The element of distance traveled during Δt is given by

$$\Delta\mathbf{r} = \mathbf{v}\,\Delta t$$

Fig. 10–2 (a) Velocity change due to an impulse perpendicular to v. (b) Velocity change due to an impulse parallel to v. (c) Velocity change due to an arbitrary impulse.

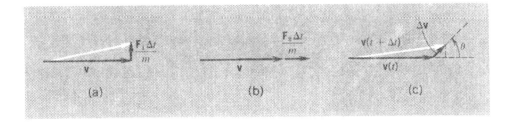

by the definition of the velocity vector. Hence the force component along **v**, multiplied by the displacement, is given by

$$|\mathbf{F}| |\Delta \mathbf{r}| \cos\theta = m |\mathbf{v}| |\Delta \mathbf{v}| \cos\theta$$

The above is a *scalar* equation; we have suppressed all reference to the effect of **F** in changing the direction of **v** and are left with information about the magnitude only. To express this more neatly we use the notation of the "dot product" (or scalar product) of two vectors:

$$\mathbf{A} \cdot \mathbf{B} = |\mathbf{A}| |\mathbf{B}| \cos\theta = AB \cos\theta$$

In this notation, we have

$$\mathbf{F} \cdot \Delta \mathbf{r} = m(\mathbf{v} \cdot \Delta \mathbf{v})$$

But we can now play a neat (and valuable) trick. Consider the quantity $\mathbf{v} \cdot \mathbf{v}$. This is a scalar, and its magnitude is just the square of $|\mathbf{v}|$, i.e., v^2 simply. But since

$$v^2 = \mathbf{v} \cdot \mathbf{v}$$

we have, by differentiation,

$$\Delta(v^2) = \Delta\mathbf{v} \cdot \mathbf{v} + \mathbf{v} \cdot \Delta\mathbf{v} = 2\mathbf{v} \cdot \Delta\mathbf{v}$$

(This last step is possible because the scalar product of two vectors is independent of the order in which the factors come—i.e., the commutative law holds.) Hence

$$\mathbf{v} \cdot \Delta\mathbf{v} = \tfrac{1}{2}\Delta(v^2)$$

It follows that we have

$$\mathbf{F} \cdot \Delta\mathbf{r} = \tfrac{1}{2} m \Delta(v^2)$$

So now, integrating over any path that the body may have followed under the action of the force, we find the relation

$$W = \int_{r_1}^{r_2} \mathbf{F} \cdot d\mathbf{r} = \tfrac{1}{2} m(v_2^2 - v_1^2) \tag{10-3}$$

which describes in general terms the relation between the work done and the change of kinetic energy.

Figure 10-3 illustrates what is involved in evaluating the work integral of Eq. (10-3), in going from a point A ($\mathbf{r}_1$) to a point B ($\mathbf{r}_2$) in a two-dimensional displacement.

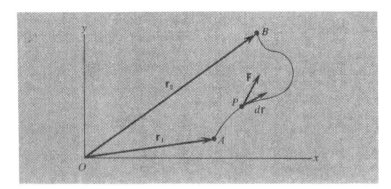

Fig. 10-3 Calculation of work done along a given path.

Equation (10–3) displays the most important property of work and energy: They are *scalar* quantities. An object moving vertically with a speed v has exactly the same kinetic energy as if it were traveling horizontally at this same speed, although its vector momentum would be quite different. This scalar property of energy will be exploited repeatedly in our future work.

WORK, ENERGY, AND POWER

This chapter is chiefly concerned with developing some general dynamical methods based upon the concepts of work and mechanical energy. The practical use of these methods will, however, involve numerical measures of these quantities in terms of appropriate units. The purpose of this section is to introduce some of these units for future reference.

Our basic unit, already introduced in Chapter 6, will be the unit of work or energy in the MKS system—the joule:

$$1 \text{ J} = 1 \text{ N-m} = 1 \text{ kg-m}^2/\text{sec}^2$$

If the CGS system of units is used, the unit of energy is the erg:

$$1 \text{ erg} = 1 \text{ dyn-cm} = 10^{-7} \text{ J}$$

Before going any further, we shall introduce a seeming diversion —the concept of *power*, defined as the rate of doing work:

$$\text{power} = \frac{dW}{dt}$$

(For a mechanical system, putting $dW = \mathbf{F} \cdot d\mathbf{r}$, we have power = $\mathbf{F} \cdot \mathbf{v}$.) Power is a concept (and a quantity) of great practical importance, because the time that it takes to perform some given amount of work may be a vital consideration. For example, a small electric motor may be just as capable of driving a hoist as a big one (given, perhaps, a few extra gear wheels), but it may be quite unacceptable because the job would take far too long. Power is essentially a practical engineering concept; we shall not be using it in our development of the principles of dynamics. But one of our accepted measures of work is often expressed in terms of a unit of power—the *watt*. In terms of mechanical quantities,

$$1 \text{ W} = 1 \text{ J/sec}$$

i.e.,

$$1 \text{ W-sec} = 1 \text{ J}$$

The most familiar use of the watt is, of course, electrical, through the relation watts = volts × amperes, but it is important to realize that it is not a specifically electrical quantity.[1] A convenient energy unit for domestic purposes (especially one's electricity bill) is the kilowatt-hour (kWh):

$$1 \text{ kWh} = 3.6 \times 10^6 \text{ J}$$

In chemical and thermal calculations the standard unit is the Calorie, defined as the amount of energy required to raise 1 kg of water from 15 to 16°C:

$$1 \text{ Cal} = 4.2 \times 10^3 \text{ J}$$

In atomic and nuclear physics, energy measurements are usually expressed in terms of the electron volt (eV) or its related units keV (10^3), MeV (10^6), and GeV (10^9). The electron volt is the amount of energy required to raise one elementary charge through 1 V of electric potential difference:

$$1 \text{ eV} = 1.6 \times 10^{-19} \text{ J}$$

Finally, as Einstein first suggested and as innumerable observa-

---

[1]The other familiar unit of power is of course the horsepower. About this we shall say only that 1 hp = 746 W, so that (as a very rough rule of thumb) one can say that it takes about 1 kw to drive a 1 hp electric motor.

Most of the power levels of practical importance can be conveniently described using power units based on the watt—microwatts (10^{-6} W, typical of very weak radio signals), milliwatts (10^{-3} W), kilowatts (10^3 W), and megawatts (10^6 W, useful as a unit in generating-plant specifications). It is worth noticing, for comparison, that the sun's power output is about 3.8×10^{26} W!

TABLE 10-1: THE ENERGY OF THINGS

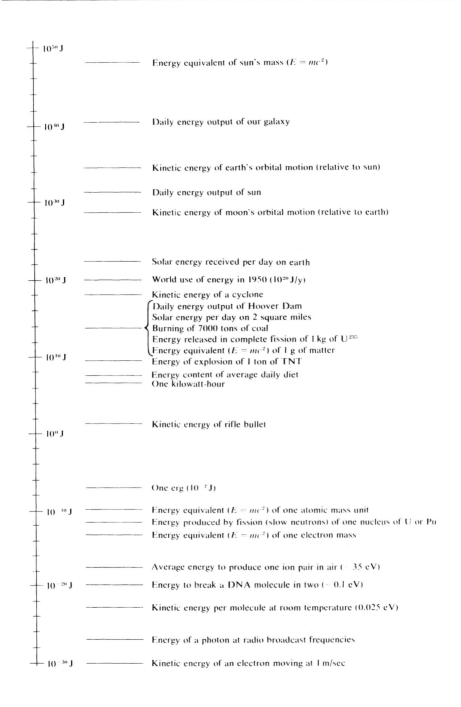

tions have confirmed, there is an equivalence between what we customarily call mass and what we customarily call energy. In classical mechanics these are treated as entirely separate concepts, but it is perhaps worth quoting this equivalence here, so that we have our selection of energy measures all in one place:

1 kg of mass is equivalent to 9×10^{16} J

For the sake of interest, we show in Table 10-1 some representative physical examples involving different orders of magnitude of energy, all expressed in terms of our basic mechanical unit, the joule.

GRAVITATIONAL POTENTIAL ENERGY

We shall begin with a simple and very familiar problem. An object has been thrown vertically upward and is moving under gravity, losing speed as it goes higher. Let us take a y axis, positive upward, and suppose that the object passes the horizontal level $y = y_1$ with velocity v_1 and reaches $y = y_2$ with velocity v_2 (Fig. 10-4). Then the purely kinematic description of the motion is given by

$$v_2^2 = v_1^2 + 2a(y_2 - y_1)$$

with

$$a = -g$$

leading to the familiar result[1]

$$v_2^2 = v_1^2 - 2g(y_2 - y_1)$$

Now let us consider this in terms of work and energy. The change of the kinetic energy (K) between y_1 and y_2 is given by

$$K_2 - K_1 = \tfrac{1}{2}mv_2^2 - \tfrac{1}{2}mv_1^2$$

Using the preceding kinematic equation this can be written

$$K_2 - K_1 = -mg(y_2 - y_1)$$

Now the quantity $-mg$ is just the constant gravitational force,

---

[1] Remember, we have defined g as a *positive* number equal to the magnitude of the local gravitational acceleration in any units that may be chosen.

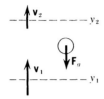

Fig. 10-4 Velocity change associated with vertical motion under gravity.

F_g, that acts on the object as it moves up from y_1 to y_2 (see Fig. 10-4). Thus we see that the change of K is precisely equal to the work done by the gravitational force, in conformity with the general result expressed by Eq. (10-3):

$$K_2 - K_1 = F_g(y_2 - y_1)$$

We are going to rewrite this equation in a different way, such that the quantities referring to the position $y = y_1$ appear on one side, and the quantities referring to the position $y = y_2$ appear on the other. We shall express the result in what may seem at first to be a clumsy fashion, but the reason will quickly appear. Our new statement of the result is

$$K_2 + (-F_g y_2) = K_1 + (-F_g y_1) \qquad (10\text{-}4)$$

What we have done here is to deliberately frame the mathematical statement in such a way that the *sum* of two quantities has the same value at two different positions. That is the formal basis of the statement of conservation of energy. We *define* the *potential energy* $U(y)$ at any given value of y through the equation

$$U(y) = -F_g y \qquad (10\text{-}5)$$

(thus making $U = 0$ at $y = 0$).[1] Notice particularly that $U(y)$ is the *negative* of the work done by the gravitational force. Substituting this definition of the potential energy into Eq. (10-4) we thus get

$$E = K_2 + U_2 = K_1 + U_1 \qquad (10\text{-}6)$$

where E is the *total mechanical energy*. Putting $F_g = -mg$ in Eq. (10-5), we have the well-known result

$$U(h) = mgh$$

for an object at height h above the ground (or other horizontal level that is defined as the zero of potential energy). For this case

---

[1] We shall use the symbol U for potential energy throughout this book. The symbol V is widely used also, but we shall avoid it here because we so often use v's (large and small) to denote velocities.

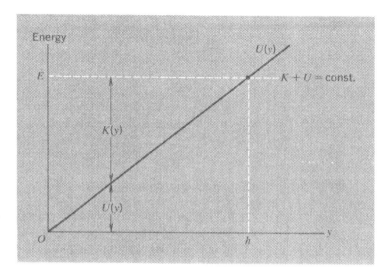

Fig. 10-5 Energy diagram for vertical motion above a horizontal surface.

the energy-conservation statement can be simply represented as in Fig. 10-5. We shall have more to say about such graphs shortly.

You will undoubtedly be familiar with another way of interpreting a potential energy such as $U(h)$ in the last equation. It represents exactly the amount of work that *we* would have to do in order to raise an object through a distance h, against the gravitational pull, without giving it any kinetic energy. In order to achieve this, we must supply an *external* force, F_{ext}; if this is insignificantly greater than mg, the object will move upward with negligibly small acceleration ($\mathbf{F}_{net} \approx 0$), thus arriving at some higher level (Fig. 10-6) with almost zero velocity.[1] If the object

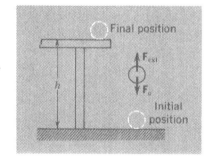

Fig. 10-6 Use of an external force to move an object in the direction of increasing gravitational potential energy.

[1]To make it even more precise, we could apply a force just a shade bigger than mg for a brief time at the beginning, to get the object moving, then change to F_{ext} *exactly* equal to mg for most of the trip (the object thus continuing to move upward at constant velocity under zero net force), and finally let F_{ext} become a shade less than mg for a brief time just before the end, so that the object finishes up at rest at height h above its starting point.

is subsequently released, then the work done on it by the gravitational force is given to the object as kinetic energy (corresponding to traveling from $y = h$ to $y = 0$ in Fig. 10-5). Many devices (pile drivers are a particularly clear example) operate in precisely this way.

MORE ABOUT ONE-DIMENSIONAL SITUATIONS

Equation (10-6) is a compact statement of the conservation of total mechanical energy in any one-dimensional problem for which the force acting on an object, due to its environment, depends only on the object's position. To derive this energy-conservation result more generally, suppose that the environment supplies a force $F(x)$ that varies with position x in any arbitrary way—but has a unique value at any given value of x. The work done by this force as an object moves from x_1 to x_2 is given by

$$W = \int_{x_1}^{x_2} F(x)\, dx$$

Equating this work to the change of kinetic energy, we have

$$K_2 - K_1 = \int_{x_1}^{x_2} F(x)\, dx \tag{10-7}$$

In order to cast this into the form of a conservation statement, we introduce an arbitrary reference point x_0 and express the work integral as follows:

$$\int_{x_1}^{x_2} F(x)\, dx = \int_{x_0}^{x_2} F(x)\, dx - \int_{x_0}^{x_1} F(x)\, dx \tag{10-8}$$

Substituting this in Eq. (10-7), we have

$$K_2 + \left[-\int_{x_0}^{x_2} F(x)\, dx \right] = K_1 + \left[-\int_{x_0}^{x_1} F(x)\, dx \right]$$

We then define the potential energy $U(x)$ at any point x by the following equation:

$$U(x) - U(x_0) = -\int_{x_0}^{x} F(x)\, dx \tag{10-9}$$

Notice once again the minus sign on the right. The potential energy at a point, relative to the reference point, is always defined as the *negative* of the work done by the force as the object moves from the reference point to the point considered. The value of $U(x_0)$, the potential energy at the reference point itself, can be set equal to zero if we please, because in any actual problem we are concerned only with *differences* of potential energy between one point and another, and the associated changes of kinetic energy.

In obtaining Eq. (10-9) we took the force as the primary quantity and the potential energy as the secondary one. Increasingly, however, as one goes deeper into mechanics, potential energy takes over the primary role, and force becomes the derived quantity—literally so, indeed, because by differentiation of both sides of Eq. (10-9) with respect to x, we obtain

$$F(x) = -\frac{dU}{dx} \tag{10-10}$$

This inversion of the roles is not just a formal one (although it does prove to be valuable theoretically) for there are many physical situations in which one's only measurements are of energy differences between two very distinct states, and in which one has no direct knowledge of the forces acting. The electronic work function of a metal, for example, and the dissociation energy of a molecule, represent the only directly observable quantities in these processes of removing a particle to infinity from some initial location. How the force varies from point to point may not be known well—perhaps not at all.

We shall often be spelling out the kinetic energy K in terms of m and v, so that the equation of energy conservation in one dimension is written as follows:

$$\tfrac{1}{2}mv^2 + U(x) = E \tag{10-11}$$

Suppose we choose any particular value of x. Then Eq. (10-11) becomes a quadratic equation for v with two equal and opposite roots:

$$v(x) = \pm \left(\frac{2[E - U(x)]}{m}\right)^{1/2} \tag{10-12}$$

This is the expression of a familiar result, which we can discuss in terms of motion under gravity. If an object were observed to pass a certain point, traveling vertically upward with speed v, then (to the approximation that air resistance could be ignored) it would be observed, a little later, to pass the same point, traveling *downward* at the same speed v. The direction of the velocity has been reversed, but there has been no loss or gain of kinetic energy. In such a case the force is said to be *conservative*. We can see from Eq. (10–12) that this result will hold as long as $U(x)$ is a unique function of x. It means that a particle, after passing through any given point at any speed, will be found to have the same kinetic energy every time it passes through that point again.

Under what conditions does the force have this conservative property? It will certainly *not* be conservative if $F(x)$ depends on the direction of motion of the object to which it is applied. Consider, for example, the addition of a resistive force to the gravitational force in the vertical motion of an object. As the object goes upward through a certain point, the net force on it (downward) is greater than F_g. After it reaches its highest point and begins moving down, the net force on it (again downward) is less than F_g. Hence the net *negative* work done on it as it rises is numerically greater than the net *positive* work as it descends. Thus on balance negative work has been done and the kinetic energy as the object passes back through the designated point is less than initially. The crucial feature is, indeed, that the net work done by $F(x)$ should be *zero* over any journey beginning and ending at any given value of x; only if this condition is satisfied can one define a potential-energy function. It might seem that an equivalent condition is that F be a unique function of position. In one-dimensional situations this is correct. In two- and three-dimensional situations, however, as we shall see later (Chapter 11), the condition that **F** be a single-valued function of **r** is necessary but not sufficient. The condition of zero net work over any *closed* path defines a conservative force in *all* circumstances and should be remembered as a basic definition.

THE ENERGY METHOD FOR ONE-DIMENSIONAL MOTIONS

The use of energy diagrams, such as that of Fig. 10–5, provides an excellent way of obtaining a complete, although perhaps qualitative, picture of possible motions in a one-dimensional system.

Frequently the information so obtained suffices for obtaining physical insight into situations for which analytic solutions are complicated or even unobtainable. In fact, even when analytic solutions can be obtained in terms of unfamiliar functions, they often are of little help in revealing the essential physical characteristics of the motion. The general scheme is as follows: We

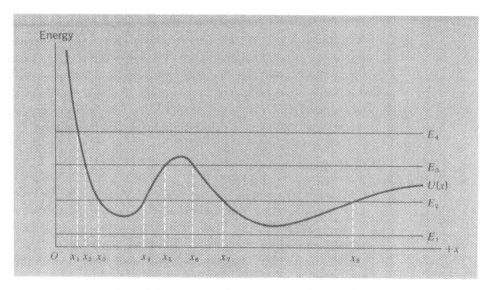

Fig. 10-7 Hypothetical energy diagram for a one-dimensional system.

plot $U(x)$ as a function of x, and on the same plot draw horizontal lines corresponding to different total energies. In Fig. 10-7 is shown such a potential-energy curve and several values of the total energy.

The kinetic energy K of a particle is equal to $(E - U)$, i.e., to the vertical distance from one of the lines of constant energy to the curve $U(x)$ at any point x. For a low energy E_1, $U(x)$ is greater than E for all values of x; this would simply imply a negative value of K and hence an imaginary value of v. Such a situation has no place in classical mechanics—although it must not be discarded so lightly when one comes to atomic and nuclear systems requiring the use of quantum mechanics. For a higher total energy E_2, the motion can occur in two regions, between x_3 and x_4 or between x_7 and x_8. These represent two quite separate situations, because a particle cannot escape from one region to the other as long as its energy is held at the value E_2. One way of seeing this is, of course, in terms of the impossible negative value of K between x_4 and x_7. But there is another way which is valuable as an example of how one "reads" such an energy diagram.

Suppose that our particle, with total energy E_2, is at the point $x = x_3$ at some instant. Its potential energy $U(x_3)$ at this point is equal to the total energy, for this is where the curve of $U(x)$ and the line $E = E_2$ intersect. Thus the particle has zero kinetic energy and hence is instantaneously at rest. However, there *is* a force on it:

$$F(x) = -\frac{dU}{dx}$$

At $x = x_3$, dU/dx is negative and hence $F(x_3)$ is positive—i.e., in the $+x$ direction. Thus the particle accelerates to the right. The force on it, and hence its acceleration, decreases as the slope of the $U(x)$ curve decreases, falling to zero at the value of x at which $U(x)$ is minimum. At this point the speed of the particle is a maximum, and as it moves further in the $+x$ direction (with $dU/dx > 0$) it now experiences a force in the $-x$ direction. The diagram displays all this information before us, and shows the kinetic energy $E_2 - U$ continuing to decrease as the particle approaches x_4. Finally, at x_4 itself, the velocity has fallen to zero —but there is still a force acting in the $-x$ direction. What happens? The particle picks up speed again, traveling to the left, until it reaches x_3 with its velocity reduced to zero. This whole cycle of motion will continue to repeat itself indefinitely as long as the total energy does not decrease. We have, in short, a *periodic* motion, of which we can discern many of the principal features without solving a single equation—just by seeing what the energy diagram has to tell us. The motion between x_7 and x_8 is likewise periodic.

We can dispose of the other possibilities more briefly, having indicated the method. For a still higher energy E_3, two kinds of motion are possible; either a periodic motion between x_2 and x_5, or the unbounded motion of a particle coming in from large values of x, speeding up as it passes x_8, then slowing down and reversing its direction of motion at x_6, moving off to the right and duplicating all the changes of speed on the way in. Finally, for the still larger energy E_4, the only possible motion is unbounded; a particle coming in from large values of x, speeding up, slowing down, speeding up, slowing down again, and reversing its direction of motion at x_1, after which it proceeds inexorably in the direction of ever-increasing x. For each of these motions, the speed at any point can be obtained graphically by measuring the vertical distance from the appropriate line of constant energy to the corresponding point on the potential-energy curve.

Caution: The curve of $U(x)$ in Fig. 10–7 is almost *too* graphic. It tends to conjure up a picture of a particle sliding down the slopes and up the peak like a roller coaster. Do not forget that it is a *one*-dimensional motion that is the subject of the analysis. The vertical scale is energy, and has nothing necessarily to do with altitude.

After this general introduction, let us consider some specific examples of one-dimensional motions as analyzed by the energy method.

SOME EXAMPLES OF THE ENERGY METHOD

Bouncing ball

Suppose that a ball, moving along a vertical line, bounces repeatedly on a horizontal floor. Let us first imagine that there is no dissipation (loss) of mechanical energy, so that this energy remains constant at some value E.

We shall use y to denote the position of the center of gravity (CG) of the ball, and take $y = 0$ to be defined by the first contact of the ball with the floor. We shall take this configuration to correspond to $U = 0$. For $y > 0$ the potential energy of the ball is given by

$$U(y) = mgy \qquad (y > 0)$$

Now $y = 0$ does not, in any real physical situation, represent the

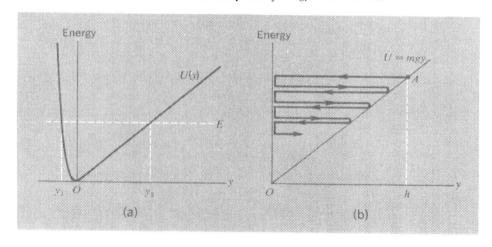

Fig. 10–8 (a) Energy diagram for a ball bouncing vertically. (b) Idealization of (a) to represent a situation in which the impact at $y = 0$ is completely rigid but in which there is some dissipation of energy at each bounce.

lowest point reached by the CG of the ball. The floor does not exert any force on the ball until it (the floor) has been compressed slightly. An equivalent remark can be made about the ball. Thus the ball certainly moves into the region $y < 0$. As it does this, however, it experiences a positive (upward) force that increases extremely rapidly as y becomes more negative, and completely overwhelms the (negative) gravitational force that exists, of course, at all values of y. This large positive force gives rise to a very steep increase of $U(y)$ with y for $y < 0$ [see Fig. 10-8(a)].

For any given value of the total energy, therefore, the ball oscillates between positions y_1 and y_2 as shown in the figure. The motion is periodic—that is, there is some well-defined time T between successive passages *in the same direction* of the ball through any given point.

Now in practice y_1 may be numerically very small compared to y_2. For instance, if a steel ball bounces on a glass plate, we might easily have y_1 of the order of 0.01 cm and y_2 of the order of 10 cm. Thus for many purposes we can approximate the plot of $U(y)$ against y for $y < 0$ by a vertical line, coinciding with the energy axis of Fig. 10-8(a). This represents the physically unreal property of perfect rigidity—an arbitrarily large force is called into play for zero deformation. However, if we can justifiably use this approximation, then we have a simple quantitative description of the situation. The motion is confined to $y \geq 0$ and is defined by

$$\tfrac{1}{2}mv_y^2 + mgy = E \qquad (y > 0) \qquad (10\text{-}13)$$

The maximum height h is, of course, defined by putting $v_y = 0$:

$$h = \frac{E}{mg} \qquad (10\text{-}14)$$

To find the period T of the motion we can calculate the time for the ball to travel from $y = 0$ to $y = h$ and then double it. That is, we have the following relation:

$$\frac{T}{2} = \int_{y=0}^{h} \frac{dy}{v_y} \qquad (10\text{-}15)$$

because the elementary contribution dt to the time of flight is equal to dy divided by the speed v_y at any given point.

Now from Eq. (10-13) we have

$$v_y = \pm \left[\frac{2(E - mgy)}{m} \right]^{1/2}$$

Taking the positive root, to correspond to upward motion, we have, from Eq. (10–15),

$$T = 2\int_0^h \left[\frac{m}{2(E - mgy)} \right]^{1/2} dy$$

$$= \sqrt{\frac{2}{g}} \int_0^h \frac{dy}{[(E/mg) - y]^{1/2}}$$

We can simplify this by noting that E/mg is just the maximum height h [Eq. (10–14)]. Thus we have

$$T = \sqrt{\frac{2}{g}} \int_0^h \frac{dy}{(h - y)^{1/2}}$$

This is an elementary integral (change the variable to $w = h - y$) yielding the result

$$T = 2\sqrt{\frac{2h}{g}}$$

You will, of course, recognize the correctness of this result from the simple kinematic problem of an object falling with constant acceleration, and could reasonably object that this is another of those cases in which we have used a sledgehammer to kill a fly. But it is the *method* that you should focus attention on, and perhaps the use of a familiar example will facilitate this. It should not be forgotten that most motions involve *varying* accelerations, so that the standard kinematic formulas for motion with constant acceleration do not apply. But Eq. (10–15), in which v is defined at any point by the energy equation, can be used for any one-dimensional motion and can be integrated numerically if necessary.

Before leaving this example, let us use it to illustrate one other instructive feature of the energy diagram. We know that the total mechanical energy of a bouncing ball does not in fact stay constant but decreases quite rapidly. Although there is little loss of energy while the ball is in flight, there is a substantial loss at each bounce. Figure 10-8(b) shows how this behavior can be displayed on the energy diagram. Starting at the point A, the history of the whole motion is obtained by following the arrows. The successive decreases in the maximum height of bounce, and the inevitable death of the motion at $y = 0$, are quite apparent by inspection of the figure.

Mass on a spring

There are very many physical systems—not just ordinary mechanical systems, but also atomic systems, and even electrical ones—that can be analyzed by analogy with a mass on a spring. The reason for this lies in two features:

1. A mass typifies the property of inertia, which has its analogues in diverse systems and which acts as a repository of kinetic energy.

2. A spring represents a means of storing potential energy according to a particular law of force that has its counterparts in all kinds of physical interactions.

We have already studied this problem—the problem of the harmonic oscillator—in some detail (in Chapter 7) as an application of Newton's law, but it is well worth analyzing the problem again from the standpoint of energy conservation—in part as an illustration of this method but chiefly because the description in terms of energy opens the way to a far wider range of situations. Not only does it provide a pattern for the handling of more complex oscillatory problems in classical mechanics; it also supplies the foundation for formulating equivalent problems in quantum theory.

Our starting point will again be the restoring force of an ideal spring as described by Hooke's law:

$$F = -kx \tag{10-16}$$

where x is the position of the free end of the spring relative to its relaxed position, k the "force constant" of the spring, measurable in N/m, and the negative sign gives the direction of the force, opposite to the displacement of the free end. No real spring obeys this law over more than a limited range. The properties of a real spring can be expressed by a graph such as Fig. 10–9(a), which represents the force $F(x)$ exerted *by the spring* as a function of its extension x. Within the linear range the potential energy stored in the spring is, according to Eq. (10–9),

$$U(x) = -\int_0^x F(x)\,dx = +\int_0^x kx\,dx = \frac{kx^2}{2} \tag{10-17}$$

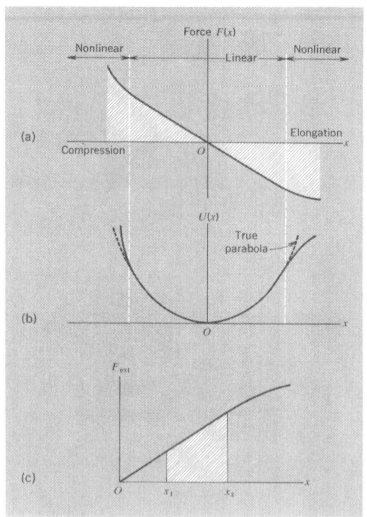

Fig. 10-9 (a) Restoring force versus displacement for a spring. (b) Potential-energy diagram associated with (a). (c) Graph of applied force versus extension in a static deformation of a spring.

where we have chosen $U = 0$ for $x = 0$, i.e., when the spring is relaxed. Figure 10–9(b) shows this potential energy plotted against x. Since the potential-energy change can be calculated as the work done by a force F_{ext} just sufficient to overcome the spring force itself, the increase of potential energy in the spring for any given increase of extension can be obtained as the area, between given limits, under a graph of F_{ext} against x [see Fig. 10–9(c)]. F_{ext} can be measured as the force needed to maintain the spring at constant extension for various values of x. Outside the linear region (whose boundaries are indicated by dashed lines) $F(x)$ can be integrated graphically so as to obtain numerically the potential

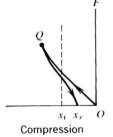

Fig. 10-10 Mechanical hysteresis.

energy for an arbitrary displacement.

Before we examine motion under this spring force, let us consider briefly what would happen if the spring employed in our example were made of lead, for example. As we compress such a spring from its original length, it will exert a force that behaves very much in the way shown in Fig. 10–9(a) and by the line OQ in Fig. 10–10. However, if we now remove the external agency, the spring will not return to its original length but will acquire a permanent deformation represented by point x_r in Fig. 10–10. Clearly, such a spring exerts a force that is *not* conservative. The value of the force depends not only on the compression but also on the past history. At the value x_1 of the compression in Fig. 10–10, we find two values of the force, one as the spring is shortened and the other as it is released. This type of behavior is called *hysteresis* and results in dissipation of mechanical energy.

Let us now consider in more detail the way in which the use of the energy diagram helps us to analyze the straight-line motion of a harmonic oscillator. In Fig. 10–11 are shown the potential energy $U = \frac{1}{2}kx^2$ plotted against x, and two different total energies, E_1 and E_2.

For a given energy E_1, as we have already discussed, the vertical distance from the horizontal line E_1 to the curve $U = \frac{1}{2}kx^2$ for any value of x is equal to the kinetic energy of the particle

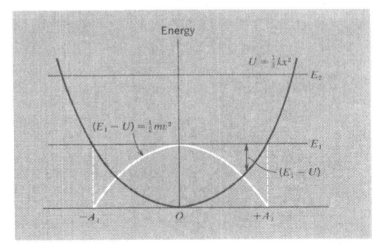

Fig. 10-11 Energy diagram for a spring that obeys Hooke's law.

at x. This is maximum at $x = 0$; at this point all the energy is kinetic and the particle attains its maximum speed. The kinetic energy and hence the particle speed decreases for positions on either side of the equilibrium position O and is reduced to zero at the points $x = \pm A_1$. For values of x to the right of $+A_1$ or to the left of $-A_1$, $(E_1 - U)$ becomes negative; v^2 is negative and there exists no real value of v. This is the region into which the particle never moves (at least in classical mechanics); thus the positions $x = \pm A_1$ are *turning points* of the motion, which is clearly oscillatory.

The amplitude A_1 of the motion is determined by the total energy E_1. Since the kinetic energy $K_1 (= E_1 - U)$ is zero at $x = \pm A_1$, we have

$$\tfrac{1}{2} k A_1^2 = E_1$$

or

$$A_1 = \left(\frac{2E_1}{k}\right)^{1/2} \tag{10-18}$$

For a larger energy, E_2, the amplitude is larger in the ratio $(E_2/E_1)^{1/2}$, but the qualitative features of the motion are the same.

It is interesting to note that the general character of the motion as inferred from the energy diagram would be the same for any potential-energy curve that has a minimum at $x = 0$ and is symmetrical about the vertical axis through this point. All motions of this sort are periodic but differ one from the other in detail, e.g., the dependence of speed on position and the dependence of the period on amplitude. Suppose that the period of the motion is T. Then for any *symmetrical* potential-energy diagram, we can imagine this time divided up into four equal portions, any one of which contains the essential information about the motion. For suppose that, at $t = 0$, the particle is traveling through the point $x = 0$ in the positive x direction. Let its velocity at this instant be called v_m—it is the biggest velocity the particle will have during its motion. At $t = T/4$ the particle is at its maximum positive displacement ($x = +A$ in Fig. 10-11), and $v = 0$. It then retraces its steps, reaching $x = 0$ after a further time $T/4$ and passing through this point with $v = -v_m$. In two further intervals $T/4$ it goes to its extreme negative displacement ($x = -A$) and at $t = T$ is once again passing through the point

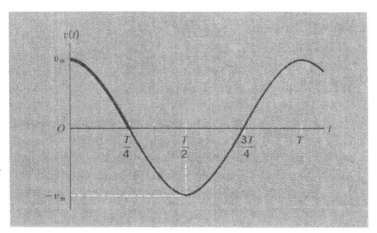

Fig. 10–12 Sinusoidal variation of velocity with time for a particle subjected to a restoring force proportional to displacement. The motion during the first quarter-period suffices to define the rest of the curve.

$x = 0$ with $v = v_m$. This sequence will repeat itself indefinitely. Furthermore, knowing the symmetry of the problem, we could construct the complete graph of v against t from a detailed graph for the first quarter-period alone (see Fig. 10–12, in which the basic quarter period is drawn with a heavier line than the rest).

In the next section we shall go beyond this rather general examination of the mass-spring system, including nonlinear restoring forces, and shall redevelop the rest of the detailed results that apply to the ideal harmonic oscillator. Before doing that, however, we shall consider one more simple example that illustrates the usefulness of the energy method.

Dynamics of a catapult

The catapult is an ancient and effective device for launching stones or other missiles with quite high velocities, by converting the potential energy of a stretched elastic cord into the kinetic energy of a mass. In Fig. 10–13(a) we show the essentials of the arrange-

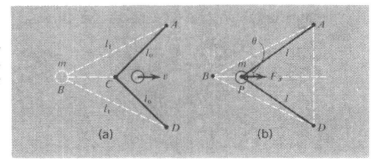

Fig. 10–13 (a) Initial and final stages of the launching of an object by a catapult. (b) Intermediate state, showing the instantaneous force acting.

ment. An elastic cord, of natural (unstretched) length $2l_0$, is attached to fixed supports at the points A and D ($AC = CD = l_0$). An object of mass m is placed at the midpoint of the cord and drawn back to point B. When it is released, the object travels along the line BC, and at the point C it begins its free flight with speed v.

Let us assume that the cord develops a tension proportional to its increase of length. Then we can use Eq. (10-17) to calculate the stored energy when the mass is at the initial position, B. Remembering that there are two segments of cord, each of initial length l_0 and stretched length l_1, we have

$$U = 2 \times \tfrac{1}{2}k(l_1 - l_0)^2 = k(l_1 - l_0)^2$$

In the idealization that the cord has negligible mass and thus does not drain off any of the elastic energy for its own motion, we can equate the final kinetic energy of the projectile to the initial potential energy as given above. Thus we have

$$\tfrac{1}{2}mv^2 = k(l_1 - l_0)^2$$

and so

$$v = \left(\frac{2k}{m}\right)^{1/2}(l_1 - l_0)$$

This example displays particularly well the advantage of making the calculation with the help of the scalar quantity, energy, instead of the vector quantity, force. For suppose we wanted to calculate the final velocity of the mass by direct application of $F = ma$. Then, as indicated in Fig. 10-13(b), we should have to consider the state of affairs at an arbitrary instant when the mass was at some point P between B and C such that $AP = l$. The tension in each half of the cord would be $k(l - l_0)$, and the instantaneous acceleration would be obtained by resolving these tension forces along the line BC:

$$m\frac{dv}{dt} = F_x = 2k(l - l_0)\cos\theta$$

This equation for dv/dt would then have to be integrated between the points B and C. Or, alternatively, we could calculate the total work done by the varying force F_x along the path BC. In either case, this is a far harder road to the final result than is the direct application of conservation of mechanical energy.

THE HARMONIC OSCILLATOR BY THE ENERGY METHOD

We shall now return to the analysis of the oscillatory motion of an object attached to a spring that obeys Hooke's law. The basic energy equation for a mass on a spring with a restoring force proportional to displacement is

$$\tfrac{1}{2}mv^2 + \tfrac{1}{2}kx^2 = E \tag{10-19}$$

where E is some constant value of the total energy. Since $v = dx/dt$, this can be rewritten as

$$\tfrac{1}{2}m\left(\frac{dx}{dt}\right)^2 + \tfrac{1}{2}kx^2 = E \tag{10-20}$$

Equation (10–19) already gives us v as a function of x, but to have a full description of the motion we must solve Eq. (10–20) so as to obtain x (and hence v) as functions of t.

Our way of dealing with Eq. (10–20) will appeal to the knowledge of trigonometric functions and their derivatives that one develops at an early stage of any calculus course. We start out by dividing the equation throughout by E. Then we get

$$\frac{m}{2E}\left(\frac{dx}{dt}\right)^2 + \frac{k}{2E}x^2 = 1 \tag{10-21}$$

We notice that this is a sum of two terms involving the square of a variable (x) and the square of its derivative (with respect to t). The sum is equal to 1. Now we can relate this to a very familiar relationship involving trigonometric functions: If $s = \sin \varphi$, then

$$\frac{ds}{d\varphi} = \cos \varphi$$

and

$$\left(\frac{ds}{d\varphi}\right)^2 + s^2 = \cos^2 \varphi + \sin^2 \varphi = 1 \tag{10-22}$$

Equations (10–21) and (10–22) are exactly similar in form! We must be able to match them, term by term:

$$\frac{m}{2E}\left(\frac{dx}{dt}\right)^2 \equiv \left(\frac{ds}{d\varphi}\right)^2$$

$$\frac{k}{2E}x^2 \equiv s^2 \tag{10-23}$$

The second of these is satisfied by putting

$$x = \left(\frac{2E}{k}\right)^{1/2} s = \left(\frac{2E}{k}\right)^{1/2} \sin \varphi \tag{10-24}$$

What is φ? We can find it by evaluating dx/dt by differentiation of both sides of Eq. (10-24) with respect to t:

$$\frac{dx}{dt} = \left(\frac{2E}{k}\right)^{1/2} \cos \varphi \frac{d\varphi}{dt}$$

But the first equation of (10-23) is satisfied by putting

$$\frac{dx}{dt} = \left(\frac{2E}{m}\right)^{1/2} \frac{ds}{d\varphi} = \left(\frac{2E}{m}\right)^{1/2} \cos \varphi$$

Comparing these two expressions for dx/dt we find the following condition on φ:

$$\frac{d\varphi}{dt} = \left(\frac{k}{m}\right)^{1/2} = \omega \tag{10-25}$$

where ω is the angular velocity (also called the circular frequency) that we met in our previous solution of the harmonic oscillator problem (Ch. 7, p. 226), starting from $F = ma$.

Integrating the last equation with respect to t we thus get

$$\varphi = \omega t + \varphi_0$$

where φ_0 is the initial phase. Substituting this expression for φ back into Eq. (10-24) then gives us

$$x = \left(\frac{2E}{k}\right)^{1/2} \sin(\omega t + \varphi_0) \tag{10-26}$$

If we note that $(2E/k)^{1/2}$ is equal to the amplitude A of the motion [Eq. (10-18)] we arrive finally at the same equation for $x(t)$ that we found in Chapter 7 [Eq. (7-42)].

[If you have some prior knowledge of differential equations, you may regard our method of solution above as being rather cumbersome. You may prefer to proceed at once to the recognition that Eq. (10-20) leads to the relationship

$$\left(\frac{dx}{dt}\right)^2 = \omega^2(A^2 - x^2)$$

and hence to the following solution by direct integration:

$$\frac{dx}{dt} = \omega(A^2 - x^2)^{1/2}$$

$$\omega\, dt = \frac{dx}{(A^2 - x^2)^{1/2}}$$

$$\omega t + \varphi_0 = \sin^{-1}\left(\frac{x}{A}\right)$$

and so $x = A \sin(\omega t + \varphi_0)$ as before.]

Equations (10-25) and (10-26) tell us something very remarkable indeed: *The period of a harmonic oscillator*, as typified by a mass on a spring, *is completely independent of the energy or amplitude of the motion*—a result that is not true of periodic oscillations under any other force law. The physical consequences of this are tremendously important. We depend heavily on the use of vibrating systems. If the frequency ν (defined as the number of complete oscillations per second, i.e., $1/T$ or $\omega/2\pi$) varied significantly with the amplitude for a given system, the situation would become vastly more complicated. Yet most vibrating systems behave, to some approximation, as harmonic oscillators with properties as described above. Let us see why.

SMALL OSCILLATIONS IN GENERAL

There are many situations in which an object is in what we call *stable equilibrium*. It is at rest at some point—under no net force—but if displaced in any direction it experiences a force tending to return it to its original position. Such a force, unless it has pathological properties (such as a discontinuous jump in value for some negligible displacement) will have the kind of variation with position shown in Fig. 10-14(a). The normal resting position is marked as x_0. This force function can be integrated to give the potential-energy graph of Fig. 10-14(b). One

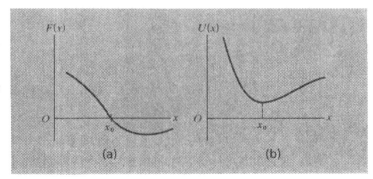

Fig. 10-14 (a) Variation of force with displacement on either side of the equilibrium position in a one-dimensional system. (b) Potential-energy curve associated with (a).

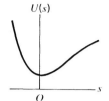

Fig. 10-15 Potential-energy curve of Fig. 10-14(b) referred to an origin located at the equilibrium position.

can then form a mental picture of the object sitting at the bottom, as it were, of the potential-energy hollow, the minimum of which is at $x = x_0$.

Now we can fit any curve with a polynomial expansion. Let us do this with the potential-energy function—but let us do it with reference to a new origin chosen at the point x_0, by putting

$$x = x_0 + s$$

where s is the displacement from equilibrium. The potential-energy curve, now appearing as in Fig. 10-15, can be fitted by the following expansion:

$$U(s) = U_0 + c_1 s + \tfrac{1}{2} c_2 s^2 + \tfrac{1}{3} c_3 s^3 + \cdots \qquad (10\text{-}27)$$

(The numerical factors are inserted for a reason that will appear almost immediately.)

The force as a function of s is obtained from the general relation

$$F(s) = -\frac{dU}{ds}$$

so that we have

$$F(s) = -c_1 - c_2 s - c_3 s^2 \cdots$$

However, by definition, $F(s) = 0$ at $s = 0$; this is the equilibrium position. Hence $c_1 = 0$, and so our equation for F becomes

$$F(s) = -c_2 s - c_3 s^2 \cdots \qquad (10\text{-}28)$$

Now, whatever the relative values of the constants c_2 and c_3, there will always be a range of values of s for which the term in s^2 is much less than the term in s, for the ratio of the two is equal to $c_3 s / c_2$, which can be made as small as we please by choosing s small enough. A similar argument applies, even more strongly, to all the higher terms in the expansion. Hence, unless our potential-energy function has some very special properties (such as having $c_2 = 0$) we can be sure that for sufficiently small oscillations it will be just like the potential-energy function of a spring that obeys Hooke's law. We can write

$$U(s) \approx \tfrac{1}{2}c_2 s^2 \qquad (10\text{-}29)$$

which means that the effective spring constant k for the motion is equal to the constant c_2. We shall discuss a specific application of this analysis—molecular vibration—later in this chapter.

THE LINEAR OSCILLATOR AS A TWO-BODY PROBLEM

So far, in all our discussions of potential energy, we have analyzed the problems as though we had a single object exposed to given forces. A statement of our calculation on an object near the earth's surface could well be in the form: *The potential energy of an object of mass m raised to a height h above the earth's surface is mgh*. Such a statement is perfectly legitimate for situations in which the mass of a particle is very small compared to the mass of the object (or objects) with which it interacts. In such a case, the center of mass of the system is effectively determined by the position of the larger mass. A frame of reference anchored to this larger mass is both a zero-momentum frame and a fixed frame of reference. This is the case for the earth and an ordinary object moving near its surface. It is also the case for interactions between any two objects if one of them is rigidly attached to the earth. One must remember, however, that, strictly speaking, one is analyzing a *two-body system* (the earth and the object which is raised): *mgh* is the increase of potential energy *of the system* when the *separation* between the earth and the object of mass *m* is increased by an amount *h*. In other words, the potential energy is a property of the two objects jointly; it cannot be associated with one or the other individually. If one has two interacting particles of comparable mass, both will accelerate and gain or lose kinetic energy as a result of the interaction between them. It is to this basic two-body aspect of the potential-energy problem that we now turn.

Suppose that we have two particles, of masses m_1 and m_2, connected by a spring of negligible mass aligned parallel to the x axis (Fig. 10-16). Let the particles be at positions x_1 and x_2, as shown, referred to some origin O. If the spring is effectively massless, the forces on it at its two ends must be equal and opposite (otherwise it would have infinite acceleration) and hence, accepting the equality of action and reaction in the contacts between the masses and the spring, the forces exerted on the masses are also equal and opposite. Thus, denoting the force

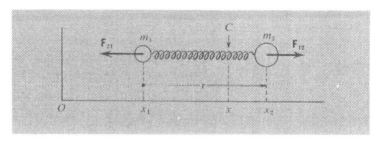

Fig. 10–16 System of two masses connected by a spring, showing the separate coordinates and forces.

exerted on mass 2 *by the spring* as F_{12}, the force F_{21} exerted on mass 1 by the spring is equal to $-F_{12}$.

We shall relate the changes in kinetic energy of the masses to the changes in stored potential energy in the spring.

The potential energy of the spring

First, suppose that m_1 moves a distance dx_1 while m_2 moves dx_2. The work done by the spring is given by

$$dW = F_{12}\,dx_2 + F_{21}\,dx_1$$
$$= F_{12}(dx_2 - dx_1) \quad \text{(since } F_{21} = -F_{12})$$
$$= F_{12}\,d(x_2 - x_1)$$

Clearly the difference $x_2 - x_1$, rather than x_1 and x_2 separately, defines the elongation of the spring (and hence the energy stored in it). Let us introduce a special coordinate, r, to denote this:

$$r = x_2 - x_1$$

Then

$$dW = F_{12}\,dr \quad \text{(work done } by \text{ spring)} \tag{10-30}$$

The change of potential energy of the spring is equal to $-dW$. Introducing the potential energy function $U(r)$ we have

$$dU = -F_{12}\,dr$$
$$U(r) = -\int F_{12}\,dr \tag{10-31}$$

The kinetic energy of the masses

Our discussion of two-body systems in Chapter 9 suggests clearly that we should introduce the center of mass of the system and refer the motions of the individual masses to it. This allows us, as we have seen, to consider the dynamics in the *CM* frame without reference to the motion of the system as a whole. By Eq. (9–29) we have

$$K = K' + \tfrac{1}{2}M\bar{v}^2$$

where K' is the total kinetic energy of the two masses as measured in the CM frame. Denoting the velocities relative to the CM by v_1' and v_2' as usual, we have

$$K' = \tfrac{1}{2}m_1 v_1'^2 + \tfrac{1}{2}m_2 v_2'^2 \tag{10-32}$$

We have seen (pp. 338–339) that it is very convenient to express K' in terms of the *relative velocity*, v_r, and the *reduced mass*, μ, of the two particles:

$$v_r = v_2' - v_1'$$

$$\mu = \frac{m_1 m_2}{m_1 + m_2}$$

From the definition of the CM (zero-momentum) frame, we have

$$m_1 v_1' + m_2 v_2' = 0$$

Using this, together with the equation for v_r, we find

$$v_1' = -\frac{m_2}{m_1 + m_2} v_r \qquad v_2' = \frac{m_1}{m_1 + m_2} v_r$$

Substituting these values into Eq. (10-32) one arrives once again at the result expressed by Eq. (9-30a):

$$K' = \frac{1}{2}\frac{m_1 m_2}{m_1 + m_2} v_r^2 = \tfrac{1}{2}\mu v_r^2 \tag{10-33}$$

We shall be considering the changes of kinetic energy of the masses as related to the work done on them by the spring. On the assumption that no external forces are acting, we have $\bar{v} = $ const., in which case

$$dK = dK'$$

The motions

At this point we can assemble the foregoing results and equate the change of kinetic energy to the work done by the spring. We evaluated dW [in Eq. (10–30)] in terms of laboratory coordinates, although, as we saw (and could have predicted), the result depends only on $x_2 - x_1$, which is equal to $x_2' - x_1'$—both being equal to the relative coordinate r. Likewise, as we have just seen, $dK = dK'$. We can, in fact, put

$$dK' = F_{12}\, dr \quad \text{(work done } by \text{ spring)}$$

Integrating,

$$K' = \int F_{12}\, dr + \text{const.}$$

And now, with the help of Eq. (10-31), we can write this as a statement of the total mechanical energy E' in the CM frame:

$$K' + U(r) = E' \tag{10-34}$$

For the specific case of a spring of spring constant k and natural length r_0, we can put

$$r = r_0 + s$$
$$U(s) = \tfrac{1}{2}ks^2$$

Also

$$v_r = \frac{dr}{dt} = \frac{ds}{dt}$$

Thus the equation of conservation of energy [Eq. (10-34)] becomes

$$\tfrac{1}{2}\mu \left(\frac{ds}{dt}\right)^2 + \tfrac{1}{2}ks^2 = E' \tag{10-35}$$

where

$$\mu = \frac{m_1 m_2}{m_1 + m_2}$$

This is exactly of the form of the linear oscillator equation; its angular frequency ω and its period T are given by

$$\omega = \left(\frac{k}{\mu}\right)^{1/2} \qquad T = 2\pi \left(\frac{\mu}{k}\right)^{1/2} \tag{10-36}$$

It is to be noted that the reduced mass μ is less than either of the individual masses, so that for a given spring the period is shorter in free oscillation than if one of the masses is clamped tight.

COLLISION PROCESSES INVOLVING ENERGY STORAGE

With the help of the analysis developed in the last section, we can gain further insight into certain inelastic or explosive collisions of the type discussed in Chapter 9.

We shall introduce the problem by imagining a little mechanical gadget that could be constructed without much trouble. The gadget is a spring equipped with a buffer that slides along a guide and becomes locked in place if the spring is compressed by more than a certain amount [see Fig. 10–17(a)]. Suppose that this device (assumed to be of negligible mass) is attached to an object of mass m_2, and that an object of mass m_1 collides with it. For simplicity, let us take m_2 to be initially stationary, and let m_1 approach it with a speed u_1.

If u_1 is small, the collision is perfectly elastic. The spring will be compressed a little when m_1 strikes it, but it will return to its original unstretched condition, and at this instant the collision process comes to an end. Throughout the time that the spring is at all compressed, the mass m_2 is subjected to an accelerative force to the right and m_1 is subjected to a decelerative force to the left. A certain positive amount of work is done on m_2 and an equal amount of negative work is done on m_1. The total kinetic energy is the same after the collision as beforehand, but it has been reapportioned.

If u_1 is increased, the situation is finally reached in which the maximum compression of the spring just brings it to the locking position. At this instant m_1 and m_2 are at rest relative to one another, and because the spring is prevented from pushing them apart again, this is the way they remain. In other words, they would continue to move on as a single composite object of mass $m_1 + m_2$; the collision has suddenly become completely inelastic.

Fig. 10–17 (a) Collision involving one object with an energy-storage device. (b) Same collision in the center-of-mass frame. The collision is elastic if the total kinetic energy in this frame is less than the work needed to compress the spring by the critical amount.

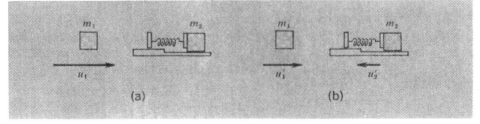

What defines the critical condition at which the collision changes from elastic to inelastic? We can most easily discuss this from the standpoint of the CM frame [Fig. 10–17(b)]. In this frame the masses have a total kinetic energy given (cf. the previous section) by

$$K' = \tfrac{1}{2}\mu v_r^2$$

with

$$\mu = \frac{m_1 m_2}{m_1 + m_2} \qquad v_r = -u_1 \qquad (\text{because } u_2 = 0)$$

If we denote the initial kinetic energy of m_1 in the Lab frame as K_1, we can put

$$K' = \frac{m_2}{m_1 + m_2} K_1$$

Now at the critical value of K_1 all the energy K' is used up in compressing the spring to the locking position; this requires a well-defined amount of work equal to the energy, U_0, stored in the spring in this configuration. Thus we can put

$$K' = \frac{m_2}{m_1 + m_2} K_1 = U_0$$

or

$$K_1 = \frac{m_1 + m_2}{m_2} U_0 \qquad (10\text{–}37)$$

If U_0 is given, the above equation defines a *threshold value* of K_1, at which inelastic collisions become possible and below which the collisions can only be elastic.

If we consider still higher values of K_1, we obtain situations in which the spring is compressed beyond its locking point but then returns to that critical length and stops. During this partial reextension it pushes m_1 and m_2 apart, giving them a final relative velocity and a kinetic energy of relative motion in the CM frame. The collision is still inelastic but only partially so. A well-defined amount (U_0) of the originally kinetic energy of the colliding masses has been locked up in the spring, and the dynamics of the collision can be analyzed in these terms. Our analysis in Chapter 9 of inelastic and explosive collisions was, in fact, made precisely in this way. The quantity Q, representing the value of $K_f - K_i$ that we introduced there (p. 347) is in this case simply equal to $-U_0$.

The purely classical dynamical situation described above is closely paralleled by many collision processes in atomic and nuclear physics, in which an incident particle (e.g., an electron or a proton) strikes a stationary target particle that has various sharply defined states of higher internal energy than its normal "ground state." The sharpness and discreteness of these characteristic states is understood in terms of quantum theory, but all that we need here is the knowledge of their existence. If a study is made of the distribution of kinetic energies of the electrons or protons after they have collided with target particles of a given species, the results give information about the excited energy levels involved. Some examples of this kind of analysis are shown in Fig. 10–18. The higher the bombarding energy of the incident particle, the larger the number of excited states that can be stimulated and detected in this way. Figure 10–18(a) shows data from the scattering of electrons by helium atoms, and Fig. 10–18(b) shows some results for the scattering of protons by the nucleus boron 10. In the former case the target particle is so massive that the available kinetic energy in the CM (K') is insignificantly different from the electron energy K_1. In the latter case, however, the center-of-mass effect is considerable and one must use Eq. (10–37) to infer the excitation energy U_0 from the threshold value of K_1. One important difference between these atomic or nuclear scattering processes and the classical ones is that perfectly elastic collisions may still occur, with a certain probability, even after the threshold energy for inelastic processes has been exceeded. This is an example of the general feature of quantum mechanics that one only has relative probabilities of events when several outcomes are possible.

To return briefly to the classical situations once again, one can of course imagine mechanisms that would provide for an increase rather than a decrease of kinetic energy as the result of a collision. One could, for example, have a spring already compressed that would be released if a trip mechanism were activated by the initial impact. Such a process—an explosive collision—would, like the inelastic collision with energy storage, require a certain minimum threshold energy to make it go. Something very comparable to this happens with chemical explosive systems that require the firing of a detonator. It is possible to do a quantitative experiment in which a simple percussion cap (as used in a child's toy pistol) is mounted on a mass that is free to recoil, and is detonated by the impact of an incident object (e.g., on an air track).

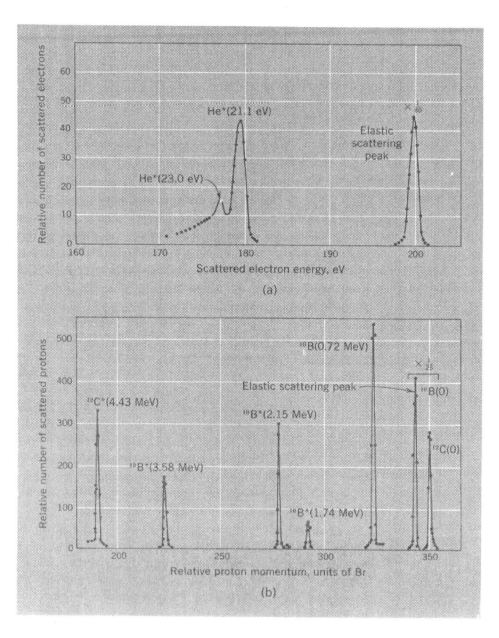

Fig. 10-18 Experimental results on elastic and inelastic collision processes, showing the production of characteristic excited states of sharply defined energy: (a) Scattering of electrons by neutral helium atoms. [After L. C. Van Atta, Phys. Rev., **38**, 876 (1931).] (b) Scattering of protons by nuclei of boron 10. Some carbon was also present as an unavoidable contaminant. [After C. K. Bockelman, C. P. Browne, W. W. Buechner, and A. Sperduto, Phys. Rev., **92**, 665 (1953).]

The threshold kinetic energy required under these conditions is significantly greater than if the cap is mounted on an unyielding support.

THE DIATOMIC MOLECULE[1]

The diatomic molecule, as typified by such molecules as HCl or Cl_2, is a system to which the methods developed in this chapter can be very effectively applied. In such molecules, the atomic nuclei play the role of point masses (which of course they are to an extremely good approximation, being so tiny in comparison to an atomic diameter) and the outer electron structure of the atoms plays the role of a spring system that can store potential energy. We thus have the physical basis for characteristic vibrations of the nuclei along the line joining them, and this will be a possible mode of internal motion of the molecule. Before discussing the oscillations as such, we must consider the shape and energy scale of the potential energy curve for a molecule of this type.

We begin with the knowledge that in a diatomic molecule there is a fairly well defined distance between the nuclei of the two atoms. This is called the bond length and is always of the order of 1 angstrom (10^{-10} m). The fact that such an equilibrium distance exists implies that the potential-energy function $U(r)$ has a minimum at the separation r_0 equal to the bond length. One way of defining such a potential, while still leaving plenty of room for empirical adjustment of the constants to experimental data, is to assume the following potential-energy function:

$$U(r) = \frac{A}{r^a} - \frac{B}{r^b} \qquad (10\text{-}38)$$

where A and B are positive constants, and a and b are suitably chosen positive exponents with $a > b$. The term A/r^a is a positive potential energy that falls away rapidly with increasing distance and represents a repulsive force. The term $-B/r^b$ is a negative potential energy, also falling off with increasing r but less rapidly than the first term, and it represents an attractive force. These functions, and the $U(r)$ curve obtained by adding them, are shown in Fig. 10-19.

Now from Eq. (10-38) we can define the equilibrium dis-

---

[1]The remainder of this chapter can be omitted without loss of continuity, but it does introduce some quite interesting features in addition to the atomic physics as such.

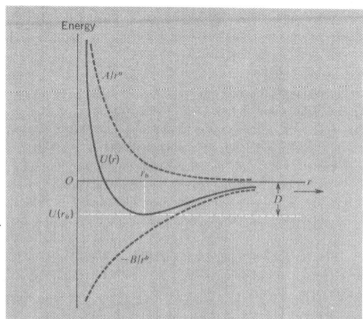

Fig. 10-19 Simple model of the potential-energy diagram of a diatomic molecule, defining an equilibrium separation r_0 of the nuclei.

tance; it is the distance at which $dU/dr = 0$:

$$\frac{dU}{dr} = -\frac{aA}{r^{a+1}} + \frac{bB}{r^{b+1}} \tag{10-39}$$

At $r = r_0$, we have

$$\frac{aA}{r_0^a} = \frac{bB}{r_0^b} \tag{10-40}$$

If r_0 is experimentally known, this defines one connection between the parameters A, B, a, and b. Also, at $r = r_0$, the potential energy is given by

$$U(r_0) = \frac{A}{r_0^a} - \frac{B}{r_0^b}$$

and substituting for A/r_0^a from Eq. (10-40), this gives

$$U(r_0) = -\frac{B}{r_0^b}\left(1 - \frac{b}{a}\right)$$

The energy $U(r_0)$ is a measurable quantity, for if we are to dissociate the molecule completely we must add a quantity of energy, the dissociation energy D, equal and opposite to $U(r_0)$ (cf. Fig. 10-19). It is typically a few eV. Thus we can put

$$D = \frac{B}{r_0^b}\left(1 - \frac{b}{a}\right) \tag{10-41}$$

Equations (10–40) and (10–41) can be used to narrow down the choice of parameters—we have two equations and four unknowns. Clearly, however, we must either appeal to some further, independent information, or make some specific assumptions, or both, to have a quantitatively defined situation.

The molecular spring constant

The quantity we really need to know, for the purpose of considering molecular vibrations, is the effective spring constant of the molecule for small displacements from equilibrium. We discussed this situation in general terms earlier in this chapter (p. 395), and arrived at the result [Eq. (10–29)]

$$U(s) \approx \tfrac{1}{2} C_2 s^2$$

where

$$s = r - r_0$$

The constant c_2 can be identified with spring constant k and can be deduced from the given form of $U(r)$:

$$k = \left(\frac{d^2 U}{ds^2}\right)_{s=0} = \left(\frac{d^2 U}{dr^2}\right)_{r=r_0} \tag{10-42}$$

Applying this to the present problem, we shall differentiate both sides of Eq. (10–39) with respect to r:

$$\frac{d^2 U}{dr^2} = \frac{a(a+1)A}{r^{a+2}} - \frac{b(b+1)B}{r^{b+2}} \tag{10-43}$$

This looks complicated, but for the particular value $r = r_0$ we can reduce it to one term with the help of Eq. (10–40):

$$\left(\frac{d^2 U}{dr^2}\right)_{r=r_0} = \frac{a+1}{r_0}\frac{bB}{r_0^{b+1}} - \frac{b(b+1)B}{r_0^{b+2}}$$

Therefore,

$$k = \frac{(a-b)bB}{r_0^{b+2}} \tag{10-44}$$

This still looks rather forbidding, but if you compare Eq. (10–44) with Eq. (10–41) you will see that the equation for the dissociation energy D contains a very similar combination of factors, and taking D as known we can greatly simplify our statement of the value of k, as follows. From Eq. (10–41) we have

$$\frac{(a-b)B}{r_0{}^b} = aD$$

Substituting this result in Eq. (10–44) we find

$$k = \frac{abD}{r_0{}^2} \tag{10-45a}$$

Beyond this point we cannot go, even with a knowledge of D and r_0, without assigning a value to the product ab which can be taken to play the role of a single adjustable parameter, so that we content ourselves with a final semiempirical equation:

$$k = C\frac{D}{r_0{}^2} \tag{10-45b}$$

where C is an empirical constant ($=ab$) greater than unity.

Fig. 10–20 Empirical potential-energy diagram for the molecule HCl, based on the Morse potential, with energies expressed in wave numbers (cm^{-1}).

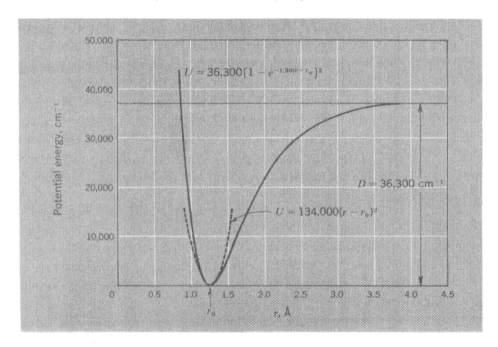

Actually a quite different analytic form of the potential function finds favor with spectroscopists. It is known as the Morse potential (after P. M. Morse). It has the same general features as the one we have used (see Fig. 10–20) but is based on a difference of exponentials rather than of simple powers of r. Like the one we have used, it has adjustable constants that can be deduced from spectroscopic data. Indeed, once the theory of molecular vibrations has been set up, one uses observations to feed back the numerical values into the theoretical formulas. It is a two-way traffic, in other words. Notice the unfamiliar units (cm^{-1}) on the ordinate of Fig. 10–20; we shall be coming back to them shortly.

The molecular vibrations

The period of vibration of a diatomic molecule with atoms of masses m_1 and m_2 is given by combining Eqs. (10–36) and (10–45b):

$$T = 2\pi r_0 \left(\frac{\mu}{CD}\right)^{1/2} \tag{10-46}$$

where μ = reduced mass = $m_1 m_2/(m_1 + m_2)$. The frequency ν, in vibrations per second, proves to be a more interesting quantity:

$$\nu = \frac{1}{2\pi r_0}\left(\frac{CD}{\mu}\right)^{1/2} \tag{10-47}$$

Let us see what this equation might suggest with an actual molecule. We shall take carbon monoxide (CO) for which the following data apply:

$$m_1(^{12}\text{C}) = 12 \text{ amu} = 2.0 \times 10^{-26} \text{ kg}$$
$$m_2(^{16}\text{O}) = 16 \text{ amu} = 2.7 \times 10^{-26} \text{ kg}$$
$$r_0 \approx 1.1 \text{ Å} = 1.1 \times 10^{-10} \text{ m}[1]$$
$$D \approx 10 \text{ eV} = 1.6 \times 10^{-18} \text{ J}[1]$$

First we have

$$\mu = \frac{m_1 m_2}{m_1 + m_2} = \frac{12}{28} \times 2.7 \times 10^{-26} \text{ kg} = 1.16 \times 10^{-26} \text{ kg}$$

---

[1] These values come from a tabulation of molecular constants in the *Smithsonian Physical Tables* (published by the Smithsonian Institution, Washington, D.C.)

Therefore,

$$D/\mu \approx 1.4 \times 10^8 \text{ m}^2/\text{sec}^2$$

whence

$$\nu \approx 1.7 \times 10^{13} C^{1/2} \text{ sec}^{-1}$$

If we ignored the factor $C^{1/2}$, we would have

$$\nu \approx 1.7 \times 10^{13} \text{ sec}^{-1}$$

Since molecular vibrations are studied mainly through spectroscopy and the measurement of wavelengths of absorbed or emitted radiation, let us calculate the wavelength λ corresponding to ν:

$$\lambda = \frac{c}{\nu} = \frac{3 \times 10^8}{1.7 \times 10^{13}} \approx 2 \times 10^{-5} \text{ m}$$

The visible spectrum extends from about 4.5×10^{-7} m to 7×10^{-7} m, so the wavelength we have calculated would be well into the infrared. That is exactly where we find the spectral lines associated with molecular vibrations. Actually there is a fundamental vibrational line of CO with a wavelength of about 4.7×10^{-6} m—about one quarter of the value we have just calculated. We would have obtained this value by putting $C^{1/2} \approx 4$ in Eq. (10–47). This value of C could correspond, for example, to having $a = 5$ and $b = 3$ in the original expression for $U(r)$ [Eq. (10–38)]. This specific form of potential would result from a short-range repulsive force varying as $1/r^6$, and a longer-range attractive force varying as $1/r^4$. We should emphasize, however, that our model of the potential is a very crude one and should not be regarded as a source of precise information about intramolecular forces. Our purpose in introducing it is simply to illustrate the general way in which vibrations about an equilibrium configuration can be analyzed.

There is a feature—a vital feature—that we have ignored; this is that molecular vibration, and the radiation associated with it, is quantized. The energy of a quantum of the radiation is $h\nu$ (where h is Planck's constant) and is provided by a transition between two sharply defined energy states of the molecule:

$$h\nu = E_1 - E_2 \tag{10-48}$$

We shall not go into this here, except to point out that the smallest possible jump between successive vibrational energy levels does, as it happens, correspond exactly to the classical oscillator frequency we have calculated. It is the basic quantum nature of the process, as expressed in Eq. (10–48), that makes the frequency ν rather than the wavelength λ the important quantity. Since, however, spectroscopists measure wavelengths in the first instance, and since ν is inversely proportional to λ, there is a custom of giving results in terms of λ^{-1} simply. To convert this to an equivalent energy measurement we must use the relation, applying to any photon, that

$$E = h\nu = hc\lambda^{-1}$$

This scheme of units has been extended still further, and used in spectroscopy as a measure of energies generally, whether or not photon emission is involved. For example, in Fig. 10–20, the dissociation energy D of HCl is given as

$$D = 36{,}300 \text{ cm}^{-1} = 3.63 \times 10^6 \text{ m}^{-1}$$

Using the above relation, and substituting $h = 6.62 \times 10^{-34}$ J-sec, we have in this case

$$\begin{aligned}D &= (6.62 \times 10^{-34})(3 \times 10^8)(3.63 \times 10^6) \\ &= 7.2 \times 10^{-19} \text{ J} \\ &= 4.5 \text{ eV (approx.)}\end{aligned}$$

PROBLEMS

10–1 A particle of mass m, at rest at $t = 0$, is subjected to a force F whose value at $t = 0$ is F_0 and which decreases linearly with time, becoming zero at $t = T$. What is the kinetic energy of the particle at $t = T$?

10–2 An object of mass 5 kg is acted on by a force that varies with the position of the object as shown. If the object starts out from rest at the point $x = 0$, what is its speed (a) at $x = 25$ m, and (b) at $x = 50$ m?

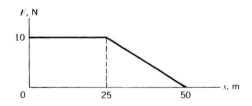

10-3 (a) A particle of mass m, initially at rest, is acted upon by a force F which increases linearly with time: $F = Ct$. Deduce the relationship between F and the particle's position x, and graph your result.

(b) How is the graph of F versus x altered if the particle has initial velocity v_0?

10-4 (a) Since antiquity man has made use of the "mechanical advantage" of simple machines, defined as the ratio of the load that is to be raised to the applied force that can raise it. By analyzing the work–energy relation $dW = \mathbf{F} \cdot d\mathbf{s}$, and appealing to conservation of energy, give a reasoned definition of "mechanical advantage."

(b) In terms of your definition, calculate the particular "mechanical advantage" offered by each of the devices shown.

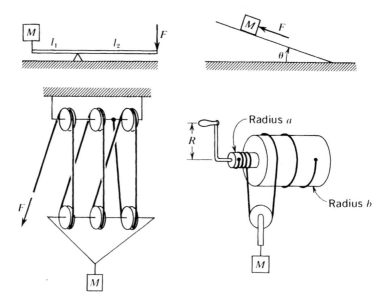

10-5 (a) The Stanford linear accelerator ("SLAC") delivers energy to electrons at the rate of approximately 200 keV/ft. How does this compare to the energy per unit length imparted to electrons by a cathode-ray tube? By a television tube?

(b) The effects of relativity are extremely pronounced in this example (after 2 miles of travel, the electrons in SLAC attain a velocity that is within $10^{-7}\%$ that of light). How far would the electrons travel in attaining the speed of light if their kinetic energy were given by the classical value $KE = \tfrac{1}{2}mv^2$?

10-6 A railroad car is loaded with 20 tons of coal in a time of 2 sec while it travels through a distance of 10 m beneath a hopper from which the coal is discharged.

(a) What average extra force must be applied to the car during this loading process to keep it moving at constant speed?
(b) How much work does this force perform?
(c) What is the increase in kinetic energy of the coal?
(d) Explain the discrepancy between (b) and (c).

10–7 A car is being driven along a straight road at constant speed v. A passenger in the car hurls a ball straight ahead so that it leaves his hand with a speed u *relative to him*.

(a) What is the gain of energy of the ball as measured in the reference frame of the car? Of the road?

(b) Relate the answers in (a) to the work done by the passenger and by the car. Satisfy yourself that you understand exactly what forces are acting on what objects, over what distances.

10–8 A common device for measuring the power output of an engine at a given rate of revolution is known as an absorption dynamometer. A small friction brake, called a Prony brake, is clamped to the output shaft of the engine (as shown), allowing the shaft to rotate, and is held in position by a spring scale a known distance R away.

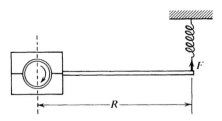

(a) Derive an expression for the horsepower of the engine in terms of the quantities R, F (the force recorded at the spring scale), and ω (the angular velocity of the shaft).

(b) You will recognize the product of the force F times its lever arm R as the torque exerted by the engine. Does your relation between torque and horsepower agree with the data published for automobile engines? Explain why there may be discrepancies.

10–9 An electric pump is used to empty a flooded basement which measures 30 by 20 ft and is 15 ft high. In a rainstorm, the basement collected water to a depth of 4 ft.

(a) Find the work necessary to pump the water out to ground level.

(b) Supposing that the pump was driven by a 1-hp motor with a 50% efficiency, how long did the operation take?

(c) If the depth of basement below ground were 50 ft instead of 15 ft, how much work would be required of the pump, again for a 4-ft flood?

10–10 (a) Look up the current world records for shot put, discus, and javelin. The masses of these objects are 7.15, 2.0, and 0.8 kg, respectively. Ignoring air resistance, calculate the minimum possible kinetic energy imparted to each of these objects to achieve these record throws.

   (b) What force, exerted over a distance of 2 m, would be required to impart these energies?

   (c) Do you think that the answers imply that air resistance imposes a serious limitation in any of these events?

10–11 A perverse traveler walks *down* an ascending escalator, so as to remain always at the same vertical level. Does the motor driving the escalator have to do more work than if the man were not there? Analyze the dynamics of this situation as fully as you can.

10–12 The Great Pyramid of Gizeh when first erected (it has since lost a certain amount of its outermost layer) was about 150 m high and had a square base of edge length 230 m. It is effectively a solid block of stone of density about 2500 kg/m^3.

   (a) What is the total gravitational potential energy of the pyramid, taking as zero the potential energy of the stone at ground level?

   (b) Assume that a slave employed in the construction of the pyramid had a food intake of about 1500 Cal/day and that about 10% of this energy was available as useful work. How many man-days would have been required, at a minimum, to construct the pyramid? (The Greek historian Herodotus reported that the job involved 100,000 men and took 20 years. If so, it was not very efficient.)

10–13 It is claimed that a rocket would rise to a greater height if, instead of being ignited at ground level (A), it were ignited at a lower level (B) after it had been allowed to slide from rest down a frictionless chute—see the figure. To analyze this claim, consider a simplified model in which the body of the rocket is represented by a mass M, the fuel is represented by a mass m, and the chemical energy released in the burning of the fuel is represented by a compressed spring between M and m which stores a definite amount of potential energy, U, sufficient to eject m suddenly with a velocity V relative to M. (This

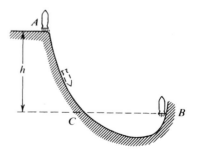

corresponds to instantaneous burning and ejection of all the fuel—i.e., an explosion.) Then proceed as follows:

(a) Assuming a value of g independent of height, calculate how high the rocket would rise if fired directly upward from rest at A.

(b) Let B be at a distance h vertically lower than A, and suppose that the rocket is fired at B after sliding down the frictionless chute. What is the velocity of the rocket at B just before the spring is released? just after the spring is released?

(c) To what height *above* A will the rocket rise now? Is this higher than the earlier case? By how much?

(d) Remembering energy conservation, can you answer a skeptic who claimed that someone had been cheated of some energy?

(e) If you are ambitious, consider a more realistic case in which the ejection of the fuel is spread out over some appreciable time. Assume a constant rate of ejection during this time.

10-14 A neutral hydrogen atom falls from rest through 100 m in vacuum. What is the order of magnitude of its kinetic energy *in electron volts* at the bottom? (1 eV = 1.6×10^{-19} J. Avogadro's number = 6×10^{23}.)

10-15 A spring exerts a restoring force given by

$$F(x) = -k_1 x - k_2 x^3$$

where x is the deviation from its unstretched length. The spring rests on a frictionless surface, and a frictionless block of mass m and initial velocity v hits a spike on the end of the spring and sticks (see the figure). How far does the mass travel, after being impaled, before it comes to rest? (Assume that the mass of the spring is negligible.)

10-16 A particle moves along the x axis. Its potential energy as a function of position is as shown in the figure. Make a careful freehand sketch of the force $F(x)$ as a function of x for this potential-energy curve. Indicate on your graph significant features and relationships.

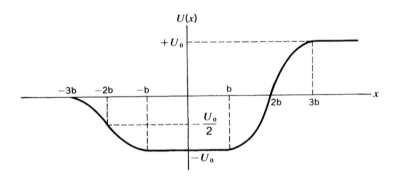

10-17 A uniform rope of mass *M* and length *L* is draped symmetrically over a frictionless horizontal peg of radius *R*. It is then disturbed slightly and begins to slide off the peg. Find the speed of the rope at the instant it leaves the peg completely.

10-18 Two masses are connected by a massless spring as shown.

(a) Find the minimum downward force that must be exerted on m_1 such that the entire assembly will barely leave the table when this force is suddenly removed.

(b) Consider this problem in the time-reversed situation: Let the assembly be supported above the table by supports attached to m_1. Lower the system until m_2 barely touches the table and then release the supports. How far will m_1 drop before coming to a stop? Does knowledge of this distance help you solve the original problem?

(c) Now that you have the answer, check it against your intuition by (1) letting m_2 be zero and (2) letting m_1 be zero. Especially in the second case, does the theoretical answer agree with your common sense? If not, discuss possible sources of error.

10-19 A particle moves in a region where the potential energy is given by

$$U(x) = 3x^2 - x^3 \qquad (x \text{ in m}, U \text{ in J})$$

(a) Sketch a freehand graph of the potential for both positive and negative values of *x*.

(b) What is the maximum value of the total mechanical energy such that oscillatory motion is possible?

(c) In what range(s) of values of *x* is the force on the particle in the positive *x* direction?

10-20 A highly elastic ball (e.g., a "Superball") is released from rest a distance *h* above the ground and bounces up and down. With each bounce a fraction *f* of its kinetic energy just before the bounce is lost. Estimate the length of time the ball will continue to bounce if $h = 5$ m and $f = \frac{1}{10}$.

10-21 The elastic cord of the catapult shown in the diagram has a total relaxed length of $2l_0$; its ends are attached to fixed supports a distance $2b$ apart.

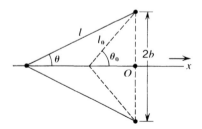

(a) Show that if the tension developed in the cord is proportional to the increase in its length, the component of force in the x direction is

$$F_x = 2kb \left(\frac{1}{\sin \theta} - \frac{1}{\sin \theta_0} \right) \cos \theta$$

(b) Noting that the position of the stone in the catapult is given by $x = -b \cot \theta$, derive the expression for the work done in moving a distance dx. Then integrate this expression between θ_0 and θ to find the total work done in extending the catapult and compare with the result that is obtainable directly by considering the energy stored in the stretched cord (p. 392). (After carrying out the calculation in the way prescribed above, you will better appreciate the advantages that may come from the use of energy conservation instead of a work integral that involves the force explicitly.)

10-22 A perfectly rigid ball of mass M and radius r is dropped upon a deformable floor which exerts a force proportional to the distance of deformation, $F = ky$.

(a) Make a graph of the potential energy of the ball as a function of height y. (Take $y = 0$ as the undisturbed floor level.)

(b) What is the equilibrium position of the center of the ball when the ball is simply resting on the floor? (Note that this corresponds to the minimum of the potential-energy curve.)

(c) By how much is the period of M increased over its period of bouncing on a perfectly rigid floor?

10-23 A particle of mass $m = 2$ kg on a frictionless table is at rest at time $t = 0$ at position $x = 0$. A force of magnitude $F_x = 4 \sin(\pi t)$ newtons acts on the particle from $t = 0$ to $t = 2$ sec.

(a) Plot the resulting acceleration a_x, velocity v_x, and position x as functions of t in this period.

(b) What is the total work done by the force during the period $t = 0$ to $t = 1$ sec?

10-24 A spring of negligible mass exerts a restoring force given by

$$F(x) = -k_1 x + k_2 x^2$$

(a) Calculate the potential energy stored in the spring for a displacement x. Take $U = 0$ at $x = 0$.

(b) It is found that the stored energy for $x = -b$ is twice the stored energy for $x = +b$. What is k_2 in terms of k_1 and b?

(c) Sketch the potential-energy diagram for the spring as defined in (b).

(d) The spring lies on a smooth horizontal surface, with one end fixed. A mass m is attached to the other end and sets out at $x = 0$ in the positive x direction with kinetic energy equal to $k_1 b^2 / 2$. How fast is it moving at $x = +b$?

(e) What are the values of x at the extreme ends of the range of oscillation? [Use your graph from part (c) for this.]

10–25 An object of mass m, moving from the region of negative x, arrives at the point $x = 0$ with speed v_0. For $x \geq 0$ it experiences a force given by

$$F(x) = -\alpha x^2$$

How far along the $+x$ axis does it get?

10–26 The potential energy of a particle of mass m as a function of its position along the x axis is as shown. (The discontinuous jumps in the value of U are not physically realistic but may be assumed to approximate a real situation.) Calculate the period of one complete oscillation if the particle has a total mechanical energy E equal to $3U_0/2$.

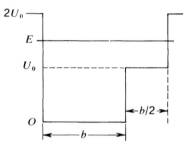

10–27 Consider an object of mass m constrained to travel on the x axis (perhaps by a frictionless guide wire or frictionless tracks), attached to a spring of relaxed length l_0 and spring constant k which has its other end fixed at $x = 0$, $y = l_0$ (see the figure).

(a) Show that the force exerted on m in the x direction is

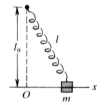

$$F_x = -kx\left[1 - \left(1 + \frac{x^2}{l_0^2}\right)^{-1/2}\right]$$

(b) For small displacements ($x \ll l_0$), show that the force is proportional to x^3 and hence

$$U(x) \simeq Ax^4 \qquad (x \ll l_0)$$

What is A in terms of the above constants?

(c) The period of a simple harmonic oscillator is independent of its amplitude. How do you think the period of oscillation of the above motion will depend on the amplitude? (An energy diagram may be helpful.)

10–28 Consider a particle of mass m moving along the x axis in a force field for which the potential energy of the particle is given by $U = Ax^2 + Bx^4$ ($A > 0$, $B > 0$). Draw the potential-energy curve and, arguing from the graph, determine something about the dependence of the period of oscillation T upon the amplitude x_0. Show that, for amplitudes sufficiently small so that Bx^4 is always very small compared to Ax^2, the approximate dependence of period upon amplitude is given by

$$T = T_0 \left(1 - \frac{3B}{4A} x_0^2\right)$$

10–29 A particle of mass m and energy E is bouncing back and forth between vertical walls as shown, i.e., over a region where $U = 0$. The potential energy is slightly changed by introducing a tiny rectangular hump of height ΔU ($\ll E$) and width Δx. Show that the period of oscillation is changed by approximately $(m/2E^3)^{1/2}(\Delta U \Delta x)$. [It is worth noting that the effect of the small irregularity in potential energy depends simply on the product of ΔU and Δx, not on the individual values of these quantities. This is typical of such small disturbing effects—known technically as perturbations.]

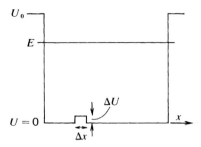

10–30 Two blocks of masses m and $2m$ rest on a frictionless horizontal table. They are connected by a spring of negligible mass, equilibrium length L, and spring constant k. By means of a massless thread connecting the blocks the spring is held compressed at a length $L/2$. The whole system is moving with speed v in a direction *perpendicular* to the length of the spring. The thread is then burned. In terms of m, L, k, and v find

  (a) The total mechanical energy of the system.
  (b) The speed of the center of mass.
  (c) The maximum relative speed of the two blocks.
  (d) The period of vibration of the system.

How do the quantities of parts (a) through (d) change if the initial velocity v is along, rather than perpendicular to, the length of the spring?

10–31 The mutual potential energy of a Li^+ ion and an I^- ion as a function of their separation r is expressed fairly well by the equation

$$U(r) = \frac{-Ke^2}{r} + \frac{A}{r^{10}}$$

where the first term arises from the Coulomb interaction, and the values of its constants in MKS units are

$$K = 9 \times 10^9 \text{ N-m}^2/\text{C}^2, \quad e = 1.6 \times 10^{-19} \text{ C}$$

The equilibrium distance r_0 between the centers of these ions in the LiI molecule is about 2.4 Å. On the basis of this information,

(a) How much work (in eV) must be done to tear these ions completely away from each other?

(b) Taking the I^- ion to be fixed (because it is so massive), what is the frequency ν (in Hz) of the Li^+ ion in vibrations of very small amplitude? (Calculate the effective spring constant k as the value of d^2U/dr^2 at $r = r_0$—see p. 407. Take the mass of the Li^+ as 10^{-26} kg.)

10-32 (a) If in addition to the van der Waals attractive force, which varies as r^{-7}, two identical atoms of mass M experience a repulsive force proportional to r^{-l} with $l > 7$, show that

$$U(r) = \frac{-A}{r^6} + \frac{B}{r^n} \quad (n > 6)$$

and graph your result versus r.

(b) Calculate the equilibrium separation r_0 in this molecule in terms of the constants by requiring

$$\left.\frac{dU(r)}{dr}\right|_{r=r_0} = 0$$

(c) The dissociation energy D of the molecule should be equal to $-U(r_0)$. What is its value in terms of A, n, and r_0?

(d) Calculate the frequency of small vibrations of the molecule about the equilibrium separation r_0. Show that it is given by the mass M, the constant n, the equilibrium separation, and the dissociation energy of the molecule, as follows:

$$\omega^2 = \frac{12nD}{Mr_0^2}$$

10-33 The potential energy of an ion in a crystal lattice of alternately charged ions may be written

$$U(r) = \frac{-Ae^2}{r} + \frac{B}{r^n}$$

where A, B, and n are constants, e is the elementary charge, and r is the distance between closest neighbors in the lattice, called the lattice constant. This potential arises from the $1/r^2$ force of electrostatic attraction, together with a short-range repulsive force as two adjacent ions are brought close together.

(a) Make a graph of $U(r)$ versus r. Making sure to identify r correctly, what can you say about the stability of the crystal structure?

(b) Calculate the lattice constant r_0 in terms of A, B, e, and n, from the equilibrium condition

$$\left.\frac{dU(r)}{dr}\right|_{r=r_0} = 0$$

(This defines the inter-ionic distance for which the energy of the lattice as a whole is minimized.)

(c) Show that the binding energy per mole of molecules is

$$u = -N_0 U(r_0) \qquad N_0 = \text{Avogadro's number} \\ \simeq 6 \times 10^{23} \text{ mole}^{-1}$$

Using the result of part (b), determine u in terms of r_0.

(d) For NaCl crystal, the equilibrium lattice spacing r_0 is 2.8 Å, and the exponent n is about 10. The constant A is equal to 1.7 (dimensionless) if r is expressed in centimeters and e is given in esu ($e = 4.8 \times 10^{-10}$ esu). Using these values, calculate a theoretical value for u and compare with the experimental value of 183 kcal/mole.

A technique succeeds in mathematical physics, not by a clever trick, or a happy accident, but because it expresses some aspect of a physical truth.

                    O. G. SUTTON, *Mathematics in Action* (1954)

11

Conservative forces and motion in space

EXTENDING THE CONCEPT OF CONSERVATIVE FORCES

THROUGHOUT CHAPTER 10 we consistently applied one important simplification or restriction, by confining our discussion to motion in one dimension only. This clearly prevented us from studying some of the most interesting and important problems in dynamics. In the course of the present chapter we shall free ourselves of this restriction and in the process show the energy method of analysis to still greater advantage.

To begin the discussion, let us consider a problem in motion under gravity near the earth's surface. Suppose we have two very smooth tubes connecting two points, A and B, at different levels in the same vertical plane (Fig. 11-1). A small particle, placed in either of these tubes and released from rest at A, slides down and emerges at B. If the tubes are effectively frictionless, the forces exerted by them on the particle are always at right angles to the particle's motion. Hence these forces do no work

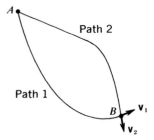

Fig. 11-1 Alternative paths between two given points for motion in a vertical plane.

on the particle; whatever changes may occur in its kinetic energy cannot be ascribed to them. What these forces *do* achieve is to compel the particle to follow a particular path so that it emerges at B traveling in a designated direction. If it follows path 1, it emerges with a velocity v_1 as shown; if it follows path 2, it emerges with a velocity v_2. Of course the energy of the particle does change as it moves along either tube; the gravitational force is doing work on it. But we observe a very interesting fact: Although the directions of the velocities v_1 and v_2 are quite different, their magnitudes are the same. *The kinetic energy given to the particle by the gravitational force is the same for all paths beginning at A and ending at B.*

How does this come about? It is not difficult to see. As the particle travels through some element of displacement ds the work dW done on it is given by

$$dW = |\mathbf{F}_g| |d\mathbf{s}| \cos\theta = \mathbf{F}_g \cdot d\mathbf{s}$$

where θ is the angle between the directions of $\mathbf{F}_g$ and $d\mathbf{s}$. But the force $\mathbf{F}_g$ is a constant force (i.e., the same at any position) with the following components:

$$F_x = 0 \qquad F_y = -mg$$

(we take the y axis as positive upward). The element of displacement has components (dx, dy). Now from the basic definition of dW as $|\mathbf{F}| |d\mathbf{s}| \cos\theta$ we have

$$dW = F_x\, dx + F_y\, dy$$

To see this, consider any two vectors **A** and **B** in the xy plane, making angles α and β, respectively, with the x axis (Fig. 11-2). Then if the angle between them is θ we have, by a standard trigonometric theorem, that

$$\cos\theta = \cos(\beta - \alpha) = \cos\beta \cos\alpha + \sin\beta \sin\alpha$$

The scalar product S ($= \mathbf{A} \cdot \mathbf{B}$) is thus given by

$$S = |\mathbf{A}| |\mathbf{B}| (\cos\beta \cos\alpha + \sin\beta \sin\alpha)$$

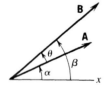

Fig. 11-2 Basis for obtaining the scalar product $\mathbf{A} \cdot \mathbf{B}$ in terms of individual components, as may be convenient in calculating the work $\mathbf{F} \cdot d\mathbf{s}$ in an arbitrary displacement.

But

$$|\mathbf{A}|\cos\alpha = A_x \qquad |\mathbf{A}|\sin\alpha = A_y$$

and similarly for the components of **B**. Thus we have

$$\mathbf{A}\cdot\mathbf{B} = A_x B_x + A_y B_y \qquad \text{(two dimensions only)}$$

More generally, if the vectors also have nonzero z components,

$$\mathbf{A}\cdot\mathbf{B} = A_x B_x + A_y B_y + A_z B_z$$

Thus in general we shall have

$$dW = \mathbf{F}\cdot d\mathbf{s} = F_x\,dx + F_y\,dy + F_z\,dz \qquad (11\text{-}1)$$

In the present two-dimensional problem, with $F_x = 0$ and $F_y = -mg$, this gives us

$$dW = -mg\,dy$$

Hence for a change of vertical coordinate from y_1 to y_2, regardless of the change of x coordinate or of the particular path taken, we have a change of kinetic energy given by

$$K_2 - K_1 = \int dW = -mg(y_2 - y_1) \qquad (11\text{-}2)$$

which exactly reproduces the result that we derived in Chapter 10 for purely vertical motion and which permits us again to define a gravitational potential energy $U(y)$ equal to *mgy*.

This result makes for a great extension of our energy methods to situations where, in addition to gravity, we have so-called "forces of constraint," which control the path of an object but, because they act always at right angles to its motion, do nothing to change its energy. Let us consider some specific examples.

ACCELERATION OF TWO CONNECTED MASSES

We shall begin with a problem which, if tackled by the direct application of Newton's laws, requires us to write a statement of $\mathbf{F} = m\mathbf{a}$ for each of two objects separately. The problem is to find the magnitude of the acceleration of two connected masses, m_1 and m_2, moving on smooth planes as shown in Fig. 11–3. The accelerations of the masses are different vectors, $\mathbf{a}_1$ and $\mathbf{a}_2$,

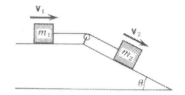

Fig. 11-3 Motion of two connected masses—a simple example of the use of energy-conservation methods.

even though they have the same magnitude a. Thus a single statement of $\mathbf{F} = m\mathbf{a}$ will not suffice. Using the scalar quantity energy, however, we can exploit the fact that the *magnitudes* of the displacements and their time derivatives are common to both masses. Thus we have $|\mathbf{v}_1| = |\mathbf{v}_2| = v$, so that the total kinetic energy of the system is given simply by

$$K = \tfrac{1}{2}(m_1 + m_2)v^2$$

and its change in a time dt is given by

$$dK = (m_1 + m_2)v\,dv$$
$$= (m_1 + m_2)va\,dt$$

Now $v\,dt$ is the distance moved by each mass parallel to the surface on which it rides, and for m_2 the distance $v\,dt$ down the slope means a change dy in vertical coordinate equal to $-v\,dt\sin\theta$ (y positive upward). The associated change dU in gravitational potential energy is thus

$$dU = -m_2 gv\,dt\sin\theta$$

But, given conservation of total mechanical energy, assuming friction to be absent, we have

$$dK + dU = 0$$

Thus

$$(m_1 + m_2)va\,dt - m_2 gv\,dt\sin\theta = 0$$

whence the familiar result

$$a = \frac{m_2}{m_1 + m_2} g\sin\theta$$

OBJECT MOVING IN A VERTICAL CIRCLE

Suppose that a particle P of mass m is attached to one end of a rod of negligible mass and length l, the other end of which

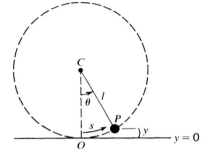

Fig. 11-4 Motion of a simple pendulum.

is pivoted freely at a fixed center C. Let us take an origin O at the normal resting position of the object (Fig. 11-4). Then the position of the object is conveniently described in terms of the single angular coordinate θ, or, if we prefer, by the displacement s along the circular arc of its path from O (assuming the rod to be of invariable length).[1]

If the angular displacement is θ, the y coordinate of the object is given by

$$y = l(1 - \cos\theta)$$

and hence its potential energy by

$$U(\theta) = mgl(1 - \cos\theta) \tag{11-3}$$

Using our basic energy-conservation statement we have

$$K_1 + U_1 = K_2 + U_2$$

for any two points on the path. Substituting for K in terms of m and v and for U by Eq. (11-3), we have

$$\tfrac{1}{2}mv_1^2 + mgl(1 - \cos\theta_1) = \tfrac{1}{2}mv_2^2 + mgl(1 - \cos\theta_2)$$

Therefore,

$$v_2^2 = v_1^2 + 2gl(\cos\theta_2 - \cos\theta_1) \tag{11-4}$$

Clearly, if $\theta_2 > \theta_1$, then $v_2 < v_1$.

If the object were started out at the lowest point with a velocity v_0, we should have

---

[1] In a sense, therefore, the motion is one-dimensional, even though the one dimension is not a straight line. But we shall not stress this aspect of such a system at this point.

$$v_2{}^2 = v_0{}^2 - 2gl(1 - \cos\theta_2)$$

With the help of this result we can answer such questions as: What initial velocity is needed for the object to reach the top of the circle? and: What position does it reach if started out with less than this velocity?

Notice the great advantage that the energy method has over the direct use of $\mathbf{F} = m\mathbf{a}$ in this problem. The velocity of the object is changing in both magnitude and direction, it has radial and transverse components of acceleration, there is an unknown push or pull on the object from the rod—yet none of these things need be considered in calculating the speed v at any given θ (or y). Once v has been found by Eq. (11-4) [or perhaps by going back to Eq. (11-2) if that is more convenient], then one can proceed to deduce the acceleration components and other things.

There may be subtleties in such problems, however. Suppose, for example, that instead of a more or less rigid rod we had a string to constrain the object. This is now a constraint that works only one way; it can pull radially inward but it cannot push radially outward. One may have a situation in which the velocity is not great enough to take the object to the top of the

Fig. 11-5 (a) Path of the bob of a simple pendulum launched with insufficient velocity to maintain a tension in the string up to $\theta = \pi$. (b) Stroboscopic photograph of a ball falling away from a circular channel at the point where the contact force becomes zero. (Photograph by Jon Rosenfeld, Education Research Center, M.I.T.)

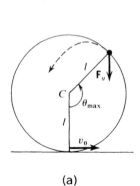

(a)

(b)

circle (although it might be possible with a rigid rod); the object will fall away in a parabolic path [Fig. 11–5(a)]. The breakaway point is reached when the tension in the string just falls to zero and the component of $\mathbf{F}_g$ along the radius of the circle is just equal to the mass m times the requisite centripetal acceleration v^2/l. An exactly similar situation can arise if an object moves along a circular track made of grooved metal, and Fig. 11–5(b) shows a stroboscopic photograph of an object falling away from such a track at the point where the normal reaction supplied by the track has fallen to zero. The angular position, θ_m, at which this occurs [cf. Fig. 11–5(a)] is defined by a statement of Newton's law:

$$mg \cos(\pi - \theta_m) = \frac{mv^2}{l} \quad \left(\text{assumes } \frac{\pi}{2} < \theta_m \leq \pi\right)$$

where

$$v^2 = v_0^2 - 2gl(1 - \cos \theta_m)$$

This leads to the result

$$\cos \theta_m = \frac{2}{3} - \frac{v_0^2}{3gl}$$

We can thus deduce that a particle that starts out from O with v_0 less than $\sqrt{5gl}$ will fail to reach the top of the circle, whereas with a rigid rod to support it an initial speed of $2\sqrt{gl}$ would suffice. Notice, therefore, that the energy-conservation principle should not be used blindly; one must always be on the alert as to whether Newton's law can be satisfied at every stage with the particular constraining forces available.

AN EXPERIMENT BY GALILEO

It was Galileo, in his *Dialogues Concerning Two New Sciences* (1638), who first clearly stated the result that the speed attained by an object descending under gravity depends only on the vertical distance traveled. He applied this result to the uniformly accelerated motion of an object down smooth planes of different slopes, as indicated in Fig. 11–6(a), which is based on a diagram from Galileo's own book. He could not, however, directly demonstrate the correctness of this proposition using inclined planes as such; instead, he performed a very clever experiment

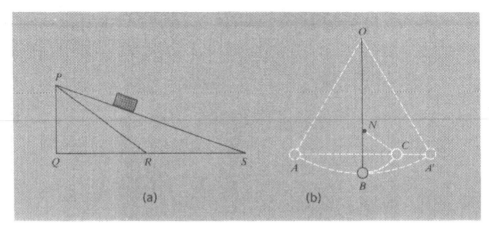

Fig. 11-6 (a) Speed attained by a block sliding down a frictionless plane depends only on the vertical distance traveled, not on the slope. (b) Galileo's pendulum experiment to demonstrate the properties of motion on idealized inclined planes.

with a mass swinging in a circular arc, and applied his remarkable scientific insight to what he observed in this situation. Here is an account of the experiment, taken with only minor changes of notation from Galileo's own description; the clarity and modernity of his presentation is quite striking:

> Imagine this page to represent a vertical wall, with a nail driven into it; and from the nail let there be suspended a lead bullet of one or two ounces by means of a fine vertical thread, OB, say from four to six feet long [see Fig. 11-6(b), based on Galileo's own sketch]. On this wall draw a horizontal line AA', at right-angles to the vertical thread OB (which hangs about two finger-breadths in front of the wall). Now bring the thread OB with the attached ball into position OA and set it free; first it will be observed to descend along the arc ABA' . . . till it almost reaches the horizontal AA', a slight shortage being caused by the resistance of the air and the string. From this we may rightly infer that the ball in its descent through the arc AB acquired a momentum on reaching B which was just sufficient to carry it through a similar arc BA' to the same height.
>
> Having repeated this experiment many times, let us now drive a nail into the wall close to the perpendicular OB, say at N, so that it projects out some five or six finger-breadths in order that the thread, again carrying the bullet through the arc AB, may strike upon the nail N when the bullet reaches B, and thus compel it to traverse the arc BC, described about N as

center ... Now, gentlemen, you will observe with pleasure that the ball swings to the point C in the horizontal line AA', and you would see the same thing happen if the obstacle were placed at some lower point. ...[1]

In this way Galileo convinced himself that the speed acquired by an object in descending any vertical distance is sufficient to carry it up through an equal vertical distance by a different path. He then added to this the reversibility of the motion along any given arc and could reach his main conclusion that the speed attained by his suspended object must be the same whether it descends along the arc AB or the arc CB, or any other arc beginning on the level AA' and with its lowest point at B. Finally, he could visualize such motion as taking place on a continuous succession of inclined planes of different slopes, and so formulate his proposition about uniformly accelerated motion along different inclines. This was an enormously important result, because he was then able to make inferences about free fall under gravity by observing the motion of an object rolling down an inclined plane with a very gentle slope. This permitted him to stretch the time scale of the whole phenomenon ($a = g \sin \theta$!) and make quantitative measurements of distance against time, using as his clock a water-filled container with a hole in it that he could open and close with his finger—a brilliantly simple device.

MASS ON A PARABOLIC TRACK

Suppose we bend a piece of very smooth metal track so that it has the shape of a parabola curving symmetrically upward with its vertex at point O (Fig. 11–7). Let the equation of the parabola be

$$y = \tfrac{1}{2}Cx^2 \tag{11-5}$$

where C is a constant.

Imagine an object of mass m, free to slide along the track with negligible friction. Its gravitational potential energy, referred to an arbitrary zero at O, is given by

$$U(y) = mgy$$

[1] Quoted from the English translation of *Dialogues Concerning Two New Sciences* (H. Crew and A. de Salvio, translators), Dover Publications, New York.

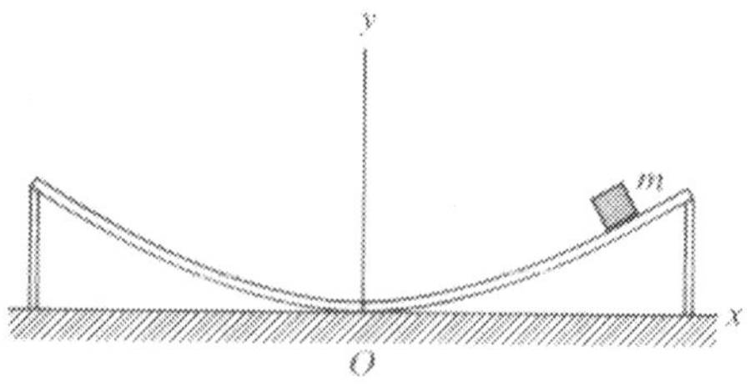

Fig. 11-7 Motion of a particle on a parabolic track in a vertical plane.

which, by Eq. (11-5), we can alternatively write as a function of x:

$$U(x) = \tfrac{1}{2}mgCx^2 = \tfrac{1}{2}kx^2 \tag{11-6}$$

where k is a constant of the same dimensions (newtons per meter) as a spring constant. At any point (x, y) the mass has a speed v, necessarily along the track, and by conservation of energy we have

$$\tfrac{1}{2}mv^2 + U = E \quad (= \text{const.})$$

Denoting the x and y components of $\mathbf{v}$ by v_x and v_y, and using Eq. (11-6), we have

$$\tfrac{1}{2}mv_x^2 + \tfrac{1}{2}mv_y^2 + \tfrac{1}{2}kx^2 = E \tag{11-7}$$

This has a marked resemblance to the energy-conservation equation of the linear harmonic oscillator. Only the presence of the term $\tfrac{1}{2}mv_y^2$ spoils things. This suggests to us that if we had a situation in which v_y were very small compared to v_x, the motion of the mass on this track would closely approximate that of a harmonic oscillator. What do we need to achieve this? Common sense tells us more or less immediately that if the track is only very slightly curved, so that $y \ll x$ at each point, the vertical motion is very small compared to the horizontal and approximately harmonic motion will occur.

Such motion can be achieved and beautifully demonstrated with a "linear air track"[1] in which a metal glider rides on a cushion of air blown up through holes in the track. Because of the low friction, the oscillation can take place even if the curvature is extremely small. To take an example, here is a specification of the shape of the track in an actual demonstration (the setting was by machine screws at 1-ft spacings; hence the lapse from MKS):

[1] R. B. Runk, J. L. Stull, and O. L. Anderson, *Am. J. Phys.*, **31**, 915 (1963).

x, ft	±1	±2	±3	±4	±5	±6	±7
y, in.	$\frac{1}{64}$	$\frac{1}{16}$	$\frac{9}{64}$	$\frac{1}{4}$	$\frac{25}{64}$	$\frac{9}{16}$	$\frac{49}{64}$

Taking any pair of these values, one establishes the value of the constant C in Eq. (11–5), e.g.,

$$x = 4 \text{ ft} = 1.22 \text{ m}$$
$$y = \tfrac{1}{4} \text{ in.} = 6.35 \times 10^{-3} \text{ m}$$

Thus

$$C = 2y/x^2 = 8.55 \times 10^{-3} \text{ m}^{-1}$$

What would be the expected period of the motion? For mass m and "spring" constant k, we have $T = 2\pi(m/k)^{1/2}$, and, from Eq. (11–6), $k = mgC$. Therefore,

$$m/k = (gC)^{-1} \approx 11.9 \text{ sec}^2$$
$$T = 2\pi\sqrt{11.9} \approx 21.6 \text{ sec}$$

The measured period was within a few tenths of a second of this and independent of amplitude to the accuracy of the measurement.

In this example (and the others of motion under gravity) the moving object, in its two-dimensional motion, really does ride a contour that corresponds in form to the graph of $U(x)$ against x. These are true two-dimensional motions; they are physically distinct from the one-dimensional situations discussed in Chapter 10. To appreciate the difference between them, consider two quite realizable physical systems. The first one [Fig. 11–8(a)] is a mass m attached to a horizontal spring and resting on a smooth table. The second [Fig. 11–8(b)] is an equal

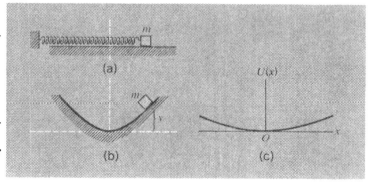

Fig. 11-8 (a) Mass attached to a horizontal spring. (b) Same mass on a smooth parabolic track. (c) Arrangements (a) and (b) can be made to have identical variations of potential energy with x, but the motions are different.

mass m free to slide on a parabolic track. The strength of the spring and the curvature of the track are adjusted so that the systems have the same parabolic potential energy curve of $U(x)$ against x, as shown in Fig. 11-8(c). If projected horizontally from the origin with the same speed v_0, both masses will have the same *magnitude* of velocity at any other value of x. If set into oscillation, both systems will be periodic. But their periods will be different. The mass on the spring will have the same period at all amplitudes; the other system will show a progressive change of period with amplitude (which way do you think it will be?). It is only in special situations, such as that with the very slightly curved track, that the two motions become essentially the same.

THE SIMPLE PENDULUM

In discussing the simple pendulum, we are returning to the problem of an object moving in a vertical circle. This time, however, we shall go into it more carefully. It is an important type of physical system, over and above its use in clocks, and it is not quite so simple as its traditional name implies. The simplicity is primarily in its structure—idealized as a point mass on a massless, rigid rod.

As we saw earlier [Eq. (11-3)], the potential energy, $U(\theta)$, expressed in terms of the angle that the supporting rod makes with the vertical, can be written in the form

$$U = mgl(1 - \cos\theta) = 2mgl \sin^2\left(\frac{\theta}{2}\right) \tag{11-8}$$

If we plot this expression for U as a function of θ, we get Fig. 11-9.

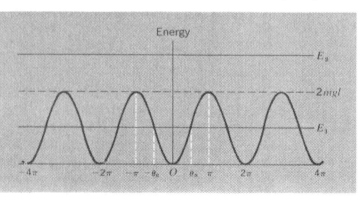

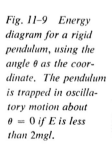

Fig. 11-9 Energy diagram for a rigid pendulum, using the angle θ as the coordinate. The pendulum is trapped in oscillatory motion about $\theta = 0$ if E is less than 2mgl.

In Fig. 11-9, θ can take all values from $-\infty$ to $+\infty$. However, all *positions* of the pendulum in space are described by values of θ between $-\pi$ and $+\pi$. Any value of θ outside this range corresponds to one and only one angle θ inside the range. The latter is obtained by adding to or subtracting from the former a whole-number multiple of 2π.

Two kinds of motion are possible, depending on whether E is less than or greater than $2mgl$ (referred to a zero of energy defined by an object at rest at the lowest position of the pendulum bob).

If the total energy is sufficiently great (e.g., as for E_2 in the figure), there are no turning points in the motion; θ increases (or decreases) without limit, corresponding to continued rotation of the pendulum rather than oscillation. The speed of the pendulum bob is maximum at $\theta = 0$ (or $\pm 2\pi$, $\pm 4\pi$, etc.) and minimum at $\theta = \pm\pi$ (or $\pm 3\pi$, $\pm 5\pi$, etc.). It is clear from Fig. 11-9 that to produce such rotational motion the pendulum must have a kinetic energy at least equal to $2mgl$ at the lowest point of the swing.

If $E < 2mgl$, say E_1, the motion is oscillatory and the angle θ changes from $+\theta_0$ to $-\theta_0$ and back again (Fig. 11-9). However, the motion is not harmonic except for sufficiently small amplitudes. In the neighborhood of O, one can very nearly match the potential-energy curve of Fig. 11-9 by a parabola (Fig. 11-10), so that for these small amplitudes the oscillations are just those of the linear oscillator. One way of justifying this statement is to recall the general argument, made in Chapter 10, that almost any symmetrical potential-energy curve can be approximated by a parabola over some limited range of small displacements. Another way is to note that in Eq. (11-8), if θ

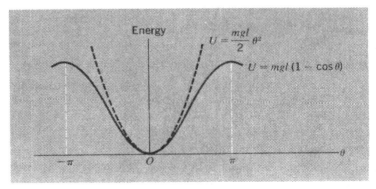

Fig. 11-10 Potential-energy diagram of a simple pendulum.

is small, one can set $\sin(\theta/2)$ approximately equal to $\theta/2$, so that we have

$$U(\theta) \approx \tfrac{1}{2}mgl\theta^2 \tag{11-9}$$

A third way depends upon using the same approximation that Newton used when he described the moon as a falling object. Applying it to the present problem, and taking the origin at O, the circular path is given by the equation

$$x^2 + (l - y)^2 = l^2$$

(see Fig. 11-11). Therefore,

$$y^2 - 2ly + x^2 = 0$$
$$y = l - (l^2 - x^2)^{1/2}$$
$$= l - l\left(1 - \frac{x^2}{l^2}\right)^{1/2}$$

If $x^2 \ll l^2$, we can use the binomial expansion to yield the approximate result

$$y \approx \frac{1}{2}\frac{x^2}{l} \tag{11-10}$$

[Compare this with Eq. (11-5); you will then realize what the constant C in that equation represents. In the actual experiment described, the motion was like that of a simple pendulum nearly 400 ft long!]

Equation (11-10) suggests a somewhat different approximation for small oscillations of the pendulum, describing its motion in terms of its horizontal displacement instead of its angular one. But whatever analysis we adopt, it is clear that the period must depend on the amplitude. We can also argue qualitatively which way it varies. Looking at Fig. 11-10, we can say that the curve

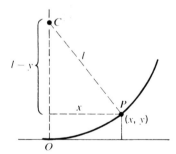

Fig. 11-11 Geometry of displacements of a simple pendulum.

of $U(\theta)$ for the pendulum describes a kind of spring that gets "softer" at large extensions. Compared to the parabolic behavior that would hold for an ideal spring, the restoring force is relatively less at larger displacements. Thus one might guess that the motion becomes more sluggish and hence that the period increases. Certainly there is one extreme case that leaves no room for doubt. If the energy is such that the pendulum just exactly arrives at $\theta_0 = \pi$, there is no restoring force at all; the pendulum would sit upside down indefinitely—although in practice this is of course an unstable equilibrium. Even at amplitudes short of this the increase of period is drastic. If you are interested, the section after next describes the analysis of larger-amplitude motion. But first we shall make a detailed study of the small-angle approximation.

THE PENDULUM AS A HARMONIC OSCILLATOR

The speed of the pendulum bob at any point is equal to $l\, d\theta/dt$, so that the energy-conservation equation is

$$\tfrac{1}{2}ml^2\left(\frac{d\theta}{dt}\right)^2 + U(\theta) = E \quad \text{(exact)}$$

Using the approximate expression for $U(\theta)$ from Eq. (11–9), we have

$$\tfrac{1}{2}ml^2\left(\frac{d\theta}{dt}\right)^2 + \tfrac{1}{2}mgl\theta^2 = E \quad \text{(approx.)}$$

or

$$\left(\frac{d\theta}{dt}\right)^2 + \frac{g}{l}\theta^2 = \text{const.} \tag{11-11}$$

By now we have met this form of equation several times and can identify $(g/l)^{1/2}$ as being the quantity ω that defines the period of the oscillation:

$$T = \frac{2\pi}{\omega} = 2\pi\left(\frac{l}{g}\right)^{1/2} \tag{11-12}$$

[*Caution:* It is the angular displacement θ itself that undergoes a simple harmonic variation, described by the equation

$$\theta(t) = \theta_0 \sin(\omega t + \varphi_0) \tag{11-13}$$

This can be confusing; the actual angular displacement of the pendulum is described in terms of the sine of the purely mathematical phase angle ($\omega t + \varphi_0$). The actual angular velocity of the pendulum is $d\theta/dt$—not the ω in ($\omega t + \varphi_0$), which serves merely to define the periodicity.]

The behavior of the simple pendulum as a close approximation to a harmonic oscillator is so important that we shall comment on it further. First, let us take a moment to consider the derivation of its equation of motion by a direct application of Newton's law. We can do this in two different ways, according to whether we analyze the problem in terms of linear or angular displacements.

Linear motion

We consider the horizontal force acting on the pendulum bob when it is a horizontal distance x from its equilibrium position [Fig. 11-12(a)]. If the tension in the suspending string is F_T, we have

$$m \frac{d^2 x}{dt^2} = -F_T \sin \theta = -F_T \frac{x}{l}$$

If the vertical component of acceleration of the bob is negligibly small, we can, however, put

$$F_T \cos \theta \approx F_g = mg$$

For small angles θ, $\cos \theta \approx 1 - \theta^2/2 \approx 1$, so that $F_T \approx mg$ and the equation of horizontal motion becomes

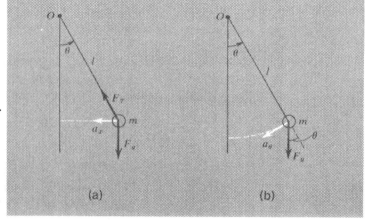

Fig. 11-12 Basis for analyzing the motion of a simple pendulum (a) in terms of horizontal displacements, and (b) in terms of angular displacements.

$$m\frac{d^2x}{dt^2} \approx -\frac{mg}{l}x$$

leading once again to Eq. (11–12) for the period T.

Angular motion

In this case we consider the transverse acceleration a_θ ($= l\,d^2\theta/dt^2$), which is produced by the component of $\mathbf{F}_g$ in the direction of the tangent to the circular arc along which the pendulum bob moves (see Fig. 11–12b). The magnitude of the tension force $\mathbf{F}_T$ (which actually varies with time) does not enter into this treatment of the problem. We have, simply,

$$ml\frac{d^2\theta}{dt^2} = -F_g \sin\theta = -mg\sin\theta$$

For small θ, $\sin\theta \approx \theta$, and so we have

$$ml\frac{d^2\theta}{dt^2} \approx -mg\theta$$

which leads to Eq. (11–13) for θ as a function of time. It can be seen that the analysis in terms of the angular variables is a "cleaner" treatment than the other, involving only the one approximation $\sin\theta \approx \theta$.

The isochronous behavior of the simple pendulum—the fact that its period is almost completely independent of amplitude over a wide range—provides a striking example of this remarkable and unique property of systems governed by restoring forces proportional to the displacement from equilibrium. Suppose we had two identical pendulums, each with a string of length 30 ft, suspended from a high support. Each would have a period of about 6 sec. And if we set one swinging with an amplitude of only an inch or two, so that its motion was almost imperceptible, and set the other swinging with an amplitude of 5 ft, so that it swept through the central position at a speed of about 5 ft/sec, the difference in their periods would be too small to put them significantly out of step in less than a hundred swings or so.

This same isochronous property of the pendulum was of great and perhaps crucial help to Newton, Huygens, et al. when they made their fundamental experiments on collision phenomena (cf. Chapter 9). All these experiments were done with masses suspended from equal strings. As long as the small-amplitude

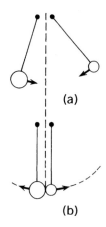

Fig. 11-13 (a) Two pendulums of equal length are released from arbitrary positions at the same instant. (b) The isochronous property of the pendulum ensures that the collision occurs when both masses are at the bottom points of their swing.

approximation is valid, two masses released at the same instant from arbitrary positions [Fig. 11-13(a)] will reach their lowest points at the same instant [Fig. 11-13(b)], so that the collision occurs when each mass is traveling horizontally with its maximum velocity—the magnitude of which is given by the equations of simple harmonic motion.[1] Even for angular amplitudes that are large enough to require the use of the exact equation for v_{max},

$$v_{max}^2 = 2gl(1 - \cos\theta_0)$$

the colliding masses will reach their lowest points at almost the same instant as long as θ_0 is less than 90°, which it would have to be for an object attached to a string rather than to a rigid rod. The next section briefly discusses the detailed dependence of period on amplitude for a simple pendulum.

THE PENDULUM WITH LARGER AMPLITUDE[2]

To find how the period of a simple pendulum departs from its ideal small-amplitude value, we write the equation for conservation of energy in the exact form

$$\left(\frac{d\theta}{dt}\right)^2 + 2\omega_0^2(1 - \cos\theta) = 2\omega_0^2(1 - \cos\theta_0) \quad (11\text{-}14)$$

where $\omega_0^2 = g/l$ and θ_0 is the maximum angle of deflection from the vertical.

The period of oscillation is then given by

$$T(\theta_0) = \frac{2}{\omega_0\sqrt{2}} \int_{-\theta_0}^{\theta_0} \frac{d\theta}{(\cos\theta - \cos\theta_0)^{1/2}} \quad (11\text{-}15)$$

---

[1] You can easily verify that the magnitude of this velocity is proportional, in the small-amplitude approximation, to the horizontal distance through which a mass is initially drawn aside before being released. This makes analysis of the experiments very simple.

[2] This section may be omitted without loss of continuity.

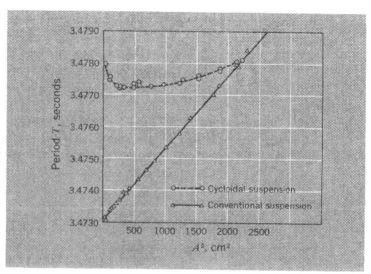

Fig. 11-14 Precise measurements on the period of a simple pendulum, showing the quadratic increase of period with amplitude, and the close approach to isochronism obtained by clamping the top end of the supporting wire between cycloidal jaws, so that the free length shortens as the amplitude increases.

For small amplitudes, we of course have a period T_0 equal to $2\pi/\omega_0$, as discussed in the previous section. The integral of Eq. (11-15) cannot be carried out exactly; one has to resort to numerical methods or to a series expansion of the integrand which gives, as a next approximation to the period, the following result:

$$T(\theta_0) \approx T_0\left[1 + \tfrac{1}{4}\sin^2\left(\frac{\theta_0}{2}\right)\right] \tag{11-16}$$

If θ_0 is not too large, another acceptable form of this result, in terms of the horizontal amplitude A ($= l\sin\theta_0$) is the formula

$$T(A) \approx T_0\left(1 + \frac{A^2}{16l^2}\right) \tag{11-17}$$

Figure 11-14 shows the results of some precision measurements that verify Eq. (11-17).[1] The pendulum had a length of about 3 m, and the greatest amplitude used was about 0.5 m, so that values of θ_0 up to about 10° (≈ 0.17 rad) are represented. It is worth noting that the over-all change of T from the least to the greatest values of θ_0 studied is less than 2 parts in 1000; the individual points are accurate to about 1 part in 10^5, and the validity of Eq. (11-17) is very nicely demonstrated over this range of θ_0. The graph also shows the results of using a special

[1] M. K. Smith, *Am. J. Phys.*, **32**, 632 (1964).

pendulum support that shortens the pendulum as θ increases, in such a way that the tendency for T to increase with θ_0 is almost completely compensated. Ideally, the support should be shaped so that the path of the pendulum is a portion of a cycloid curve. (Incidentally, the method of making an exactly isochronous pendulum, through the use of a cycloidal suspension, was first discovered by Huygens. In 1673 he published a great book, the *Horologium Oscillatorium*, in which many important dynamical results were first presented within the framework of the very practical problems of clock design.)

UNIVERSAL GRAVITATION: A CONSERVATIVE CENTRAL FORCE

The problems of energy conservation that we have discussed so far have involved only the familiar force of gravity near the earth's surface—a force which, as experienced by any given object, has effectively the same magnitude and the same direction in a given locality, regardless of horizontal or vertical displacements (within wide limits).[1] But, as we know, the basic gravitational interaction is a force varying as the inverse square of the distance between two particles and exerted along the line joining their centers. It is an example of the very important class of forces—*central* forces—which are purely radial with respect to a given point, the "center of force". It has the further property of being spherically symmetric; that is, the magnitude of the force depends only on the radial distance from the center of force, and not on the direction. We shall show that all such spherically symmetric central forces are conservative, and shall then consider the special features of the $1/r^2$ force that holds for gravitation and electrostatics.

If a particle is exposed to a central force, the force vector **F** acting on it has only one component, F_r. If the force is also spherically symmetric, then F_r can be written as a function of r only:

$$F_r = f(r) \tag{11-18}$$

We shall prove that any such spherically symmetric central force is conservative by showing that the work done by the force on a

---

[1] Recall, however, that small local variations are in fact detectable by sensitive gravity survey instruments, as described in Chapter 8.

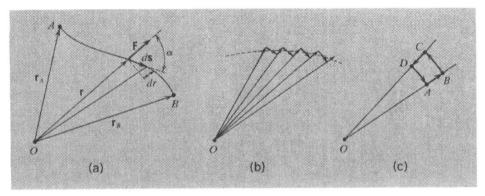

Fig. 11-15 (a) Diagram for consideration of potential-energy changes in a central force field. (b) Analysis of an arbitrary path into radial elements along which work is done, and transverse elements along which no work is done. (c) Closed path in a conservative central force field.

test particle, as the latter changes its position from a point A to another point B [Fig. 11-15(a)], does not depend on the particular path connecting A and B, but only on these end points. If this result holds, then it will be true that the particle, if it went from A to B by this path, and returned from B to A by any other path, would have *no* net work done on it by the central force and would (if no work were done on it by other forces) return to A with the same kinetic energy that it had to start with. This then corresponds exactly to the conservative property as defined for one-dimensional motions.

Let the center of force be at O [Fig. 11-15(a)], and consider the work done by the central force **F** on the test particle as it undergoes a displacement *d***s** along the path as shown. This work is given by

$$dW = \mathbf{F} \cdot d\mathbf{s} = F\,ds\cos\alpha \tag{11-19}$$

where α is the angle between the direction of **F**, i.e., the direction of **r**, and the direction of *d***s**. From the figure, however, we see that

$$ds\cos\alpha = dr$$

where dr is the change of distance from O, resulting from the displacement *d***s**. Inserting this value of $ds\cos\alpha$ into Eq. (11-19), we have

$$dW = F\,dr$$

where the magnitude of the force [$F = f(r)$] depends only on r as indicated in Eq. (11–18). We then have for the work done by **F** on the test object as it moves from A to B,

$$W = \int_{r_A}^{r_B} f(r)\, dr \tag{11-20}$$

Because this integral has a value that depends only on its limits and not on the path, we can conclude that the spherically symmetric central force is conservative. This result can be arrived at in slightly different terms by picturing the actual path as being built up from a succession of small steps, as shown in Fig. 11–15(b). One component of each step is motion along an arc at constant r, so that the force is perpendicular to the displacement and the contribution to W is zero, and the other component is a purely radial displacement so that the force and the displacement are in the same direction, resulting in the amount of work $F\,dr$ by the central force.

[A converse to the result we have just derived is the following important proposition:

A central force field that is also conservative must be spherically symmetric.

To show this, suppose that a center of force exists at the point O in Fig. 11–15(c). Imagine a closed path $ABCD$, formed by very short portions of two radial lines drawn from O, and by the two circular arcs BC and DA. Since, by definition, the force is purely central, it has no component perpendicular to the radial direction from O at any point. Thus, if we imagine a particle carried around $ABCD$, it experiences no force along BC and DA. The condition that the force be conservative thus requires equal and opposite amounts of work along AB and CD. Since these lines are of equal length, the mean magnitude of the force must be the same on each. If we imagine the lengths of these elements of path to become arbitrarily small, we conclude that the value of the force at a particular scalar value of r is independent of the direction in which the vector **r** is drawn. The most important sources of such spherically symmetric force fields are spherically symmetric distributions of mass or electric charge.]

It is important to realize that a force *may* be conservative without being central or spherically symmetric. For example, the combined gravitational effect of a pair of concentrated unequal masses, separated by some distance, has a complicated dependence on position and direction, but we know that it is conservative because it is the superposition of two individually conservative force fields of the separate masses.

Given the result expressed by Eq. (11–20), we can proceed to define a potential energy $U(r)$ for any object exposed to a spherically symmetric central force:

$$U_B - U_A = -\int_{r_A}^{r_B} f(r)\, dr \tag{11–21}$$

If the kinetic energy of the object at points A and B has the values K_A and K_B, respectively, then (if no work is done by other forces) we have

$$K_B - K_A = W$$

and

$$K_A + U_A = K_B + U_B = E \tag{11–22}$$

Thus we have established an energy-conservation statement for an object moving under the action of any central force.

For an inverse-square force, we have

$$F(r) = \frac{C}{r^2} \tag{11–23}$$

In such a case, therefore, Eq. (11–21) gives us

$$U_B - U_A = -C \int_{r_A}^{r_B} r^{-2}\, dr$$

from which we get

$$U_B - U_A = C\left(\frac{1}{r_B} - \frac{1}{r_A}\right) \tag{11–24}$$

There remains only the choice of the zero of potential energy. It is convenient to set $U = 0$ for $r = \infty$, i.e., at points infinitely far from the source at O, since the force vanishes at these points, and in colloquial terms one can say that the existence of either particle is of no consequence to the other one under these conditions. We now apply Eq. (11–24) to the case where $r_A = \infty$, $U_A = 0$, and the potential energy of the test particle becomes $U_B = C/r_B$—or, if we drop the now redundant subscript B, there follows

$$U(r) = \frac{C}{r} \qquad (11\text{-}25)$$

for the potential energy of a test particle as a function of its position.

Equation (11-25) is valid for either an attractive or a repulsive force, the constant C being negative for attractive forces and positive for repulsive forces. In particular, for a particle of mass m under the gravitational attraction of a point mass M (which we shall suppose to be fixed at the origin O) we have

$$F(r) = -\frac{GMm}{r^2} \qquad (11\text{-}26)$$

$$U(r) = -\frac{GMm}{r} \qquad (11\text{-}27)$$

Note that these last two equations refer only to the interaction between two objects that can be regarded as though they were points—i.e., their linear dimensions are small compared to their separation. As we saw in Chapter 8, however, some of the most interesting and important gravitational problems concern the gravitational forces exerted by large spherical objects such as the earth or the sun. In our earlier discussion of such problems, we saw that the basic problem is the interaction between a point particle and a thin spherical shell of material. In Chapter 8 we presented a frontal attack on this problem, going directly to an evaluation of an integral over all contributions to the net force. Now we shall approach the problem by way of a consideration of potential energies, and in the process we shall see the great value of the potential energy concept in such calculations.

A GRAVITATING SPHERICAL SHELL

Suppose, as in our earlier treatment of the problem, that we have a thin, uniform shell of matter, of radius R and mass M (Fig. 11-16). Let a particle of mass m be placed at some point P a distance r from the center of the shell. If we deal directly in terms of forces, then, as we saw, the force from material in the vicinity of a point such as A must be resolved along the line OP. In other words, a vector sum of all the force contributions is necessary. If we deal in terms of potential energy, however, we can exploit its most important property: *Potential energy is a*

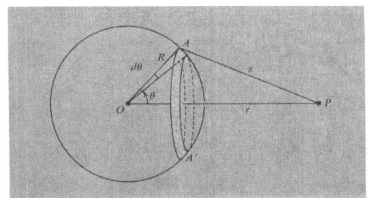

Fig. 11-16 Diagram for considering the gravitational potential energy due to a circular zone of a thin spherical shell of matter.

scalar quantity. We can just add up the contributions to U from all parts of the shell. The force on m is then obtained simply from the relation

$$F_r = -\frac{dU}{dr} \tag{11-28}$$

As in our direct calculation of the force, we take advantage of the symmetry of the system by considering the zone of material marked off on the sphere between the angles θ and $\theta + d\theta$ (Fig. 11-16). For this zone we have

$$\text{area} = 2\pi R^2 \sin\theta\, d\theta$$

$$\text{mass} = \frac{2\pi R^2 \sin\theta\, d\theta}{4\pi R^2} M$$

i.e.,

$$dM = \tfrac{1}{2} M \sin\theta\, d\theta$$

All of this material in dM is at the same distance, s, from P. Thus the contribution that dM makes to U is given, according to Eq. (11-27), by

$$dU = -\frac{Gm\, dM}{s} = -\frac{GMm}{2}\frac{\sin\theta\, d\theta}{s}$$

The total potential energy of m is thus

$$U(r) = -\frac{GMm}{2} \int \frac{\sin\theta\, d\theta}{s} \tag{11-29}$$

This integral is to be evaluated, keeping R and r constant, by

447 A gravitating spherical shell

allowing θ to sweep from zero to π (s changing accordingly) so as to include the contributions of all parts of the shell. The calculation is made simple by the fact that, in the triangle AOP, we have

$$s^2 = R^2 + r^2 - 2Rr\cos\theta$$

Differentiating both sides with respect to θ, remembering that R and r are constants for the purpose of the integration in Eq. (11–29), we have

$$2s\frac{ds}{d\theta} = 2Rr\sin\theta$$

Therefore,

$$\frac{\sin\theta\, d\theta}{s} = \frac{ds}{Rr}$$

But the left-hand side of this is just the integrand of Eq. (11–29), which thus becomes

$$U(r) = -\frac{GMm}{2Rr}\int_{\theta=0}^{\pi} ds \tag{11-30}$$

Equation (11–30) is the key result in calculating gravitational potential energies and forces due to spherical objects. In evaluating

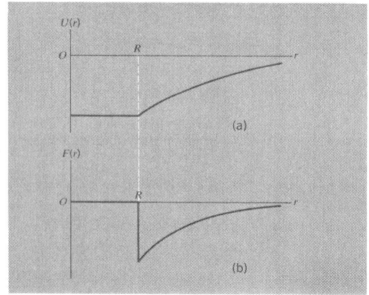

Fig. 11-17 (a) Gravitational potential energy of a point particle as a function of its distance from the center of a thin spherical shell of radius R. (b) Variation of F with r, derived from (a).

the integral of ds, however, we must distinguish two cases:

Case 1. Point P *outside* the shell (i.e., $r > R$, as in Fig. 11–16). The limits on s are defined as follows:

$\theta = 0$ giving $s_{min} = r - R$
$\theta = \pi$ giving $s_{max} = r + R$

Hence

$s_{max} - s_{min} = 2R$

$$U(r) = -\frac{GMm}{r} \quad \text{(point } outside \text{ shell)} \quad (11\text{–}31)$$

Case 2. Point P *inside* the shell (i.e., $r < R$). The limits now become

$\theta = 0$ giving $s_{min} = R - r$
$\theta = \pi$ giving $s_{max} = R + r$

Hence

$s_{max} - s_{min} = 2r$

$$U(r) = -\frac{GMm}{R} \quad \text{(point } inside \text{ shell)} \quad (11\text{–}32)$$

These two results are disarmingly alike. The forces derived from them are, however, quite different. If we proceed now to calculate the forces from Eq. (11–28), we must remember that R (the radius of the shell) is a *constant*. It is only r, the distance of the mass m from O, that can vary. Hence we have

Case 1.

$$F_r = -\frac{GMm}{r^2} \quad \text{(point } outside \text{ shell)} \quad (11\text{–}33)$$

Case 2.

$$F_r = 0 \quad \text{(point } inside \text{ shell)} \quad (11\text{–}34)$$

In Fig. 11–17 we show the graphs of potential energy and force as functions of the radial distance r of the test mass m from the center of the shell. The discontinuities at $r = R$ are easily seen.

Once we have obtained these results, it is a simple matter to consider a solid sphere of material.

A GRAVITATING SPHERE

Our primary assumption is that the sphere can be regarded as made up of a whole succession of *uniform* spherical shells, even though the density may vary with radial distance from the center.

Granted this assumption of symmetry, we can at once draw some conclusions about the gravitational force exerted by such a sphere, of total mass M and radius R.

Case 1. Observation point *outside* the sphere ($r > R$). Since each component spherical shell acts as though its whole mass were at the center, the same can be said for the sphere as a whole. Regardless of the way in which the density of the material varies with radius, we have

$$F(r) = -\frac{GMm}{r^2} \quad \text{(point outside sphere)} \quad (11\text{-}35)$$

This is the result that we have already used in Chapter 8 in discussing Newton's famous comparison of the gravitational accelerations of the moon and the apple.

Case 2. Observation point *inside* sphere ($r < R$). Here we must be more careful, but we can at once make two clear statements:

a. For all the spherical shells lying *outside* the observation point, the contribution to the force is *zero*, by Eq. (11-34).

b. For all shells lying *inside* the radius defined by the observation point, the mass is effectively concentrated at the center, by Eq. (11-33).

To specify the force in this case, therefore, we must know how much mass is enclosed within the sphere of radius r drawn through the observation point.

Case 2'. (special) The same as case 2, but with the extra proviso that the density of the sphere is the same for all r—the sphere is *homogeneous*. In this case, we know that the fraction of the mass contained within radius r is equal to r^3/R^3, because this is the ratio of the partial volume to the whole volume. Hence the amount of mass to be considered is equal to Mr^3/R^3, effectively at the center, i.e., a distance r from the position of the test mass. This then gives us

$$F(r) = -\frac{GMm}{R^3} r \quad \text{(point inside homogeneous sphere)} \quad (11\text{-}36)$$

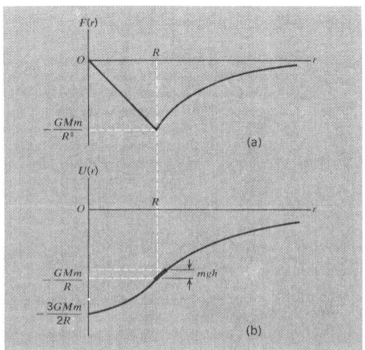

Fig. 11-18 (a) Force on a point particle as a function of its distance from the center of a uniform solid sphere of radius R. (b) Variation of mutual potential energy with r, obtained from (a) by integration. The approximately linear increase of U with r just outside r = R corresponds to gravity as observed near the earth's surface.

The combined results of Eqs. (11-35) and (11-36) for a homogeneous sphere are shown in Fig. 11-18(a). One can also construct the graph of potential energy versus r for all values of the distance of a particle of mass m from the center of the sphere. This is shown in Fig. 11-18(b). For interior points ($r < R$) we have

$$U(r) - U(R) = \frac{GMm}{R^3} \int_R^r r\, dr$$

$$= -\frac{GMm}{2R^3}(R^2 - r^2)$$

But we already know, by putting $r = R$ in Eq. (11-31), that

$$U(R) = -\frac{GMm}{R}$$

Thus we get

$$(r < R) \quad U(r) = -\frac{GMm}{2R^3}(3R^2 - r^2) \qquad (11\text{-}37)$$

In particular, at $r = 0$, we have

451   A gravitating sphere

$$U(0) = -\frac{3GMm}{2R}$$

If one imagines starting out from $r = 0$, then, as Fig. 11–18(b) shows, the potential energy increases parabolically with distance up to $r = R$ and then goes over smoothly into the continued increase of U with r that is described by Eq. (11–31).

The variations of $F(r)$ and $U(r)$ for $r > R$ in Fig. 11–18 hold good, as we have seen, for any spherically symmetrical distribution of matter; there is no requirement that the sphere should be homogeneous. On the other hand, one must be careful to remember that Eqs. (11–36) and (11–37), for $r < R$, refer only to the special case of a sphere of the same density throughout. Thus it does *not* correctly describe the variation of gravitational force with radial distance inside such objects as the earth and the sun, which have drastic variations of density with r (see Fig. 11–19). This invalidates a favorite textbook exercise: "Show that a particle would execute simple harmonic motion in a tunnel bored along a diameter of the earth." The practical impossibility of making such a tunnel may, however, be considered as a far more powerful objection.

If one is used to thinking that the potential energy of an object of mass m, a distance h above the earth's surface, is given

Fig. 11–19 (a) Radial variation of density inside the earth. (After E. C. Bullard.) (b) Calculated variation of density with radial distance inside the sun (I. Iben and Z. Abraham, M.I.T.).

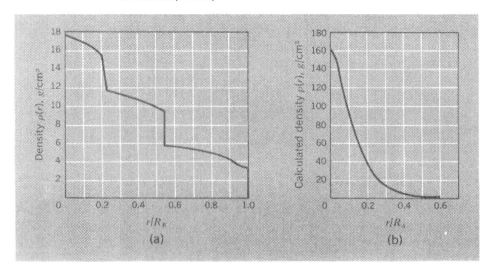

by the formula $U = mgh$, it may seem hard to reconcile this result with the result expressed by Eq. (11–31):

$$U(r) = -\frac{GMm}{r} \quad (r > R)$$

There is really no difficulty, however, once one recognizes that the zero of potential energy is arbitrary and that the simple linear formula applies only to objects raised through distances that are exceedingly small compared to the earth's radius. We have, in fact, from the more general formula [i.e., Eq. (11–31)],

$$U(R + h) - U(R) = -\frac{GMm}{R + h} + \frac{GMm}{R}$$
$$\approx \frac{GMm}{R^2} h$$

Since, however, GMm/R^2 is the gravitational force mg exerted on m, this gives us

$$U(R + h) - U(R) \approx mgh$$

By putting $U(R) = 0$ as an arbitrary reference level of potential energy, we see that $U = mgh$ is an acceptable approximation that applies to small displacements near the earth's surface. This approximately linear increase of potential energy with distance just outside a sphere is indicated on Fig. 11–18(b).

ESCAPE VELOCITIES

Suppose we have an object at the surface of a large gravitating sphere, such as a planet, and we want to shoot the object off into outer space so that it never returns. What speed must we give it? This problem is easily considered in terms of energy conservation. At the surface of the planet ($r = R$) the potential energy is given by

$$U(R) = -\frac{GMm}{R}$$

At $r = \infty$ we have $U = 0$. The particle, to reach $r = \infty$, must survive to this distance with some kinetic energy $K(\infty)$. Thus at launch it must have a kinetic energy $K(R)$ defined through the conservation equation

$$K(R) - \frac{GMm}{R} = K(\infty)$$

Therefore,

$$K(R) \geq \frac{GMm}{R}$$

The critical condition is reached if we take the equality in the above statement; the object would then just reach infinity with zero residual speed. The minimum escape speed at radius R is thus given by

$$\tfrac{1}{2}mv_0^2 = \frac{GMm}{R}$$

Therefore,

$$v_0^2 = \frac{2GM}{R} \qquad (11\text{--}38)$$

In calculating the escape velocity from the earth's surface, it may be convenient to state the result of Eq. (11–38) in another form that makes use of our knowledge that the gravitational force on a mass m at $r = R_E$ can be set equal to the magnitude of m times the local acceleration due to gravity, g:

$$\frac{GMm}{R_E^2} = mg$$

Therefore,

$$\frac{GM}{R_E} = gR_E$$

and so, from Eq. (11–38), we have

$$v_0 = (2gR_E)^{1/2} \qquad (11\text{--}39)$$

Putting in the familiar values $g \approx 9.8 \text{ m/sec}^2$, $R_E \approx 6.4 \times 10^6$ m, we have

$$v_0 \approx 11.2 \text{ km/sec}$$

Notice once again the remarkable implications of energy conservation. For the purpose of calculating *complete escape*, we need specify nothing about the direction in which the escaping object is fired. It could be radially outward, or tangentially, or anything in between; the same value of v_0 applies to all cases [Fig. 11–20(a)].

It is interesting that the magnitude of the escape speed v_0 is exactly $\sqrt{2}$ times as great as the orbital speed that a particle

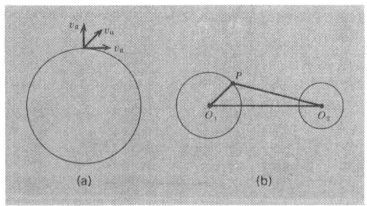

Fig. 11–20 (a) The possibility of escape from a gravitating sphere depends on the magnitude of the velocity, not on its direction. (b) The escape speed from a double-star system depends on the position of the starting point P.

would have if it could skim around the earth's surface in a circular orbit of radius R_E.

The preceding calculation assumes that the object from which the escape takes place is isolated from all other objects. More complicated escape problems would arise if, for example, we wanted to calculate the escape velocity from the surface of one member of a double-star system [Fig. 11–20(b)]. We should then have to consider where the launching point P was in relation to the centers O_1 and O_2 of the two stars. But still the *scalar* property of potential energy makes the calculation relatively simple. The potential energy of a mass m at P is just the straightforward sum of the potential energies due to the two stars separately. And again the direction of launch for complete escape at the critical velocity is immaterial (as long as the trajectory misses the other star!)

A particularly important example of this last type of system is, of course, the earth–moon combination. This is shown schematically in Fig. 11–21(a). If we consider a particle of mass m, its potential energy as a function of position along a straight line joining the centers of earth and moon is as shown by the full line in Fig. 11–21(b). Mathematically it is given by the formula

$$U(r) = -\frac{GM_E m}{r} - \frac{GM_M m}{D - r} \qquad (11\text{–}40)$$

where D is the distance between the centers of earth and moon and r is the distance of the particle from the center of the earth. This potential energy is simply the sum of the separate contributions from earth and moon.

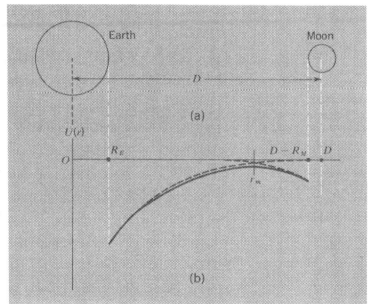

Fig. 11-21 (a) Schematic diagram of the earth-moon system. (b) Form of variation of the total potential energy of a particle along the line joining the centers of earth and moon.

The maximum value of $U(r)$ occurs at the point $r = r_m$, where $(dU/dr)_{r_m} = 0$. This is clearly also the point at which the attractive forces due to earth and moon are equal and opposite, so that we can put

$$\frac{GM_E m}{r_m^2} = \frac{GM_M m}{(D - r_m)^2}$$

This gives

$$r_m = \frac{D}{1 + (M_M/M_E)^{1/2}}$$

Now the value of M_M/M_E is about $\frac{1}{81}$, so that $r_m \approx 0.90D$. This represents a point about 24,000 miles from the moon's center. The value of $U(r)$ at this value of r is given, according to Eq. (11-40), by

$$\begin{aligned}U(r_m) &= -\frac{GM_E m}{0.9D} - \frac{GM_M m}{0.1D} \\ &= -\frac{GM_E m}{0.9D}\left(1 + \frac{9M_M}{M_E}\right) \\ &\approx -\frac{1.23 GM_E m}{D}\end{aligned}$$

If a spacecraft is to reach the moon from the earth along the direct line considered, it would have to surmount this potential energy hump. If, further, this were to be done by simply coasting after an initial rocket blast at the point of departure from the earth's surface, the necessary minimum initial speed at the earth would be defined by the following equation:

$$\tfrac{1}{2}mv_0^2 - \frac{GM_E m}{R_E} = U(r_m) = -\frac{1.23 GM_E m}{D}$$

This then gives us

$$v_0^2 = \frac{2GM_E}{R_E}\left(1 - \frac{1.23 R_E}{D}\right)$$

The factor $2GM_E/R_E$ is just the square of the speed needed for complete escape from the earth, as given by Eq. (11-39) and evaluated numerically as 11.2 km/sec. Putting $R_E/D \approx \tfrac{1}{60}$ in the above expression for v_0^2, one finds $v_0 \approx 11.1$ km/sec. At this minimum velocity the traveling object would just barely surmount the potential "hill" between earth and moon and would then fall toward the moon, reaching its surface with a speed of about 2.3 km/sec. (You should check these results for yourself.)

MORE ABOUT THE CRITERIA FOR CONSERVATIVE FORCES[1]

In our discussions of forces and potential energy, we have pointed out that the fundamental criterion for a force to be conservative is that the net work done by the force be zero over any closed path. As an aside we pointed out that in one-dimensional problems, but only in one-dimensional problems, this condition is automatically met if the force is a unique function of position. We shall now give a simple example to illustrate how, in two dimensions, this latter condition is not sufficient. A force $\mathbf{F}(x, y)$ that is a unique function of x and y may nevertheless be nonconservative.

Our example is this: Suppose that a particle, at any point in the xy plane, finds itself exposed to a force $\mathbf{F}$ given by

$$\begin{aligned} F_x &= -ky \\ F_y &= +kx \end{aligned} \qquad (11\text{-}41)$$

where k is a constant and x and y are the coordinates of the

[1] This section may be omitted without loss of continuity.

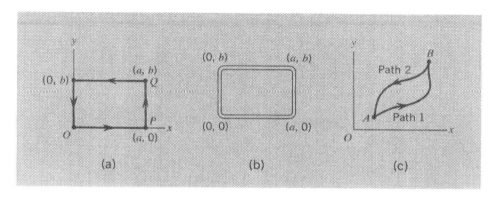

Fig. 11-22 (a) Rectangular path in a plane. (b) Smooth tube shaped to force a particle to follow the path shown in (a). (c) Closed path made up of two different paths between the given points A and B. If the force is conservative, the net work is zero.

particle. Such a force evidently depends only on the position of the particle. Now let us calculate the work done by this force on the particle if the latter moves counterclockwise around the closed path shown in Fig. 11-22(a). It may well be objected that the particle would never follow this path under the action of the force **F** and nothing else, but (as we did in our first approach to potential energy at the beginning of Chapter 10) we can imagine that **F** is exactly balanced by another force, supplied for example by a spring, so that the object can be freely moved around without any net work being done. Or, alternatively, we can imagine a very smooth pipe, as in Fig. 11-22(b), which compels the particle to travel along the sides of a rectangle and yet does no work on the particle because the force exerted by the pipe is always at right angles to the particle's motion. We should also add the proviso, in this case, that the particle has enough kinetic energy to carry it around the path even if **F** is taking energy away from it. Now for the calculation.

Starting at O, let the particle move along the x axis a distance a to the point P with coordinates $x = a$, $y = 0$. The work done by **F** during this motion is

$$W_1 = \int_{x=0}^{a} F_x \, dx = -ky \int_0^a dx = 0$$

since $y = 0$ everywhere on this portion of the closed path. Next, the particle moves from P to $Q(x = a, y = b)$. The work done by **F** along this path is

$$W_2 = \int_{y=0}^{b} F_y\, dy = +ka \int_0^b dy = kab$$

For the path from Q to R $(0, b)$, the work done is

$$W_3 = \int_{x=a}^{0} F_x\, dx = -kb \int_a^0 dx = +kab$$

and finally, the work W_4 done as the particle moves from R back to the origin O is zero, since x and hence F_y is zero everywhere on this path:

$$W_4 = 0$$

The total work done in the round trip is therefore

$$W = W_1 + W_2 + W_3 + W_4 = 2kab \neq 0$$

and the force **F** given by Eq. (11–41) is *not* conservative, although it depends only on position. If k were positive, the kinetic energy of the particle would be increased by $2kab$ for each complete circuit taken in this direction. On the other hand, if the particle traveled clockwise, its kinetic energy would be *decreased* by this amount each time. This may seem—and indeed is—a very artificial example; nevertheless, forces having precisely this nonconservative property [although not the same analytic forms as described by Eq. (11–40)] play an important role in physics, especially in electromagnetism.

There is another way of looking at the analysis of such a situation. The physical criterion for the force to be conservative is that it should do no net work, either positive or negative, as the particle to which it is applied makes a complete circuit around the path of Fig. 11–22(a) in either direction. Let us consider a more general situation [Fig. 11–22(c)] in which a particle travels from a point A to a point B along one path and returns from B to A along a different path (again imagine that constraining forces, doing no work, are applied as necessary). We shall assume that the force we are studying (i.e., not including any constraining forces) is a function of the position of the particle only. If this force is conservative, we then have

$$W = \int_{A}^{B} \mathbf{F} \cdot d\mathbf{s} + \int_{B}^{A} \mathbf{F} \cdot d\mathbf{s} = 0$$
$$\text{Path 1} \qquad \text{Path 2}$$

It follows from this that

$$\int_A^B \mathbf{F}\cdot d\mathbf{s} \bigg|_{\text{Path 1}} = -\int_B^A \mathbf{F}\cdot d\mathbf{s} \bigg|_{\text{Path 2}}$$

But now, if the force is a function only of position, we can interchange the limits of the integral on the right and reverse its sign. Hence, if **F** is conservative, we must have

$$\int_A^B \mathbf{F}\cdot d\mathbf{s} \bigg|_{\text{Path 1}} = \int_A^B \mathbf{F}\cdot d\mathbf{s} \bigg|_{\text{Path 2}}$$

If this condition is satisfied we can put

$$W_{AB} = \int_A^B \mathbf{F}\cdot d\mathbf{s} \qquad (11\text{-}42a)$$

without reference to the particular path. And if W_{AB} is the same for all paths from A to B, we can conclude that the force is conservative and that the potential-energy function can be defined through the equation

$$U_B - U_A = -\int_A^B \mathbf{F}\cdot d\mathbf{s} \qquad (11\text{-}42b)$$

In evaluating work integrals such as that in Eq. (11-42a), one may wish to resolve the force **F** into its components F_x and F_y, and resolve the element of path $d\mathbf{s}$ into its components dx and dy, thus getting (for a two-dimensional problem)

$$W = \int_{x_1}^{x_2} F_x\, dx + \int_{y_1}^{y_2} F_y\, dy$$

If one uses this equation, however, one should always remember that, basically, *the integrals are defined as being taken along an actual designated path of the particle.* This might seem to be obvious, because the force must necessarily be applied wherever the particle is. However, there are many situations in which an object experiences a force wherever it happens to be (gravitational force, for example). One can then set up a purely *mathematical* statement that defines a value of **F** for any given x and y. In general each component of **F** is a function of both x and y. And unless one already knows that the force is conservative, it is

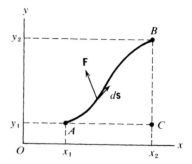

Fig. 11-23 *Consideration of work done along a specified path between two points.*

essential to follow the given path, as in Fig. 11-23, and not take the simpler but perhaps unjustified course of (for example) finding the integral of F_x from x_1 to x_2 along the line AC ($y = y_1 = $ const.), followed by the integral of F_y from y_1 to y_2 along the line CB ($x = x_2 = $ const.).

It is perhaps worth ending this discussion with the remark that, in many situations, one may use the concept of potential energy even when additional nonconservative forces act on the particle. For example, a satellite moving in the gravitational field of the earth may be subject to the frictional drag of the earth's upper atmosphere. This drag is a dissipative force, nonconservative, and the total mechanical energy of the motion will not be constant but will decrease as the motion proceeds. Nevertheless, one may still properly talk of the gravitational potential energy that the satellite possesses at any given point.

FIELDS[1]

The forces that we have labeled gravitational, electric, and magnetic are "action-at-a-distance" forces. No apparent physical contact between interacting objects comes into play. For this kind of interaction, the idea of a *field* of force is most useful. To introduce this concept, consider the gravitational attraction of the earth for a particle outside it. The pull of the earth depends both on the mass of the attracted particle and on its location relative to the center of the earth. This attractive force divided by the mass of the particle being pulled depends only on the earth and the location of the attracted object. We can therefore assign to each point of space a vector, of magnitude equal to

---

[1]This section may be omitted without loss of continuity.

the earth's pull on a particle divided by its mass, and of direction identical with that of the attractive force. Thus we imagine a collection of vectors throughout space, in general different in magnitude and direction at each point of space, which define the gravitational attraction of the earth for a test particle located at any arbitrary position. The totality of such vectors is called a field, and the vectors themselves are called the field strengths or intensities of the field.

In this example, the gravitational field strength **g** at a point P is

$$\mathbf{g} = \frac{\mathbf{F}}{m}$$

The magnitude of this field is measured in units of acceleration (e.g., m/sec^2) and is given explicitly by the law of gravitation as

$$\mathbf{g} = -G\frac{M}{r^2}\frac{\mathbf{r}}{r} = -G\frac{M}{r^2}\mathbf{e}_r \tag{11-43}$$

where M is the mass of the earth, **r** the position vector of the point P relative to the center of the earth, and $\mathbf{e}_r$ the unit vector in the direction of **r**. Equation (11-43) is valid at all points outside the surface of an ideal spherical earth.

One can generalize this for the field produced by an arbitrary distribution of matter, for which the field strength **g** as a function of position describes quantitatively the gravitational field. The gravitational force exerted on an object of mass m by this field is then given by $\mathbf{F} = m\mathbf{g}$.

This field description of forces is especially useful in specifying electromagnetic forces. The electric field produced by a charged particle or by a collection of such charged particles is described by the electric field strength or intensity $\mathbf{\mathcal{E}}$, where

$$\mathbf{\mathcal{E}} = \frac{\mathbf{F}}{q}$$

F is the vector force acting on a positive test charge of magnitude q, and $\mathbf{\mathcal{E}}$ depends on position. For electric fields produced by charges at rest, the situation is similar to the gravitational case.

As an aid to visualizing the character of a force field, use is frequently made of the concept of a *field line*. Starting from an arbitrary point, we draw an infinitesimal line element in the direction of the field at that point. Being thus brought to a neighboring point in the field, we draw another line element in the direction of the field at the new point, and so on. In the limit of making the line elements vanishingly short, we obtain a smooth curve, the tangent to which, at any point, is the direction of the field at that point. This construct—which is the field line—has no real existence, but a vivid picture of the properties of a force field can be obtained by drawing a whole collection of such lines.

Hand-in-hand with the concept of a line of force goes that of a line or surface of constant potential energy. We shall therefore discuss this briefly in the next section.

EQUIPOTENTIAL SURFACES AND THE GRADIENT OF POTENTIAL ENERGY[1]

Besides its utility in connection with the dynamics of conservative systems, the concept of potential energy enables us to describe conservative fields of force quantitatively in a relatively simple fashion. The reason is that we can use the simple *scalar* field of potential energy to calculate the relatively complicated *vector* field of force. This calculation proceeds as follows. Suppose we know the potential energy U of a test particle at each point of space; i.e., we have a relation of the form $U = f(x, y, z)$, where the single-valued function $f(x, y, z)$ depends on the particular field of force under consideration. If we wish to know at what points of space the test particle will have a given value of potential energy, say U_0, we set $U = U_0$ and obtain an equation of the form

$$f(x, y, z) = \text{const.} = U_0$$

This is the equation of a surface, and this surface is called an energy *equipotential* surface. There exists a whole family of these equipotential surfaces, one for each value of U_0. Since, by definition, it requires no work to move our test particle from one point to another on the same equipotential surface, it follows that *the lines of force are everywhere perpendicular to the equipotentials.*

We should draw attention to a distinction between two quantities here. Just as we define *field* as the *force per unit mass* (or charge, etc.) so we define *potential* (φ) as the *potential energy per unit mass* (or charge, etc.). To take the specific case of gravi-

[1] This section may be omitted without loss of continuity.

tation, we thus have the following paired quantities: gravitational force **F** (newtons) with gravitational potential energy U (joules), and gravitational field **g** (m/sec^2) with gravitational potential φ (m^2/sec^2).

The complete array of field lines and equipotential surfaces (or, in two dimensions, equipotential lines) provides a graphic picture of a complete field pattern. In two dimensions we see two sets of curves which, however complicated their appearance may be, are everywhere orthogonal to one another. The gravitational field of a spherical object has a simple pattern, in which the field lines are radial lines and the equipotentials are a set of concentric spheres [Fig. 11-24(a)]. The field pattern due to two spheres close together is far less simple [see Fig. 11-24(b)]. (For one way of constructing such a diagram, see Problem 11-26.) In making drawings of the equipotentials, it is often convenient

Fig. 11-24 (a) Equipotentials and field lines due to a sphere with a $1/r^2$ force law. (b) Equipotentials and field lines due to a system of two nearby spheres, of masses 2M and M.

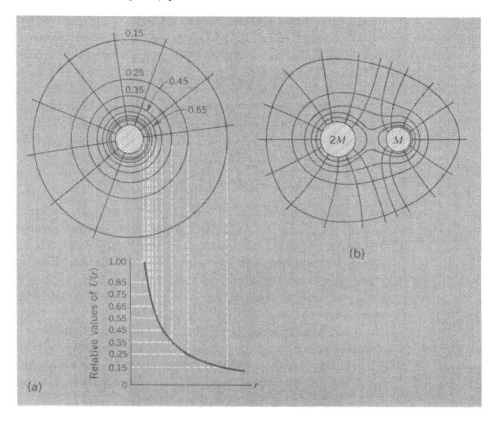

to draw them for equal successive increments of the potential (or potential energy); this makes the picture very informative, just like a contour map—which in effect it is (see Fig. 11-24).

To obtain a complete specification of the force field, we must be able to get the magnitude of the force vector, as well as its direction, at every point of space. Consider a test particle at a point P and let it be displaced an amount $d\mathbf{s}$ to a neighboring point. Its potential energy will change by an amount

$$dU = -\mathbf{F} \cdot d\mathbf{s} = -F_s \, ds$$

This may be written in the form

$$F_s = -\frac{dU}{ds} \tag{11-44}$$

In words, the component of the force $\mathbf{F}$ in any direction equals the negative rate of change of potential energy with position *in that direction*. The spatial derivative on the right-hand side of Eq. (11-44) is called a *directional* derivative, because its value depends on the direction in which $d\mathbf{s}$ is chosen at the point P. (Strictly speaking, we should be using the notation of *partial* derivatives: $F_s = -\partial U/\partial s$.) If we move from P to a neighboring point on the same equipotential surface as that on which P lies, then dU/ds is zero for this direction. If, however, we move to a neighboring point not on the same equipotential surface as that containing P, dU/ds will be different from zero. That particular direction for which dU/ds has its *maximum* value at a given point defines the direction of the line of force at that point, and the magnitude of this maximum value of dU/ds is the magnitude of the vector force at the point in question.

This maximum value of dU/ds, together with its associated direction, is called the *gradient* of the potential energy; it is a vector directed at right angles to the equipotential surface. In symbols, we write

$$\mathbf{F} = -\text{grad } U \tag{11-45}$$

To help clarify the idea of the gradient, consider a conservative field in two dimensions. Here the equipotentials are lines, rather than surfaces as they would be in three dimensions. In Fig. 11-25 we show two equipotentials, one for $U = U_0$ and one for $U = U_0 + \Delta U$. Starting at P on U_0, we can move to the equipotential $U_0 + \Delta U$ by any of an infinite number of dis-

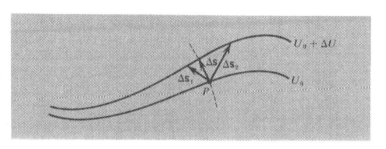

Fig. 11-25 Different paths between two neighboring equipotentials.

placements. However, for a given change of potential energy ΔU, one moves along a line of force to attain this change in the shortest possible displacement Δs. The rate of change of potential energy with position is maximum for this direction and this is the direction of the gradient of potential energy. This is indicated in Fig. 11-25, where three directions from P are shown. It is clear that Δs is shorter in length than either Δs_1 or Δs_2 and hence that $\Delta U/\Delta s$ is larger than $\Delta U/\Delta s_1$ or $\Delta U/\Delta s_2$.

MOTION IN CONSERVATIVE FIELDS

We now turn our attention to the problem of the *motion* of particles in a conservative field of force. We have of course already discussed a number of problems involving motion under gravity, but here we shall try to indicate the value of the energy method as it applies to more complex situations.

If the only force acting on a particle is that due to the field, the law of conservation of mechanical energy provides a first step in solving for the motion. This law, expressed as

$$\tfrac{1}{2}mv^2 + U(x, y, z) = E \tag{11-46}$$

where $v^2 = v_x^2 + v_y^2 + v_z^2$, does not, however, contain the whole story. It is the result of combining three statements of Newton's law as applied to the three independent coordinate directions, and the synthesis of these three vector equations into one scalar equation involves the discarding of information that in principle is still available to us. Given that U is a conservative potential, we can always find its gradient and hence the vector force [Eq. (11-45)]. With this, plus the initial conditions (values of $\mathbf{r}_0$ and $\mathbf{v}_0$ at $t = 0$) everything that one needs for a solution of the problem is provided.

Our interest here, however, lies in taking advantage of the

energy-conservation equation as far as possible, and supplementing it with whatever other information may be necessary. This extra information may take the form of the explicit use of one or two of the Newton's law statements, as we shall see in the example about to be discussed. Or it may, in suitable circumstances, be contained in a conservation law for quantities besides energy—in particular, angular momentum. The exploitation of this latter property is one of the principal concerns of Chapter 13.

As a good example of the methods we shall consider the motion of a charged particle in a combined electric and magnetic field. Suppose we have a pair of parallel metal plates mounted inside an evacuated tube and connected to a battery as shown in Fig. 11-26, so that a uniform electric field $\mathcal{E}$ ($= V/d$) exists between the plates. The plates are placed between the poles of a magnet that produces a uniform magnetic field B in a direction perpendicular to the page. We shall assume that electrons start out with negligible energy and velocity from the lower plate. This could, for example, be arranged by giving the lower plate a photosensitive coating and shining light onto it, as in a commercial phototube.

Consider one electron of charge $q = -e$ emitted at O. It will be accelerated vertically by a constant force equal to eV/d directed along the positive y axis, and will be deflected into the path indicated in the figure. Since the magnetic deflecting force, always acting perpendicular to the particle's velocity, does no work and hence does not directly affect the particle's energy, the statement of energy conservation can be written simply as follows:

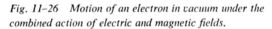

Fig. 11-26 Motion of an electron in vacuum under the combined action of electric and magnetic fields.

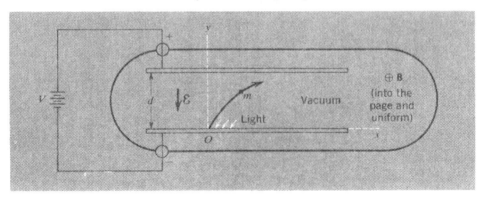

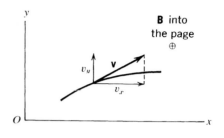

Fig. 11-27 Analysis of the magnetic force into components associated with the separate x and y components of **v**.

$$\tfrac{1}{2}m(v_x^2 + v_y^2) - \frac{eV}{d}y = 0 \qquad (11\text{-}47)$$

where we have chosen $U\ [= -(eV/d)y]$ equal to zero for $y = 0$, i.e., at the bottom plate where the kinetic energy at $y = 0$ is negligible.

As stated at the beginning of this section, we need more information to determine the motion. Since the electric force is directed along the y axis, the only x component of force on the electron is the x component of the magnetic force. To evaluate this force, think of the velocity vector of the electron at some point of its path resolved into x and y components (Fig. 11-27). The component of magnetic force associated with v_x in this situation is parallel to the y axis and does not concern us here, but the velocity component v_y gives rise to a magnetic force component F_x given by

$$F_x = ev_y B$$

and Newton's law of motion for the x component of motion is accordingly

$$ev_y B = m\frac{dv_x}{dt} \qquad (11\text{-}48)$$

Equations (11-47) and (11-48) can be solved for the motion of the particle, as follows:

Since $v_y = dy/dt$, Eq. (11-48) can be written as

$$eB\, dy = m\, dv_x$$

or, integrated,

$$mv_x = eBy \qquad (11\text{-}49)$$

since $v_x = 0$ when $y = 0$. Now we use the value of v_x from Eq. (11-49) in the energy equation (11-47) and get

$$\frac{mv_y^2}{2} + \frac{m}{2}\left(\frac{eB}{m}\right)^2 y^2 - \frac{eV}{d} y = 0 \tag{11-50}$$

This is of the form of the energy equation in one dimension; the total energy is zero and the effective potential energy $U'(y)$ is given by

$$U'(y) = -\frac{eV}{d} y + \tfrac{1}{2} m \left(\frac{eB}{m}\right)^2 y^2$$

This effective potential-energy curve is shown in Fig. 11–28. It is precisely of the form of a harmonic oscillator potential (centered on a point at a certain positive value of y) and its characteristic angular frequency ω is given by

$$\omega = \frac{eB}{m}$$

This implies a very interesting result. Equation (11–50) and Fig. 11–28 show that, as y increases from zero, the kinetic energy of the electron increases to some maximum value and then decreases again. Mathematically, Eq. (11–50) defines a maximum value of y given by putting $v_y = 0$:

$$y_{\max} = 2\frac{m}{e}\frac{V}{B^2 d}$$

If this is less than the separation d between the plates (a condition that can be obtained, as the above equation shows, by making B large enough), the y displacement of the electron will perform simple harmonic motion with the angular frequency ω and with an amplitude A equal to $\tfrac{1}{2} y_{\max}$. Taking $t = 0$ at the instant the electron leaves the lower plate, we can put

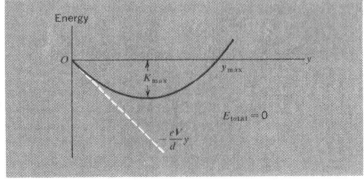

Fig. 11–28 Effective potential-energy curve for the y component of motion in the arrangement of Fig. 11–26.

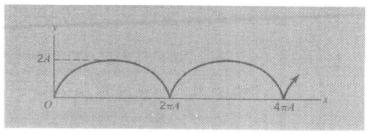

Fig. 11-29 Cycloidal path of electron between charged parallel plates in a magnetic field.

$$y = A(1 - \cos \omega t) \qquad (11\text{-}51\text{a})$$

What about the x component of the motion? If we look at Eq. (11-49), we see that v_x is proportional to y. Since y is always positive, v_x is always in the positive x direction. When one couples this with the harmonic oscillation along y, one sees that the path of the electron is a succession of rabbit-like hops, as shown in Fig. 11-29. Specifically, we have, from Eq. (11-49),

$$\frac{dx}{dt} = \frac{eB}{m} y = \omega y$$

where ω is the same angular frequency, eB/m, that we introduced earlier. Thus, substituting the explicit expression for y from Eq. (11-51a), we have

$$\frac{dx}{dt} = \omega A(1 - \cos \omega t)$$

Integrating this and putting $x = 0$ at $t = 0$ we then find that

$$x = A(\omega t - \sin \omega t) \qquad (11\text{-}51\text{b})$$

Equations (11-51a) and (11-51b), taken together, show that the path of the electron is in fact a cycloid—just as if it were on the rim of a wheel of radius A rolling along the x axis.

THE EFFECT OF DISSIPATIVE FORCES

We mentioned earlier how the conservative properties of a force field may be useful even in circumstances in which dissipative forces are also present. The slow decay of the orbits of artificial earth satellites provides a particularly interesting example of this.

We all know that a satellite placed in orbit a few hundred miles above the earth's surface will eventually come down. The

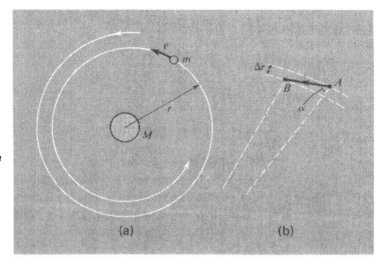

Fig. 11-30 (a) Earth satellite spiraling inward as it loses energy. (b) Small element of the path shown in (a).

descent is, however, spread over many thousands of revolutions. Thus, although the actual path is a continuous inward spiral, as shown in Fig. 11-30(a), the motion during a single revolution is very nearly a closed orbit, which we will take to be circular for simplicity. The statement of Newton's law at any stage is thus given, with extremely little error, by the usual equation for uniform circular motion:

$$F = \frac{GMm}{r^2} = \frac{mv^2}{r}$$

Thus the kinetic energy at any particular value of r is given by

$$K = \tfrac{1}{2}mv^2 = \frac{GMm}{2r} \qquad (11\text{-}52)$$

This exposes a very curious feature. As the orbital radius gets smaller, the kinetic energy increases—in other words, the satellite speeds up. This happens in spite of the fact that the motion of the satellite is continually opposed by a resistive force. If there were no such force the orbital radius and the speed would remain constant.

We know, however, that the satellite must have lost energy, and the amount of this loss is well defined. This becomes clear when we consider the potential energy:

$$U = -\frac{GMm}{r}$$

The total energy, E, when the orbit radius is r, is given by

$$E = K + U = -\frac{GMm}{2r} \tag{11-53}$$

As r decreases, E becomes more strongly negative. We may note the following relationships in any such circular orbit:

$$E = -K = \tfrac{1}{2}U \tag{11-54}$$

But still one may wonder: Why does the satellite *accelerate* in the face of a resistive force? Figure 11–30(b) suggests the answer. The line AB represents a small segment, of length Δs, of the path of the satellite. The starting point A lies on a circle of radius r; the end point B lies on a circle of radius $r + \Delta r$ (Δr being actually negative). We greatly exaggerate the magnitude of Δr, for the sake of clarity. We then see that as the satellite travels from A to B, it feels a component of the gravitational attraction along its path. If the resistive force is $R(v)$, the total force acting on the satellite in the direction of Δs is given by

$$F = \frac{GMm}{r^2} \cos\alpha - R(v)$$

Thus, in the distance Δs, the gain of kinetic energy, equal to the work $F\,\Delta s$, is given by

$$\Delta K = F\,\Delta s = \frac{GMm}{r^2}\Delta s \cos\alpha - R(v)\,\Delta s$$

In the first term on the right, we can substitute

$$\Delta s \cos\alpha = -\Delta r$$

In the second term on the right, we can substitute

$$\Delta s = v\,\Delta t$$

If we also put $\Delta K = mv\,\Delta v$, we arrive at the following equation:

$$mv\,\Delta v = -\frac{GMm}{r^2}\Delta r - R(v)v\,\Delta t$$

However, by differentiation of both sides of Eq. (11–52) we have

$$mv\,\Delta v = -\frac{GMm}{2r^2}\Delta r$$

Using this result, we can substitute for the value of $-(GMm/r^2)\Delta r$

in the previous equation, and we get

$$mv\,\Delta v = 2mv\,\Delta v - R(v)v\,\Delta t$$

Hence

$$\frac{\Delta v}{\Delta t} = +\frac{R(v)}{m}$$

This is a most intriguing result. The positive sign is not an error. The rate of *increase* of speed of the satellite is directly proportional to the magnitude of the resistive force that opposes its motion! The seeming paradox is resolved when we recognize that the resistance, by changing the orbital path from a circle to an inward spiral, acts as an agent that allows the gravitational force to do *positive* work on the satellite, the amount of which is numerically twice as great as the amount of negative work done by the resistive force itself.

GAUSS'S LAW[1]

The particular case of the inverse-square field of force due to a point mass or charge is so important that we will append here a brief discussion of Gauss's law, which is a compact and powerful statement of the inverse-square law. Since we are primarily concerned with mechanics here, we shall discuss the problem in terms of the gravitational rather than the electric field. We begin with a very elementary observation. As we have seen, the gravitational field **g** due to a point mass M is given by the equation

$$\mathbf{g}(\mathbf{r}) = -\frac{GM}{r^2}\mathbf{e}_r$$

This field acts at every point of the surface of a sphere of radius r centered on M. If we use the convention that the positive direction is radially outward, then the radial component of **g** multiplied by the total surface area of the sphere gives us a quantity ϕ defined as follows:

$$\phi = 4\pi r^2 g_r = -4\pi GM$$

We shall call ϕ the *flux* of the gravitational field **g** through the surface of the sphere.

[1] This section may be omitted without loss of continuity.

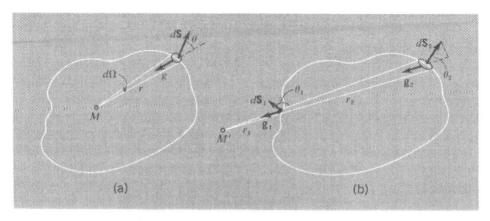

Fig. 11–31 (a) *Gaussian surface enclosing a particle of mass M. There is a net gravitational flux due to M.* (b) *Gaussian surface not enclosing M'. The net flux due to M' is zero.*

Now a noteworthy property of ϕ is that it is independent of r. If we had two spheres of different radii, r_1 and r_2, centered on M, the flux of **g** would be the same through both. We can extend this result to the case of a closed surface of any shape surrounding the mass [Fig. 11–31(a)]. If at an arbitrary point on this surface the outward normal to the surface makes an angle θ with **r**, as shown, we can resolve **g** into orthogonal components, parallel to the surface and along the normal. If we think of the flux literally as a kind of *flow* of the **g** field across the surface, it is only the normal component that contributes. Multiplying this by the element of area dS, we thus have

$$d\phi = -\frac{GM}{r^2} \cos\theta \, dS$$

Now $dS \cos\theta$ can be equated to the projection of dS perpendicular to **r**, and the quotient $dS \cos\theta / r^2$ is the element of solid angle, $d\Omega$, subtended at M by dS. Thus we can put

$$d\phi = -GM \, d\Omega$$

The total flux of **g** through the surface, as defined in this way, is then obtained by integrating over all the contributions $d\Omega$. But this means simply including the complete solid angle, 4π. Thus we have

$$\phi = \int d\phi = -GM \int d\Omega = -4\pi GM$$

exactly as for a sphere. If we consider the situation for a point mass M' *outside* the surface, the result is quite different. This time a cone of small angle $d\Omega$ intersects the surface twice; the contribution to the flux is given by

$$d\phi = -GM'\left(\frac{dS_1 \cos \theta_1}{r_1^2} + \frac{dS_2 \cos \theta_2}{r_2^2}\right)$$

It is not hard to verify that the two terms inside the parentheses are equal to $-d\Omega$ and $+d\Omega$, respectively, so that $d\phi = 0$, from which it follows that the total flux ϕ is also zero in this case.

One can formalize these calculations a little by representing an element of surface area as a vector, $d\mathbf{S}$, with a magnitude equal to dS and a direction along the outward normal, as shown in Fig. 11–31. The element of flux, $d\phi$, is then defined as equal to the scalar product $\mathbf{g} \cdot d\mathbf{S}$. It is then apparent that, in Fig. 11–31(b), $\mathbf{g}_1 \cdot d\mathbf{S}_1$ and $\mathbf{g}_2 \cdot d\mathbf{S}_2$ are of opposite sign.

Suppose now that we have a number of masses, m_1, m_2, m_3, ..., inside a surface of some kind, as in Fig. 11–32(a). Their gravitational fields combine to produce a resultant gravitational field $\mathbf{g}$ at any point P:

$$\mathbf{g} = \mathbf{g}_1 + \mathbf{g}_2 + \mathbf{g}_3 + \cdots$$

The total gravitational flux is then given by

$$\phi = \int \mathbf{g} \cdot d\mathbf{S}$$

and we see that this is simply the sum of the contributions from the individual masses:

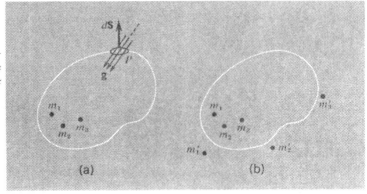

Fig. 11–32 (a) Basis of calculating the gravitational flux due to an arbitrary collection of masses inside a given surface. (b) The addition of external masses does not change the net flux, although it will alter the local field intensities.

$$\phi = \int \mathbf{g}_1 \cdot d\mathbf{S} + \int \mathbf{g}_2 \cdot d\mathbf{S} + \int \mathbf{g}_3 \cdot d\mathbf{S} + \cdots$$
$$= -4\pi G(m_1 + m_2 + m_3 + \cdots)$$

Thus this total flux depends only on the total enclosed mass, regardless of how it may be distributed:

$$\phi = -4\pi G M_{\text{total}} \tag{11-55}$$

If, as in Fig. 11-32(b), we suppose that additional masses m_1', m_2', ..., are placed *outside* the surface, our previous calculation shows that these contribute nothing to the total gravitational flux through the surface. Thus Eq. (11-55) emerges as a completely general result; it is known as Gauss's law (or theorem), and its validity depends completely on the fact that the law of force is an exact inverse-square law. An exact parallel to it holds for electrostatics, in which context Gauss, in fact, first developed it.

If a physical system has certain obvious symmetries, Gauss's theorem leads at once to important conclusions about field strengths. We shall consider a few examples, some of which we have already treated by other methods.

APPLICATIONS OF GAUSS'S THEOREM

Field outside a sphere

Suppose we want to know the gravitational field at a point P, outside an isolated sphere of mass M at a distance r from its center [Fig. 11-33(a)]. If the mass distribution within the sphere

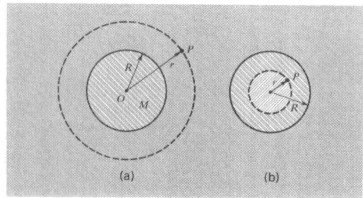

Fig. 11-33 Use of Gauss's law to calculate the gravitational field of a solid sphere (a) at exterior points, and (b) at interior points.

is symmetrical—i.e., if the density of matter is the same at all points at the same distance from the center of the sphere—then the gravitational field **g** itself has the same strength at all points on a spherical surface of radius r. Thus, if we draw a "Gaussian surface" in the form of a sphere of radius r, the total gravitational flux is given by

$$\phi = 4\pi r^2 g_r = -4\pi GM$$

whence

$$g_r = -\frac{GM}{r^2} \quad \text{at once}$$

The result is of course entirely familiar, but the point to notice is that, once we have Gauss's general theorem, the process of carrying out an explicit integration over the mass distribution, as we did in Chapter 8, becomes unnecessary.

Field inside a sphere

An exactly similar treatment holds for the other familiar problem of the gravitational field *inside* a spherically symmetric mass distribution [Fig. 11–33(b)]. Gauss's theorem tells us at once that, if we imagine a spherical surface of radius $r < R$, the material outside this surface contributes nothing to the flux of **g** through the surface. Thus, if the mass enclosed within the surface of radius r is $m(r)$, we can at once deduce that

$$g_r = -\frac{Gm(r)}{r^2}$$

The assumption of spherical symmetry throughout the system, outside as well as inside the radius r, is, however, essential. Otherwise the value of **g** may be different at different points on a spherical surface, and even though the *total* flux still has no contribution from exterior material, the field strength at individual points will be affected. Thus we must assume, as in the first example, that the sphere is effectively isolated, i.e., far from any outside masses.

Field due to a flat sheet

Suppose we have an infinite sheet of matter, of surface density σ (mass per unit area). Symmetry in this case tells us that the field must be everywhere normal to the sheet and that it has the same

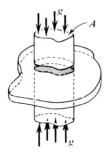

Fig. 11-34 Use of Gauss's law to calculate the gravitational field of a flat sheet of matter.

magnitude on the two sides (Fig. 11-34). We can construct a Gaussian surface for this problem in the form of a right cylinder, with ends of arbitrary shape but the same area A. Then there is no flux of **g** through the curved sides of the cylinder. The enclosed mass is just σA, so the total gravitational flux passing through the surface is $-4\pi G\sigma A$. But, by Gauss's theorem, this is equal to the field strength g multiplied by the total area of the two ends. Thus we have

$$2Ag = -4\pi G\sigma A$$

Therefore,

$$g = -2\pi G\sigma \qquad (11\text{-}56)$$

The strength of the field is thus entirely independent of the distance from the sheet of matter, assuming this to be indeed of unlimited extent.

In terms of gravitational systems, this example is not very realistic. But the result is highly relevant in electrostatics, where distributions of electric charge over large plane areas are frequently met. The parallel-plate capacitor is a prime example, and the constant field at all points between the charged plates in the Millikan experiment, for example, can at once be understood in the light of the above calculation.

PROBLEMS

11-1 A particle of mass m slides without friction on a wire that is curved into a vertical circle (see the figure). The position of the particle can be described in terms of θ, or in terms of the arc length s, or by the vertical distance h that it has fallen.

(a) Consider the force that acts on the particle along the direction of its path at any point, and show that the work done on the particle as it moves through an arc length ds is given by

$$dW = F_s\, ds = mgR\sin\theta\, d\theta$$

(b) By integrating this expression from $\theta = 0$ to $\theta = \theta_0$, show that the work done is $W = mgR(1 - \cos\theta_0)$, and express this result in terms of the vertical distance h that the particle has descended.

11-2 Two blocks, of masses m and $2m$, rest on two frictionless planes

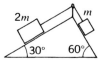

inclined at angles of 60° and 30°, respectively, and are connected by a string through an eyelet of negligible friction (see the figure). Using energy methods, find the magnitude of the acceleration of the masses, and deduce the tension in the connecting string.

11-3 Show that if a mass on the end of a string is allowed to swing down in a circular arc from a position in which the string is initially horizontal (and the mass is at rest), then at the lowest point of the swing the tension in the string is three times as great as when the mass simply hangs there. (This result is quoted by Huygens at the end of his book on pendulum clocks, published in 1673.)

11-4 A pendulum bob of mass m, at the end of a string of length l, starts from rest at the position shown in the figure, with the string at 60° to the vertical. At the lowest point of the arc the bob strikes a previously stationary block, of mass nm, that is on a frictionless horizontal surface. The collision is perfectly elastic.

    (a) What is the speed of the bob just before the impact occurs?
    (b) What is the tension in the string at this instant?
    (c) What velocity is given to the block by the impact?
    (d) In the oscillations of the pendulum after the collision, what maximum angle θ to the vertical does the string make? (Obtain your answer in the form $\cos\theta = \cdots$.)

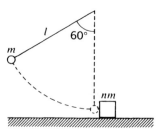

11-5 A student is assigned a "conservation of momentum" experiment in which two masses, m_1 and m_2, are suspended from strings of length l. Mass m_1 is pulled aside a horizontal distance x_i and released. It strikes m_2 and sticks to it, after which the combined mass swings out to some final position x_f. The student objects that momentum is obviously not conserved in this process. It is zero initially, increases, and then decreases again to zero. The student also claims that energy is not conserved. Analyze carefully what actually goes on in this experiment, so that you are in a position to answer the student's objections.

11-6 A small ball of putty, of mass m, is attached to a string of length l fastened to an upright on a wooden board resting on a horizontal table (see the figure). The combined mass of the board and upright is M. The friction coefficient between the board and the table

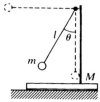

479 Problems

is μ. The ball is released from rest with the string in a horizontal position. It hits the upright in a completely inelastic collision. While the ball swings down, the board does not move.

(a) How far does the board move after the collision?

(b) What is the minimum value that μ must have to prevent the board from moving to the left while the ball swings down? Assume that $m \ll M$, which gives the critical condition at $\theta \simeq 45°$.

11-7 An object of mass m slides on a frictionless loop-the-loop apparatus. The object is released from rest at a height h above the top of the loop (see the figure).

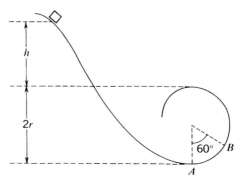

(a) What is the magnitude and direction of the force exerted on the object by the track as the object passes through the point A?

(b) Draw an isolation diagram showing the forces acting on the object at point B, and find the magnitude of the radial acceleration at that point.

(c) Show that the object must start from $h \geq r/2$ to successfully complete the loop.

(d) For $h < r/2$ the object will begin to fall away from the track before reaching the top of the circle. Show that this happens at a position such that $3 \cos \alpha = 2 + 2h/r$, where α is the angular distance from the (upward) vertical.

11-8 A ball of mass m hangs at the end of a string of length l. It is struck so as to start out horizontally from this position with a speed v_0.

(a) What must v_0 be if the ball is just able to travel through a complete circular path?

(b) If $v_0 = \sqrt{4gl}$ (so that the ball could just rise through the height $2l$ if projected vertically), what angle does the string make with the vertical when the ball begins to fall away from its circular path?

(c) In situation (b), analyze the subsequent motion, assuming no losses of total mechanical energy.

11-9 A mass slides down a very smooth curved chute as shown. At what horizontal distance from the end of the chute does it hit the ground?

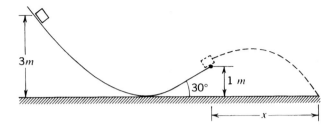

11-10 A daredevil astronomer stands at the top of his observatory dome (see figure) wearing roller skates, and starts with negligible velocity to coast down over the dome surface.

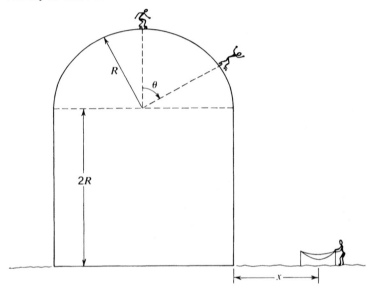

(a) Neglecting friction, at what angle θ does he leave the dome's surface?

(b) If he were to start with an initial velocity v_0, at what angle θ would he leave the dome?

(c) For the observatory shown, how far from the base should his assistant position a net to break his fall, in situation (a)? Evaluate your answer for $R = 8$ m, and use $g \approx 10$ m/sec^2.

11-11 A pendulum of length L is held at an angle α from the vertical. A peg is located in the path of swing of the string as shown in the figure. The pendulum is released from rest.

(a) If $\alpha = 60°, \beta = 30°$, and $r = L\sqrt{3}/4$, what is the maximum angle with the vertical that the bob will reach?

(b) Show that the string will buckle during the bob's ascent when $0 < (r \cos \beta - L \cos \alpha) < 3(L - r)/2$. At what angle θ does this buckling occur?

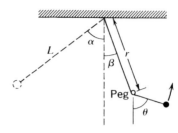

(c) A carnival skill game involves such a pendulum and peg apparatus. A prize is awarded for causing the bob to strike the peg. In the game, L, r, and β are fixed; the contestant releases the bob from rest at any desired α. Find the value of α for which a prize will be won. (Express your answer in the form $\cos \alpha = \cdots$.)

11-12 A frictionless airtrack is deformed into the parabolic shape $y = \frac{1}{2}Cx^2$. An object oscillates along the track in almost perfect simple harmonic motion.

(a) If the period is 30 sec, by what distance is the track at $x = \pm 2$ m higher than at $x = 0$?

(b) If the amplitude of oscillation is 2 m, how long does it take for the object to pass from $x = -0.5$ m to $x = +1.5$ m traveling in the positive x direction? (The description of SHM as a projection of uniform circular motion is useful for this.)

11-13 A pendulum clock keeps perfect time at ground level. Approximately how many seconds *per week* would it gain or lose at a height of 20 m above this level? (Assume a simple pendulum, with $T = 2\pi\sqrt{l/g}$. Earth's radius = 6.4×10^6 m. 1 week $\simeq 6 \times 10^5$ sec.)

11-14 Two identical, and equally charged, ping-pong balls are hung from the same point by strings of length L (see the figure).

(a) Given that the mass of each is m and that the equilibrium position is as shown in the figure, what is the charge on each? (Recall that the electrostatic force between two charges at a separation r is given by $F = kq_1q_2/r^2$.)

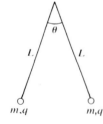

(b) Suppose that the balls are displaced slightly toward each other by the same amount. Make appropriate approximations and describe the subsequent motion. Find an expression for the frequency of oscillations.

(c) On the other hand, if the ping-pong balls are displaced slightly in the same direction and by the same amount, what will the subsequent motion be? In this second case, does it matter whether they are charged or merely held apart by a massless rod?

11-15 (a) Show that free fall in the earth's gravitational field from infinity results in the same velocity at the surface of the earth that would be achieved by a free fall from a height $H = R_E$ (= radius of earth) under a constant acceleration equal to the value of g at the earth's surface.

(b) Show that the speed at the earth's surface of an object dropped from height h ($h \ll R_E$) is given approximately by

$$\frac{dr}{dt} = v = (2gh)^{1/2}\left(1 - \frac{1}{2}\frac{h}{R_E}\right)$$

(c) Verify the statement in the text (p. 454) that the speed needed for escape from the surface of a gravitating sphere is $v_0\sqrt{2}$, where v_0 is the speed of a particle skimming the surface of the sphere in circular orbit.

11-16 A physicist plans to determine the mass of the earth by observing the change in period of a pendulum as he descends a mine shaft. He knows the radius R of the earth, and he measures the density ρ_s of the crustal material that he penetrates, the distance h he descends, and the fractional change ($\Delta T/T$) in the pendulum's period.

(a) What is the mass of the earth in terms of these measurements and the earth's radius?

(b) Suppose that the mean over-all density of the earth is twice the mean density of the portion above 3 km. (This supposition is actually rather accurate.) How many seconds per day would a pendulum clock at the bottom of a deep mine (3 km) gain, if it had kept time accurately at the surface?

11-17 The figure shows a system of two uniform, thin-walled, concentric spherical shells. The smaller shell has radius R and mass M, the larger has radius $2R$ and mass $2M$, and the point P is their common center. A point mass m is situated at a distance r from P.

(a) What is the gravitational force $F(r)$ these shells exert on m in each of the three ranges of r: $0 < r < R$, $R < r < 2R$, $r > 2R$?

(b) What is the gravitational potential energy of the point mass m when it is at P? (Take the potential energy to be zero when m is infinitely far from P.)

(c) If the particle is released from rest very far away from the spheres, what is its speed when it reaches P? (Assume that the particle can pass freely through the walls of the shells.)

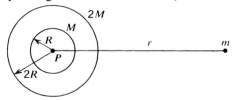

11-18 Assume the moon to be a sphere of uniform density with radius 1740 km and mass 7.3×10^{22} kg. Imagine that a straight smooth tunnel is bored through the moon so as to connect *any two*

points on its surface.

(a) Show that the motion of objects along this tunnel under the action of gravity would be simple harmonic.

(b) Calculate the period of oscillation.

(c) Compare this with the period of a satellite traveling around the moon in a circular orbit at the moon's surface.

11-19 The escape speed from the surface of a sphere of mass M and radius R is $v_e = (2GM/R)^{1/2}$; the mean speed $\bar{v}$ of gas molecules of mass m at temperature T is about $(3kT/m)^{1/2}$, where k ($= 1.38 \times 10^{-23}$ J/°K) is Boltzmann's constant. Detailed calculation shows that a planetary atmosphere can retain for astronomical times (10^9 years) only those gases for which $\bar{v} \lesssim 0.2 v_e$. Using the data below, find which, if any, of the gases H_2, NH_3, N_2, and CO_2 could be retained for such periods by the earth; by the moon; by Mars. $M_E = 6.0 \times 10^{24}$ kg, $R_E = 6.4 \times 10^3$ km, $T_E = 250°$K. For the moon, $M = M_E/81$, $R = 0.27 R_E$, and $T = T_E$. For Mars, $M = 0.11 M_E$, $R = 0.53 R_E$, and $T = 0.8 T_E$.

11-20 A double-star system consists of two stars, of masses M and $2M$, separated by a distance D center to center. Draw a graph showing how the gravitational potential energy of a particle would vary with position along a straight line that passes through the centers of both stars. What can you infer about the possible motions of a particle along this line?

11-21 A double-star system is composed of two identical stars, each of mass M and radius R, the centers of which are separated by a distance of $5R$. A particle leaves the surface of one star at the point nearest to the other star and escapes to an effectively infinite distance.

(a) Ignoring the orbital motion of the two stars about one another, calculate the escape speed of the particle.

(b) Assuming that the stars always present the same face to one another (like the moon to the earth), calculate the orbital speed of the point from which the particle is emitted. How would the existence of this orbital motion affect the escape problem?

11-22 The earth and the moon are separated by a distance $D = 3.84 \times 10^8$ m. Ignoring their motion about their common center of mass, but taking into account both of their gravitational fields, answer the following questions:

(a) How much work must be done on a 100-kg payload for it to reach a height 1000 km above the earth's surface in the direction of the moon?

(b) Calculate the difference in potential energy for this mass between the moon's surface and the earth's surface.

(c) Calculate the necessary initial kinetic energy to get the payload to the moon. It must be greater than the potential difference given in (b) because the payload must overcome a "potential hill." Find the location of the top of this hill and compute the potential difference from the earth to it.

11-23 Two stars, each of mass M, orbit about their center of mass. The radius of their common orbit is r (their separation is $2r$). A planetoid of mass m ($\ll M$) happens to move along the axis of the system (the line perpendicular to the orbital plane which intersects the center of mass), as shown in the figure.

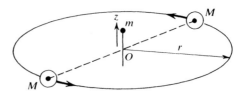

(a) Calculate directly the force exerted on the planetoid if it is displaced a distance z from the center of mass.

(b) Calculate the gravitational potential energy as a function of this displacement z and use it to verify the result of part (a).

(c) Find approximate expressions for the potential energy and the force in the cases $z \gg r$ and $z \ll r$.

(d) Show that if the planetoid is displaced slightly from the center of mass, simple harmonic motion occurs. Compare the period T_P of this oscillation with the orbital period T_0 of the binary system.

11-24 A frictionless wire is stretched between the origin and the point $x = a$, $y = b$ in a horizontal plane. A bead on the wire starts out from the origin and moves under the action of a force that varies with position. Find the kinetic energy with which the bead arrives at the point (a, b) if

(a) The components of the force are $F_x = k_1 x$, $F_y = k_2 y$.

(b) The components of the force are $F_x = k_1 y$, $F_y = k_2 x$
(with $k_1, k_2 > 0$).

In one case the force is conservative, in the other case it is not. By considering a different path—e.g., from $(0, 0)$ to $(a, 0)$ and then from $(a, 0)$ to (a, b)—verify which is which.

11-25 The nuclear part of the interaction of two nucleons (protons or neutrons) is described pretty well by a potential $V(r) = -\lambda e^{-r/r_0}/r$ when the separation r is greater than 1 F (1 F = 10^{-15} m). In this expression, $r_0 = 1.4$ F and $\lambda = 70$ MeV-F. (The potential was proposed by H. Yukawa and is named after him.)

(a) Find an expression for the nuclear force that acts on (each of) two nucleons separated by $r > 1$ F.

(b) Evaluate this force for two protons 1.4 F apart, and compare this with the repulsive Coulomb force at that separation.

(c) Estimate the separation at which the nuclear force has

dropped to 1% of its value at $r = 1.4$ F. What is the Coulomb force at this separation?

(d) The last result indicates why the Yukawa potential is not part of our macroscopic experience. Note that r_0 characterizes the range of distance over which the interaction is important. In contrast to the nuclear force, gravitational and Coulomb forces have been called forces of infinite range. Indicate why such a description is appropriate.

11-26 In Fig. 11-24(b) is shown a plot of gravitational equipotentials and field lines for two spheres of masses $2M$ and M. One way of developing this diagram is as follows. We know that the gravitational potential at any point P is given by

$$\varphi_P = -\frac{G(2M)}{r_1} - \frac{G(M)}{r_2}$$

where r_1 and r_2 are measured from the centers of the spheres. An equipotential is defined by $\varphi_P = $ constant. In order to construct such equipotentials, draw a set of circles, with the sphere of mass M as center, such that the values of $1/r_2$ are in arithmetic progression— e.g., $1/r_2 = 1, 1.5, 2, 2.5, 3, 3.5, \ldots$. Then with the sphere of mass $2M$ as center, draw a set of circles with radii r_1 twice as great as the values of r_2. Since $\varphi_P \sim (2/r_1) + (1/r_2)$, the corresponding circles in each pair represent equal contributions to φ by $2M$ and M. The two sets of circles have a number of intersections, and each intersection corresponds to a well-defined value of φ_P. For instance, with $10r_2 = 2.85$ units and $10r_1 = 10$ units, we have $\varphi_P \sim (2/1.0) + (1/0.285) = 5.5$. But $\varphi_P \sim 5.5$ is obtained also at the intersections of other pairs of circles as follows: $r_2 = 0.33$, $r_1 = 0.80$; $r_2 = 0.40$, $r_1 = 0.67$; $r_2 = 0.67$, $r_1 = 0.50$. Joining these intersections having the same value of $(2/r_1) + (1/r_2)$ gives us an equipotential, and other equipotentials can be constructed in the same way. Once the equipotentials are obtained, the field lines can be drawn as lines everywhere normal to the equipotentials.

11-27 A science fiction story concerned a space probe that was retrieved after passing near a neutron star. The unfortunate monkey that rode in the probe was found to be dismembered. The conclusion reached was that the strong gravity gradient (dg/dr) which the probe had experienced had pulled the monkey apart. Given that the probe had passed within 200 km of the surface of a neutron star of mass 10^{33} g (about half the sun's mass) and radius 10 km, was this explanation reasonable? (The probe was in free fall at the time.)

11-28 A straight chute is to be used to transport articles a given horizontal distance l. The vertical drop of the chute can be freely chosen. The articles are to arrive at the top of the chute with negligible velocity and the chute is to be chosen such that the transit time is a minimum.

(a) If the surface is frictionless, what is the angle of the chute for which the time is minimized?

(b) What is the corresponding angle if the coefficient of friction is μ?

(c) If a playground slide were designed to give minimum duration of ride for given horizontal displacement, and if the coefficient of friction of the child-slide surface is 0.2, what angle would the slide make with the horizontal? (Ignore the curved portion at the lower end of the slide.)

(d) If the optimization problem of (a) is encountered and if curved chutes are allowed, can you guess roughly what form the best design would have?

11-29 The magnetic force exerted on a particle of charge q and mass m traveling in a uniform magnetic field $\mathbf{B}$ (which is constant in time) is given by $\mathbf{F} = kq\mathbf{v} \times \mathbf{B}$ (where the constant $k = 1$ in the MKS system).

(a) Show that the work done by such a force on the particle is zero for arbitrary particle motion.

(b) Situations arise in which a moving charged particle is slowed by a force of constant magnitude F_D and of direction opposite to the instantaneous velocity. (For example, the ionization of atoms along the track of a charged particle produces an almost constant energy loss per unit distance, corresponding to a constant retarding force.) Show that if a particle has speed v_0 at $t = 0$, then for $t > 0$ its speed will be given by $v(t) = v_0 - F_D t/m$ until $t = mv_0/F_D$, after which time the particle will remain motionless. Note that the presence of a magnetic field will not affect this result.

(c) Under the action of a magnetic force alone, a particle with an initial velocity v_0 (normal to $\mathbf{B}$) describes a circle of radius $r = mv_0/kqB$ with a circular frequency $\omega = kqB/m$. Show that a particle subject also to the force described in (b) moves inward along a spiral and that the number of circuits it makes in the spiral before it comes to rest is given by $(kqBv_0/2\pi F_D)$.

11-30 (a) Obtain an expression for the gravitational field due to a thin disk (thickness d, radius R, and density ρ) at a distance h above the center of mass of the disk. In calculating the field, assume the surface density $\sigma = \rho d$ to be concentrated in a disk of negligible thickness a distance h beneath the test point. This will give an accurate result whenever $h \gg d$, and reduces the complexity of the calculation. Note that for $R \to \infty$, the calculation agrees with the prediction of Gauss's theorem, as it must.

(b) Express the field obtained in part (a) as a fraction of $2\pi G\sigma$ for the cases $R = 2h$, $R = 5h$, and $R = 25h$.

(c) How many seconds per year would a pendulum clock gain

when suspended with the bob 5 cm above a lead floor 1 cm thick, if it keeps correct time in the absence of the floor?

11-31 Any mass of matter has a gravitational "self-energy" arising from the gravitational attraction among its parts.

(a) Show that the gravitational self-energy of a uniform sphere of mass M and radius R is equal to $-3GM^2/5R$. You can do this either by calculating the potential energy of the sphere directly, integrating over the interactions between all possible pairs of thin spherical shells within the sphere (and remembering that each shell is counted twice in this calculation) or, perhaps more simply, by imagining the sphere to be built up from scratch by the addition of successive layers of matter brought in from infinity.

(b) Calculate the order of magnitude of the gravitational self-energy of the earth. Check where this lies on the scale of energies displayed in Table 10-1.

Part III
Some special topics

When first studying mechanics one has the impression that everything in this branch of science is simple, fundamental and settled for all time. One would hardly suspect the existence of an important clue which no one noticed for three hundred years. The neglected clue is connected with one of the fundamental concepts of mechanics—that of mass.

A. EINSTEIN AND L. INFELD,
The Evolution of Physics (1938)

12

Inertial forces and noninertial frames

IMAGINE THAT YOU are sitting in a car on a very smooth road. You are holding a heavy package. The car is moving, but you cannot see the speedometer from where you sit. All at once you get the feeling that the package, instead of being just a dead weight on your knees, has begun to push backward horizontally on you as well. Even though the package is not in contact with anything except yourself, the effect is as if a force were being applied to it and transmitted to you as you hold it still with respect to yourself and the car. If you did not restrain the package in this way, it would in fact be pushed backward. You notice that this is what happens to a mascot that has been hanging at the end of a previously vertical string attached to the roof of the car.

How do you interpret these observations? If you have any previous experience of such phenomena, you will have no hesitation in saying that they are associated with an increase of velocity of the car—i.e., with a positive acceleration. Even if this were your first experience of this type, but if you had a well-developed acquaintance with Newton's laws, you could reach the same conclusion. An acceleration of the car calls for an acceleration of everything connected with it; the acceleration of the package requires, through $F = ma$, a force of the appropriate size supplied by your hands. Nonetheless, it does feel just as if the package itself is somehow subjected to an extra force—a "force of inertia"—that comes into play whenever the effort is made to change the state of motion of an object.

These extra forces form an important class. They can be held responsible for such phenomena as the motion of a Foucault pendulum, the effects in a high-speed centrifuge, the so-called g forces on an astronaut during launching, and the preferred direction of rotation of cyclones in the northern and southern hemispheres. These forces are unique, however, in the sense that one cannot trace their origins to some other physical system, as was possible for all the forces previously considered. Gravitational, electromagnetic, and contact forces, for example, have their origins in other masses, other charges, or the "contact" of another object. But the additional forces that make their appearance when an object is being accelerated have no such physical objects as sources. Are these inertial forces real or not? That question, and the answer to it, is bound up with the choice of reference frame with respect to which we are analyzing the motion. Let us, therefore, begin this analysis with a reminder of dynamics from the standpoint of an unaccelerated frame.

MOTION OBSERVED FROM UNACCELERATED FRAMES

An unaccelerated reference frame belongs to the class of reference frames that we have called *inertial*. We saw, in developing the basic ideas of dynamics in Chapter 6, that a unique importance and interest attaches to these frames, in which Galileo's law of inertia holds. We have seen how, if one such frame has been identified, any other frame having an arbitrary constant velocity relative to the first is also inertial, and our inferences about the forces acting on an object are the same in both.

To a good first approximation, as we know, the surface of the earth defines an inertial frame. So also, therefore, does any system moving at constant speed over the earth. Galileo himself was the first person to present a clear recognition of this fact, and one aspect of it that he discussed is useful as a starting point for us now. In his *Dialogue Concerning the Two World Systems*, in which he advocated the Copernican view of the solar system in preference to the Ptolemaic, Galileo pointed out that a rock, dropped from the top of the mast of a ship, always lands just at the foot of the mast, whether or not the ship is moving. Galileo argued from this that the vertical path of a falling object does not compel one to the conclusion that the earth is stationary.

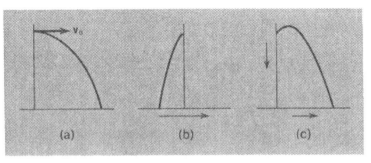

Fig. 12-1 (a) Parabolic trajectory under gravity, as observed in the earth's reference frame. The initial velocity v_0 is horizontal. (b) Same motion observed from a frame with a horizontal velocity greater than v_0. (c) Same motion observed from a frame having both horizontal and vertical velocity components.

The comparison here is between an object falling from rest relative to the earth and another object falling from rest relative to the ship. If we considered only an object that starts from rest relative to a moving ship, its path would be vertical in the ship's frame and parabolic in the earth's frame. More generally, if we considered an object projected with some arbitrary velocity relative to the earth, its subsequent path would have diverse shapes as viewed from different inertial frames (see Fig. 12–1) but all of them would be parabolic, and all of them, when analyzed, would show that the falling object had the vertical acceleration, **g**, resulting from the one force $\mathbf{F}_g$ ($= m\mathbf{g}$) due to gravity. Let us now contrast this with what one finds if the reference frame itself has an acceleration.

MOTION OBSERVED FROM AN ACCELERATED FRAME

Suppose that an object is released from rest in a reference frame that has a constant horizontal acceleration with respect to the earth's surface. Let us consider the subsequent motion as it appears with respect to the earth and with respect to the accelerating frame. We shall take the direction of the positive x axis in the direction of the acceleration and will set up two rectangular coordinate systems: system S, at rest relative to the earth, and S', fixed in the accelerating frame (Fig. 12–2). Take

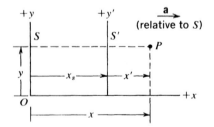

Fig. 12-2 Relationship of coordinates of a particle in two frames that are in accelerated relative motion.

the origins of the frames to coincide at $t = 0$, and suppose that the velocity of S' with respect to S at this instant is equal to v_0. The vertical axes of the two systems are taken as positive upward, and the object is released at $t = 0$ from a point for which $x = x' = 0$, $y = y' = h$.

What will the trajectories in S and S' look like? For an observer in S, we already know the answer. To him, the object is undergoing free fall with initial horizontal velocity v_0 [Fig. 12-3(a)]. Thus we have

$$\text{(As observed in } S\text{)} \quad \begin{cases} x = v_0 t \\ y = h - \tfrac{1}{2}gt^2 \end{cases}$$

These two equations uniquely define the position of the object at time t, but to describe the motion as observed in S' we must express the results in terms of the coordinates x' and y' as measured in S'. To transform to the S' frame, we substitute

$$x' = x - x_s$$
$$y' = y$$

where x_s is the separation along the x axis of the origins of S and S' (see Fig. 12-2). We know that

$$x_s = v_0 t + \tfrac{1}{2}at^2$$

Substituting these values we find

$$\text{(As observed in } S'\text{)} \quad \begin{cases} x' = v_0 t - (v_0 t + \tfrac{1}{2}at^2) = -\tfrac{1}{2}at^2 \\ y' = h - \tfrac{1}{2}gt^2 \end{cases}$$

Thus the path of the particle as observed in S' is a *straight line* given by the equation

$$x' = -\frac{a}{g}(h - y')$$

Fig. 12-3 (a) Parabolic trajectory of a particle under gravity, as observed in the earth's reference frame S. (b) Same motion observed in a frame S' that has a constant horizontal acceleration.

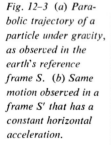
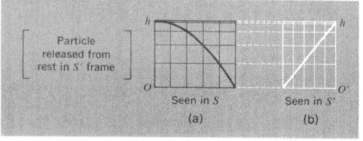

This is shown in Fig. 12-3(b). In the accelerated frame, the object appears to have not only a constant downward component of acceleration due to gravity, but also a constant horizontal component of acceleration in the $-x$ direction which causes the particle to follow a nonvertical straight-line path. [A similar simple example is the monkey-shooting drama described in Chapter 3 (p. 104). The outcome becomes almost self-evident if we choose to describe the events in the rest frame of the falling monkey. In this frame the bullet just follows a straight-line path directly toward the monkey, while the ground accelerates upward at 9.8 m/sec^2.]

There is no mystery about the unfamiliar motion represented by Fig. 12-3(b). It is a direct kinematic consequence of describing the normal free-fall motion from a frame that is itself accelerated. We could perfectly well use this path, described by measurements made entirely within S', to discover the acceleration of this frame, provided that the direction of the true vertical were already known. However, a greater interest attaches to learning about the acceleration through dynamic methods. That is the concern of the next section.

ACCELERATED FRAMES AND INERTIAL FORCES

From what has been said, it is clear that inertial frames have a very special status. All inertial frames are *equivalent* in the sense that it is impossible by means of dynamical experiments to discover their motions in any absolute sense—only their relative motions are significant. Out of this dynamical equivalence comes what is called the Newtonian principle of relativity:

> **There is no dynamical observation that leads us to prefer one inertial frame to another. Hence, no dynamical experiment will tell us whether we have a constant velocity through space.**

As we have just seen, however, a relative acceleration between two frames *is* dynamically detectable. As observed in accelerating frames, objects have unexpected accelerations. It follows at once, since Newton's law establishes a link between force and acceleration, that we have a quantitative basis for calculating the magnitude of the inertial force associated with a measured acceleration. Conversely, and more importantly, we have a dynamical basis for inferring the magnitude of an acceleration from the inertial force associated with it. This is the underlying principle of all the instruments known as accelerometers. They function because of the inertial property of some physical mass.

To make the analysis explicit, consider the motion of a particle P with respect to two reference frames like those considered in the last section and shown in Fig. 12-2: an inertial frame S and an accelerated frame S'. We then have, once again,

$$x = x' + x_s$$
$$y = y'$$

The velocity components of P as measured in the two frames are thus given by

$$u_x = u'_x + v_s$$
$$u_y = u'_y$$

where $v_s = dx_s/dt$ at any particular instant. If S' has a constant acceleration a, we can put $v_s = v_0 + at$, but the condition of constant acceleration is not at all necessary to our analysis.

Taking the time derivatives of the instantaneous velocity components, we then get

$$a_x = a'_x + a_s$$
$$a_y = a'_y$$

where a_s is the instantaneous acceleration of the frame S'. Although we have chosen to introduce the calculation in terms of Cartesian components, it is clear that a single vector statement relates the acceleration **a** of P, as measured in S, to its acceleration **a**' as measured in S' together with the acceleration $\mathbf{a}_s$ of S' itself:

$$\mathbf{a} = \mathbf{a}' + \mathbf{a}_s \qquad (12\text{-}1)$$

Multiplying Eq. (12-1) throughout by m, we recognize the left-hand side as giving the real (net) force, **F**, that is acting on the particle, since this defines the true cause of its acceleration as measured in an inertial frame. That is, in the S frame,

$$\mathbf{F} = m\mathbf{a} \qquad (12\text{-}2)$$

but, using Eq. (12-1), this gives us

$$\mathbf{F} = m\mathbf{a}' + m\mathbf{a}_s \qquad (12\text{-}3)$$

We now come to the crucial question: How do we interpret Eq. (12-3) from the standpoint of observations made within the accelerated frame S' itself?

Newton's viewpoint— that the net force on an object is the cause of accelerated motion ($\mathbf{F}_{net} = m\mathbf{a}$)—is so deeply ingrained in our thinking that we are strongly motivated to preserve this relationship at all times. When we observe an object accelerating, we interpret this as due to the action of a net force on the object. Can we achieve a mathematical format that has the appearance of $\mathbf{F}_{net} = m\mathbf{a}$ for the present case of an accelerated frame of reference? Yes. By transferring all terms but $m\mathbf{a}'$ to the left and treating these terms as forces that act on m, and have a resultant $\mathbf{F}'$, which is of the correct magnitude to produce just the observed acceleration $\mathbf{a}'$:

$$\mathbf{F}' = \mathbf{F} - m\mathbf{a}_s = m\mathbf{a}' \tag{12-4}$$

The net force in the S' frame is thus made up of two parts: a "real" force, $\mathbf{F}$, with components F_x and F_y, and a "fictitious" force equal to $-m\mathbf{a}_s$, which has its origin in the fact that the frame of reference itself has the acceleration $+\mathbf{a}_s$. An important special case of Eq. (12-4) is that in which the "real" force $\mathbf{F}$ is zero, in which case the particle, as observed in S', moves under the action of the inertial force $-m\mathbf{a}_s$ alone.

The result expressed by Eq. (12-4) is not merely a mathematical trick. From the standpoint of an observer in the accelerating frame, the inertial force is actually present. If one took steps to keep an object "at rest" in S', by tying it down with springs, these springs would be observed to elongate or contract in such a way as to provide a counteracting force to balance the inertial force. To describe such a force as "fictitious" is therefore somewhat misleading. One would like to have some convenient label that distinguishes inertial forces from forces that arise from true physical interactions, and the term "pseudoforce" is often used. Even this, however, does not do justice to such forces as experienced by someone who is actually in the accelerating frame. Probably the original, strictly technical name, "inertial force," which is free of any questionable overtones, remains the best description.

As an illustration of the way in which the same dynamical situation may be described from the different standpoints of an inertial frame, on the one hand, and an accelerated frame, on the other, consider a simple pendulum suspended from the roof of a car. The mass of the bob is m. In applying $\mathbf{F} = m\mathbf{a}$ from the standpoint of a frame of reference S attached to the earth

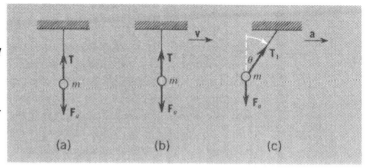

Fig. 12-4 Forces acting on a suspended mass in (a) a stationary car, (b) a car moving at constant velocity, and (c) a car undergoing a positive acceleration.

(assumed nonrotating), one can draw isolation diagrams for the possible motions of the car as shown in Fig. 12-4. In each case, there are just two (real) forces acting on the bob: F_g, the force of gravity, and **T**, the tension in the string. Cases (a) and (b) do not involve acceleration and the application of $\mathbf{F} = m\mathbf{a}$ is trivial. In (c), the bob undergoes acceleration toward the right and the string hangs at an angle with some increase in its tension (from T to T_1). The isolation diagram of Fig. 12-5(a) leads us to apply $\mathbf{F} = m\mathbf{a}$ as follows:

Horizontal component: $T_1 \sin \theta = ma$
Vertical component: $T_1 \cos \theta - mg = 0$

In the S' frame, however, because of the acceleration of the frame, there will be an additional force of magnitude ma in the direction opposite to the acceleration of the frame. Figure 12-5(b) shows an isolation diagram for the bob as seen in S'. The bob is in equilibrium. Here, application of $\mathbf{F}' = m\mathbf{a}'$ gives (because $\mathbf{a}' = 0$):

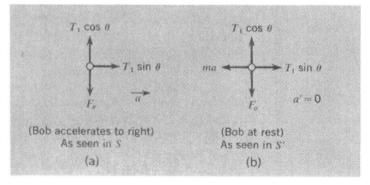

Fig. 12-5 Forces on an object that is at rest relative to an accelerated car (a) as judged in an inertial frame, and (b) as judged in the accelerated frame.

$$T_1 \sin\theta - ma = 0$$
$$T_1 \cos\theta - mg = 0$$

Thus the equilibrium inclination of the pendulum is defined by the condition

$$\tan\theta = \frac{a}{g} \qquad (12\text{–}5)$$

ACCELEROMETERS

The result expressed by Eq. (12–5) provides the theoretical basis for a simple accelerometer. If we have first established the true vertical direction, representing $\theta = 0$, the observation of the angle of inclination of a pendulum at any subsequent time tells us the value of a through the equation

$$a = g \tan\theta$$

For example, if a passenger in an airplane lets his tie, or a keychain, hang freely from his fingers during the takeoff run, he can make a rough estimate of the acceleration, which is usually almost constant [Fig. 12–6(a)]. If he also records the time from the beginning of the run to the instant of takeoff, he can obtain

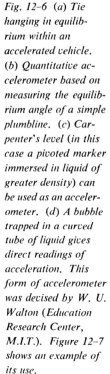

Fig. 12-6 (a) Tie hanging in equilibrium within an accelerated vehicle. (b) Quantitative accelerometer based on measuring the equilibrium angle of a simple plumbline. (c) Carpenter's level (in this case a pivoted marker immersed in liquid of greater density) can be used as an accelerometer. (d) A bubble trapped in a curved tube of liquid gives direct readings of acceleration. This form of accelerometer was devised by W. U. Walton (Education Research Center, M.I.T.). Figure 12–7 shows an example of its use.

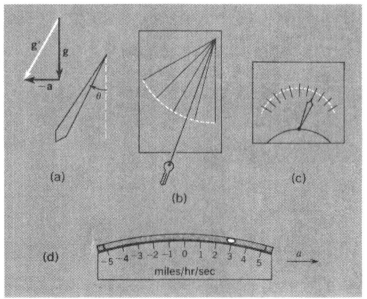

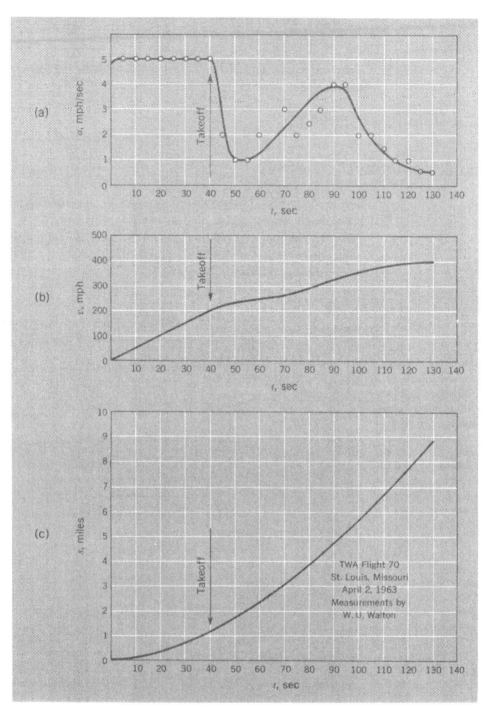

Fig. 12-7 (a) Record obtained with the accelerometer of Fig. 12-6(d) before and after takeoff of a commercial jet aircraft. The accelerometer was held so as to record the horizontal component of acceleration only. Note the sharp decrease in a at takeoff. (b) Graph of velocity versus time, obtained by graphical integration of (a). (c) Graph of distance versus time, obtained by graphical integration of (b).

a fairly good estimate of the length of the run and the takeoff speed. If he is more ambitious, he can go armed with a card, as in Fig. 12-6(b), already marked out as a goniometer (= angle measurer) or even directly calibrated in terms of acceleration measured in convenient units (e.g., mph per second).[1] Another simple accelerometer is obtainable readymade in the form of a carpenter's level made of a small pivoted float that is completely immersed in a liquid [Fig. 12-6(c)]. All these devices make use of the fact that the natural direction of a plumbline in an accelerated frame is defined by the combination of the gravitational acceleration vector **g** and the negative of the acceleration **a** of the frame itself.

A quite sensitive accelerometer of this same basic type, with the further advantages of a quick response and a quick attainment of equilibrium (without much overshoot or oscillation) can be made by curving a piece of plastic tubing into a circular arc and filling it with water or acetone until only a small bubble remains [see Fig. 12-6(a)]. Figure 12-7(a) shows the record of acceleration versus time as obtained with such an accelerometer during the takeoff of a jet aircraft. Figures 12-7(b) and (c) show the results of numerically integrating this record so as to obtain the speed and the total distance traveled.

Accelerometers of a vastly more sophisticated kind can be made by using very sensitive strain gauges, with electrical measuring techniques, to record in minute detail the deformations of elastic systems to which a mass is attached. Figure 12-8 shows in schematic form the design of such an instrument. If the object on which the accelerometer is mounted undergoes an acceleration, the inertial force experienced by the pendulum bob begins to deflect it. This, however, unbalances slightly an electrical capacitance bridge in which the pendulum forms part of two of the capacitors, as shown. An error signal is obtained which is used both to provide a measure of the acceleration and to drive a coil that applies a restoring force to the pendulum. Such an accelerometer unit may have a useful range from about 10^{-5} "g" to more than $10\,g$.

---

[1] A book entitled *Science for the Airplane Passenger* by Elizabeth A. Wood (Ballantine, New York, 1969) has such a goniometer on its back cover and a discussion of its use in the text. The book also describes a host of other ways in which airplane passengers can discover or apply scientific principles during their travels.

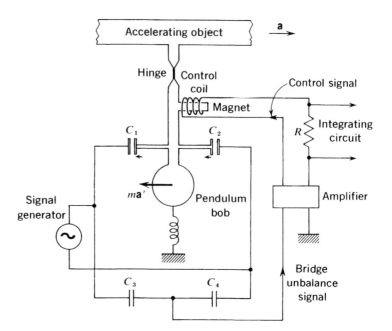

Fig. 12-8 Electromechanical accelerometer system.

ACCELERATING FRAMES AND GRAVITY

In all our discussions of accelerated frames, we have assumed that the observers know "which way is up"—i.e., they know the direction and magnitude of the force of gravity and treat it (as we have done) as a real force, whose source is the gravitating mass of the earth. But suppose our frame of reference to be a completely enclosed room with no access to the external surroundings. What can one then deduce about gravity and inertial forces through dynamical experiments wholly within the room?

We shall suppose once again that there is an observer in a frame, S, attached to the earth. This observer is not isolated; he is able to verify that the downward acceleration of a particle dropped from rest is along a line perpendicular to the earth's surface and hence is directed toward the *center* of the earth.[1] He is able to draw the orthodox conclusion that this acceleration is due to the gravitational attraction from the large mass of the earth. Our second observer is shut up in a room that defines the frame S'. Initially it is known that the floor of his room is

[1] We are still ignoring the rotation of the earth, which causes this statement to be not quite correct. A falling object does *not* fall exactly parallel to a plumb-line. We shall come back to this when we discuss rotating reference frames.

horizontal and that its walls are vertical. In subsequent measurements, however, the observer in S' finds that a plumbline hangs at an angle to what he had previously taken to be the vertical, and that objects dropped from rest travel parallel to his plumbline. The observers in S and S' report their findings to one another by radio. The observer in S' then concludes that he has three alternative ways of accounting for the component of force, parallel to the floor, that is now exerted on all particles as observed in *his* frame:

1. In addition to the gravitational force, there is an inertial force in the $-x$ direction due to the acceleration of his frame in the $+x$ direction.

2. His frame is not accelerating, but a large massive object has been set down in the $-x$ direction outside his closed room, thus exerting an additional gravitational force on all masses in his frame.

3. His room has been tilted through an angle θ *and* an extra mass has been placed beneath the room to increase the net gravitational force. (This is close to being just a variant of alternative 2.)

In supposing that all three hypotheses work equally well to explain what happens in S', we must assume that the additional massive object, postulated in alternatives 2 and 3, produces an effectively uniform gravitational field throughout the room.

From dynamical experiments made entirely within the closed room, there is no way to distinguish among these hypotheses. The acceleration of the frame of reference produces effects that are identical to those of gravitational attraction. Inertial and gravitational forces are both proportional to the mass of the object under examination. The procedures for detecting and measuring them are identical. Moreover, they are both describable in terms of the properties of a *field* (an acceleration field) that has a certain strength and direction at any given point. An object placed in this field experiences a certain force without benefit of any contact with its surroundings. Is all this just an interesting parallel, or does it have a deeper significance?

Einstein, after pondering these questions, concluded that there was indeed something fundamental here. In particular, the completely exact proportionality (as far as could be determined) between gravitational force and inertial mass suggested to him that no physical distinction could be drawn, at least within a

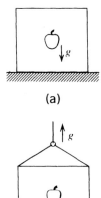

Fig. 12-9 (a) Apple falling inside a box that rests on the earth. (b) Indistinguishable motion when the apple is inside an accelerated box in outer space.

limited region, between a gravitational field and a general acceleration of the reference frame (see Fig. 12-9). He announced this—his famous principle of equivalence—in 1911.[1] The proportionality of gravitational force to inertial mass now becomes an exact necessity, not an empirical and inevitably approximate result. It is also implied that *anything* traversing a gravitational field must follow a curved path, because such a curvature would appear on purely kinematic and geometrical grounds if we replaced the gravitational field by the equivalent acceleration of our own reference frame. In particular, this should happen with rays of light (see Fig. 12-10). With the help of these ideas Einstein proceeded to construct his general theory of relativity, which (as we pointed out in Chapter 8) is primarily a geometrical theory of gravitation.

Fig. 12-10 Successive stages in the path of a horizontally traveling object as observed within an enclosure accelerating vertically upward. This illustrates the equivalence of gravity and a general acceleration of the reference frame.

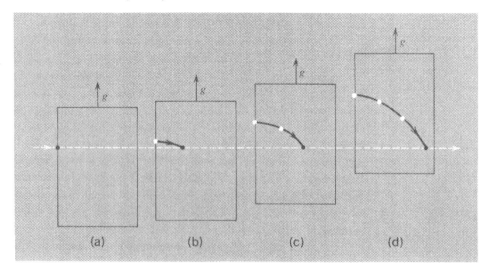

[1] A. Einstein, *Ann. Phys.* (4) **35,** 898 (1911), reprinted in translated form in *The Principle of Relativity* (W. Perrett and G. B. Jeffery, translators), Methuen, London, 1923 and Dover, New York, 1958.

CENTRIFUGAL FORCE

We shall now consider a particular kind of inertial force that always appears if the motion of a particle is described and analyzed from the standpoint of a *rotating* reference frame. This force—the centrifugal force—is familiar to us as the force with which, for example, an object appears to pull on us if we whirl it around at the end of a string.[1] To introduce it, we shall consider a situation of just this kind.

Suppose that a "tether ball" is being whirled around in horizontal circular motion with constant speed (Fig. 12-11). We shall analyze the motion of the ball as seen from two viewpoints: a stationary frame S, and a rotating frame S' that rotates with the same (constant) rotational speed as the ball. For convenience, we align the coordinate systems with their z and z' axes (as well as origins) coincident. The rotational speed of S' relative to S will be designated ω (in rad/sec). Figure 12-11 shows the analysis with respect to these two frames. The essential conclusions are these:

1. From the standpoint of the stationary (inertial) frame, the ball has an acceleration $(-\omega^2 r)$ toward the axis of rotation. The force, F_r, to cause this acceleration is supplied by the tethering cord, and we must have

(In S) $F_r = -m\omega^2 r$

2. From the standpoint of a frame that rotates so as to keep exact pace with the ball, the acceleration of the ball is zero. We can maintain the validity of Newton's law in the rotating frame if, in addition to the force F_r, the ball experiences an inertial force F_i, equal and opposite to F_r, and so directed radially outward:

(In S') $\begin{cases} F'_r = F_r + F_i = 0 \\ F_i = m\omega^2 r \end{cases}$

The force F_i is then what we call the centrifugal force.

The magnitude of the centrifugal force can be established experimentally by an observer in the rotating frame S'. Let him hold a mass m stationary (as seen in his rotating frame) by

---

[1] The name "centrifugal" comes from the Latin: *centrum*, the center, and *fugere*, to flee.

Procedure	Viewed from stationary frame S	Viewed from rotating frame S'
Pictorial sketch of problem	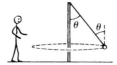 Ball is observed to move with speed v in a circle of radius r (angular speed ω)	 Boy turns around at the same angular speed ω as the ball; from his point of view, the ball is at rest
"Isolate the body." Draw all forces that act on the ball		 T = tension in cord mg = force of gravity F_i = inertial force due to viewing the problem from a rotating frame
For ease of calculation we resolve forces into components in mutually perpendicular directions		
We now analyze the problem in terms of $\mathbf{F} = m\mathbf{a}$	Vertical Direction Because there is no vertical acceleration, we conclude that the net vertical force must be zero; hence $T \cos \theta = mg$ Horizontal Direction The object is traveling in a circle, therefore accelerating; the net force (i.e., the sum of all three forces) is horizontal toward the center of the circle, and must be equal in magnitude to mv^2/r; hence  This force is directed radially inward	Vertical Direction Because there is no vertical acceleration, we conclude that the net vertical force must be zero; hence: $T \cos \theta = mg$ Horizontal Direction The object is "at rest," therefore the sum of all the forces on it must be zero; hence F_i is equal in magnitude to $T \sin \theta$. From the analysis in the left column, it is given by $$F_i = \frac{mv^2}{r} = m\omega^2 r$$ and is directed outward. We call F_i the centrifugal force

Fig. 12–11 *Motion of a suspended ball, which is traveling in a horizontal circle, as analyzed from the earth's reference frame and from a frame rotating with the ball.*

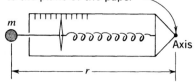

Fig. 12-12 *Measurement of the force needed to hold an object at rest in a rotating reference frame.*

attaching it to a spring balance (Fig. 12-12). If the mass is at any location except on the axis of rotation, the spring balance will show that it is exerting on the mass an inward force proportional to m and r. If the observer in S' is informed that his frame is rotating at the rate of ω rad/sec, he can confirm that this force is equal to $m\omega^2 r$. The observer explains the extension of the spring by saying that it is counteracting the outward centrifugal force on m which is present in the rotating frame. Furthermore, if the spring breaks, then the net force on the mass is just the centrifugal force and the object will at that instant have an outward acceleration of $\omega^2 r$ in response to this so-called "fictitious" force. Once again the inertial force is "there" by every criterion we can apply (except our inability to find another physical system as its source).

The magnitude of the centrifugal force is given, as we have seen, by the equation

$$F_{\text{centrifugal}} = m\frac{v^2}{r} = m\omega^2 r \quad \text{(radially outward)} \quad (12\text{-}6)$$

A nice example of our almost intuitive use of this force, under conditions in which there is nothing to balance it, is provided by situations such as the following: We have been washing a piece of straight tubing, and we want to get it dry on the inside. As a first step we get rid of the larger drops of water that are sitting on the inside walls. And we do this, not by shaking the tube longitudinally, but by whirling it in a circular arc [Fig. 12-13(a)]. The analysis of what happens as we begin this rotation gives a particularly clear picture of the difference between the descriptions of the process in stationary and rotating frames. It also provides us with a different way of deriving the formula for the centrifugal force itself.

Suppose that a drop, of mass m, is sitting on the inner wall of the tube at a point A [Fig. 12-13(b)], a distance r from the axis of rotation. Assume that the tube is very smooth, so that

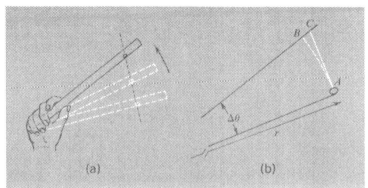

Fig. 12-13 (a) Shaking a drop of water out of a tube. (b) Analysis of initial motion in terms of centrifugal forces.

the drop encounters no resistance if it moves along the tube. The drop must, however, be carried along in any transverse movement of the tube resulting from the rotation. Then if the tube is suddenly set into motion and rotated through a small angle $\Delta\theta$, the drop, receiving an impulse normal to the wall of the tube at A, moves along the straight line AC. This, however, means that it is now further from the axis of rotation than if it had been fixed to the tube and had traveled along the circular arc AB. We have, in fact,

$$BC = r \sec \Delta\theta - r$$

Now

$$\sec \Delta\theta = (\cos \Delta\theta)^{-1} \approx [1 - \tfrac{1}{2}(\Delta\theta)^2]^{-1}$$
$$\approx 1 + \tfrac{1}{2}(\Delta\theta)^2$$

Therefore,

$$BC \approx \tfrac{1}{2} r (\Delta\theta)^2$$

We can, however, express $\Delta\theta$ in terms of the angular velocity ω and the time Δt: $\Delta\theta = \omega \Delta t$. Thus we have

$$BC = \tfrac{1}{2} \omega^2 r (\Delta t)^2$$

This is then recognizable as the radial displacement that occurs in time Δt under an acceleration $\omega^2 r$. Hence we can put

$$a_{\text{centrifugal}} = \omega^2 r$$

and so

$$F_{\text{centrifugal}} = m\omega^2 r$$

Notice, then, that what is, in fact, a small transverse displacement in a straight line, with no real force in the radial direction, appears in the frame of the tube as a small, purely radial displacement under an unbalanced centrifugal force. The physical fact that the drop is moved outward along the tube is readily understood in terms of either description. (We should add, however, that our analysis as it stands does only apply to the initial step of the motion. Once the drop has acquired an appreciable radial velocity, things become more complicated.)

The term "centrifugal force" is frequently used incorrectly. For example, one may read such statements as "The satellite does not fall down as it moves around the earth because the centrifugal force just counteracts the force of gravity and hence there is no net force to make it fall." Any such statement flouts Newton's first law—A body with no net force on it travels in a *straight* line.... For if the satellite is described as *moving* in a curved path around the earth, it must also have an unbalanced force on it. The only frame in which the centrifugal force does balance the gravitational force is the frame in which the satellite appears not to move at all. One can, of course, consider the description of such motions with respect to a reference frame rotating at some arbitrary rate different from that of the orbiting object itself. In this case, however, the centrifugal contribution to the inertial forces represents only a part of the story, and the simple balancing of "real" and centrifugal forces does not apply. In particular, let us reemphasize that in a nonrotating frame of reference there is no such thing as centrifugal force. The long-standing confusion that leads people to use the term "centrifugal force" incorrectly has driven at least one author to extreme vexation. In an otherwise sober and quite formal text the author writes: "There is no answer to these people. Some of them are good citizens. They vote the ticket of the party that is responsible for the prosperity of the country; they belong to the only true church; they subscribe to the Red Cross drive—but they have no place in the Temple of Science; they profane it."[1]

CENTRIFUGES

The laboratory centrifuge represents an immensely important and direct application of the dynamical principle of centrifugal force. The basic arrangement of a simple type of centrifuge is

[1] W. F. Osgood, *Mechanics*, Macmillan, New York, 1937.

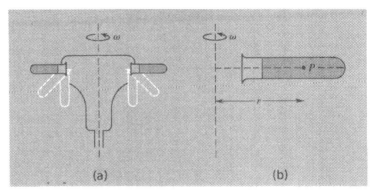

Fig. 12-14 (a) Vertical section through a simple centrifuge. (b) Analysis of radial sedimentation in terms of centrifugal forces.

shown in Fig. 12-14(a). Carefully balanced tubes of liquid are suspended on smooth pivots from a rotor. When the rotor is made to spin at high speed, the tubes swing upward and outward into almost horizontal positions and may be maintained in this orientation for many hours on end. At any point P in one of the tubes [Fig. 12-14(b)], distance r from the axis of rotation, there is an effective gravitational field of magnitude $\omega^2 r$, which may be made very much greater than g. For example, if $r = 15$ cm and the rotor spins at 25 rps ($\omega = 50\pi$ sec^{-1}), the value of $\omega^2 r$ is about 4000 m/sec^2 or 400 g. Small particles in suspension in the liquid will be driven toward the outward (bottom) end of the tube much more quickly than they would ever be under the action of gravity alone.

The basis for calculating the drift speed is the formula for the resistive force to motion through a fluid at low speeds, which we first met in Chapter 5. For a spherical particle of radius r and speed v, this force is proportional to the product rv. If the medium is water, the approximate magnitude of the force is given by

$$R(v) \approx 0.02 rv$$

where R is in newtons, r in meters, and v in m/sec. A steady value of v is attained when this force just balances the driving force associated with the effective gravitational field strength, g'. In calculating this driving force it is important to allow for buoyancy effects—i.e., Archimedes' principle. If the density of the particle is ρ_p and the density of the liquid is ρ_l, the driving force is given by

$$F = \frac{4\pi}{3}(\rho_p - \rho_l)r^3 g'$$

This can be more simply expressed if we introduce the true mass m of the particle ($= 4\pi\rho_p r^3/3$), in which case we can put

$$F = \left(1 - \frac{\rho_l}{\rho_p}\right) mg'$$

To take a specific example, suppose that we have an aqueous suspension of bacterial particles of radius 1 μ, each with a mass of about 5×10^{-15} kg and a density about 1.1 times that of water. If we take for g' the value 400 g calculated earlier, we find

$$F \approx 2 \times 10^{-12} \text{ N}$$

We thus obtain a drift speed given by

$$v \approx \frac{2 \times 10^{-12}}{2 \times 10^{-2} \times 10^{-6}} \approx 10^{-4} \text{ m/sec}$$

This represents a settling rate of several centimeters per hour, which makes for effective separation in reasonable times, whereas under the normal gravity force alone one would have only a millimeter or two per day.

The above example represents what one may regard as a more or less routine type of centrifugation, but in 1925 the Swedish chemist T. Svedberg opened up a whole new field of research when, by achieving centrifugal fields thousands of times stronger than g, he succeeded in measuring the molecular weights of proteins by studying their radial sedimentation. The type of machine he developed for this purpose was appropriately named the *ultracentrifuge*, and Svedberg succeeded in producing centrifugal fields as high as about 50,000 g. The physicist J. W. Beams has taken the technique even further through his development of magnetic suspensions, in vacuum, that dispense with mechanical bearings altogether. The rotor simply spins in empty space, with carefully controlled magnetic fields to hold it at a constant vertical level against the normal pull of gravity. By such methods Beams has produced centrifugal fields equivalent to about $10^6 \, g$ in a usable centrifuge and fields as high as $10^9 \, g$ in tiny objects (e.g., spheres of 0.001-in. diameter). The limitation is set by the bursting speed of the rotor; this defines a maximum value of ω proportional to $1/r$ (see Chapter 7, p. 208). Since the centrifugal field g' is equal to $\omega^2 r$, and the limiting speed sets an upper limit to ωr, it may be seen that the attainable value of g' varies as $1/r$.

The technique of ultracentrifuge methods has been brought

to an extraordinary pitch of refinement. It has become possible to determine molecular weights to a precision of better than 1% over a range from about 10^8 (virus particles) down to as low as about 50. The possibility of measuring the very low molecular weights by this method is particularly impressive. Beams has pointed out that in a solution of sucrose in water, the calculated rate of descent of an individual sucrose molecule, of mass about 340 amu and radius about 5 Å, would (according to the kind of analysis we gave earlier) be less than 1 mm in 100 years under normal gravity. (A rate as slow as this becomes in fact meaningless because, as Beams points out, it would be completely swamped by random thermal motions.) If a field of $10^5 g$ is available, however, the time constant of the sedimentation process is reduced to the order of 1 day or less, which brings the measurement well within the range of possibility.[1]

This whole subject of centrifuges and centrifugation is a particularly good application of the concept of inertial force, because the phenomena are so appropriately described in terms of static or quasistatic equilibrium in the rotating frame.

CORIOLIS FORCES

We have seen how the centrifugal force, $m\omega^2 r$, exerted on a particle of a given mass m in a frame rotating at a given angular velocity ω, depends only on the distance r of the particle from the axis of rotation. In general, however, another inertial force appears in a rotating frame. This is the *Coriolis force*,[2] and it depends only on the velocity of the particle (not on its position). We shall introduce this force in a simple way for some specific situations. Later, by introducing *vector* expressions for rotational motion, we shall develop a succinct notation that gives both the centrifugal and Coriolis forces in a form valid in three dimensions using any type of coordinate system.

The need to introduce the Coriolis force is easily shown by comparing the straight-line motion of a particle in an inertial frame S with the motion of the same particle as seen in a rotating frame S'.

---

[1] For further reading on this extremely interesting subject, see T. Svedberg and K. O. Pedersen, *The Ultracentrifuge*, Oxford University Press, New York, 1960, and J. W. Beams, "High Centrifugal Fields," *Physics Teacher* **1**, 103 (1963).
[2] G. Coriolis, *J. de l'Ecole Polytechnique*, Cahier **24**, 142 (1835).

Suppose that S' is a coordinate system attached to a horizontal circular table that rotates with constant angular speed ω. Let the vertical axis of rotation define the z' axis and suppose that the table surface (in the $x'y'$ plane) has no friction. A string fastened to the origin holds a particle on the y' coordinate axis at a radial distance r'_0 from the axis of rotation. Thus, in the S' frame, the particle is at rest in equilibrium under the combined forces of the tension in the string and the centrifugal force. (The vertical force of gravity and the normal force of the table surface always add to zero and need not concern us further.)

The same particle is viewed from an inertial frame S which coincides with S' at $t = 0$. In this stationary frame, the particle travels with uniform speed $v_\theta = \omega r_0$ in a circle of constant radius r_0 $(= r'_0)$ under the single unbalanced force of the tension in the string. There is, of course, no centrifugal force in this inertial frame.

At $t = 0$ the string breaks. In S the particle then travels

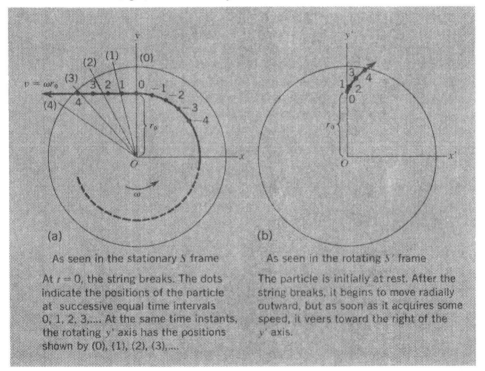

Fig. 12–15 *Two different descriptions of the motion of an object that is initially tethered on a rotating disk and begins motion under no forces at $t = 0$.*

(a) As seen in the stationary S frame

At $t = 0$, the string breaks. The dots indicate the positions of the particle at successive equal time intervals 0, 1, 2, 3,.... At the same time instants, the rotating y' axis has the positions shown by (0), (1), (2), (3),....

(b) As seen in the rotating S' frame

The particle is initially at rest. After the string breaks, it begins to move radially outward, but as soon as it acquires some speed, it veers toward the right of the y' axis.

515 *Coriolis forces*

in a straight line with constant speed $v = \omega r_0$ as shown in Fig. 12-15(a). To find the motion in S', we compare *in the stationary frame* the positions of the y' axis and the corresponding locations of the particle at successive equally-spaced times. We discover that the particle, as observed in S', not only moves radially outward, but also moves farther and farther to the right of the single radial line formed by the rotating y' axis. This result is plotted in Fig. 12-15(b). To explain this motion as observed in the rotating frame, it is necessary to postulate, in addition to the centrifugal force, a sideways deflecting force. This deflecting force is the *Coriolis* force. In the course of the following discussion, we shall determine its magnitude and show that it always acts at right angles to any velocity $\mathbf{v}'$ in the S' frame.

We can find the magnitude of the Coriolis force by investigating another simple motion in these two frames. Suppose that, instead of the situation just described, we make a particle follow a radially outward path in the rotating frame at *constant* velocity $\mathbf{v}'_r$. In this frame there must be no net force on the particle. Hence we shall have to supply some real (inward) force to counteract the varying (outward) centrifugal force as the particle moves. We shall not concern ourselves with these radial components but will concentrate our attention only on the transverse-force components. In this way we can remove from consideration the distortion of the trajectory by the centrifugal force, which is purely radial.

How does the motion appear in the two frames? Figure 12-16(b) shows the straight-line path of the object in the rotating frame. But the path of the object in the stationary frame is a curved line AB as shown in Fig. 12-16(a). In S the transverse velocity v_θ ($= \omega r$) is greater at B than at A, because the radial distance from the axis is greater at B. Hence there must be a *real* transverse force to produce this increase of velocity seen in the stationary frame. This real force might be provided, for example, by a spring balance.

What does this motion look like in the rotating frame? In S' the object moves outward with constant speed and hence has no acceleration [see Fig. 12-16(b)]. This means, as we have said, that there can be no net force on the object in the rotating frame. But since an observer in S' sees the spring balance exerting a *real* sideways force on the object in the $+\theta$ direction, he infers that there is a counteracting *inertial* force in the $-\theta$ direction to balance it. This is the Coriolis force.

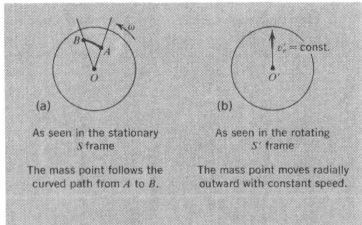

Fig. 12-16 (a) Laboratory view of the path of a particle that moves radially outward on a rotating table. (b) The motion as it appears in the rotating frame itself.

As seen in the stationary S frame

The mass point follows the curved path from A to B.

As seen in the rotating S' frame

The mass point moves radially outward with constant speed.

To determine its magnitude, let OA and OB in Fig. 12-17 be successive positions of the same radial line at times separated by Δt. Let OC be the bisector of the angle $\Delta\theta$. The velocity perpendicular to OC changes by the amount Δv_θ during Δt, where

$$\Delta v_\theta = [\omega(r + \Delta r)\cos(\Delta\theta/2) + v_r \sin(\Delta\theta/2)] \\ - [\omega r \cos(\Delta\theta/2) - v_r \sin(\Delta\theta/2)]$$

For small angles we can put the cosine equal to 1 and the sine equal to the angle, which leads to the following very simple expression for Δv_θ:

$$\Delta v_\theta \approx \omega\,\Delta r + v_r\,\Delta\theta$$

The transverse acceleration a_θ is thus given by

$$a_\theta = \omega(\Delta r/\Delta t) + v_r(\Delta\theta/\Delta t)$$

But

$$\Delta r/\Delta t = v_r \quad \text{and} \quad \Delta\theta/\Delta t = \omega$$

Hence

$$a_\theta = 2\omega v_r$$
$$F_\theta = 2m\omega v_r$$

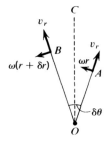

Fig. 12-17 Basis of calculating the Coriolis force for a particle moving radially at constant speed with respect to a rotating table.

This gives us the *real* force needed to cause the *real* acceleration as judged in S. But as observed in the rotating frame S', there is no acceleration and no net force. Hence the existence of the Coriolis force, equal to $-2m\omega v'_r$, is inferred. (Note that $v_r = v'_r$.) This inertial force is in the negative θ' direction, opposite to the spring force, and is at right angles to the direction of motion of the particle.

$$F'_\theta(\text{Coriolis}) = -2m\omega v'_r \qquad (12\text{-}7)$$

An important feature, which you should verify for yourself, is that if we had considered a radially *inward* motion (v'_r negative), then we would have inferred the existence of a Coriolis force acting in the *positive* θ' direction. In both cases, therefore, the Coriolis force acts to deflect the object in the same way with respect to the direction of the velocity $\mathbf{v}'$ itself—to the right if the frame S' is rotating counterclockwise, as we have assumed, or to the left if S' rotates clockwise. It turns out, in fact, as we shall prove later, that even in the case of motion in an arbitrary direction the Coriolis force is always a *deflecting* force, exerted at right angles to the direction of motion as observed in the rotating frame.

The Coriolis force is very real from the viewpoint of the rotating frame of reference. If you want to convince yourself of the reality of this "fictitious" force, ride a rotating merry-go-round and try walking a radial line outward or inward. (Proceed cautiously—the Coriolis force is so unexpected and surprising that it is easy to lose one's balance!)

DYNAMICS ON A MERRY-GO-ROUND

As we have just mentioned, the behavior of objects in motion within a rotating reference frame can run strongly counter to one's intuitions. It is not too hard to get used to the existence of the centrifugal force acting on an object at rest with respect to the rotating frame, but the combination of centrifugal and Coriolis effects that appear when the object is set in motion can be quite bewildering, and sometimes entertaining. Suppose, for example, that a man stands at point A on a merry-go-round [Fig. 12-18(a)] and tries to throw a ball to someone at B (or perhaps aims for the bull's-eye of a dart board placed there). Then the thrown object mysteriously veers to the right and misses its target every time. One can blame part of this, of course, on the centrifugal force itself. However, it is to be noted that

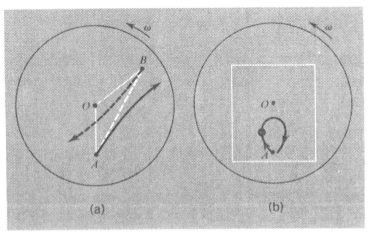

Fig. 12-18 (a) Trajectories of objects as they appear to observers on a rotating table. (b) An object projected on a frictionless rotating table can return to its starting point.

since the magnitude of the centrifugal force is $m\omega^2 r$ and that of the Coriolis force is $2m\omega v'$, the ratio of these two forces is proportional to $v'/\omega r$. Thus if v' is made much greater than the actual peripheral speed of the merry-go-round, the peculiarities of the motion are governed almost entirely by the Coriolis effects. If this condition holds, the net deflection of a moving object will always be to the right with respect to $\mathbf{v'}$ on a merry-go-round rotating counterclockwise. Thus if the positions A and B in Fig. 12-18(a) are occupied by two people trying to throw a ball back and forth, each will have to aim to the left in order to make a good throw.

An extreme case of this kind of behavior can cause an object to follow a continuously curved path that brings it back to its starting point, although it is not subjected to any real forces at all. This phenomenon has been demonstrated in the highly entertaining and instructive film, *Frames of Reference*.[1] A dry-ice puck, launched at point A on a tabletop of plate glass [Fig. 12-18(b)], can be caused by a skilled operator to follow a trajectory of the kind indicated.

GENERAL EQUATION OF MOTION IN A ROTATING FRAME[2]

The goal of this discussion will be to relate the time derivatives of the displacement of a moving object as observed in a sta-

[1] "*Frames of Reference*," by J. N. P. Hume and D. G. Ivey, Education Development Center, Newton, Mass., 1960.
[2] This section may be omitted by a reader who is willing to take on trust its final results—that the total inertial force in a rotating frame is the combination of the centrifugal force with a Coriolis force corresponding to a generalized form of Eq. (12-7).

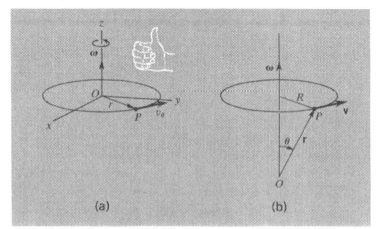

Fig. 12-19 Use of angular velocity as a vector to define the linear velocity of a particle on a rotating table: $v = \omega \times r$.

tionary frame S and in a rotating frame S'. To set the stage, we shall introduce the idea that angular velocity may be represented as a vector.

Consider first a point P on a rotating disk [Fig. 12–19(a)]. It has a purely tangential velocity, v_θ, in a direction at right angles to the radius OP. We can describe this velocity, in both magnitude and direction, if we define a *vector* according to the same convention that we introduced for torque in Chapter 4. That is, if the fingers of the *right* hand are curled around in the sense of rotation, keeping the thumb extended as shown in the figure, then ω is represented as a vector, of length proportional to the angular speed, in the direction in which the thumb points. Thus with ω pointing along the positive z direction, one is defining a rotation that carries each point such as P from the positive x direction toward the positive y direction. The rotation of the disk is in this case counterclockwise as seen from above.

The velocity of P is now given by the vector (cross) product of ω with the radius vector r:

$$v = \omega \times r \tag{12-8}$$

This vector-product expression is valid in three dimensions also, if the position vector r of P is measured from any point on the axis of rotation, as shown in Fig. 12–19(b). The radius of the circle in which P moves is $R = r \sin \theta$. Thus we have $v = v_\theta = \omega r \sin \theta$, in a direction perpendicular to the plane defined by ω and r. That is precisely what Eq. (12-8) gives us.

Next, we consider how the change of *any* vector during a

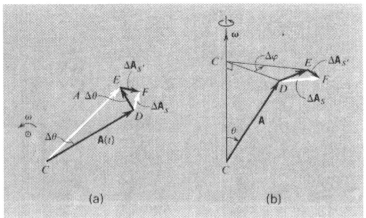

Fig. 12-20 (a) Change of a vector, analyzed in terms of its change as measured on a rotating table, together with the change due to rotation of the table itself. (b) Similar analysis for an arbitrary vector referred to any origin on the axis of rotation.

small time interval Δt can be expressed as the vector sum of two contributions:

1. The change that would occur if it were simply a vector of constant length embedded in the rotating frame S'.
2. The further change described by its change of length and direction as observed in S'.

In Fig. 12-20(a) we show this analysis for motion confined to a plane. The vector **A** at time t is represented by the line CD. If it remains fixed with respect to a rotating table, its direction at time $t + \Delta t$ is given by the line CE, where $\Delta\theta = \omega\,\Delta t$. Thus its change due to the rotation alone would be represented by DE, where $DE = A\,\Delta\theta = A\omega\,\Delta t$. From the standpoint of frame S' this change would not be observed. There might, however, be a change represented by the line EF; we shall denote this as $\Delta\mathbf{A}_{S'}$—the change of **A** as observed in S'. The vector sum of DE and EF, i.e., the line DF, then represents the true change of **A** as observed in S. We therefore denote this as $\Delta\mathbf{A}_S$.

In Fig. 12-20(b) we show the corresponding analysis for three dimensions. The length of DE is now equal to $A \sin\theta\,\Delta\varphi$; its direction is perpendicular to the plane defined by **ω** and **A**. Since $\Delta\varphi = \omega\,\Delta t$, we can put

vector displacement $DE = (\boldsymbol{\omega} \times \mathbf{A})\,\Delta t$

The displacement $\Delta\mathbf{A}_{S'}$ may be in any direction with respect to DE, but the two again combine to give a net displacement DF which is to be identified with $\Delta\mathbf{A}_S$. Thus we have

$$\Delta \mathbf{A}_S = \Delta \mathbf{A}_{S'} + (\boldsymbol{\omega} \times \mathbf{A})\Delta t$$

We can at once proceed from this to a relation between the rates of change of **A** as observed in S and S', respectively:

$$\left(\frac{d\mathbf{A}}{dt}\right)_S = \left(\frac{d\mathbf{A}}{dt}\right)_{S'} + \boldsymbol{\omega} \times \mathbf{A} \tag{12-9}$$

This is a very powerful relation because **A** can be any vector we please.

First, we shall choose **A** to be the position vector **r**. Then $(d\mathbf{A}/dt)_S$ is the true velocity, **v**, as observed in S, and $(d\mathbf{A}/dt)_{S'}$ is the apparent velocity, **v'**, as observed in S'. Thus we immediately have

$$\mathbf{v} = \mathbf{v}' + \boldsymbol{\omega} \times \mathbf{r} \tag{12-10}$$

Next, we shall choose **A** to be the velocity **v**:

$$\left(\frac{d\mathbf{v}}{dt}\right)_S = \left(\frac{d\mathbf{v}}{dt}\right)_{S'} + \boldsymbol{\omega} \times \mathbf{v} \tag{12-11}$$

Now $(d\mathbf{v}/dt)_S$ is the true acceleration, **a**, as observed in S. The quantity $(d\mathbf{v}/dt)_{S'}$ is, however, a sort of hybrid—it is the rate of change in S' of the velocity as observed in S. We can make more sense of this if we substitute for **v** from Eq. (12-10); we then have

$$\left(\frac{d\mathbf{v}}{dt}\right)_{S'} = \left(\frac{d\mathbf{v}'}{dt}\right)_{S'} + \boldsymbol{\omega} \times \left(\frac{d\mathbf{r}}{dt}\right)_{S'}$$

The two terms on the right of this equation are now quite recognizable; $(d\mathbf{v}'/dt)_{S'}$ is the acceleration, **a'**, as observed in S', and $(d\mathbf{r}/dt)_{S'}$ is just **v'**. Thus we have

$$\left(\frac{d\mathbf{v}}{dt}\right)_{S'} = \mathbf{a}' + \boldsymbol{\omega} \times \mathbf{v}'$$

Substituting this in Eq. (12-11) we thus get

$$\mathbf{a} = \mathbf{a}' + \boldsymbol{\omega} \times \mathbf{v}' + \boldsymbol{\omega} \times \mathbf{v}$$

We do not need to have both **v** and **v'** on the right-hand side, and we shall again substitute for **v** from Eq. (12-10). This gives us finally

$$\mathbf{a} = \mathbf{a}' + 2\boldsymbol{\omega} \times \mathbf{v}' + \boldsymbol{\omega} \times (\boldsymbol{\omega} \times \mathbf{r}) \tag{12-12}$$

A remark is in order regarding the last term, which involves the cross product of three vectors. According to the rules of vector algebra, the cross product inside the parentheses is to be taken first, then the other cross product performed. A nonzero answer will result for all cases where the angle formed by ω and $\mathbf{r}$ is other than 0° or 180°. Performing the cross products in the reverse (incorrect) order, however, would result in zero for all cases, regardless of the angle between these vectors.

Multiplying Eq. (12–12) throughout by the mass m of the object, we recognize the left side as the net external force on the mass as seen in the stationary system.

$$m\mathbf{a} = \mathbf{F}_{net} = m\mathbf{a}' + 2m(\omega \times \mathbf{v}') + m[\omega \times (\omega \times \mathbf{r})]$$

In the rotating frame of reference, the object m has the acceleration $\mathbf{a}'$. We may preserve the format of Newton's second law in this accelerated frame of reference by rearranging the above equation, so as to be able to write

$$\mathbf{F}'_{net} = m\mathbf{a}' \qquad (12\text{–}13a)$$

where

$$\mathbf{F}'_{net} = \underbrace{\mathbf{F}_{net}}_{\text{"real" force}} - \underbrace{\underbrace{2m(\omega \times \mathbf{v}')}_{\text{Coriolis force}} - \underbrace{m[\omega \times (\omega \times \mathbf{r})]}_{\text{centrifugal force}}}_{\text{inertial forces}} \qquad (12\text{–}13b)$$

The mathematical form of Eq. (12–13b) shows that both the Coriolis force and the centrifugal force are in a direction at right angles to the axis of rotation defined by ω. The centrifugal force, in particular, is always radially outward from the axis, as is clear if one considers the geometrical relationships of the vectors involved in the product $-\omega \times (\omega \times \mathbf{r})$, as shown in Fig. 12–21. The equation also shows that the Coriolis force

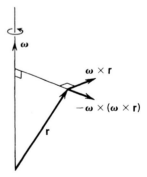

Fig. 12–21 Relation of the vectors involved in forming the centrifugal acceleration $-\omega \times (\omega \times \mathbf{r})$.

would reverse if the direction of ω were reversed, but the direction of the centrifugal force would remain unchanged.

The specification of $\mathbf{F}'$ in Eq. (12–13) can be made entirely on the basis of measurements of position, velocity, and acceleration as observed within the rotating frame itself. The centrifugal term, involving the vector $\mathbf{r}$, might seem to contradict this, but we could just as well put $\mathbf{r}'$ instead of $\mathbf{r}$, because observers in the two frames do agree on the vector position of a moving object at a given instant, granted that they use the same choice of origin.

To summarize, we have established by the above calculation that the dynamics of motion as observed in a uniformly rotating frame of reference may be analyzed in terms of the following three categories of forces:

"*Real*": $\mathbf{F}_{net}$ — This is the sum of all the "real" forces on the object such as forces of contact, tensions in strings, the force of gravity, electrical forces, magnetic forces, and so on. Only these forces are seen in a stationary frame of reference.

Coriolis: $-2m(\boldsymbol{\omega} \times \mathbf{v}')$ — The Coriolis force is a *deflecting* force always at right angles to the velocity $\mathbf{v}'$ of the mass m. If the object has no velocity in the rotating frame of reference, there is no Coriolis force. It is an inertial force *not* seen in a stationary frame of reference. Note minus sign.

Centrifugal: $-m[\boldsymbol{\omega} \times (\boldsymbol{\omega} \times \mathbf{r})]$ — The centrifugal force depends on position only and is always radially outward. It is an inertial force *not* seen in a stationary frame of reference. We could equally well write it as $-m[\boldsymbol{\omega} \times (\boldsymbol{\omega} \times \mathbf{r}')]$. Note the minus sign.

THE EARTH AS A ROTATING REFERENCE FRAME

In this section we shall consider a few examples of the way in which the earth's rotation affects the dynamical processes occurring on it.

The local value of g

If a particle P is at rest at latitude λ near the earth's surface, then as judged in the earth's frame it is subjected to the gravita-

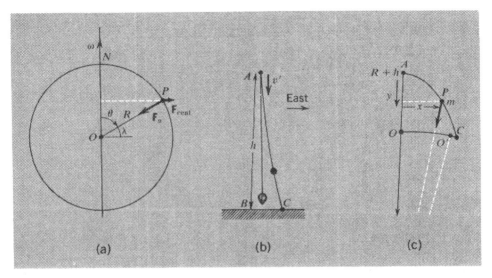

Fig. 12-22 (a) Forces on an object at rest on the earth, as interpreted in a reference frame that rotates with the earth. (b) An object falling from rest relative to the earth undergoes an eastward displacement. (c) The falling motion of (b), as seen from a frame that does not rotate with the earth.

tional force $\mathbf{F}_g$ and the centrifugal force $\mathbf{F}_{\text{cent}}$ shown in Fig. 12-22(a). The magnitude of the latter is given, according to Eq. (12-13b), by the equation

$$F_{\text{cent}} = m\omega^2 R \sin\theta = m\omega^2 R \cos\lambda$$

where R is the earth's radius. We have already discussed in Chapter 8 the way in which this centrifugal term reduces the local magnitude of g and also modifies the local direction of the vertical as defined by a plumbline. The analysis is in fact much simpler and clearer from the standpoint of our natural reference frame as defined by the earth itself. We have, as Fig. 12-22(a) shows, the following relations:

$$\begin{aligned} F'_r &= F_g - F_{\text{cent}} \cos\lambda = F_g - m\omega^2 R \cos^2\lambda \\ F'_\theta &= F_{\text{cent}} \sin\lambda = m\omega^2 R \sin\lambda \cos\lambda \end{aligned} \quad (12\text{-}14)$$

Deviation of freely falling objects

If a particle is released from rest at a point such as P in Fig. 12-22(a), it begins to accelerate downward under the action of a net force $\mathbf{F}'$ whose components are given by Eq. (12-14). As soon as it has any appreciable velocity, however, it also experiences a Coriolis force given by the equation

$$\mathbf{F}_{\text{Coriolis}} = -2m\boldsymbol{\omega} \times \mathbf{v}' \tag{12-15}$$

Now the velocity $\mathbf{v}'$ is in the plane PON containing the earth's axis. The Coriolis force must be perpendicular to this plane, and a consideration of the actual directions of $\boldsymbol{\omega}$ and $\mathbf{v}'$ shows that it is eastward. Thus if we set up a local coordinate system defined by the local plumbline vertical and the local easterly direction, as in Fig. 12–22(b), the falling object deviates eastward from a plumbline AB and hits the ground at a point C. The effect is very small but has been detected and measured in careful experiments (see Problem 12–24).

To calculate what the deflection should be for an object falling from a given height h, we use the fact that the value of v' to be inserted in Eq. (12-15) is extremely well approximated by the simple equation of free vertical fall:

$$v' = gt$$

where v' is measured as positive downward. Thus if we label the eastward direction as x', we have

$$m\frac{d^2 x'}{dt^2} = (2m\omega \cos \lambda)gt$$

Integrating this twice with respect to t, we have

$$x' = \tfrac{1}{3}g\omega t^3 \cos \lambda \tag{12-16}$$

For a total distance of vertical fall equal to h, we have $t = (2h/g)^{1/2}$, which thus gives us

$$x' = \frac{2\sqrt{2}}{3} \frac{\omega \cos \lambda}{g^{1/2}} h^{3/2} \tag{12-17}$$

Inserting approximate numerical values ($\omega = 2\pi$ day$^{-1} \approx 7 \times 10^{-5}$ sec^{-1}), one finds

$$x' \approx 2 \times 10^{-5} h^{3/2} \cos \lambda \qquad (x' \text{ and } h \text{ in m})$$

Thus, for example, with $h = 50$ m at latitude 45°, one has $x' \approx 5$ mm, or about $\tfrac{1}{4}$ in.

It is perhaps worth reminding oneself that the effects of inertial forces can always be calculated, if one wishes, from the standpoint of an inertial frame in which these forces simply do not exist. In the present case, one can begin by recognizing that a particle held at a distance h above the ground has a higher eastward velocity than a point on the ground below. For simplicity, let us consider how this operates at the equator ($\lambda = 0$). Figure 12–22(c) shows the trajectory of the falling object as seen in a nonrotating frame. The object has an initial horizontal velocity given by

$$v_{0x} = \omega(R + h)$$

After a time t it has traveled a horizontal distance x given, very nearly, by $\omega R t$. With the object now at P (see the figure) the gravitational force acting on it has a very small component in the negative x direction. We have, in fact,

$$F_x \approx -\frac{x}{R} F_y \approx -mg\omega t$$

Hence

$$\frac{d^2 x}{dt^2} \approx -g\omega t$$

Integrating once, we have

$$\frac{dx}{dt} \approx v_{0x} - \tfrac{1}{2}g\omega t^2$$

Substituting the value $v_{0x} = \omega(R + h)$, this gives, as a very good approximation,

$$\frac{dx}{dt} = \omega(R + h) - \tfrac{1}{2}g\omega t^2$$

Integrating a second time, we have

$$x = \omega(R + h)t - \tfrac{1}{6}g\omega t^3$$

However, the point O at the earth's surface is also moving, with a constant speed of ωR. Thus, when the falling object hits the ground at C, the point O has reached O', where $OO' = \omega R t$. Hence we have

$$x' = O'C \approx \omega h t - \tfrac{1}{6}g\omega t^3$$

If we substitute $h = \tfrac{1}{2}g t^2$, we at once obtain the result given by

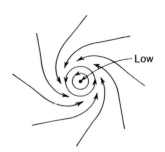

Fig. 12–23   Formation of a cyclone in the northern hemisphere, under the action of Coriolis forces on the moving air masses.

Eq. (12–16) for $\lambda = 0$. [Or, of course, we can substitute $t = \sqrt{2h/g}$ and arrive at Eq. (12–17)].

Patterns of atmospheric circulation

Because of the Coriolis effect, air masses being driven radially inward toward a low-pressure region, or outward away from a high-pressure region, are also subject to deflecting forces. This causes most cyclones to be in a counterclockwise direction in the northern hemisphere and clockwise in the southern hemisphere. The origin of these preferred rotational directions may be seen in Fig. 12–23, which shows the motions of air in the northern hemisphere moving toward a region of low pressure. The horizontal components of the Coriolis force deflect these motions

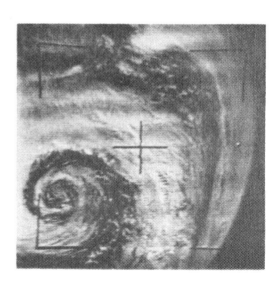

Fig. 12–24   Tiros satellite photograph of a cyclone. (Courtesy of Charles W. C. Rogers and N.A.S.A.)

toward the right. Thus, as the air masses converge on the center of the low-pressure region, they produce a net counterclockwise rotation. For air moving north or south over the earth's surface the Coriolis force is due east or due west, parallel to the earth's surface. If we consider a 1-kg mass of air at a wind velocity of 10 m/sec (about 22 mph) at 45° north latitude, a direct application of Eq. (12-15) gives us

$$F_{\text{Coriolis}} = 2m\omega v' \sin \lambda \approx (2)(1)(2\pi \times 10^{-5})$$
$$\times (10)(0.707) \approx 10^{-3} \text{ N}$$

If we had considered air flowing in from east or west, the Coriolis forces would not be parallel to the earth's surface, but their components parallel to the surface would be given by the same equation as that used above. (Verify this.)

The approximate radius of curvature of the resultant motion may be obtained from

$$F = m\frac{v^2}{R}$$

or

$$R = m\frac{v^2}{F_{\text{Coriolis}}} \approx 1 \times \frac{100}{10^{-3}} = 10^5 \text{ m (about 60 miles)}$$

As air masses move over hundreds of miles on the earth's surface, they often form huge vortices—as is dramatically shown in the Tiros weather satellite photograph in Fig. 12-24.

Occasionally one reads that water draining out of a basin also circulates in a preferred direction because of the Coriolis force. In most cases, the Coriolis force on the flowing water is negligible compared with other larger forces which are present; however, if extremely precise and careful experiments are performed, the effect can be demonstrated.[1]

The Foucault pendulum

No account of Coriolis forces would be complete without some mention of the famous pendulum experiment named after the French physicist J. B. L. Foucault, who first demonstrated in 1851 how the slow rotation of the plane of vibration of a pen-

---

[1] See, for example, the film "Bathtub Vortex," an excerpt from "Vorticity," by A. H. Shapiro, National Council on Fluid Mechanics, 1962.

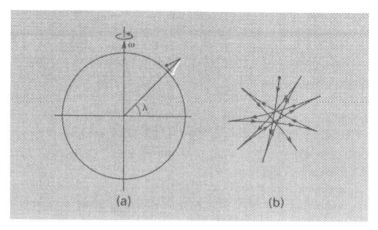

Fig. 12-25 (a) A pendulum swinging along a north–south line at latitude λ. (b) Path of pendulum bob, as seen from above. (The change of direction per swing is, however, grossly exaggerated.)

dulum could be used as evidence of the earth's own rotation.

It is easy, but rather too glib, to say that of course we are simply seeing the effect of the earth turning beneath the pendulum. This description might properly be used for a pendulum suspended at the north or south pole. One can even press things a little further and say that at a given latitude, λ [see Fig. 12-25(a)] the earth's angular velocity vector has a component $\omega \sin \lambda$ along the local vertical. This would indeed lead to the correct result—that the plane of the pendulum rotates at a rate corresponding to one complete rotation in a time $T(\lambda)$ given by

$$T(\lambda) = \frac{2\pi}{\omega \sin \lambda} = 24 \csc \lambda \quad \text{hours} \tag{12-18}$$

But the pendulum is, after all, connected to the earth via its suspending wire, and both the tension in the wire and the gravitational force on the bob lie in the vertical plane in which the pendulum is first set swinging. (So, too, is the air resistance, if this needs to be considered.) It is the Coriolis force that can be invoked to give a more explicit basis for the rotation. For a pendulum swinging in the northern hemisphere, the Coriolis force acts always to curve the path of the swinging bob to the right, as indicated in exaggerated form in Fig. 12-25(b). As with the Coriolis force on moving air, the effect does not depend on the direction of swing—contrary to the intuition most of us probably have that the rotation is likely to be more marked when the pendulum swings along a north–south line than when it swings east–west.

THE TIDES

As everyone knows, the production of ocean tides is basically the consequence of the gravitational action of the moon—and, to a lesser extent, the sun. Thus we could have discussed this as an example of universal gravitation in Chapter 8. The analysis of the phenomenon is, however, considerably helped by introducing the concept of inertial forces as developed in the present chapter.

The feature that probably causes the most puzzlement when one first learns about the tides is the fact that there are, at most places on the earth's surface, two high tides every day rather than just one. This corresponds to the fact that, at any instant, the general distribution of ocean levels around the earth has two bulges. On the simple model that we shall discuss, these bulges would be highest at the places on the earth's surface nearest to and farthest from the moon [Fig. 12–26(a)]. While the earth performs its rotation during 24 hr, the positions of the bulges would remain almost stationary, being defined by the almost constant position of the moon. Thus, if one could imagine the earth completely girdled by water, the depth of the water as measured from a point fixed to the earth's solid surface would pass through two maxima and two minima in each revolution. A better approximation to the observed facts is obtained by considering the bulges to be dragged eastward by friction from the land and the ocean floor, so that their equilibrium positions with respect to the moon are more nearly as indicated in Fig. 12–26(b).

To conclude these preliminary remarks, we may point out that the bulges are, in fact, also being carried slowly eastward all the time by the moon's own motion around the earth. This motion (one complete orbit relative to the fixed stars every

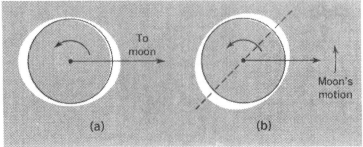

Fig. 12-26 (a) Double tidal bulge as it would be if the earth's rotation did not displace it. The size of the bulge is enormously exaggerated. (b) Approximate true orientation of the tidal bulges, carried eastward by the earth's rotation.

27.3 days) has the consequence that it takes more than 24 hr for a given point on the earth to make successive passages past a particular tidal bulge. Specifically, this causes the theoretical time interval between successive high tides at a given place to be close to 12 hr 25 min instead of precisely 12 hr (see Problem 2-15). For example, if a high tide is observed to occur at 4 P.M. one day, its counterpart next day would be expected to occur at about 4:50 P.M.

Now let us consider the dynamical situation. The first point to appreciate is the manner in which the earth as a whole is being accelerated toward the moon by virtue of the gravitational attraction between them. With respect to the CM of the earth–moon system (inside the earth, at about 3000 miles from the earth's center), the earth's center of mass has an acceleration of magnitude a_C given by Newton's laws:

$$M_E a_C = \frac{GM_E M_m}{r_m^2}$$

i.e.,

$$a_C = \frac{GM_m}{r_m^2} \qquad (12\text{-}19)$$

where M_m and r_m are the moon's mass and distance. What may not be immediately apparent is that every point in the earth receives this *same* acceleration from the moon's attraction. If

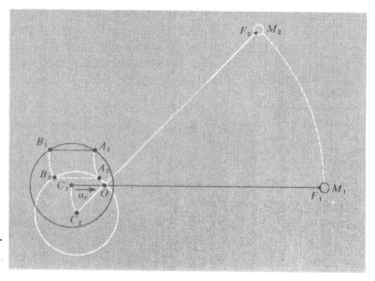

Fig. 12-27 The orbital motion of the earth about the moon does not by itself involve any rotation of the earth; the line A_1B_1 is carried into the parallel configuration A_2B_2.

one draws a sketch, as shown in Fig. 12-27, of the arcs along which the earth's center and the moon travel in a certain span of time, one is tempted to think of the earth–moon system as a kind of rigid dumbbell that rotates as a unit about the center of mass, O. It is true that the moon, for its part, does move so that it presents always the same face toward the earth, but with the earth itself things are different. If the earth were not rotating on its axis, every point on it would follow a circular arc identical in size and direction to the arc C_1C_2 traced out by the earth's center. The line A_1B_1 would be translated into the parallel line A_2B_2. The earth's intrinsic rotation about its axis is simply superposed on this general displacement and the associated acceleration.

This is where noninertial frames come into the picture. The dynamical consequences of the earth's orbital motion around the CM of the earth–moon system can be correctly described in terms of an inertial force, $-ma_C$, experienced by a particle of mass m wherever it may be, in or on the earth. This force is then added to all the other forces that may be acting on the particle.

In the model that we are using—corresponding to what is called the *equilibrium theory* of the tides—the water around the earth simply moves until it attains an equilibrium configuration that remains stationary from the viewpoint of an observer on the moon. Now we know that for a particle at the earth's center, the centrifugal force and the moon's gravitational attraction are equal and opposite. If, however, we consider a particle on the earth's surface at the nearest point to the moon [point A in Fig. 12-28(a)], the gravitational force on it is greater than the centrifugal force by an amount that we shall call f_0:

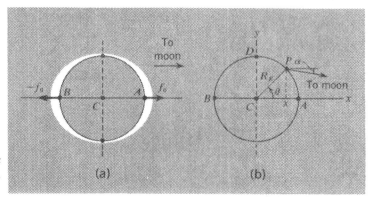

Fig. 12-28 (a) Difference between centrifugal force and the earth's gravity at the points nearest to and farthest from the moon. (b) Tide-producing force at an arbitrary point P, showing existence of a transverse component.

$$f_0 = \frac{GM_m m}{(r_m - R_E)^2} - \frac{GM_m m}{r_m^2}$$

Since $R_E \ll r_m$ ($R_E \approx r_m/60$), we can approximate this expression as follows:

$$f_0 = \frac{GM_m m}{r_m^2}\left[\left(1 - \frac{R_E}{r_m}\right)^{-2} - 1\right]$$

i.e.,

$$f_0 \approx \frac{2GM_m m}{r_m^3} R_E \tag{12-20}$$

By an exactly similar calculation, we find that the tide-producing force on a particle of mass m at the farthest point from the moon [point B in Fig. 12-28(a)] is equal to $-f_0$; hence we recognize the tendency for the water to be pulled or pushed away from a midplane drawn through the earth's center (see the figure).

By going just a little further we can get a much better insight into the problem. Consider now a particle of water at an arbitrary point P [Fig. 12-28(b)]. Relative to the earth's center, C, it has coordinates (x, y), with $x = R_E \cos\theta$, $y = R_E \sin\theta$. The tidal force on it in the x direction is given by a calculation just like those above:

$$f_x \approx \frac{2GM_m m}{r_m^3} x = \frac{2GM_m m}{r_m^3} R_E \cos\theta \tag{12-21}$$

This yields the results already obtained for the points A and B if we put $\theta = 0$ or π. In addition to this force parallel to the line joining the centers of the earth and the moon there is also, however, a transverse force, because the line from P to the moon's center makes a small angle, α, with the x axis, and the net gravitational force, $GM_m m/r^2$, has a small component perpendicular to x, given by

$$f_y = -\frac{GM_m m}{r^2} \sin\alpha \quad \text{(with } r \approx r_m\text{)}$$

Now we have

$$\tan\alpha = \frac{y}{r_m - x}$$

Since α is a very small angle [$\leq \tan^{-1}(R_E/r_m)$, which is about 1°] we can safely approximate the above expression:

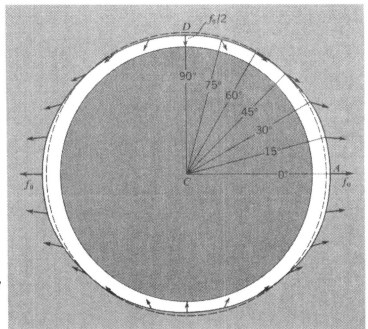

Fig. 12-29 Pattern of tide-producing forces around the earth. The circular dashed line shows where the undisturbed water surface would be.

$$\alpha \approx \frac{y}{r_m} = \frac{R_E \sin\theta}{r_m}$$

The component f_y of the tidal force is then given by

$$f_y \approx \frac{GM_m m}{r_m^3} y = -\frac{GM_m m}{r_m^3} R_E \sin\theta \qquad (12\text{-}22)$$

We see that this transverse force is greatest at $\theta = \pi/2$, at which point it is equal to half the maximum value (f_0) of f_x. Using Eqs. (12-21) and (12-22) together, we can develop an over-all picture of the tide-producing forces, as shown in Fig. 12-29. This shows much more convincingly how the forces act in such directions as to cause the water to flow and redistribute itself in the manner already qualitatively described.

TIDAL HEIGHTS; EFFECT OF THE SUN[1]

How high ought the equilibrium tidal bulge to be? If you are familiar with actual tidal variations you may be surprised at the

---

[1]This section goes well beyond the scope of the chapter as a whole but is added for the interest that it may have.

result. The equilibrium tide would be a rise and fall of less than 2 ft. We can calculate this by considering that the work done by the tidal force in moving a particle of water from D to A (Fig. 12–29) is equivalent to the increase of gravitational potential energy needed to raise the water through a height h against the earth's normal gravitational pull.[1] The distance h is the difference of water levels between A and D. Now, using Eqs. (12–21) and (12–22) we have

$$dW = f_x\, dx + f_y\, dy$$
$$= \frac{GM_m m}{r_m^3}(2x\, dx - y\, dy)$$
$$W_{DA} = \frac{GM_m m}{r_m^3}\left[\int_0^{R_E} 2x\, dx - \int_{R_E}^0 y\, dy\right]$$
$$= \frac{3GM_m m}{2r_m^3} R_E^2$$

Setting this amount of work equal to the gain of gravitational potential energy, mgh, we have

$$h = \frac{3GM_m R_E^2}{2gr_m^3} \tag{12-23}$$

The numerical values of the relevant quantities are as follows:

$$G = 6.67 \times 10^{-11} \text{ m}^3/\text{kg-sec}^2$$
$$M_m = 7.34 \times 10^{22} \text{ kg}$$
$$r_m = 3.84 \times 10^8 \text{ m}$$
$$R_E = 6.37 \times 10^6 \text{ m}$$
$$g = 9.80 \text{ m/sec}^2$$

Substituting these in Eq. (12–23) we find

$$h \approx 0.54 \text{ m} \approx 21 \text{ in.}$$

The great excess over this calculated value in many places (by factors of 10 or even more) can only be explained by considering the problem in detailed dynamical terms, in which the accumulation of water in narrow estuaries, and resonance effects, can completely alter the scale of the phenomenon. The value that we have calculated should be approximated in the open sea.

The last point that we shall consider here is the effect of the sun. Its mass and distance are as follows:

[1] Technically, this condition corresponds to the water surface being an energy equipotential.

$$M_s = 1.99 \times 10^{30} \text{ kg}$$
$$r_s = 1.49 \times 10^{11} \text{ m}$$

If we directly compare the gravitational forces exerted by the sun and the moon on a particle on the earth, we discover that the sun wins by a large factor:

$$\frac{F_s}{F_m} = \frac{M_s/r_s^2}{M_m/r_m^2} = \frac{M_s}{M_m}\left(\frac{r_m}{r_s}\right)^2 \approx 180$$

What matters, however, for tide production is the amount by which these forces *change* from point to point over the earth. This is expressed in terms of the *gradient* of the gravitational force:

$$F(r) = \frac{GMm}{r^2}$$

$$f = \Delta F = -\frac{2GMm}{r^3}\Delta r \qquad (12\text{–}24)$$

Putting $M = M_m$, $r = r_m$, and $\Delta r = \pm R_E$, we obtain the forces $\pm f_0$ corresponding to Eq. (12–20).

We now see that the comparative tide-producing forces due to the sun and the moon are given, according to Eq. (12–24), by the following ratio:

$$\frac{f_s}{f_m} = \frac{M_s/r_s^3}{M_m/r_m^3} = \frac{M_s}{M_m}\left(\frac{r_m}{r_s}\right)^3 \qquad (12\text{–}25)$$

Substituting the numerical values, one finds

$$\frac{f_s}{f_m} \approx 0.465$$

This means that the tide-raising ability of the moon exceeds that of the sun by a factor of about 2.15. The effects of the two combine linearly—and, of course, vectorially, depending on the relative angular positions of the moon and the sun. When they are on the same line through the earth (whether on the same side or on opposite sides) there should be a maximum tide equal to 1.465 times that due to the moon alone. This should happen once every 2 weeks, approximately, when the moon is new or full. At intermediate times (half-moon) when the angular positions of sun and moon are separated by 90°, the tidal amplitude should fall to a minimum value equal to 0.535 times that of the moon. The ratio of maximum to minimum values is thus about 2.7.

THE SEARCH FOR A FUNDAMENTAL INERTIAL FRAME

The phenomena that we have discussed in this chapter seem to leave us in no doubt that the acceleration of one's frame of reference can be detected by dynamical means. They suggest that a very special status does indeed attach to inertial frames. But how can we be sure that we have identified a true inertial frame in which Galileo's law of inertia holds exactly?

We saw at the very beginning of our discussion of dynamics that the earth itself represents a good approximation to such a frame for many purposes, especially for dynamical phenomena whose scale in distance and time is small. But we have now seen abundant evidence that a laboratory on the earth's surface is accelerated. If the laboratory is at latitude λ (see Fig. 12–30), each point in it is accelerating toward the earth's axis of rotation with an acceleration given by

$$a_\lambda = \omega^2 R \cos \lambda$$

with

$$\omega = 2\pi/86{,}400 \text{ sec}^{-1}$$
$$R = 6.4 \times 10^6 \text{ m}$$

This gives

$$a_\lambda = 3.4 \times 10^{-2} \cos \lambda \text{ m/sec}^2$$

This acceleration of a frame of reference tied to the earth is, as we know, not the simplest case of an accelerated frame. The linearly accelerated frames with which we began this chapter are much more readily analyzed. It was, however, the phenomena associated with rotating frames that led Newton to his belief in absolute space and in the absolute character of accelerations. Near the beginning of the *Principia* he describes a celebrated

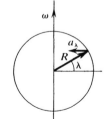

Fig. 12–30 Acceleration toward the earth's axis by virtue of its rotation.

Bucket rotating
water stationary

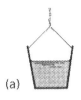

(a)

Bucket and
water rotating
together

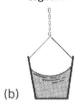

(b)

Bucket stationary
water rotating

(c)

Fig. 12–31 Main features of the experiment that Newton quoted as evidence of the absolute character of rotation and the associated acceleration.

experiment that he made with a bucket of water. It is an experiment that anyone may readily repeat for himself. The bucket is hung on a strongly twisted rope and is then released. There are three key observations, depicted in Fig. 12–31:

1. At first the bucket spins rapidly, but the water remains almost at rest, before the viscous forces have had time to set it rotating. The water surface is flat, just as it was before the bucket was released.

2. The water and the bucket are rotating together; the water surface has become concave (see Problem 12–18).

3. The bucket is suddenly stopped, but the body of water continues to rotate, and its surface remains curved.[1]

Clearly, said Newton, the relative motion of the bucket and the water is not the factor that determines the curvature of the water surface. It must be the absolute rotation of the water in space, and its attendant acceleration, that is at the bottom of the phenomenon. And with the help of $F = ma$, we can account for it quantitatively.

Newton's argument is a powerful one. He could point to further evidence in support of his views in the bulging of the earth itself by virtue of its rotation. The equatorial diameter of the earth is greater than the polar diameter by about 1 part in 300. It seems almost obvious, even without detailed calculation, that this is closely tied to the fact that a_λ/g is about $\frac{1}{300}$ at the equator and is zero at the poles (although the detailed calculation is, in fact, a bit messy).

Newton did not stop here, of course, He held the key of universal gravitation. Even a nonrotating earth would not be an inertial frame, because the whole earth is accelerating toward the sun.

For this system we have

$$\omega = 2\pi/(3.16 \times 10^7) \text{ sec}^{-1}$$
$$R = 1.49 \times 10^{11} \text{ m}$$
$$a_2 = \omega^2 R = 5.9 \times 10^{-3} \text{ m/sec}^2$$

[1] Newton does not suggest that he actually performed this third step, but it represents a natural completion of the experiment as one might perform it for oneself.

If we could conceive of an object that was immune to the gravitational attraction of the sun, it would not obey the law of inertia as observed from a reference frame attached to the earth. From Newton's standpoint the acceleration is real and absolute and is linked to the existence of a well-defined gravitational force provided by the sun.

That was about the end of the road as far as Newton was concerned. For him the system of the stars provided the arena in which the motions that he so brilliantly analyzed took place. A reference frame attached to these fixed stars could be taken to constitute a true inertial system, even though it might not coincide with the absolute space in which he believed.

Today, thanks to the work of astronomers, we know a good deal about the motions of some of those "fixed" stars. We have come to be aware of our involvement in a general rotation of our Galaxy. The sun would appear to be making a complete circuit of the Galaxy in about 2.5×10^8 years at a radial distance of about 2.5×10^4 light-years from the center. For this motion we would have

$$\omega \approx 2\pi/(8 \times 10^{15}) \text{ sec}^{-1}$$
$$R \approx 2.4 \times 10^{20} \text{ m}$$
$$a_3 \approx 10^{-10} \text{ m/sec}^2$$

It looks as though this acceleration can be reasonably accounted for by means of Newton's law of universal gravitation, if we regard the solar system as having a centripetal acceleration under the attraction of all the stars lying within its orbit. But no dynamical experiments that we do on earth require us to take into account this extremely minute effect—or, even, for most purposes, the revolution of the earth about the sun. (The rotation of the earth on its own axis is, however, an important consideration—and indeed an important aid in such matters as gyroscopic navigation.) Figure 12-32 schematizes the three rotating frames in which we find ourselves (we ignore here the acceleration caused by the moon).

But we still have not found an unaccelerated object to which we can attach our inertial frame of reference. In fact, we could extend this tantalizing search even further. There is some evidence that galaxies themselves tend to cluster together in groups containing a few galaxies to perhaps thousands. Our local group consists of about 10 galaxies. Although individual galaxies could have rather complex motions with respect to each other,

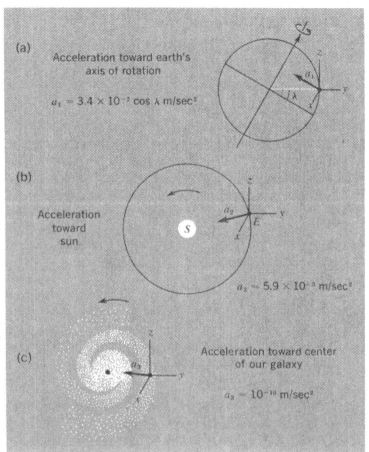

Fig. 12-32 Accelerations of any laboratory reference frame attached to the earth's surface.

this group is believed to have a more or less common motion through space.

So where are the "fixed" stars or other astronomical objects to which we can attach our inertial frame of reference? It appears that referring to the "fixed stars" is not a solution and contains an uncomfortable element of metaphysics (although we frequently use this phrase as a shorthand designation for the establishment of an inertial frame). This does not mean that the astronomical search for an inertial frame has been without value. For, at least up to the galactic level, it would seem that apparent departures from the law of inertia can be traced to identifiable accelerations of the reference frame in which motions are observed. However, the quest is incomplete, and so it seems likely to remain. Ultimately, therefore, we rely on an operational definition based upon local dynamical experiments and observation. We *define* an inertial frame to be one in which, experimentally, Galileo's law of inertia holds. The very existence of the inertial property remains, however, a deep and fascinating problem, and we shall end the chapter with a few remarks about this most fundamental feature of dynamics.

SPECULATIONS ON THE ORIGIN OF INERTIA

Not everyone accepted Newton's view that the phenomena associated with rotating objects demonstrated the absolute character of acceleration. The philosopher–bishop, George Berkeley, was perhaps the first person to argue[1] that all motions, including rotational ones, only have meaning as motions relative to other objects. The circling of two spheres around their center of mass could not, he said, be imagined in a space that was otherwise empty. Only when we introduce the background represented by the stars do we have a basis for recognizing the existence of such motion.

About 150 years later (in 1872) the German philosopher Ernst Mach presented the same idea in much more cogent form. He wrote:

> For me, only relative motions exist... and I can see, in this regard, no distinction between rotation and translation. Obviously it does not matter if we think of the earth as turning round on its axis, or at rest while the fixed stars revolve around it... But if we think of the earth at rest and the fixed stars revolving around it, there is no flattening of the earth, no Foucault's experiment and so on—at least according to our usual conception of the law of inertia. Now one can solve the difficulty in two ways. Either all motion is absolute, or our law of inertia is wrongly expressed... I prefer the second way. The law of inertia must be so conceived that exactly the same thing results from the second supposition as from the first. By this it will be evident that in its expression, regard must be paid to the masses of the universe.[2]

[1] In his tract *De Motu*, written in 1717, 30 years after the publication of Newton's *Principia*.
[2] E. Mach, *History and Root of the Principle of the Conservation of Energy*, (2nd ed.), Barth, Leipzig (1909). English translation of the 2nd edition by P. Jourdain, Open Court Publishing Co., London, 1911. Actually the first sentence of the quotation is taken from Mach's classic book, *The Science of Mechanics*, first published in 1883.

Thus was born the profound and novel idea—subsequently to become famous as *Mach's principle*—that the inertial property of any given object depends upon the presence and the distribution of other masses. Einstein himself accepted this idea and took it as a central principle of cosmology.

If one admits the validity of this point of view, then one sees that the whole basis of dynamics is involved. For consider the method that we described in Chapter 9 (p. 319) for finding the ratio of the inertial masses of two objects. This ratio is given as the negative inverse ratio of the accelerations that they produce by their mutual interaction:

$$\frac{m_1}{m_2} = -\frac{a_2}{a_1}$$

This looks very simple and straightforward, but it is clear that our ability to attach specific values to the individual accelerations, as distinct from the total *relative* acceleration, depends completely on our having identified a reference frame in which these accelerations can be measured. For this purpose the physical background provided by other objects is essential.

In looking critically at the phenomena of rotational motion, Mach attacked some intuitive notions that are much more deep-seated than any that we have in connection with straight-line motion. He considered the evidence provided by Newton's rotating-bucket experiment which we discussed in the last section. It is quite clear that the curvature of the water surface is related overwhelmingly to the existence of rotation relative to the vast amount of distant matter of the universe. When that relative rotation is stopped, the water surface becomes flat. When the bucket rotates and the water remains still (both relative to the fixed stars), the shape of the water surface remains unaffected. But, said Mach, that may be only a matter of degree. "No one," he wrote, "is competent to say how the experiment would turn out if the sides of the vessel increased in thickness and mass until they were ultimately several leagues thick." His own belief was that this rotation of a monster bucket would in fact generate the equivalent of centrifugal forces on the water inside it, even though this water had no rotational motion in the accepted sense.

This is a startling idea indeed. Let us present it in a slightly different context. We know that the act of giving an object an acceleration **a**, with respect to the inertial frame defined by the fixed stars, calls into play an inertial force, equal to $-m\mathbf{a}$, that expresses the resistance of the object to being accelerated. In

Mach's view we are equally entitled (indeed, compelled) to accept a description of the phenomenon in a frame always attached to the object itself. In this frame the rest of the universe has the acceleration $\mathbf{a}'$ ($= -\mathbf{a}$) and the inertial force $m\mathbf{a}'$ that the object experiences must be ascribable to the acceleration of the other masses.

This then brings us to the quantitative question: If a mass M, at distance r, is given the acceleration $\mathbf{a}$ relative to a given object, what contribution does it make to the total inertial force $m\mathbf{a}$ that the object experiences? Since we know that the force is proportional to m, we can argue on the grounds of symmetry and relativity that it must be proportional to M also. But at this point we enter a more speculative realm. A very suggestive analogy is provided by electromagnetic interactions. If two electric charges, q_1 and q_2, are separated by a distance r, we know that the static force exerted by q_1 on q_2 is given by

$$F_{12} = \frac{kq_1q_2}{r^2}$$

where k is a constant that depends on the particular choice of units. If, however, the charge q_1 is given the acceleration a there is an additional force that comes into play, directly proportional to a and inversely proportional to the distance:

$$F'_{12} = \frac{kq_1q_2a}{c^2r}$$

where c is the speed of light. Since this force falls off more slowly with distance than the static interaction, it can survive in appreciable magnitude at distances at which the static $1/r^2$ force has become negligible. This is, in fact, the basis of the electromagnetic radiation field by which signals can be transmitted over large distances.

Suppose now that we assume an analogous situation for gravitational interactions. The basic static law of force is known to be

$$F_{12} = \frac{GMm}{r^2}$$

The force on m associated with an acceleration of M would then be given by

$$F'_{12} = \frac{GMma}{c^2r} \tag{12-26}$$

On this basis we can estimate the relative magnitudes of the contributions from various masses of interest—the earth, the sun, our own Galaxy, and the rest of the universe. All we have to do is to calculate the values of M/r for these objects. The results are shown in Table 12-1, using numbers to the nearest power of 10 only. (The value of M for the universe as a whole is the somewhat speculative value quoted in Chapter 1.) We

TABLE 12-1: RELATIVE CONTRIBUTIONS TO INERTIA

Source	M, kg	r, m	M/r, kg/m	M/r (relative)
Earth	10^{25}	10^7	10^{18}	10^{-8}
Sun	10^{30}	10^{11}	10^{19}	10^{-7}
Our Galaxy	10^{41}	10^{21}	10^{20}	10^{-6}
Universe	10^{52}	10^{26}	10^{26}	1

see that, according to this theory, the effect of a nearby object, even one as massive as the earth itself, would be negligible compared to the effect of the universe at large.

The *total* inertial force called into existence if everything in the universe acquires an acceleration a with respect to a given object would be obtained by summing the forces F'_{12} of Eq. (12-26) over all masses other than m itself:

$$F_{\text{inertial}} = ma \sum \frac{GM}{c^2 r}$$

This, however, should be identical with what we know to be the magnitude of the inertial force as directly given by the value of ma. Thus the theory would require the following identity to hold:

$$\sum_{\text{universe}} \frac{GM}{c^2 r} = 1 \qquad (12\text{-}27)$$

It is clear from Table 12-1 that even such a large local mass as our Galaxy represents only a minor contribution; what we are involved with is a summation over the approximately uniform distribution of matter represented by the universe as a whole. If we regard it as a sphere, centered on ourselves, of mean density ρ and radius R_U ($\approx 10^{10}$ light-years $= 10^{26}$ m), we would have

$$\sum_{\text{universe}} \frac{M}{r} \to \int_0^{R_U} \frac{4\pi \rho r^2 \, dr}{r} = 2\pi \rho R_U^2$$

The total mass is however given by

$$M_U = \frac{4\pi}{3} \rho R_U^3$$

Thus we have, on this simple picture (based on Euclidean geometry)

$$\sum_{\text{universe}} \frac{M}{r} = \frac{3}{2} \frac{M_U}{R_U} \approx 10^{26} \text{ kg/m}$$

Using the values $G \approx 10^{-10}$ N-m^2/kg^2 and $c^2 \approx 10^{17}$ m^2/sec^2, we would then have

$$\sum_{\text{universe}} \frac{GM}{c^2 r} \approx 10^{-1}$$

Taking into account the uncertainties in our knowledge of the distribution of matter throughout space, many would say that the factor of about 10 that separates the above empirical value from the theoretical value (unity) called for by Eq. (12-27) is not significant. The result is intriguing, to say the least, and many cosmologists have accepted as fundamentally correct this development from the primary ideas espoused by Mach and Einstein.[1]

PROBLEMS

12-1 A single-engine airplane flies horizontally at a constant speed v. In the frame of the aircraft, each tip of the propeller sweeps out a circle of radius R at the rate of n revolutions per second. Obtain an equation for the path of a tip of the propeller as viewed from the earth.

12-2 A person observes the position of a post from the origin of a reference frame (S') rigidly attached to the rim of a merry-go-round, as shown in the figure. The merry-go-round (of radius R) is rotating with angular velocity ω, the distance of the post from the axis of the merry-go-round is D, and at $t = 0$, the coordinates of P in S' are $x' = D - R$, $y' = 0$ (equivalently, $r' = D - R$, $\theta' = 0$).

(a) Find the coordinates $r'(t)$, $\theta'(t)$ of the post; also give the corresponding $x'(t)$ and $y'(t)$.

---

[1] For further reading on this fascinating topic, see, for example, R. H. Dicke, "The Many Faces of Mach," in *Gravitation and Relativity* (ed. H.-Y. Chin and W. F. Hoffmann, eds.), W. A. Benjamin, New York, 1964; N. R. Hanson, "Newton's First Law," and P. Morrison, "The Physics of the Large," both in *Beyond the Edge of Certainty* (R. G. Colodny, ed.), Prentice-Hall, Englewood Cliffs, N.J., 1965; D. W. Sciama, *The Unity of the Universe*, Doubleday, New York, 1961, and *The Physical Foundations of General Relativity*, Doubleday, New York, 1969.

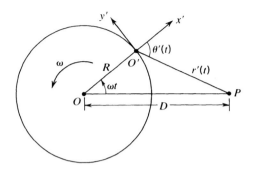

(b) By differentiating the results of (a), obtain the velocity and acceleration of the post in both Cartesian and polar coordinates.

(c) Make a plot of the path of the post in S'.

12-3 A boy is riding on a railroad flatcar, on level ground, that has an acceleration a in the direction of its motion. At what angle with the vertical should he toss a ball so that he can catch it without shifting his position on the car?

12-4 A railroad train traveling on a straight track at a speed of 20 m/sec begins to slow down uniformly as it enters a station and comes to a stop in 100 m. A suitcase of mass 10 kg having a coefficient of sliding friction $\mu = 0.15$ with the train's floor slides down the aisle during this deceleration period.

(a) What is the acceleration of the suitcase (with respect to the ground) during this time?

(b) What is the velocity of the suitcase just as the train comes to a halt?

(c) The suitcase continues sliding for a period after the train has stopped. When it comes to rest, how far is it displaced from its original position on the floor of the train?

12-5 A man weighs himself on a spring balance calibrated in newtons which indicates his weight as $mg = 700$ N. What will he read if he repeats the observation while riding an elevator from the first to the twelfth floors in the following manner?

(a) Between the first and third floors the elevator accelerates at the rate of 2 m/sec².

(b) Between the third and tenth floors the elevator travels with the constant velocity of 7 m/sec.

(c) Between the tenth and twelfth floors the elevator decelerates at the rate of 2 m/sec².

(d) He then makes a similar trip down again.

(e) If on another trip the balance reads 500 N, what can you say of his motion? Which way is he moving?

12-6 If the coefficient of friction between a box and the bed of a truck is μ, what is the maximum acceleration with which the truck can climb a hill, making an angle θ with the horizontal, without the box's slipping on the truck bed?

12-7 A block of mass 2 kg rests on a frictionless platform. It is attached to a horizontal spring of spring constant 8 N/m, as shown in the figure. Initially the whole system is stationary, but at $t = 0$ the platform begins to move to the right with a constant acceleration of 2 m/sec.^{2} As a result the block begins to oscillate horizontally relative to the platform.

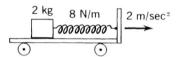

(a) What is the amplitude of the oscillation?
(b) At $t = 2\pi/3$ sec, by what amount is the spring longer than it was in its initial unstretched condition?

12-8 A plane surface inclined 37° ($\sin^{-1}\frac{3}{5}$) from the horizontal is accelerated horizontally to the left (see the figure). The magnitude of the acceleration is gradually increased until a block of mass m, originally at rest with respect to the plane, just starts to slip up the plane. The static friction force at the block-plane surface is characterized by $\mu = \frac{4}{5}$.

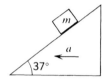

(a) Draw a diagram showing the forces acting on the block, just before it slips, in an inertial frame fixed to the floor.
(b) Find the acceleration at which the block begins to slip.
(c) Repeat part (a) in the noninertial frame moving along with the block.

12-9 A nervous passenger in an airplane at takeoff removes his tie and lets it hang loosely from his fingers. He observes that during the takeoff run, which lasts 30 sec, the tie makes an angle of 15° with the vertical. What is the speed of the plane at takeoff, and how much runway is needed? Assume that the runway is level.

12-10 A uniform steel rod (density = 7500 kg/m^3, ultimate tensile strength 5×10^8 N/m^2) of length 1 m is accelerated along the direction of its length by a constant force applied to one end and directed away from the center of mass of the rod. What is the maximum allowable acceleration if the rod is not to break? If this acceleration is exceeded, where will the rod break?

12-11 (a) A train slowed with deceleration a. What angle would the liquid level of a bowl of soup in the dining car have made with the horizontal? A child dropped an apple from a height h and a distance d from the front wall of the dining car. What path did the apple take as observed by the child? Under what conditions would the apple have hit the ground? The front wall?

(b) As a reward for making the above observations, the parents bought the child a helium-filled balloon at the next stop. For fun, they asked him what would happen to the balloon if the train left the station with acceleration a'. Subsequently, they were surprised to find his predictions correct. What did the precocious child answer?

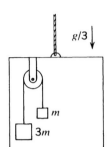

12-12 An elevator has a *downward* acceleration equal to $g/3$. Inside the elevator is mounted a pulley, of negligible friction and inertia, over which passes a string carrying two objects, of masses m and $3m$, respectively (see the figure).

(a) Calculate the acceleration of the object of mass $3m$ *relative to the elevator*.

(b) Calculate the force exerted on the pulley by the rod that joins it to the roof of the elevator.

(c) How could an observer, completely isolated inside the elevator, explain the acceleration of m in terms of forces that he himself could measure with the help of a spring balance?

12-13 In each of the following cases, find the equilibrium position as well as the period of small oscillations of a pendulum of length L:
(1) In a train moving with acceleration a on level tracks.
(2) In a train free-wheeling on tracks making an angle θ with the horizontal.
(3) In an elevator falling with acceleration a.

12-14 The world record for the 16-lb hammer throw is about 70 m. Assuming that the hammer is whirled around in a circle of radius about 2 m before being let fly, estimate the magnitude of the pull that the thrower must be able to withstand.

12-15 (a) A man rides in an elevator with vertical acceleration a. He swings a bucket of water in a vertical circle of radius R. With what angular velocity must he swing the bucket so that no water spills?

(b) With what angular frequency must the bucket be swung if the man is on a train with horizontal acceleration a? (The plane of the circle is again vertical and contains the direction of the train's acceleration.)

12-16 Consider a thin rod of material of density ρ rotating with constant angular velocity ω about an axis perpendicular to its length.

(a) Show that if the rod is to have a constant stress S (tensile force per unit area of cross section) along its length, the cross-sectional area must decrease exponentially with the square of the distance from the axis:

$$A = A_0 e^{-kr^2} \quad \text{where } k = \rho\omega^2/2S$$

[Consider a small segment of the rod between r and $r + \Delta r$, having a mass $\Delta m = \rho A(r)\, \Delta r$, and notice that the difference in tensions at its ends is $\Delta T = \Delta(SA)$.]

(b) What is the maximum angular velocity ω_{max} in terms of ρ, S_{max}, and k?

(c) The ultimate tensile strength of steel is about 10^9 N/m^2. Estimate the maximum possible number of rpm of a steel rotor for which the "taper constant" $k = 100$ m^{-2} ($\rho = 7500$ kg/m^3).

12-17 A spherically shaped influenza virus particle, of mass 6×10^{-16} g and diameter 10^{-5} cm, is in a water suspension in an ultracentrifuge. It is 4 cm from the vertical axis of rotation, and the speed of rotation is 10^3 rps. The density of the virus particle is 1.1 times that of water.

(a) From the standpoint of a reference frame rotating with the centrifuge, what is the effective value of "g"?

(b) Again from the standpoint of the rotating reference frame, what is the net centrifugal force acting on the virus particle?

(c) Because of this centrifugal force, the particle moves radially outward at a small speed v. The motion is resisted by a viscous force given by $F_{res} = 3\pi\eta v d$, where d is the diameter of the particle and η is the viscosity of water, equal to 10^{-2} cgs units (g/cm/sec). What is v?

(d) Describe the situation from the standpoint of an inertial frame attached to the laboratory.

[In (b) and (c), account must be taken of buoyancy effects. Think of the ordinary hydrostatics problem of a body completely immersed in a fluid of different density.]

12-18 (a) Show that the equilibrium form of the surface of a rotating body of liquid is parabolic (or, strictly, a paraboloid of revolution). This problem is most simply considered from the standpoint of the rotating frame, given that a liquid cannot withstand forces tangential to its surface and will tend toward a configuration in which such forces disappear. It is instructive to consider the situation from the standpoint of an inertial frame also.

(b) It has been proposed that a parabolic mirror for an astronomical telescope might be formed from a rotating pool of mercury. What rate of rotation (rpm) would make a mirror of focal length 20 m?

12-19 To a first approximation, an object released anywhere within an orbiting spacecraft will remain in the same place relative to the spacecraft. More accurately, however, it experiences a net force proportional to its distance from the center of mass of the spacecraft. This force, as measured in the noninertial frame of the craft, arises from the small variations in both the gravitational force and the centrifugal force due to the change of distance from the earth's center. Obtain an expression for this force as a function of the mass, m, of the object, its distance ΔR from the center of the spacecraft, the radius R of the spacecraft's orbit around the earth, and the gravitational acceleration g_R at the distance R from the earth's center.

12–20 A circular platform of radius 5 m rotates with an angular velocity $\omega = 0.2$ rad/sec. A man of mass 100 kg walks with constant velocity $v' = 1$ m/sec along a diameter of the platform. At time $t = 0$ he crosses the center and at time $t = 5$ sec he jumps off the edge of the platform.

(a) Draw a graph of the centrifugal force felt by the man as a function of time in the interval $t = 0$ to $t = 5$ sec.

(b) Draw a similar graph of the Coriolis force. For both diagrams, give the correct vertical scale (in newtons).

(c) Show on a sketch the direction of these forces, assuming the platform to rotate in a clockwise direction as seen from above.

12–21 On a long-playing record (33 rpm, 12 in.) an insect starts to crawl toward the rim. Assume that the coefficient of friction between its legs and the record is 0.1. Does it reach the edge by crawling or otherwise?

12–22 A child sits on the ground near a rotating merry-go-round. With respect to a reference frame attached to the earth the child has no acceleration (accept this as being approximately true) and experiences no force. With respect to polar coordinates fixed to the merry-go-round, with origin at its center:

(a) What is the motion of the child?

(b) What is his acceleration?

(c) Account for this acceleration, as measured in the rotating frame, in terms of the centrifugal and Coriolis forces judged to be acting on the child.

12–23 The text (p. 516) derives the Coriolis force in the transverse (θ) direction by considering the motion of an object along a radial line in the rotating frame. Correspondingly, if one considers an object that is moving *transversely* in the rotating frame, one can obtain the net *radial* force due to Coriolis and centrifugal effects. Consider a particle on a frictionless turntable rotating with angular velocity ω. The particle is initially at rest relative to the turntable, at a distance r from the axis of rotation.

(a) Set up a fixed coordinate system S with axes x transversely and y radially and with its origin O at the position of the particle at $t = 0$ (see the figure). Set up another coordinate system S', with origin O' and axes x' and y', which rotates with the turntable and which coincides with S at $t = 0$. Show that at a later time t the coordinates of a given point as measured in S' and S are related by the following equations, where $\theta = \omega t$:

$$x' = x \cos\theta + y \sin\theta + r \sin\theta$$
$$y' = y \cos\theta - x \sin\theta - r(1 - \cos\theta)$$

(b) Suppose that, at $t = 0$, the particle is given a velocity v'

relative to O' in the x' direction. Its subsequent motion will be along the x direction at the constant velocity $v' - \omega r$ relative to O. Use this to obtain its coordinates x' and y' at a later time t.

(c) Making the approximations for the case $\omega t \ll 1$, show that for small values of t one can put $y' \simeq \frac{1}{2} a'_r t^2$, where $a'_r = \omega^2 r - 2\omega v'$. This corresponds to the required combination of centrifugal and Coriolis accelerations.

(d) If you are feeling ambitious, apply the same kind of analysis for an initial velocity in an arbitrary direction.

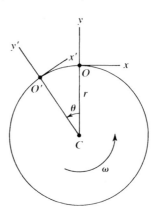

12-24 In an article entitled "Do Objects fall South?" [*Phys. Rev.*, **16**, 246 (1903)], Edwin Hall reported the results of nearly 1000 trials in which he allowed an object to fall through a vertical distance of 23 m at Cambridge, Mass. (lat. 42° N). He found, on the average, an eastward deflection of 0.149 cm and a southerly deflection of 0.0045 cm.

(a) Compare the easterly deflection with what would be expected from Eq. (12–17).

(b) Consider the fact that the development of an eastward component of motion relative to the earth would indeed lead in turn to a southerly component of Coriolis force. Without attempting any detailed analysis, estimate the order of magnitude of the ratio of the resulting southerly deflection to the predominant easterly deflection. Do you think that an explanation of Hall's results on southerly deflection can be achieved in these terms?

12-25 Calculate the Coriolis acceleration of a satellite in a circular polar orbit as observed by someone on the rotating earth. Obtain the direction of this acceleration throughout the orbit, thereby explaining why the satellite always passes through the poles even though it is subjected to the Coriolis force. Is there a similar force on a satellite in an equatorial orbit?

12-26 Imagine that a frictionless horizontal table, circular in shape and of radius R, is fitted with a perfectly elastic rim, and that a dry-ice puck is launched from a point on the rim toward the center. The puck bounces back and forth across the table at constant speed v, but because of the Coriolis force it does not quite follow a straight-line path along a diameter. Consider the rate at which the path of the puck gradually turns with respect to the table, and compare the result with that for a Foucault pendulum at the same latitude, λ.

12-27 In the text (p. 536) the height of the equilibrium tide is calculated by considering the work done by the tide-producing force in carrying a particle of water from point D to point A (see the figure).

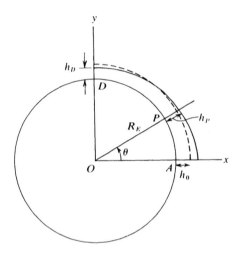

By considering the work from D to an intermediate point P, one can obtain a general expression for the elevation or depression $h(\theta)$ of the water at an arbitrary point, relative to what the water level would be in the absence of the tide-producing force. The calculation involves two parts, as follows:

(a) Evaluate the work integral of the tide-producing force from $D(x = 0, y = R_E)$ to $P(x = R_E \cos \theta, y = R_E \sin \theta)$ for a particle of water of mass m. Equating this to the difference of gravitational potential energies, $mg(h_P - h_D)$, one gets an expression for the difference $h_P - h_D$.

(b) The total volume of water is a constant. Hence, if h_0 represents the water depth in the absence of the tide-producing force, we must have

$$\int_0^{\pi/2} 2\pi R_E^2 [h(\theta) - h_0] \sin \theta \, d\theta = 0$$

Putting the results of (a) and (b) together, you should be able to verify that the deviation of the water level from its undisturbed state is proportional to $3 \cos^2 \theta - 1$.

What makes planets go around the sun? At the time of Kepler some people answered this problem by saying that there were angels behind them beating their wings and pushing the planets around an orbit. As you will see, the answer is not very far from the truth. The only difference is that the angels sit in a different direction and their wings push inwards.

R. P. FEYNMAN, *The Character of Physical Law* (1965)

13
Motion under central forces

WE HAVE ALREADY seen, especially in Chapter 8, how the motion of objects under the action of forces directed toward some well-defined center is one of the richest areas of study in mechanics. Twice in the history of physics the analysis of such motions has been linked with fundamental advances in our understanding of nature—through the explanation of planetary motions, on the macroscopic scale, and through Rutherford's studies of alpha-particle scattering, which gave man his first clear view of the subatomic world. Up until this point we have limited ourselves to the study of circular orbits, and it is remarkable how much can be learned on that basis. But now we shall begin a more general analysis of motion under the action of central forces.

BASIC FEATURES OF THE PROBLEM

As we saw in Chapter 11 (p. 444), a central force field that is also conservative must be spherically symmetric, and some of the most important fields in nature (notably electrical and gravitational) are precisely of this type. The frequent occurrence of spherically symmetric models to describe physical reality is closely linked to the basic assumption that space is isotropic and is the intuitively natural starting point in building theoretical models of various kinds of dynamical systems.

We shall begin with the specific problem of the motion of a single particle of mass m in a spherically symmetric central field

Fig. 13-1 (a) Unit vectors associated with radial and transverse directions in a plane polar coordinate system. (b) Radial and transverse components of an elementary vector displacement Δr.

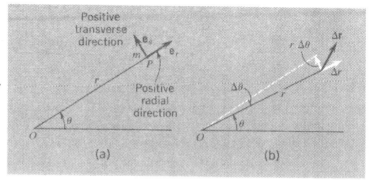

of force. Initially, at least, we shall assume that the object responsible for this central field is so massive that it can be regarded as a fixed center that defines a convenient origin of coordinates for the analysis of the motion.

The first thing to notice is that the path of the moving particle will lie in a fixed plane that passes through the center of force. This plane is defined by the initial velocity vector $\mathbf{v}_0$ of the particle and the initial vector position $\mathbf{r}_0$ of the particle with respect to the center of force. Since the force acting on the particle is in this plane, and since there is no component of initial velocity perpendicular to it, the motion must remain confined to this plane of $\mathbf{r}_0$ and $\mathbf{v}_0$. To analyze the motion we must first pick an appropriate coordinate system. Because the force $\mathbf{F}$ is a function of the scalar distance r only and is along the line of the vector $\mathbf{r}$ (positively or negatively), it is clearly most convenient to work with the plane polar coordinates (r, θ), as indicated in Fig. 13-1(a). This means that we shall be making use of the acceleration vector expressed in these coordinates. In Chapter 3 (p. 108) we calculated this vector for the particular case of circular motion (r = constant). Now we shall develop the more general expression that embraces changes of both r and θ.

Using the unit vectors $\mathbf{e}_r$ and $\mathbf{e}_\theta$ as indicated in Fig. 13-1(a) we have

$$\mathbf{r} = r\mathbf{e}_r \tag{13-1}$$

$$\mathbf{v} = \frac{d\mathbf{r}}{dt} = \frac{dr}{dt}\mathbf{e}_r + r\frac{d\theta}{dt}\mathbf{e}_\theta \tag{13-2}$$

This equation for $\mathbf{v}$ is readily constructed by recognizing that a general infinitesimal change in position, $\Delta\mathbf{r}$, is obtained by combining a radial displacement of length Δr (at constant θ) and a transverse displacement of length $r\,\Delta\theta$ (at constant r), as indicated in Fig. 13–1(b). Alternatively, one can just differentiate both sides of Eq. (13–1) with respect to time, remembering that $d(\mathbf{e}_r)/dt = (d\theta/dt)\mathbf{e}_\theta$ [see Eq. (3–17a)].

We now proceed to differentiate both sides of Eq. (13–2) with respect to t:

$$\mathbf{a} = \frac{d\mathbf{v}}{dt} = \frac{d^2r}{dt^2}\mathbf{e}_r + \frac{dr}{dt}\frac{d}{dt}\mathbf{e}_r + \frac{dr}{dt}\frac{d\theta}{dt}\mathbf{e}_\theta + r\frac{d^2\theta}{dt^2}\mathbf{e}_\theta$$
$$+ r\frac{d\theta}{dt}\frac{d}{dt}\mathbf{e}_\theta$$

Substituting $d(\mathbf{e}_r)/dt = (d\theta/dt)\mathbf{e}_\theta$, and $d(\mathbf{e}_\theta)/dt = -(d\theta/dt)\mathbf{e}_r$, the expression for $\mathbf{a}$ can be rewritten as follows:

$$\mathbf{a} = \left[\frac{d^2r}{dt^2} - r\left(\frac{d\theta}{dt}\right)^2\right]\mathbf{e}_r + \left[r\frac{d^2\theta}{dt^2} + 2\frac{dr}{dt}\frac{d\theta}{dt}\right]\mathbf{e}_\theta \qquad (13\text{--}3)$$

It will be convenient to extract from this the separate radial and transverse components of the total acceleration:

$$a_r = \frac{d^2r}{dt^2} - r\left(\frac{d\theta}{dt}\right)^2 \qquad (13\text{--}4)$$

$$a_\theta = r\frac{d^2\theta}{dt^2} + 2\frac{dr}{dt}\frac{d\theta}{dt} \qquad (13\text{--}5)$$

The statement of Newton's law in plane polar coordinates can then be made in terms of these separate acceleration components:

$$F_r = m\left[\frac{d^2r}{dt^2} - r\left(\frac{d\theta}{dt}\right)^2\right] \qquad (13\text{--}6)$$

$$F_\theta = m\left[r\frac{d^2\theta}{dt^2} + 2\frac{dr}{dt}\frac{d\theta}{dt}\right] \qquad (13\text{--}7)$$

The above two equations provide a basis for the solution of any problem of motion in a plane, referred to an origin of polar coordinates. We shall, however, consider their application to central forces in particular.

THE LAW OF EQUAL AREAS

In the case of any kind of conservative central force, we have $F_r = F(r)$ simply, and $F_\theta = 0$. The second of these immediately implies that $a_\theta = 0$. Substituting the specific expression for a_θ

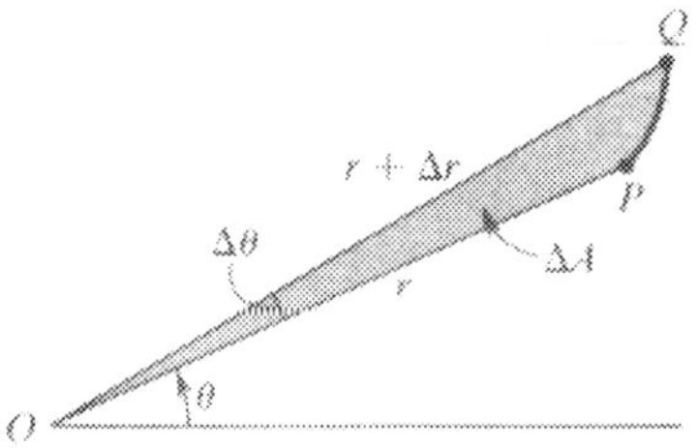

Fig. 13-2 *Illustrating the basis of calculating areal velocity (area swept out per unit time by the radius vector).*

from Eq. (13–5), we have

$$r\frac{d^2\theta}{dt^2} + 2\frac{dr}{dt}\frac{d\theta}{dt} = 0 \qquad (13\text{–}8)$$

This equation contains a somewhat veiled statement of a simple geometrical result—that the vector **r** sweeps out area at a constant rate. One way of seeing this is to multiply Eq. (13–8) throughout by r:

$$r^2\frac{d^2\theta}{dt^2} + 2r\frac{dr}{dt}\frac{d\theta}{dt} = 0$$

The left-hand side may then be recognized as the derivative with respect to t of the product $r^2\, d\theta/dt$:

$$\frac{d}{dt}\left(r^2\frac{d\theta}{dt}\right) = r^2\frac{d^2\theta}{dt^2} + 2r\frac{dr}{dt}\frac{d\theta}{dt}$$

If we integrate this expression, we therefore have

$$r^2\frac{d\theta}{dt} = \text{const.} \qquad (13\text{–}9)$$

Now in Fig. 13–2 we show the area ΔA (shaded) swept out by **r** in a short time Δt. It is the triangle POQ (we take PQ to be indistinguishable from a straight line if it is short enough) and we have

$$\Delta A = \tfrac{1}{2}r(r + \Delta r)\sin\Delta\theta$$

The rate at which area is being swept out, instantaneously, is the limit of $\Delta A/\Delta t$ for $\Delta t \to 0$. Since, as we approach this limit, $\Delta r/r \to 0$ and $\sin\Delta\theta \to \Delta\theta$, we arrive at the result

$$\frac{dA}{dt} = \frac{1}{2}r^2\frac{d\theta}{dt} \qquad (13\text{–}10\text{a})$$

Thus we recognize the constant on the right-hand side of Eq.

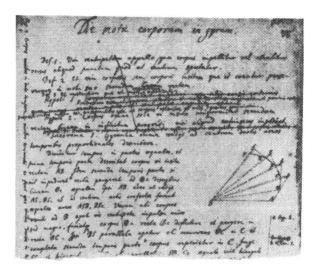

Fig. 13-3 (a) A portion of Newton's manuscript, De Motu, *showing the basis of his proof of the law of equal areas for a central force. (b) Enlarged copy of Newton's diagram. (From J. Herivel,* The Background to Newton's Principia, *Oxford University Press, London, 1965.)*

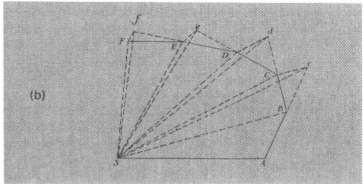

(13-9) as twice the rate (a *constant* rate) at which the radius vector **r** sweeps out area, and we therefore have

$$\text{(Any central force)} \qquad \frac{dA}{dt} = \frac{1}{2} r^2 \frac{d\theta}{dt} = \text{const.} \qquad (13\text{-}10b)$$

The result expressed by Eq. (13-10b) was first discovered by Kepler in his analysis of planetary motions (of which more later). It was stated by him in what is known as his second law (although it was actually the first chronologically). Newton understood it on the same dynamical grounds as we have discussed above, i.e., as a feature of the motion of an object acted on by any kind of force that is directed always to the same point. He visualized the action of such a force as a succession of small kicks or impulses, which in the limit would go over into a continuously applied influence. He set out this view of the process in a tract written in 1684 (about 2 years before the *Principia*).[1] Figure 13-3(a) is a reproduction of a small fragment of the work, indicating Newton's approach to the problem. With the help of an enlarged version of his sketch [Figure 13-3(b)] we can more readily follow Newton's argument, which as usual was geometrical.

Newton imagines an object traveling along AB and then receiving an impulse directed toward the point S. As a result it now travels along the line BC instead of Bc. Similar impulses carry it to D, E, and F. To make things quantitative, Newton visualizes the displacement BC as being, in effect, the combination of the displacement Bc, equal to AB, that the object would have undergone if it had continued for an equal length of time with its original velocity, together with the displacement cC parallel to the line BS along which the impulse was applied. This at once yields the law of areas by a simple argument: The triangles SAB and SBc are equal, having equal bases (AB and Bc) and the same altitude. The triangles SBc and SBC are equal, having a common base (SB) and lying between the same parallels. Hence $\triangle SAB = \triangle SBC$.

THE CONSERVATION OF ANGULAR MOMENTUM

We give a more modern and more fundamental slant to the law of areas by expressing it in terms of the conservation of *orbital angular momentum*. If a particle at P [Fig. 13-4(a)] is acted on by a force $\mathbf{F}$, we have

$$\mathbf{F} = m\mathbf{a} = m\frac{d\mathbf{v}}{dt}$$

Let us now form the vector (cross) product of the position vector $\mathbf{r}$ with both sides of this equation:

$$\mathbf{r} \times \mathbf{F} = \mathbf{r} \times m\frac{d\mathbf{v}}{dt} \tag{13-11}$$

The left-hand side is the torque $\mathbf{M}$ due to $\mathbf{F}$ about O.

[1]This tract, called *De Motu* (Concerning the Motion of Bodies), contains many of the important results that were later incorporated in the *Principia*.

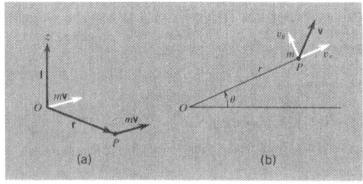

Fig. 13-4 (a) *Vector relationship of position, linear momentum, and orbital momentum.* (b) *Analysis of the velocity into radial and transverse components.*

We now introduce the *moment of the momentum*, **l**, of the particle with respect to O:

$$\mathbf{l} = \mathbf{r} \times m\mathbf{v} = \mathbf{r} \times \mathbf{p} \tag{13-12}$$

Thus **l** bears the same relation to the linear momentum, **p**, as the moment of the force, **M**, does to **F**. It is a vector, as shown in Fig. 13-4(a), perpendicular to the plane defined by **r** and **v**. If we calculate the rate of change of **l**—i.e., the derivative of **l** with respect to t—we have

$$\frac{d\mathbf{l}}{dt} = \frac{d\mathbf{r}}{dt} \times m\mathbf{v} + \mathbf{r} \times m\frac{d\mathbf{v}}{dt}$$

$$= \mathbf{v} \times m\mathbf{v} + \mathbf{r} \times m\frac{d\mathbf{v}}{dt}$$

However, the value of $\mathbf{v} \times m\mathbf{v}$ is zero, because it is the cross product of two parallel vectors. The second term on the right of the above equation is, by Eq. (13-11), equal to the torque or moment of the applied force about O. Thus we have the simple result

$$\mathbf{M} = \mathbf{r} \times \mathbf{F} = \frac{d\mathbf{l}}{dt} \tag{13-13}$$

This, then, relates the moment of **F** to the rate of change of the moment of momentum, or orbital angular momentum, of the particle about the origin O. If, now, we take **F** to be a central force directed radially toward or away from O, the value of **M** is zero. In this case, therefore, we have

$$\frac{d\mathbf{l}}{dt} = 0$$

and

$$\mathbf{l} = \mathbf{r} \times \mathbf{p} = \text{const.} \quad (13\text{-}14)$$

This, then, is the statement of the conservation of orbital angular momentum for motion under a central force—or, of course, under zero force.

If we look at the situation in the plane of the motion [Fig. 13-4(b)] the vector $\mathbf{l}$ points upward from the page, and its scalar magnitude l is given by

$$l = rmv_\theta = mr^2 \frac{d\theta}{dt} \quad (13\text{-}15a)$$

This means that the constant rate of sweeping out area, as expressed by Eqs. (13-10a) and (13-10b), is given quantitatively by

$$\frac{dA}{dt} = \frac{1}{2} r^2 \frac{d\theta}{dt} = \frac{l}{2m} \quad (13\text{-}15b)$$

The result expressed by Eqs. (13-14) and (13-15) is a valuable one. As we have pointed out before, if we find quantities that remain constant throughout some process—i.e., they are "conserved"—these become powerful tools in the analysis of phenomena. Angular momentum is a conserved quantity of this sort which is particularly valuable in the analysis of central field problems. It should be noted that the conservation of angular momentum depends only on the absence of a torque and is independent of the conservation or nonconservation of mechanical energy. A nice example of this last feature is the speeding up of an object that whirls around at the end of a gradually shortening string (Fig. 13-5). If the decrease of r in one revolution is small compared to r itself, the velocity $\mathbf{v}$ is almost perpendicular to $\mathbf{r}$ at each instant, and the angular momentum (rmv_θ) is very nearly equal to mvr. The tension $\mathbf{T}$ in the string that pulls the object inward exerts no torque about O, and so we have $l = $ const. In a change of r from r_1 to r_2 we thus have

$$mv_1 r_1 = mv_2 r_2$$

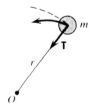

Fig. 13-5 Small portion of the path of an object that is being whirled around O at the end of a string the length of which is being steadily shortened.

or

$$v_2 = \frac{r_1}{r_2} v_1$$

Thus, if r_2 is less than r_1, there is a gain of kinetic energy given by

$$K_2 - K_1 = \tfrac{1}{2} m v_1^2 \left(\frac{r_1^2}{r_2^2} - 1 \right)$$

The work equivalent to this gain of kinetic energy must be provided by the tension in the string. In an equilibrium orbit we would have

$$T = \frac{mv^2}{r}$$

Substituting $v = l/mr$, this becomes

$$T = \frac{l^2}{mr^3} = \frac{mv_1^2 r_1^2}{r^3}$$

The work done by T in a change from r to $r + dr$ is $-T\,dr$ (since **T** acts radially inward). Thus the total work between r_1 and r_2 is given by

$$W = -mv_1^2 r_1^2 \int_{r_1}^{r_2} \frac{dr}{r^3} = \tfrac{1}{2} m v_1^2 r_1^2 \left(\frac{1}{r_2^2} - \frac{1}{r_1^2} \right)$$

It may be seen that this equals the value of $K_2 - K_1$ already calculated. The problem has a good deal in common with that of the earth satellite spiraling inward, although in that case conservation of orbital angular momentum does not apply, because the air resistance represents a transverse force providing a negative torque (see discussion on pp. 470–473).

ENERGY CONSERVATION IN CENTRAL FORCE MOTIONS

If we are dealing with a conservative central force (which was not the case in the example analyzed above) we can write a statement of the conservation of total mechanical energy in the following form:

$$\frac{m}{2}[v_r^2 + v_\theta^2] + U(r) = E \tag{13-16}$$

In addition, we have the law of conservation of angular momentum as given by Eq. (13-15a):

$$mrv_\theta = l$$

The quantities E and l are thus "constants of the motion." $U(r)$ is the potential energy of the particle m in the central field. The orbital-momentum equation allows us to reduce Eq. (13–16) to the *form* of a problem of a particle moving in one dimension only under the action of a conservative force. This is the key to the method of handling central field problems. From Eq. (13–15a) we take the value of v_θ and use this value in Eq. (13–16). There follows

$$\frac{mv_r^2}{2} + \frac{l^2}{2mr^2} + U(r) = E \qquad (13\text{-}17)$$

which describes the radial part of the motion. This is of the form

$$\frac{mv_r^2}{2} + U'(r) = E \qquad (13\text{-}18)$$

where

$$U'(r) = U(r) + \frac{l^2}{2mr^2}$$

The quantity $U'(r)$ plays the role of an *equivalent* potential energy for the one-dimensional radial problem. The additional term $l^2/2mr^2$ takes complete account, as far as the radial motion is concerned, of the fact that the position vector **r** in the actual motion is continuously changing its direction. But it is to be noted that nothing that depends explicitly on the angular coordinate θ or its time derivatives appears in Eq. (13–18).

The term $l^2/2mr^2$ is often referred to as the "centrifugal" potential energy, because the force represented by the negative gradient of this potential energy is given by

$$F_{\text{centrifugal}} = -\frac{d}{dr}\left(\frac{l^2}{2mr^2}\right) = \frac{l^2}{mr^3}$$

Putting $l = mr^2(d\theta/dt)$, this becomes

$$F_{\text{centrifugal}} = mr\left(\frac{d\theta}{dt}\right)^2$$

which is identical with the centrifugal force $m\omega^2 r$ in a frame rotating at an angular velocity ω equal to the instantaneous value of $d\theta/dt$.

It must be remembered, of course, that the "centrifugal potential energy" is in fact a portion of the kinetic energy of the particle: that part due to the component of its motion transverse to the instantaneous direction of the radius vector. The circumstance that this kinetic-energy contribution can be expressed as a function of radial position alone enables us to treat the radial motion as an independent one-dimensional problem.

The potential-energy function $U(r)$ in Eq. (13–17) can be any function of radius. In developing this discussion we shall first consider the general properties of motion in any central field for which $U(r)$ approaches zero as r increases. Subsequently we shall take up the important special case of an inverse-square force field in which the potential energy of a particle is inversely proportional to r.

USE OF THE EFFECTIVE POTENTIAL-ENERGY CURVES

The general properties of the motion of a particle in a central field can be most readily obtained by using the energy-diagram method of Chapter 10 applied to Eq. (13–17) or Eq. (13–18). There are, however, two significant differences between the use of the energy method for one-dimensional motion and its use for any two-dimensional motion which can be reduced to two one-dimensional motions, as in the case of central fields. First, in one dimension, the energy alone determines the general character of the motion in a conservative field. In the central-field case, however, specification of the energy is not enough. The angular momentum l of the particle must also be specified, and the motion depends on both parameters E and l, instead of on E alone. This shows up clearly in the energy diagram, because the plot of $U' = U + (l^2/2mr^2)$ will be different for different values of l. We have, then, a whole family of curves of equivalent potential energy, corresponding to different values of the angular momentum, and must consider what happens to the motion with a given energy for the different values of l.

In addition, to translate the results obtained for the radial motion by this scheme into the actual motion of the particle, we must remember that while the radial coordinate r is changing with time, so is θ. The changes in r are accompanied by a simultaneous rotation of the vector **r**, and the actual orbit of the particle depends on both. The rotation of the vector **r** is not uniform except in the special case of circular motion, for which the length of the position vector **r** does not change with time.

The angular momentum l and the energy E define the basic dynamical conditions and are related to the position r_0 and velocity v_0 at an arbitrary time t_0 by

$$E = \frac{mv_0^2}{2} + U(r_0)$$

$$l = mr_0(v_\theta)_0$$
(13-19)

Since we are dealing with a two-dimensional problem, we need a total of four initial conditions to specify the situation completely. A statement of the vectors $\mathbf{r}_0$ and $\mathbf{v}_0$ (two components each) fulfills this purpose.

Fig. 13-6 (a) *True potential energy $U(r)$, and several "centrifugal potential energy" curves belonging to different values of l.* (b) *Resultant effective potential energy $U'(r)$ for different orbital momenta.*

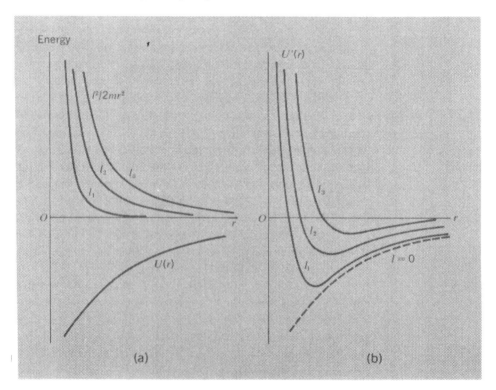

Now let us look at a typical energy diagram. In Fig. 13-6(a) are plotted the curves of $l^2/2mr^2$ against r for several different values of the angular momentum, and $U(r)$ against r for an attractive potential that rises with increasing distance from $r = 0$. $U(r) \to 0$ as $r \to \infty$. Figure 13-6(b) shows the effective potential-energy curves $U'(r)$ obtained by combining the single function $U(r)$ with each of the separate centrifugal potential curves. (To obtain curves with minima as shown, it is necessary for the variation of U with r to be less rapid than $1/r^2$.) Let us now consider how to use such curves to draw valuable qualitative conclusions about the possible motions.

In Figure 13-7 we show an effective potential-energy curve drawn for a particular value of l. We then see that, given this value of l, no physically meaningful situation can exist for any value of the total energy less than the energy E_m equal to the minimum value of $U'(r)$. At this minimum possible value of E, no radial motion can occur; the motion must be circular with a radius equal to r_0. For a range of larger values of E, between E_m and zero, the radial motion will be periodic (e.g., $E = E_a$ or E_b). With the assumed form of $U(r)$, all motions with a positive value of E (e.g., E_c) are unbounded; there is a least possible value of r but no upper limit. If one chooses to regard the value of E as given, but the orbital momentum as a variable, then it may be seen [most readily from Fig. 13-6(b)] that there is a maximum permissible value of l; any value from this down to zero is allowed, and the maximum value corresponds to a circular orbit.

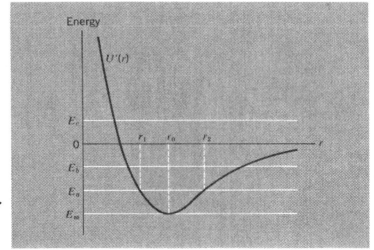

Fig. 13-7 *Effective potential-energy diagram from which the character of the radial motion for different values of the total energy can be inferred.*

BOUNDED ORBITS

With the help of the above preliminary analysis of the radial component of the motion, one can then proceed to develop some ideas about the appearance of the actual orbits in space. Suppose, for example, that we had the situation corresponding to the energy E_b in Fig. 13-7. The radial motion is bounded between certain minimum and maximum values of r. It is periodic with a certain period T_r. Thus we know that the particle moves always within the area between two circles as shown in Fig. 13-8. Furthermore, the radial distances $r_{\min}$ and $r_{\max}$ represent turning points of the radial motion. The orbit must be such that it is tangent to both these circles, because at these points the radial velocity is zero but the tangential velocity cannot be zero, given that the particle has angular momentum. Consider the particle after it has reached point A of the figure. It moves in as indicated, its trajectory becoming tangent to the inner circle at point C and continuing until it again becomes tangent to the outer circle at point B. The time it takes for this part of the motion is the radial period T_r. On the other hand, the radius vector is continually changing its direction, always in the same sense (i.e., either clockwise or counterclockwise) and will have turned through 2π after a characteristic period T_θ. In Fig. 13-8 the line OA represents the vector position of the particle at some instant, and the line OA' represents the vector position at the time T_θ later.

It is clear that the character of the orbit depends strongly on the ratio of the two periods, T_r and T_θ, of the doubly periodic

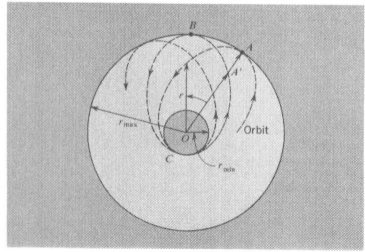

Fig. 13-8 Plan view of an orbital motion for a case in which the radial and angular periods differ. The effect is that of an approximately elliptic orbit that keeps turning (precessing) in its own plane.

motion. If the periods are commensurable (i.e., if their ratio can be expressed as the ratio of two integers), the moving particle will ultimately (after a time equal to the lowest common multiple of T_r and T_θ) find itself in exactly the same position as initially and the orbit will thus have been closed. If the two periods happen to be exactly equal, this closure will happen after only one radial period and one increment of 2π in θ. In Fig. 13-8 this would mean that the points A and B would coincide. One sees that this is a unique and (on the face of it) a thoroughly improbable situation, yet it is precisely what one has if the force is one of attraction proportional to the inverse square of r. Thus the most important forces in nature (gravitational and electrostatic) yield orbits of this remarkable character, as we shall show shortly.

If the radial and angular periods are comparable but definitely different, then one has just the kind of situation shown in Fig. 13-8. This corresponds to a case in which T_r is somewhat greater than T_θ, so that the radius vector rotates through rather more than 2π before r completes its variation from r_{max} to r_{min} and back again. In a situation like this, where the path is near to being a closed curve but is, in effect, also turning (either forward or backward), one says that the orbit is *precessing*. The study of orbital precession is important in astronomical systems.

If the radial and angular periods are incommensurable, the orbit will never close and will eventually fill up the whole region between r_{min} and r_{max}.

UNBOUNDED ORBITS

We have already pointed out that a positive total energy leads to a lower limit of r but to no upper limit in the effective potential of Fig. 13-7. This is, in fact, a rather general result, applying to any potential that tends to zero at $r = \infty$, because it corresponds to the fact that the particle has a positive kinetic energy at infinity. If $U(r)$ is a repulsive potential, everywhere positive, then (assuming that it decreases monotonically with increasing r) there are no bounded motions at all. Figure 13-9 compares the situations obtained by taking a given centrifugal potential and combining it with attractive and repulsive potential-energy curves that differ only in sign. Figure 13-10 shows what this means in terms of the trajectories that a particle of a given energy would follow in these

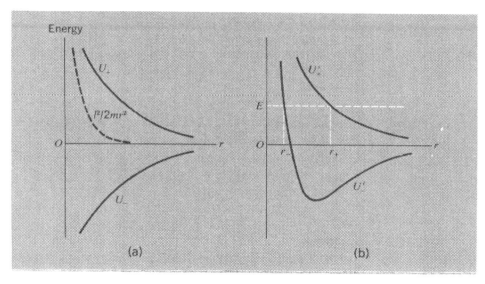

Fig. 13-9 (a) Centrifugal potential-energy curve, and two potential-energy curves, differing only in sign, that might arise from electrical interactions of like and unlike charges. (b) Effective potential-energy curves corresponding to the two cases shown in (a), indicating different distances of closest approach for a given positive total energy.

two situations. At large distances from the center of force, such that the magnitudes of $U(r)$ and $l^2/2mr^2$ are both negligible, the particle travels in a straight line with a speed v_0 equal to $(2E/m)^{1/2}$. This line of motion is offset from a parallel line through the center of force by a certain distance b that is directly related to the angular momentum; we have, in fact,

$$l = mv_0 b \tag{13-20}$$

Thus an assumption of given values of E and l in Fig. 13-9 is expressed by using the values of v_0 and b in Fig. 13-10(a), which corresponds to an attractive potential, and in Fig. 13-10(b), which corresponds to a repulsive potential. The distance b is called the *impact parameter*, and it is a very useful quantity for characterizing situations in which a particle, in an unbounded orbit, approaches a center of force from a large distance away. For a given value of v_0, the value of b completely defines the orbital angular momentum, via Eq. (13-20).

It is clear that these unbounded trajectories represent single encounters of the particle with the center of force; there is no possibility of successive returns as we have with the bounded

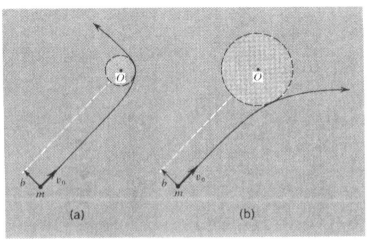

Fig. 13-10 (a) Plan view of trajectory of a particle moving around a center of attraction. The angular momentum is defined by the "impact parameter" b. (b) Corresponding trajectory with the same impact parameter, but with a repulsive center of force.

orbits of the kind shown in Fig. 13-8. The situations shown in Fig. 13-10 thus represent individual collisions or scattering processes, of the kind so important in atomic and nuclear physics. We shall return to them later.

Certain potentials may lead to the possibility of having both

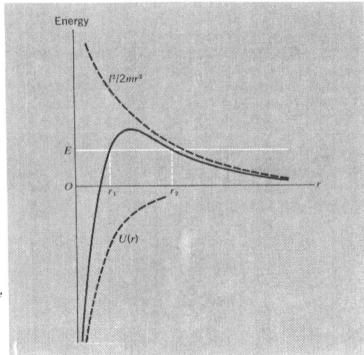

Fig. 13-11 Construction of effective potential-energy diagram for an attractive potential that varies more strongly than $1/r^2$.

bounded and unbounded motions at the same energy. This possibility exists for any attractive potential curve $U(r)$ that falls off *more* rapidly than $1/r^2$ with increasing r. An example is shown in Fig. 13-11. As long as l is not zero, the combination of $U(r)$ with the centrifugal potential gives an effective potential-energy curve that looks very much like an upside-down version of Fig. 13-7. For a positive total energy E that is less than the maximum of $U'(r)$, there are now two distinct regions of possible motion:

$$0 \leq r \leq r_1 \quad \text{bounded orbits}$$
$$r_2 \leq r \leq \infty \quad \text{unbounded orbits}$$

This is not just an academic curiosity. The effective potential-energy curve representing the interaction between, let us say, a proton and an atomic nucleus resembles Fig. 13-11 and suggests that the particle may be either trapped within r_1, or free outside r_2, with no possibility of going from one state to the other. It is, as we first mentioned in Chapter 10, one of the fascinating results of quantum mechanics that a transition through the classically forbidden region can in fact occur with a certain probability—a probability that is far too small to be significant in most cases but can become very important in atomic and nuclear systems.

CIRCULAR ORBITS IN AN INVERSE-SQUARE FORCE FIELD

As a quantitative example of the general approach discussed in the preceding sections, we shall take the case of circular or almost circular orbits under an attractive central force varying as $1/r^2$. This means that we shall mainly be discussing, from a different point of view, a situation that we have already analyzed in some detail in earlier chapters. There is merit in this, because it enables one to see more readily the relationship between the different approaches.

To be specific, let us consider the motion of a satellite of mass m in the gravitational field of a vastly more massive planet. For this case, the potential energy in the field of a planet of mass M takes the special form [see Eq. (11-31)]

$$U(r) = -\frac{GMm}{r} \tag{13-21}$$

where G is the universal gravitational constant and M the mass of the attracting object. Equation (13-17) becomes for this case

$$\frac{mv_r^2}{2} + \frac{l^2}{2mr^2} - \frac{GMm}{r} = E \tag{13-22}$$

In Fig. 13-12 are plotted the equivalent potential-energy curves

$$U'(r) = \frac{l^2}{2mr^2} - \frac{GMm}{r}$$

for two values of the angular momentum l.

A circular orbit corresponds, as we have seen, to a total energy equal to the minimum value of $U'(r)$ for a given value of l. Now from the above equation for $U'(r)$, taking l as fixed, we have

$$\frac{dU'}{dr} = -\frac{l^2}{mr^3} + \frac{GMm}{r^2}$$

Putting $dU'/dr = 0$, we thus find that

$$\frac{l^2}{mr} = GMm = \text{const.} \tag{13-23}$$

The orbital radius is therefore proportional to the square of the orbital angular momentum. This is indicated qualitatively in Fig. 13-12.

Let us now consider the energy of the motion. A circular orbit is characterized by having the radial velocity component, v_r, always equal to zero. Thus in Eq. (13-22) we can put

$$\text{(Circular orbit)} \quad E = \frac{l^2}{2mr^2} - \frac{GMm}{r}$$

From Eq. (13-23), however, we have

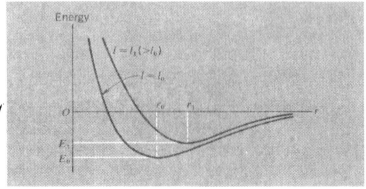

Fig. 13-12 Energy diagrams for orbits of different orbital angular momentum around a given center of force.

$$\frac{GMm}{r} = \frac{l^2}{mr^2}$$

Hence the energy can be expressed in either of the following ways:

(Circular orbit) $\quad E = -\dfrac{l^2}{2mr^2} = -\dfrac{GMm}{2r} \quad$ (13-24)

The second form shows that the orbital radius is inversely proportional to $|E|$; the first, taken together with Eq. (13-23), shows that $|E|$ is inversely proportional to l^2. These properties, too, are qualitatively indicated in Fig. 13-12.

The period of the orbit can be obtained, in this approach, with the help of the law of areas. Earlier in this chapter we established the following result [Eq. (13-15b)]:

$$\frac{dA}{dt} = \frac{l}{2m}$$

For a circular orbit, the total area A is πr^2 and the value of l is simply mvr. Thus we have $dA/dt = vr/2$ and hence $T_\theta = \pi r^2/(vr/2) = 2\pi r/v$—hardly a surprising result! To express T_θ in more interesting terms we can make use of Eq. (13-23), putting $l = mvr$:

$$\frac{(mvr)^2}{mr} = GMm$$

giving

$$v = \left(\frac{GM}{r}\right)^{1/2}$$

Using this explicit expression for v as a function of r, we then arrive once again at the equation that expresses Kepler's third law:

$$T_\theta = \frac{2\pi}{(GM)^{1/2}} r^{3/2} \quad (13\text{-}25)$$

SMALL PERTURBATION OF A CIRCULAR ORBIT

In the previous section we limited ourselves to redeveloping some familiar results. But now let us do something new, which exploits the insights provided by the effective potential-energy method.

Imagine that a satellite is describing a circular orbit of energy E_0 and radius r_0 around a spherical planet whose center

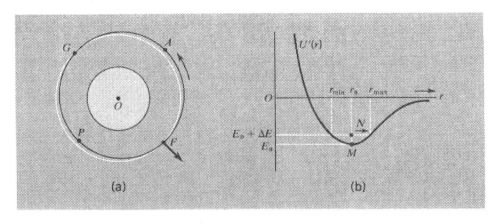

Fig. 13-13 (a) Result of applying a radial impulse to an object that is initially in a circular orbit with its center at the earth's center, O. The impulse is applied at F, and the subsequent path is the almost-circular path FAGP. (b) Effective radial potential-energy diagram for the analysis of the motion shown in (a).

is at O [Fig. 13-13(a)]. Suppose now that the satellite, when at the point F, is given a small radial impulse, along the line OF, by means of a brief firing of one of its control jets. What happens to the orbit?

Let us examine the situation with the help of the energy diagram in Fig. 13-13(b). Initially the satellite is in the state represented by the minimum point, M, of the effective potential curve. The impulse, being purely radial, leaves the orbital momentum unchanged. Thus the function $l^2/2mr^2$, and hence the whole function $U'(r)$, is unaffected.[1] The impulse does, however, raise the total energy of the satellite slightly. Thus, immediately after the firing, the satellite finds itself in the state represented by the point N in Fig. 13-13(b); it is still at $r = r_0$, but it has a small radial velocity corresponding to the energy increment ΔE. But we know what this means! The form of $U'(r)$ in the vicinity of its minimum, M, is approximately parabolic, and simple harmonic radial oscillations will ensue. Let us calculate their period.

---

[1] We are assuming that the decrease of mass associated with the firing of the jet can be ignored. Actually, since both the kinetic energy and the gravitational potential energy associated with the satellite are proportional to m, the loss of a significant amount of mass would not affect the conclusions. What matters is that the velocity, **v**, of the satellite is increased without any change in the transverse component v_θ.

We saw in Chapter 10 how the effective spring constant, k, for oscillations around the minimum of a potential-energy curve is equal to the second derivative of U with respect to distance, evaluated at the minimum point. [See pp. 395–397, and especially Eq. (10–29).] In the present case we have

$$U'(r) = \frac{l^2}{2mr^2} - \frac{GMm}{r}$$

$$\frac{dU'}{dr} = -\frac{l^2}{mr^3} + \frac{GMm}{r^2}$$

$$\frac{d^2U'}{dr^2} = \frac{3l^2}{mr^4} - \frac{2GMm}{r^3}$$

This now is to be evaluated at $r = r_0$:

$$k = \frac{3l^2}{mr_0^4} - \frac{2GMm}{r_0^3}$$

But from Eq. (13–23) we can put $l^2/mr_0 = GMm$; using this we obtain the following value for k:

$$k = \frac{GMm}{r_0^3}$$

We can immediately write down the period for one complete cycle of these small radial oscillations:

$$T_r = 2\pi \left(\frac{m}{k}\right)^{1/2} = \frac{2\pi}{(GM)^{1/2}} r_0^{3/2} \qquad (13\text{-}26)$$

This is identical with the period of one complete sweep of the radius vector through 2π, as given by T_θ in Eq. (13–25) with $r = r_0$. For $\Delta r \ll r_0$, this angular motion takes place with an almost constant value of $d\theta/dt$. Thus when θ has advanced by $\pi/2$ from the point F, the satellite is just about at its maximum distance from O, as represented by the point A in Fig. 13–13(a). After a further quarter-period it is at G, with $r = r_0$ once again, and another quarter-period brings it to P, where it is at its minimum distance from O. The orbit closes at the original firing point F and would continue in its new form, which for small radial oscillations is, to an exceedingly good approximation, a circle of the same radius as the original orbit but with its center shifted by a distance equal to the amplitude of the oscillations. We actually have here (although we have not proved it) the conversion of a circular orbit into an elliptic one. What we *have* shown is that *small* radial oscillations, in synchronism with the

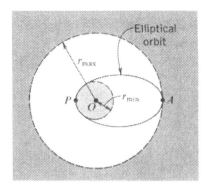

Fig. 13-14 Stationary, closed elliptic orbit ($1/r^2$ law of force) in which the periods of the radial and angular motions are equal.

angular motion, generate a closed, repeating, eccentric orbit. The same effect with a larger radial impulse, resulting in much larger (and asymmetrical) radial oscillations, can in fact lead to noticeably flattened closed orbits, as shown in Fig. 13-14, tangent to the circles of radii r_{min} and r_{max}. Our approximate (small perturbation) calculation could not, however, be taken as any guarantee of the closure property in this more extreme case—although exact analysis shows that it does hold.

If the central gravitating object were the earth, the far point A would be called the apogee, and the near point P the perigee (from the Greek *ap*, away from; *peri*, near; *geos*, earth). The prefixes ap- and peri- are used quite generally to denote far and near points of orbits around given objects [e.g., aphelion and perihelion for the sun (Helios)].

THE ELLIPTIC ORBITS OF THE PLANETS[1]

Before proceeding with the subject of orbit dynamics, we shall briefly describe the remarkable achievement of Johannes Kepler in establishing the proposition that the planetary orbits are not compounded circles but simple ellipses. This great discovery was based almost entirely on the analysis of the motion of a single planet—Mars. The behavior of Mars had puzzled and exasperated astronomers for a very long time, because the apparent irregularities in its motion were greater than those of any other planet and defied any easy analysis in terms of combinations of uniform circular motions.

[1]This section is a historical digression, included on the grounds that the question, How do we know what we know? is always worth asking and often has as much to teach us as the final answer does.

To appreciate the development of Kepler's discovery one must constantly keep in mind the fact that the primary data of observational astronomy (and this was especially true in the days before the telescope) are directions rather than distances. Although it was well understood that variations in the apparent brightness of the planets were linked to variations of distance from the earth, the precise data were only of angular positions. The whole theoretical machinery of superposed circular motions was primarily a means of reproducing the observed angular position of each planet as a function of time.

Kepler began his study of Mars at the direction of the great observational astronomer Tycho Brahe, whom he joined as an assistant in 1600. Kepler's task was to construct the actual path of Mars in space from the accumulation of original observations; it took him 6 years, and many false scents, before he arrived at the picture that is now familiar to us.

Kepler fully accepted a heliocentric model of the solar system, and (unlike Copernicus himself) he consistently held to the idea that the path of a planet must be a smooth, continuous curve of some kind around the sun. His problem was to find this curve on the basis of observations made from a laboratory—the earth—which was itself orbiting the sun in a nonuniform way. A first task was therefore to establish the path of the earth itself. Kepler attacked this problem in several ways. The one most directly based on observation was brilliant. Kepler knew the length of the Martian year to be 687 days. He used this knowledge to identify the dates on which Mars must have returned to a given point M in its orbit [Fig. 13–15(a)]. He chose this point to be one corresponding initially to a configuration in which the earth was at E_0, on a straight line between Mars and the sun (what the astronomers call *opposition*). During one Martian year, the earth travels through about 677°, or 43° less than two complete revolutions. Thus the next time Mars is at M, the earth is at E_1. Since the angular positions of the sun and Mars against the background of the stars were a matter of record, Kepler could locate the point E_1 with respect to the baseline SM. Using the same base line he was able to locate the points E_2, E_3, ... at the ends of successive Martian years. Thus a plot of the earth's orbit could be constructed.

What Kepler found was that the earth's orbit is indistinguishable from a circle in shape—but the sun is not at the center and the rate of change of the earth's angular position with respect

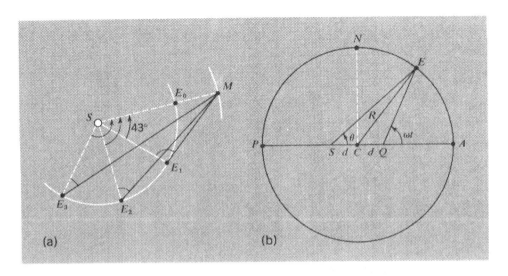

Fig. 13-15 (a) Principle of Kepler's method of charting the earth's orbit by reference to a standard position of Mars (at the ends of successive Martian years). (b) Approximate representation of a planetary orbit (with the sun at S) obtained by assuming that the line QE (from the equant point Q) rotates at constant angular speed.

to the sun (or vice versa) is not constant. To describe the earth's motion, Kepler first used a clever and effective trick that had been discovered by Ptolemy; this was to locate a fictitious center with respect to which the angular motion *was* uniform. Its use by Kepler is indicated in Fig. 13-15(b). The earth's orbit is a circle of radius R with center C. The sun is at S, a distance d from C. If one takes a point Q on the line SC produced, such that $CQ = SC = d$, then a *uniform* angular motion of the line QE causes the line SE (from sun to earth) to change its inclination θ in almost exactly the right way. The point Q was called the *equant*, because it acted as a center of equal (i.e., uniform) angular change.

Kepler found from his analysis of the earth's orbit that the ratio d/R is only about 0.018. A figure of about this size was substantiated by observations (initiated by Tycho Brahe) of the variation of the sun's measured angular diameter during the course of the year. Figure 13-16 shows the results of some more recent observations of the same effect. Thus Kepler could feel that his picture of the earth's orbit in space was substantially correct. If one introduces the aphelion and perihelion points [A and P, re-

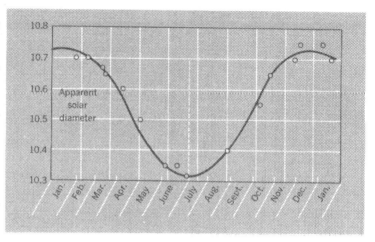

Fig. 13–16 Evidence of the change in the sun's distance during one year, as shown by the change in size of the sun's image formed by a telescope. (After a graph in Science in Secondary Schools, Ministry of Education Pamphlet No. 38, H. M. Stationery Office, London, 1960.)

spectively in Fig. 13–15(b)], then we have

$$r_{max} = SA = R + d$$
$$r_{min} = SP = R - d$$

The *eccentricity*, ϵ, of any such orbit is defined by the formula

$$\text{eccentricity } (\epsilon) = \frac{r_{max} - r_{min}}{r_{max} + r_{min}} \tag{13-27}$$

In the present case this is equal to the ratio $d/R \approx 0.018$.

We can understand why the model represented by Fig. 13–15(b) worked so well. First, as we saw in the previous section, an inverse-square law of force can lead to orbits that are closely circular but do not have the center of force at the geometrical center. Second, the use of the equant, Q, in Fig. 13–15(b) comes close to defining motion with a constant orbital angular momentum about the true center of force, S. Consider, in particular, the perihelion and aphelion points, P and A. If the line QE rotates at a constant angular velocity ω, then at A we have $v_\theta = \omega(R - d)$, and at P we have $v_\theta = \omega(R + d)$. The distances SA and SP are, however, equal to $R + d$ and $R - d$, respectively. Thus in both cases we have a rate of sweeping out area given by

$$\frac{dA}{dt} = \tfrac{1}{2}\omega(R^2 - d^2)$$

At the point N the rate is equal to $\tfrac{1}{2}\omega(R^2 + d^2)$. (Check this.) Thus for $d^2/R^2 \ll 1$ the "areal velocity" defined by this construction is very nearly constant. Since for the earth's orbit the ratio d^2/R^2 is only about 3×10^{-4} the above approximate description works very well indeed.

After this lengthy introduction we can abbreviate the rest of the story. With his knowledge of the earth's orbit Kepler could now, in a straightforward manner, construct a picture of the successive angular positions of Mars as seen from the sun [Fig. 13-17(a)]. Again he tried to represent the orbit as an eccentric circle with an equant, as in Fig. 13-15(b). But this time it did not work; there were small but significant discrepancies between the calculated angular positions and the observed positions as function of time. The discrepancies were less than 10 minutes of arc, but Kepler had such faith in the reliability and accuracy of Tycho Brahe's observations that he discarded the theory rather than try to bend the data to his ideas. Finally his prodigious analytical efforts were rewarded with two decisive insights:

1. The law of equal areas, referred to the sun as origin, made the fictitious equant unnecessary, and worked just as well (or, indeed, even better).

2. If the original circle were modified to an inscribed ellipse with the sun at one focus, as shown in Fig. 13-17(b), then (applying the law of areas to this new orbit) the agreement between theory and observation became excellent. The distance SN is equal to the radius CA of the original circle.

The circumstance that made the motion of Mars so hard to explain (but without which Kepler would probably not have been brought to his great discovery) was the large eccentricity of the orbit (about 0.09—five times greater than that of the earth, and more than 12 times greater than that of Venus). It should be

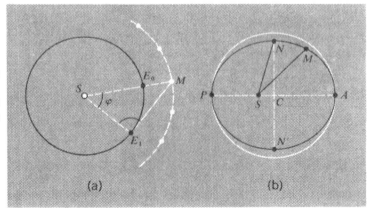

Fig. 13-17 (a) *Kepler's triangulation method for locating a point on the orbit of Mars, once the earth's orbit is known.* (b) *Kepler's discovery that the orbit of Mars is an ellipse, not an eccentric circle. (Based on one of his own diagrams in the New Astronomy.)*

noted, however, that even in the case of Mars the departure of the actual shape of the orbit from a circle is exceedingly small; the difference between the shortest diameter, NN', and the longest diameter, PA [see Fig. 13-17(b)] is less than 1 part in 200.

Kepler published the full story of his labors—the many failures as well as the final successes—in a book, *The New Astronomy* (*Astronomia Nova*), published in 1609. It is a classic of scientific discovery.[1]

Once the basic character of the planetary motions had been established, the wealth of pre-existing records made it easy to infer the orbital parameters of other planets. Table 13-1 presents a modern tabulation of such parameters, and other relevant data, for all the major planets. (The inclination, i, is the angle made by the plane of the orbit with the plane of the earth's own orbit—i.e., the ecliptic.) Figure 13-18 is a scale drawing of the orbits of the four inner planets, projected onto the plane of the ecliptic. (The angles of inclination are so small that this projection does not change the shape of the orbits detectably, and they all look like perfect circles.)

TABLE 13-1: PLANETARY ORBIT DATA

	Semimajor axis (a)		Eccentricity	Inclination	Period (T)	
	AU	10^6 km	(ϵ)	(i), deg	Years	10^8 sec
Mercury	0.387	57.9	0.206	7.00	0.241	0.076
Venus	0.723	108.2	0.007	3.39	0.615	0.194
Earth	1.000	149.6	0.017	—	1.000	0.316
Mars	1.523	227.9	0.093	1.85	1.881	0.594
Jupiter	5.203	778.3	0.048	1.31	11.862	3.75
Saturn	9.540	1427	0.056	2.50	29.46	9.31
Uranus	19.18	2869	0.047	0.77	84.02	26.6
Neptune	30.07	4498	0.009	1.78	164.77	52.1
Pluto	39.44	5900	0.249	17.2	248	78.4

[1]The book is usually referred to simply as the *New Astronomy*. Its full title, however, when translated, reads "A New Causal Astronomy or Celestial Physics Together with Commentaries on the Movements of the Planet Mars. From the Observations of Count Tycho Brahe."

Arthur Koestler's book, *The Watershed*, Doubleday (Anchor Books), New York, 1960, is a full and interesting account of Kepler's life and work. A more detailed and critical discussion of Kepler's approach to his first and second laws is an article by Curtis Wilson, "Kepler's Derivation of the Elliptical Path," *Isis*, **59**, 5-25 (1968).

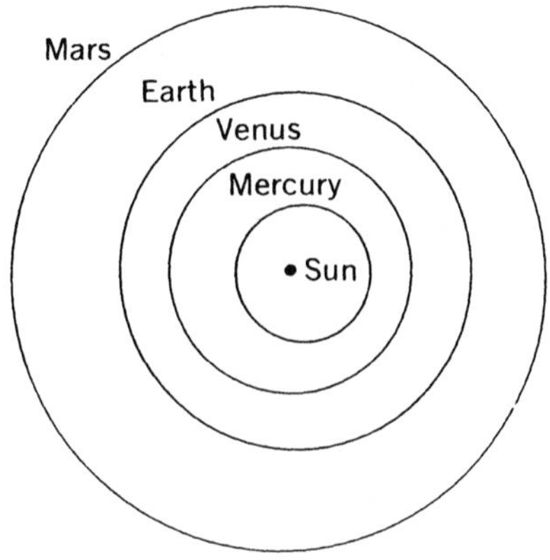

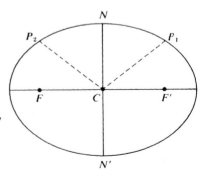

Fig. 13-18 *Scale drawing of the orbits of the four inner planets. They are all indistinguishable from circles in shape but are not concentric. All are, in fact, ellipses with the sun at one focus.*

DEDUCING THE INVERSE-SQUARE LAW FROM THE ELLIPSE

In 1689 Newton received a letter from the philosopher John Locke. Locke wrote that he had been reading the *Principia* (which had been published two years earlier) but found some of the mathematics quite beyond him. Could Newton give him a less formidable explanation of how to infer the inverse-square law from the observed elliptic orbits of the planets? Newton replied with a delightfully simple argument, which we shall use as the starting point of our own discussion. It rests on a remarkable property of these orbits—their geometrical symmetry, which exists in spite of the fact that in kinematic and dynamic terms the orbits are asymmetrical.

Consider the elliptic orbit represented by Fig. 13-19. There is a center of force at the focus, F. Located at an equal distance to the other side of the geometrical center, C, is a second focal

Fig. 13-19 *Geometrical symmetry of an ellipse with respect to reflection in its minor axis, NN'.*

point, F', but it is empty. [The equant point, Q, of Kepler's original model, as shown in Fig. 13-15(b), was at this second focus.] The ellipse has perfect reflection symmetry, not only about the major axis passing through F and F', but also about the perpendicular (minor) axis NN' drawn through C. Yet, as we have seen, the velocities at symmetrically placed points, e.g., P_1 and P_2 in the figure, are quite different; the orbiting object moves much faster at P_2 than at P_1, in accordance with the law of equal areas.

Consider now, said Newton, the motion of a planet as it passes through its perihelion point p and its aphelion point a [Fig. 13-20(a)]. Mark off the arcs pq and ab that would be traversed in equal, short times just after passing these points. The lengths of these arcs are related to the distances from the sun by the law of equal areas, using the fact that at p and a the planet's velocity is purely transverse to the radius vector from F. Hence to a very good approximation we have

$$r_1 s_1 = r_2 s_2$$

Now in the absence of any gravitational force, the planet after passing p would travel along the tangent pT. Because of the attraction, it follows the curved path pq, which is indistinguishable from a parabola over a short distance. (This is just like Newton's analysis of the motion of the moon—see Chapter 8, p. 257.) The planet "falls" through the distance Tq ($= d_1$), which is

Fig. 13-20 (a) Illustrating Newton's simple argument to deduce the inverse-square law from the motion of a planet near perihelion and aphelion. (Based on his own diagram). (b) Geometrical construction for deducing the inverse-square law from the elliptic path and the law of equal areas.

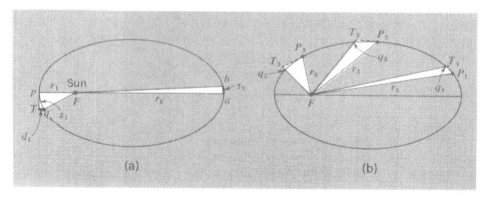

proportional to the square of the tangent pT. Remembering that the deviation Tq is, in fact, much less than pT, we can put

$$d_1 = Cs_1^2$$

where C is the constant defining the shape of the parabola. But the parabolic curvature of the ellipse near the aphelion, a, is identical, because of the symmetry, to the curvature near the perihelion, p. Thus we can also put

$$d_2 = Cs_2^2$$

where C is the same constant as before and d_2 is the deviation of the planet's path in going from a to b. It follows from these results, taken together with the law of areas, that

$$\frac{d_1}{d_2} = \frac{r_2^2}{r_1^2}$$

Since, however, the distances fallen in equal times are proportional to the accelerations and hence to the forces, it follows that the gravitational forces acting on the planet at the points p and a are also inversely as the squares of the distances r_1 and r_2.

The above argument does not demonstrate the applicability of the inverse-square law at other points around the orbit, but an extension of the same basic geometrical approach can be used to achieve this, although it no longer appeals to the geometrical symmetry of the orbit. Figure 13-20(b) indicates the method. Sectors of equal area are constructed, using various positions (P_1, P_2, P_3) as starting points. This defines the points q_1, q_2, and q_3 reached by the planet after equal time intervals. The tangent lines at P_1, P_2, and P_3 are also drawn. The radial displacements due to the gravitational force are then given by $T_1 q_1$, $T_2 q_2$, and $T_3 q_3$. With a very carefully drawn figure one can verify that these distances are in proportion to $1/r^2$. The construction does, however, have to be extremely carefully done, as the radial displacements are very small.[1]

ELLIPTIC ORBITS: ANALYTICAL TREATMENT

We begin with the fundamental "pins-and-string" definition of the ellipse, which requires that

---

[1] The method has been nicely presented in the PSSC film "Elliptic Orbits," by Albert Baez, Education Development Center, Newton, Mass., 1959. See also the paper by Baez, *Am. J. Phys.*, **28**, 254 (1960).

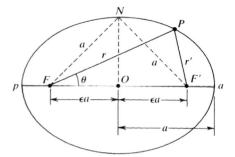

Fig. 13-21 Basic geometrical features of an elliptic orbit with a center of force at the focus F.

$$r + r' = 2a \tag{13-28}$$

as shown in Fig. 13-21, where the ellipse has a major axis equal to $2a$ and has its foci F and F' separated by a distance $2\epsilon a$. First, we need a little trigonometry, to obtain a relation between the radial and angular coordinates (r, θ) of an arbitrary point P.

Applying the law of cosines to the triangle FPF', we have

$$r'^2 = r^2 + (2\epsilon a)^2 - 4\epsilon ar \cos\theta$$

From Eq. (13-28), however, we have

$$r'^2 = 4a^2 - 4ar + r^2$$

Equating these two expressions for r'^2, we get

$$4\epsilon^2 a^2 - 4\epsilon ar \cos\theta = 4a^2 - 4ar$$

Hence

$$r(1 - \epsilon \cos\theta) = a(1 - \epsilon^2)$$

or

$$r = \frac{a(1 - \epsilon^2)}{1 - \epsilon \cos\theta} \tag{13-29}$$

We can simplify this a little by introducing the semiminor axis, b, of the ellipse, which is the distance ON in Fig. 13-21. By the geometry of the triangle FON we have

$$b^2 = a^2 - (\epsilon a)^2 = a^2(1 - \epsilon^2)$$

Using this, Eq. (13-29) can be written as

$$\frac{1}{r} = \frac{a}{b^2}(1 - \epsilon \cos\theta) \tag{13-30}$$

With the help of this last equation, together with conservation of orbital momentum as expressed in the law of areas, we can deduce that the force acting on a moving particle at any point is proportional to $1/r^2$. The essential step is to calculate the radial component of acceleration. By Eq. (13-4) we have

$$a_r = \frac{d^2r}{dt^2} - r\left(\frac{d\theta}{dt}\right)^2 \tag{13-31}$$

Now by differentiating both sides of Eq. (13-30) with respect to t, we get

$$-\frac{1}{r^2}\frac{dr}{dt} = \frac{\epsilon a}{b^2}\sin\theta\frac{d\theta}{dt}$$

However, by the law of areas we have

$$r^2\frac{d\theta}{dt} = C \tag{13-32}$$

where C is a constant (equal to twice the areal velocity). Using this, the previous equation gives us

$$\frac{dr}{dt} = -\frac{C\epsilon a}{b^2}\sin\theta$$

Differentiating a second time, we get

$$\frac{d^2r}{dt^2} = -\frac{C\epsilon a}{b^2}\cos\theta\frac{d\theta}{dt}$$

Again using Eq. (13-32) to eliminate $d\theta/dt$ we have

$$\frac{d^2r}{dt^2} = -\frac{C^2\epsilon a}{b^2}\frac{\cos\theta}{r^2}$$

To obtain a_r in Eq. (13-31), we must subtract from this the quantity $r(d\theta/dt)^2$. By Eq. (13-32) this can be written as

$$r\left(\frac{d\theta}{dt}\right)^2 = \frac{C^2}{r^3}$$

Combining this with the previous equation, we thus get

$$a_r = -\frac{C^2}{r^2}\left(\frac{\epsilon a\cos\theta}{b^2} + \frac{1}{r}\right)$$

This looks complicated, but if we look back at the original equation for $1/r$ [Eq. (13-30)] we see that

$$\frac{\epsilon a \cos\theta}{b^2} + \frac{1}{r} = \frac{a}{b^2}$$

Hence the term in parentheses in the above equation for a_r is just a geometrical constant of the ellipse, and we have

$$a_r = -\frac{C^2 a}{b^2}\frac{1}{r^2} \qquad (13\text{-}33)$$

Thus the operation of an inverse-square law is mathematically verified for any elliptic orbit known to be taking place under the action of a central force directed toward one focus. (But note the last qualification. It is perfectly possible, for example, to have an elliptic path under the action of a force directed toward the geometric center of the ellipse, but the law of force is no longer the inverse square. Can you guess what it is? See Problem 13-23.)

We can at once use Eq. (13-33) to develop another important result—Kepler's third law in a rigorous form. The parameter C is, as we have said, equal to twice the constant rate of sweeping out area. But the total area of an ellipse is given by the equation

$$A = \pi ab$$

Hence the period T of the orbit is given by

$$T = \frac{\pi ab}{C/2}$$

i.e.,

$$C = \frac{2\pi ab}{T}$$

Substituting this value in Eq. (13-33) we find

$$a_r = -\frac{4\pi^2 a^3}{T^2}\frac{1}{r^2} \qquad (13\text{-}34)$$

However, by introducing the specific law of gravitation between the orbiting mass m and a mass M fixed at F, we obtain another expression for the radial acceleration:

$$F_r = -\frac{GMm}{r^2}$$

Therefore,

$$a_r = -\frac{GM}{r^2} \qquad (13\text{-}35)$$

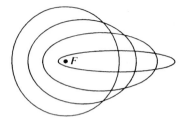

Fig. 13-22 Family of elliptic orbits of the same total energy, sharing the focus F, where the center of force is located. The major axes of the ellipses are all equal.

Equating the right-hand sides of Eqs. (13-34) and (13-35), we have the identity

$$\frac{4\pi^2 a^3}{T^2} = GM$$

whence

$$T^2 = \frac{4\pi^2 a^3}{GM} \qquad (13\text{-}36)$$

This result, although identical in form with what we identified in Chapter 8 as Kepler's third law, contains an important new feature. Previously we considered only circular orbits. Now we see that, according to Eq. (13-36), all orbits having the same major axis have the same period (for a given value of M), whether they are circular or strongly flattened. If one takes into account the fact that the gravitating mass must be at one focus, a group of elliptic orbits of the same period but different eccentricities might be as shown in Fig. 13-22.

Kepler, in stating his third law, said that the squares of the planetary periods are proportional to the cubes of the mean distances from the sun. We see now that the rather vague phrase "mean distance" takes on a very sharp and precise meaning; it is just the average of the maximum and minimum distances of a planet from the sun. This is identical with the semimajor axis, a, for from Eq. (13-29) or from Fig. 13-21, we have

$$r_{max} = a(1 + \epsilon)$$
$$r_{min} = a(1 - \epsilon)$$
$$a = \frac{r_{max} + r_{min}}{2}$$

ENERGY IN AN ELLIPTIC ORBIT

The purpose of this section is to show that, as with the orbital period, the only geometrical parameter entering into the total energy is the length of the major axis of the elliptical path.

Since the total energy is constant, we are free to evaluate it at any convenient point in the orbit. We shall choose the point a on the major axis (Fig. 13–21), for which we have

$$r = r_{\max} = a(1 + \epsilon)$$

The potential energy can be stated directly:

$$U(r_{\max}) = -\frac{GMm}{a(1 + \epsilon)} \tag{13-37}$$

The kinetic energy is a little harder to come by. Since the velocity at a is purely transverse, we have

$$K = \tfrac{1}{2}mv_\theta^2 = \tfrac{1}{2}mr^2\left(\frac{d\theta}{dt}\right)^2$$

With the help of Eq. (13–32) this becomes

$$K = \frac{mC^2}{2r^2}$$

Specifically,

$$K(r_{\max}) = \frac{mC^2}{2a^2(1 + \epsilon)^2}$$

Now the constant C is twice the rate of sweeping out area, which means, as we saw before, that $C = 2\pi ab/T$. Hence

$$C^2 = \frac{4\pi^2 a^2 b^2}{T^2} = \frac{4\pi^2 a^4(1 - \epsilon^2)}{T^2}$$

Substituting now for T^2 from Eq. (13–36) we have

$$C^2 = GMa(1 - \epsilon^2)$$

Using this expression for C^2 in the equation for $K(r_{\max})$ we find that

$$K(r_{\max}) = \frac{GMm(1 - \epsilon)}{2a(1 + \epsilon)} \tag{13-38}$$

Combining the results expressed in Eqs. (13–37) and (13–38) we obtain the following formula for the total energy of the motion:

$$E = -\frac{GMm}{2a} \tag{13-39}$$

The total energy of any elliptic orbit is thus the same as that of

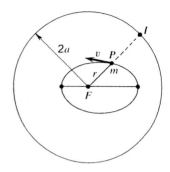

Fig. 13-23 A particle at any point P in an elliptic orbit has a speed equal to what it would acquire in falling from rest inward from the circle of radius equal to the major axis of the ellipse.

a circular orbit whose diameter is equal in length to the major axis of the ellipse.

It may be seen from Eq. (13-39) that any increase of E implies an increase in length of the major axis; the total energy remains negative but becomes numerically smaller. The negative total energy of an elliptic orbit expresses the fact that the orbiting object is bound to the center of force and cannot escape unless a positive amount of energy at least equal to $GMm/2a$ is supplied.

If we consider an object at an arbitrary point in its orbit, its gravitational potential energy is $-GMm/r$ and the total energy must, by Eq. (13-39), be equal to $-GMm/2a$. Thus we have

$$E = \tfrac{1}{2}mv^2 - \frac{GMm}{r} = -\frac{GMm}{2a} \tag{13-40}$$

The amount of kinetic energy defined by this equation can be considered with the help of a diagram (Fig. 13-23). Taking the focus F of the ellipse as center, draw a circle whose *radius* is equal to the major axis, $2a$, of the ellipse. To find the speed of the orbiting particle at a point P, imagine that the particle has been released from rest at the point I on the circle and has fallen from I to P under the gravitational attraction. Then the kinetic energy in Eq. (13-40) corresponds precisely to the change of potential energy associated with the displacement from I to P. If one could construct a device that would smoothly steer the particle into the direction of the tangent at P without changing the magnitude of the velocity, the ensuing motion would be the elliptic orbit as shown.

MOTION NEAR THE EARTH'S SURFACE

With the background provided by the foregoing analysis of elliptic orbits in general, one can recognize that all low-level

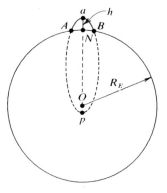

Fig. 13-24 Limited trajectory above the earth's surface, seen as a small part of an ellipse with the earth's center at the more distant focus.

trajectories at the earth's surface, hitherto regarded as parabolic paths under a uniform **g** field, are actually small portions of ellipses, as indicated in Fig. 13-24. The approximation of the trajectory by a parabola is in fact excellent, but strictly one should regard the high point of the trajectory as the apogee, a, of a very narrow ellipse. If an object is launched with a speed v from a point A on the earth's surface, the total energy is given by

$$\tfrac{1}{2}mv^2 - \frac{GMm}{R_E} = E = -\frac{GMm}{2a}$$

The major axis $2a$ is larger than the earth's radius by only a small amount, say H. We can then put

$$\tfrac{1}{2}mv^2 - \frac{GMm}{R_E} = -\frac{GMm}{R_E}\left(1 + \frac{H}{R_E}\right)^{-1}$$

$$\approx -\frac{GMm}{R_E} + \frac{GMm}{R_E^2}H$$

Hence

$$H \approx \frac{v^2}{2GM/R_E^2} = \frac{v^2}{2g}$$

This distance H is the sum of the maximum altitude h reached by the object, plus the distance Op from the earth's center to the fictitious perigee point. In the event that the object is fired vertically upward, the ellipse degenerates into a straight line and the perigee point moves into coincidence with the earth's center. In this case the value of h becomes equal to H, i.e., to $v^2/2g$, just as we would calculate directly from simple kinematics.

INTERPLANETARY TRANSFER ORBITS

A problem that used to be rather academic but has now become very practical is that of sending a spacecraft from the earth to another planet. The most efficient method is not to aim the spacecraft radially inward or outward along a radial line from the sun, but to let it coast in an elliptic orbit that joins smoothly to the orbits of the earth and the other planet in question.

To take a specific example, consider the problem of getting from the earth to Venus. We shall use the quite good approximation that the orbits of the two planets are circles with the sun at their common center. The transfer orbit would then be as shown in Fig. 13-25; it starts out tangent to the earth's orbit at E and joins the orbit of Venus tangentially at V. The length of the major axis of the transfer orbit is thus the sum of the orbital radii of the earth and Venus, which makes a total of 1.72 AU. This major axis is intermediate between those of the two planets, and it follows from Eq. (13-39) that the total energy is also intermediate between the values associated with the initial and final circular orbits. Thus to carry out the transfer the spacecraft needs to be given a sudden retardation at E and (if it is to be put in orbit with Venus) another sudden retardation at V. Let us now make this quantitative.

The speed v_0 appropriate to the earth's orbit is given by the equation

$$\frac{GM_Sm}{r_E^2} = \frac{mv_0^2}{r_E}$$

where M_S is the mass of the sun. Therefore,

$$v_0^2 = \frac{GM_S}{r_E} \tag{13-41}$$

The actual magnitude of this speed is known directly from the earth's orbit radius r_E ($= 1.49 \times 10^8$ km) and the length of the

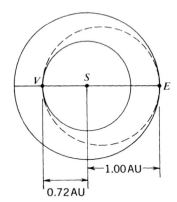

Fig. 13-25 *Interplanetary transfer orbit from the earth to Venus. The orbit is an ellipse that touches the circular orbits of the two planets.*

year, T_E (= 3.16×10^7 sec). Thus we have

$$v_0 = \frac{2\pi \times 1.49 \times 10^8}{3.16 \times 10^7} = 29.6 \text{ km/sec}$$

The speed v_1 that the spacecraft must have at E in order to follow the transfer orbit is given, according to Eq. (13-40), by the condition

$$E = -\frac{GM_S m}{1.72 r_E} = \tfrac{1}{2} m v_1^2 - \frac{GM_S m}{r_E}$$

Therefore,

$$v_1^2 = \frac{GM_S}{r_E}\left(2 - \frac{1}{0.86}\right) = 0.84 \frac{GM_S}{r_E}$$

Thus, by Eq. (13-41), we have

$$v_1 \approx 0.92 v_0 \approx 27.2 \text{ km/sec}$$

The necessary retardation is thus quite small—only about 2.5 km/sec.

Once placed in the transfer orbit, the spacecraft will take half of an orbital period to join the orbit of Venus at V. Since, by the generalized form of Kepler's third law, T^2 is proportional to the cube of the major axis, we have

$$T = \left(\frac{1.72}{2.00}\right)^{3/2} T_E = 0.80 T_E$$

Thus the journey takes about 0.40 of a terrestrial year, or about 146 days. During this time the speed of the spacecraft is continually increasing. Using the energy-conservation condition once again, the speed v_2 at V is given by

$$\tfrac{1}{2} m v_2^2 - \frac{GM_S m}{0.72 r_E} = -\frac{GM_S m}{1.72 r_E}$$

Therefore,

$$v_2^2 = \frac{GM_S}{r_E}\left(\frac{1}{0.36} - \frac{1}{0.86}\right) = 1.62 v_0^2$$
$$v_2 \approx 1.27 v_0 \approx 37.7 \text{ km/sec}$$

The orbital speed of Venus is about 34.9 km/sec. Thus an impulse sufficient to reduce the speed of the spacecraft by about 2.8 km/sec will complete the transfer operation. It may be seen, from this example, how the properties of the free orbital motions can be judiciously exploited so as to make such transfers with a relatively small expenditure of energy in rocket propulsion.

CALCULATING AN ORBIT FROM INITIAL CONDITIONS

Suppose that a particle is launched with a velocity v_0 from a point P, at a vector distance r_0 from the center of force, F (Fig. 13-26). How do we deduce the size, shape, and orientation of the subsequent orbit?

The first thing to do is to test whether the total energy is positive or negative (or perhaps, by chance, zero). Only if it is negative will we have a bounded orbit, and we shall limit our attention here to such cases, i.e., to closed elliptic orbits. From the values of v_0 and r_0 we know the total energy and hence the length of the major axis:

$$E = \tfrac{1}{2}mv_0^2 - \frac{GMm}{r_0} = -\frac{GMm}{2a} \qquad (13\text{-}42)$$

Thus we know the distance a.

Next we can use the fact that the orbital angular momentum is uniquely defined:

$$l = |\mathbf{r}_0 \times m\mathbf{v}_0| = mv_0 r_0 \sin\varphi \qquad (13\text{-}43)$$

At perigee ($r = r_1$) and apogee ($r = r_2$) the directions of $\mathbf{v}$ and $\mathbf{r}$ are orthogonal, so that we can put

$$l = mv_1 r_1 = mv_2 r_2,$$

or

$$\frac{1}{r_1} = \frac{mv_1}{l} \qquad \frac{1}{r_2} = \frac{mv_2}{l}$$

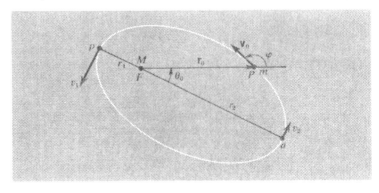

Fig. 13-26 Elliptic orbit resulting from the launching of a particle with an arbitrary velocity v_0 at a vector distance r_0 from the center of force (given that the total energy is less than zero).

We can insert the value of $1/r$ as defined by either of these equations in the expression for the total energy at apogee or perigee:

$$\tfrac{1}{2}mv_1{}^2 - \frac{GMm^2}{l}v_1 = -\frac{GMm}{2a}$$

This quadratic, when solved, has as its roots the values of both v_1 and v_2, from which we can at once deduce the values of r_1 and r_2. This then fixes the eccentricity, by the relations

$$r_1 = a(1 - \epsilon) \qquad r_2 = a(1 + \epsilon)$$

Finally, the orientation of the major axis, relative to which the initial position vector makes an angle θ_0, is determined through the general polar equation of the curve [Eq. (13–29)]:

$$r_0 = \frac{a(1 - \epsilon^2)}{1 - \epsilon \cos \theta_0}$$

Thus the orbit is completely specified, as shown in Fig. 13–26.

A FAMILY OF RELATED ORBITS

If we have a given force center and a given launching point, it is instructive to consider the variety of orbits that correspond to various possible values of the total energy. In the particular case of an attractive inverse-square force law, the situation can be illustrated with the help of Fig. 13–27. We shall suppose, for simplicity, that a particle is launched from P in a direction at right angles to the line FP from the force center F.

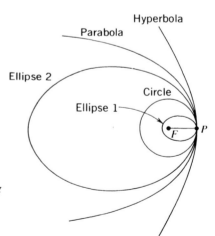

Fig. 13–27 Family of orbits of different total energy but sharing the focus F (the center of force) and the launching point P.

If the launching speed is close to zero, the particle follows an almost straight line toward F; strictly, it would be an elliptical path with an extremely small width and a major axis only very slightly longer than FP.

At a slightly higher launching speed, the orbit would resemble the orbit labeled "Ellipse 1." The launching point P would be the apogee. At some higher value of the energy the orbit would be a perfect circle with F at the center. A still further increase of energy would lead to elliptic orbits once again, but now P is the perigee point—i.e., the force center now represents the nearer of the two foci with respect to the launching point.

Ultimately the situation is reached where the total energy is precisely zero. A detailed analysis of the dynamics shows that the trajectory in this case is a parabola; the particle would move continually further from the center of force and would approach infinite distance with vanishingly small velocity.

Any further increase of launching speed produces a trajectory that is one branch of a hyperbola; the particle now approaches infinity with a significant positive kinetic energy.

One sees by this kind of evolutionary picture that there is no sharp distinction between the various trajectories shown in Fig. 13-27, despite the fact of their being mathematically different forms of conic sections. It may, in fact, be a quite difficult matter to determine, on the basis of measurements made near P, whether a given trajectory is part of an ellipse, a parabola, or a hyperbola. This is a real consideration in the analysis of the paths of comets through the solar system. The bounded orbit of a comet with a total energy only barely less than zero is almost indistinguishable from that of a hyperbolic orbit of small positive energy if the comet is visible only when it penetrates the inner regions of the solar system. An outstanding example of this is the famous comet named after Newton's friend Halley, who recognized that the comet he observed in 1682 had been recorded in previous approaches to the sun at intervals of about 76 years. With the help of Kepler's third law one can then infer that its orbit is, in fact, an ellipse with a major axis equal to $(76)^{2/3}$ times the diameter of the earth's orbit, or a distance of about 40 AU. The orbit in relation to the rest of the solar system is shown in Fig. 13-28. At its nearest approach to the sun (~0.6 AU) the comet goes inside the orbit of Venus and shines brightly; at its most distant points it goes beyond the orbit of Neptune and is quite invisible over most of its path.

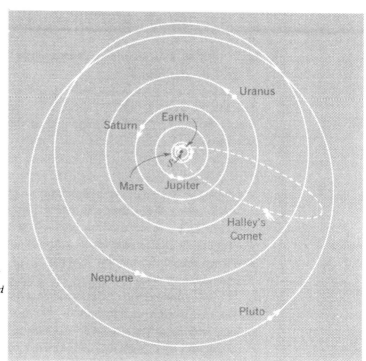

Fig. 13-28 Orbit of Halley's comet (period about 76 years) passing among the orbits of the planets.

CENTRAL FORCE MOTION AS A TWO-BODY PROBLEM

We have, of course, been treating these planetary problems and so on as two-body problems in the sense that the basis of the motion is the interaction between one object and another, but there has been something of an inconsistency. We have used the Newtonian law to define the force between two masses (M and m), but we have assumed that one mass (M) could be taken as fixed. This can be a quite good approximation for $M \gg m$ (e.g., the sun and any one of the planets), but it is never rigorously true and must fail seriously if M and m are comparable.

We have already seen an example of the correct approach in connection with a double-star system with circular orbits (p. 296); the motion of both partners is recognized, and the displacement of each is referred to the center of mass as origin, as indicated in Fig. 13-29. The particles P and Q have coordinates $\mathbf{r}_1$ and $\mathbf{r}_2$ with respect to the center of mass C; the vector distance from Q to P is $\mathbf{r}$. We then have

$$\mathbf{r}_1 = \frac{M}{M+m}\mathbf{r}$$

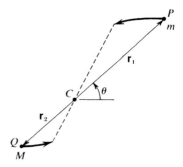

Fig. 13-29 *Binary system, in which the particles follow geometrically similar orbits about the center of mass, C, always at opposite ends of a straight line through C.*

If the particles move only under the action of a central force $F(r)$ exerted between them, it is possible to choose a reference frame in which C is fixed and the vectors $\mathbf{r}_1$ and $\mathbf{r}_2$ are always exactly opposite with a length ratio equal to M/m. The radial acceleration of P with respect to C is then given by

$$a_{r1} = \frac{M}{M+m}\left[\frac{d^2 r}{dt^2} - r\left(\frac{d\theta}{dt}\right)^2\right]$$

Thus the statement of Newton's law for the radial component of the motion of P can be written

$$F(r) = \frac{mM}{M+m}\left[\frac{d^2 r}{dt^2} - r\left(\frac{d\theta}{dt}\right)^2\right] \tag{13-44a}$$

(Note that F is a function of the *total* separation, r, of the particles.) The identical equation would arise from Newton's law applied to the other particle, Q; it involves only the relative coordinate, r, and the reduced mass μ [$= mM/(M+m)$]. Thus we can put

$$F(r) = \mu\left[\frac{d^2 r}{dt^2} - r\left(\frac{d\theta}{dt}\right)^2\right] \tag{13-44b}$$

If F is the gravitational force, $-GMm/r^2$, we see from Eq. (13-44a) that the radial equation becomes

$$\frac{d^2 r}{dt^2} - r\left(\frac{d\theta}{dt}\right)^2 = -\frac{G(M+m)}{r^2}$$

The total mass, $M + m$, thus plays the role that was occupied by M alone in the previous treatment.

The fact that the partners in a two-body system are following geometrically similar orbits about their common center of mass has been beautifully shown in a computer-generated film by

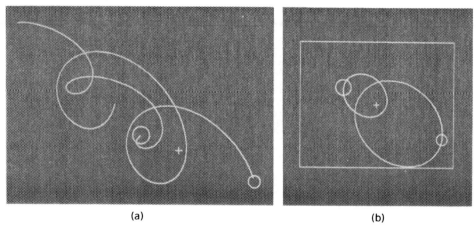

*Fig. 13-30 (a) Paths of two members of a binary system as they might appear in an arbitrary reference frame. (b) Same motions as seen in the CM frame. (From the film "Force, Mass and Motion," by F. W. Sinden, Bell Telephone Laboratories and Education Development Center Film Studio, Newton, Mass., **1965**.)*

Frank Sinden. Figure 13-30 shows two stills from the film; one is of the motions as observed in a frame in which the center of mass is itself moving, and the other refers the motions to the CM frame.

DEDUCING THE ORBIT FROM THE FORCE LAW[1]

Earlier in this chapter we showed how the operation of an inverse-square law could be inferred from the observed fact that the orbit of a planet is an ellipse with the sun at one focus. Later, however, we pointed out that parabolic and hyperbolic orbits were also possible. How do we know this? The answer is, as we said in Chapter 7, that Newton's second law can be used in two main ways. We can infer the forces from the motions, or we can infer the motions from the forces. Generally speaking, the latter is easier than the former and also leads to results of much greater generality. For example, the analysis of the path of one particular object over a limited trajectory near the earth's surface allows us to conclude that a constant vertical acceleration, g, is at work. But if we start out with the fact of this acceleration, we can quickly deduce that *all* trajectories near the earth's surface (ignoring air resistance) are parabolas or parts of parabolas. A similar situation holds for central forces, and we shall illustrate

[1]This section can be omitted without loss of continuity.

the power of the deductive use of Newton's laws in such problems.

The basic equations governing such motions are, as we have seen, the following:

$$F_r = ma_r = m\left[\frac{d^2r}{dt^2} - r\left(\frac{d\theta}{dt}\right)^2\right]$$
$$r^2\frac{d\theta}{dt} = C = \frac{l}{m} \tag{13-45}$$

The *shape* of the orbit is something that can be described without reference to the time; it is just the spatial description of the curve —given, in these problems, by r as a function of θ. Thus we shall be interested in suppressing the explicit time dependence that is represented by the derivatives d^2r/dt^2 and $d\theta/dt$. The clue to doing this is given by the second equation of (13-45): We can put

$$\frac{d\theta}{dt} = \frac{C}{r^2}$$

It proves to be very advantageous in the analysis to introduce the reciprocal of r as a variable. Calling this u we have

$$r = \frac{1}{u}$$
$$\frac{d\theta}{dt} = Cu^2 \tag{13-46}$$

Also, taking the first derivative of r with respect to t, we have

$$\frac{dr}{dt} = -\frac{1}{u^2}\frac{du}{dt}$$

Using the chain rule, we can rewrite this as follows:

$$\frac{dr}{dt} = -\frac{1}{u^2}\frac{du}{d\theta}\frac{d\theta}{dt} = -C\frac{du}{d\theta}$$

[This last step follows from Eq. (13-46).] Differentiating again, we get

$$\frac{d^2r}{dt^2} = -C\frac{d^2u}{d\theta^2}\frac{d\theta}{dt} = -C^2u^2\frac{d^2u}{d\theta^2} \tag{13-47}$$

Using Eqs. (13-46) and (13-47) gives us the following expression of Newton's law as applied to the radial component of the motion:

$$F_r = ma_r = -mC^2u^2\left(\frac{d^2u}{d\theta^2} + u\right) \tag{13-48}$$

The value of developing this particular formulation of the radial equation of motion shows up at once when we designate F_r as a specific function of r. In particular, for the case of motion under gravity with a mass M fixed at the origin, we have

$$F_r = -\frac{GMm}{r^2} = -GMmu^2$$

Substituting this in Eq. (13–48) then leads at once to the following simple equation:

$$\frac{d^2u}{d\theta^2} + u = \frac{GM}{C^2} = A \qquad (13\text{–}49)$$

where A is a constant of the motion. If we rewrite this in the form

$$\frac{d^2}{d\theta^2}(u - A) = -(u - A)$$

it is easy to see that the integrated solution can be written (with a suitable choice of the zero of θ) in the form

$$u - A = B\cos\theta$$

where B is another constant. Returning now to r as a variable, we have the following equation for the orbit:

$$\frac{1}{r} = A + B\cos\theta \qquad (13\text{–}50)$$

We shall point at once to one feature of Eq. (13–50), resulting from our particular choice of the zero of θ. This is that as θ increases from zero, in either the positive or the negative sense, the value of $1/r$ decreases and so r increases (provided that B is positive). Thus the point corresponding to $\theta = 0$ is the perigee point of the orbit; r is passing through its minimum value, which we shall call r_1 [see Fig. 13–31(a)].

Equation (13–50) has a geometrical interpretation that can be described with reference to Fig. 13–31(b). Rewriting the equation slightly, we have

$$\frac{1}{B} = d = \alpha r + r\cos\theta$$

where $d = 1/B$ and $\alpha = A/B$. If we take a line FD of length d passing through the perigee and draw an axis at right angles to this at D, then the orbit is the locus of a point P that moves so

Fig. 13-31 (a) Particle at perigee in an orbit under an inverse-square force. (b) Portion of the orbit, showing the geometrical relationships of the focus, the particle's position, and the directrix (the line DN).

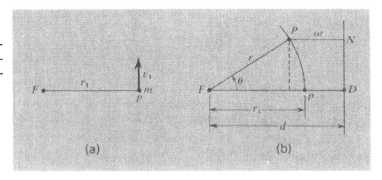

that its perpendicular distance from the line DN is a constant multiple, α, of its distance $(d - r\cos\theta)$ from the focus F. This corresponds to a general prescription for generating the various conic sections.

To interpret the result more fully, we must consider the values of the constants A and B in Eq. (13-50). The value of A is defined in Eq. (13-49):

$$A = \frac{GM}{C^2}$$

Now C is the constant value of $r^2\, d\theta/dt$. We can express this in terms of the radial distance and the speed at perigee:

$$C = r_1 v_1$$

Hence

$$A = \frac{GM}{r_1^2 v_1^2}$$

Now the potential energy and the kinetic energy at perigee are given by the following expressions:

$$U_1 = -\frac{GMm}{r_1} \qquad K_1 = \tfrac{1}{2}mv_1^2$$

This permits us to write the constant A as follows:

$$A = -\frac{U_1}{2r_1 K_1}$$

The value of B follows immediately from putting $\theta = 0$ in Eq. (13-50) itself:

$$B = \frac{1}{r_1} - A = \frac{1}{r_1}\left(1 + \frac{U_1}{2K_1}\right)$$

Now consider the difference between B and A:

$$B - A = \frac{1}{r_1}\left(1 + \frac{U_1}{K_1}\right) = \frac{E}{r_1 K_1}$$

where E is the *total* energy of the motion at every point in the orbit. This is the key to the problem, for we can now recognize three different situations that correspond to the three types of orbit discussed earlier, according to whether E is zero, negative, or positive:

$$E = 0 \quad (B = A, \alpha = 1): \quad \frac{1}{r} = A(1 + \cos\theta)$$

$$E > 0 \quad (B > A, \alpha < 1): \quad \frac{1}{r} = A(\alpha_h + \cos\theta) \qquad (13\text{-}51)$$

$$E < 0 \quad (B < A, \alpha > 1): \quad \frac{1}{r} = A(\alpha_e + \cos\theta)$$

These equations define a parabola, a hyperbola, and an ellipse, in that order. The first two equations clearly permit r to become infinitely great (the first of them as $\theta \to \pi$, the second as $\cos\theta \to -\alpha_h$). The third equation defines maximum and minimum values of r at $\theta = \pi$ and zero, respectively. Further analysis would relate the specific values of the orbit parameters to the dynamical constants of the motion—i.e., to the magnitudes of the energy and the orbital momentum.

Many problems involving force laws other than the inverse square can also be effectively attacked by the methods developed at the beginning of this section and culminating in Eq. (13–48).

RUTHERFORD SCATTERING

As another example of motion in an inverse-square central field of force, we shall consider the deflection of an electrically charged particle in the electric field of a much more massive object carrying an electrical charge of the same sign. The field is repulsive, obeys Coulomb's law, and has the equivalent potential energy

$$U'(r) = \frac{kq_1 q_2}{r} + \frac{l^2}{2mr^2} \qquad (13\text{-}52)$$

as shown plotted in Fig. 13–32(a). k is the proportionality constant in Coulomb's law and q_1 and q_2 are the electrical charges on the two particles.

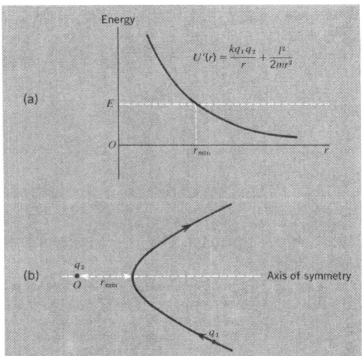

Fig. 13-32 (a) Effective radial potential-energy diagram for a particle with orbital angular momentum l in a repulsive Coulomb field. (b) Plan view of the trajectory of an alpha particle (q_1) in the neighborhood of a massive nuclear charge (q_2).

Motion is possible only for positive energies (E) and all such motions are unbounded, characterized by a distance of nearest approach, r_{min}, which depends on the energy of the moving particle. Because the particle retraces all the values of radial speed on the way out that it had on the way in, and because the angular velocity ($d\theta/dt$) of the particle depends only on its distance r from O, the trajectory will be symmetrical as shown in Fig. 13-32(b).

Historically, the understanding of this type of motion played a basic role in one of the most important experiments of this century. About 1910 Lord Rutherford and his students, especially Geiger and Marsden, performed a series of experiments on the scattering of a beam of alpha particles by thin metallic foils. These experiments showed that most of the mass of atoms is concentrated in a small positively charged nucleus. Presumably the electrons in the atom surrounded this nucleus like a cloud. This nuclear model of an atom was in sharp contrast to that previously proposed by J. J. Thomson, which was essentially a ball of distributed positive charge in which the electrons were embedded.

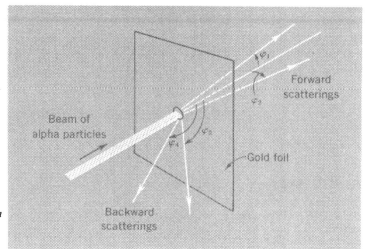

Fig. 13-33 Scattering of alpha particles by the nuclei in a thin metallic foil.

Figure 13-33 is a schematic diagram of one such experiment, using a thin gold foil as a target for the alpha particles. A collimated beam of alpha particles (helium nuclei) from a polonium source is incident on a gold foil about 4×10^{-7} m (4000 Å) in thickness. Although most of the alpha particles were observed to pass through the foil, deflected chiefly through small angles (forward scattering), a few were found to be deflected through angles greater than 90° (backward scattering). The fraction undergoing backward scattering was exceedingly small, only a few parts in a million. Rutherford reasoned that alpha particles could not be scattered backward by electrons in the gold atoms because electrons have such a small mass relative to the mass of He^{2+}. In fact, since this mass ratio is of the order of 1/7000, the scattering effect of the electrons should be quite negligible even in forward direction. The occurrence of backward scattering can be accounted for only if alpha particles pass *very close* to a mass that is *very large* compared to the mass of alpha particles. This argument led Rutherford to propose an atomic model with a very small but massive nucleus. If this is a correct model, one should be able to predict the relative number of alpha particles scattered in different directions under the influence of the Coulomb field that surrounds a much more massive positively charged gold nucleus. This was done and the quantitative experimental results agreed with the calculations. Let us see how such a calculation of scattering is obtained.

For a thin foil, an alpha particle scattered by one of the gold nuclei (there are about 6×10^{19} gold atoms/mm^3 in the foil) has a small chance of being deflected appreciably by a second nucleus before it emerges from the foil. Therefore, we can consider all scattering of the beam to be the result of single scattering processes. First, then, we must find the deflection of an alpha particle by one gold nucleus. Next, since we cannot aim each alpha particle, we must compute the relative number deflected through a certain angle, taking into account all possible "aiming errors."

Although we can use the methods of satellite orbits to obtain the deflection of an alpha particle by a gold nucleus, we shall present a much simpler and more direct argument. Consider an alpha particle (charge q_1 and mass m) moving with a speed v_0 toward a gold nucleus of charge q_2 as shown in Fig. 13-34.

Clearly, the deflection will be larger, the more nearly v_0 points at the charge q_2. The distance b in the figure is the impact parameter defined earlier in this chapter (p. 570) and is a measure of the aiming error. Since at large distances from q_2 the potential energy of m in the field of q_2 is negligibly small, the kinetic energy $\frac{1}{2}mv_0^2$ at such distances is the total energy of the motion. This total energy is conserved in the encounter. Hence the alpha particle regains its initial speed after scattering, and the only effect of the process is to change the direction of its motion by an amount equal to the angle φ in Fig. 13-34. To be sure, the alpha

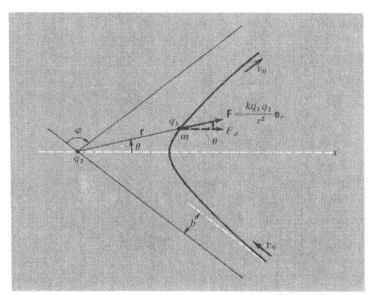

Fig. 13-34 Geometry of a Rutherford-scattering event.

particle slows down as it approaches the gold nucleus, but it regains its original speed on the way out. In addition, the angular momentum of m about q_2 is conserved and this constant angular momentum l is given by

$$l = mv_0 b = mr^2 \frac{d\theta}{dt} = \text{const.} \tag{13-53}$$

The total change of momentum $\Delta(mv)$ in the scattering process is the difference of the two vectors, each of magnitude mv_0, shown in Fig. 13-35, and is equal in magnitude to

$$\Delta(mv) = 2mv_0 \sin \frac{\varphi}{2} \tag{13-54}$$

This must be equal to the total impulse supplied by the force $\mathbf{F}$ of Fig. 13-34 during the scattering process. This impulse is the vector

$$\Delta \mathbf{p} = \int \mathbf{F}\, dt$$

From the symmetry of Fig. 13-34, we see that only the component F_x of $\mathbf{F}$ contributes to this impulse, because the perpendicular contributions from F_y at points on the trajectory below the x axis just cancel the corresponding contributions at points above the x axis. This makes sense, because $\Delta(mv)$ is parallel to x as indicated in Fig. 13-35. Thus, Newton's law of motion gives us

$$\Delta p = \int F_x\, dt = \int F \cos \theta\, dt = \Delta(mv) = 2mv_0 \sin \frac{\varphi}{2}$$

and we must evaluate the integral. Writing this integral as

$$\Delta p = \int \frac{kq_1 q_2 \cos \theta}{r^2}\, dt$$

and using Eq. (13-53), we have

$$\Delta p = \frac{kq_1 q_2}{v_0 b} \int_{\theta_1}^{\theta_2} \cos \theta\, d\theta = \frac{kq_1 q_2}{v_0 b} (\sin \theta_2 - \sin \theta_1) \tag{13-55}$$

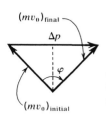

Fig. 13-35 Net dynamical result of a scattering event in terms of the impulse Δp that changes the direction but not the magnitude of the alpha-particle momentum.

θ_1 and θ_2 are the values of θ before and after scattering. From Fig. 13-34 we see that

$$\theta_1 = -\left(\frac{\pi - \varphi}{2}\right) \quad \text{and} \quad \theta_2 = +\left(\frac{\pi - \varphi}{2}\right)$$

Since $\sin[(\pi - \varphi)/2] = \cos(\varphi/2)$, Eq. (13-55) becomes

$$\Delta p = \frac{2kq_1q_2}{v_0 b} \cos \frac{\varphi}{2}$$

Equating this expression for Δp to the value of $\Delta(mv)$ according to Eq. (13-54), we find

$$\tan \frac{\varphi}{2} = \frac{kq_1q_2}{mv_0^2 b} \tag{13-56}$$

This tells us, for each value of the impact parameter b, the angle of scattering of particles of a given energy. If the beam of alpha particles is essentially monoenergetic, we can proceed to use Eq. (13-56) to calculate the relative numbers of the incident particles that are scattered through different angles φ. This involves finding the fraction of the incident alpha particles that have impact parameters between b and $b + db$ and from this the fraction scattered into the corresponding range of angles φ.

CROSS SECTIONS FOR SCATTERING

It is customary to express the relative numbers of particles scattered through different angles in terms of a quantity called the scattering cross section of the particle that does the scattering. The primary definition of a cross section is simply that it is the effective target area presented by each scattering center to the incident beam. To develop this quantitatively, consider a greatly magnified picture of a very small square of a scattering foil [Fig. 13-36(a)] and imagine that one can look right through the thickness of the foil. Each scattering center blocks out an area σ, and we shall assume that the foil is so thin that, as viewed from the front, there is no overlapping of the cross sections at various depths in the foil (see the figure).

Now the material of the foil has a certain characteristic number, n, of atoms per unit volume. Thus if the thin slice is a square of edge l and thickness Δx, the number of atoms in it is $nl^2 \Delta x$. With each atom we associate a nuclear cross section σ,

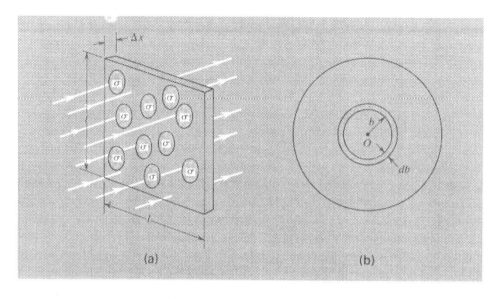

Fig. 13-36 (a) Schematic view of a portion of a scattering foil, with each nucleus presenting an effective target area σ. (b) The total target area of a nucleus can be subdivided into partial cross sections corresponding to the rings contained between neighboring values of the impact parameter, b.

so that the amount of area blocked with respect to the incoming particles is $\sigma n l^2 \Delta x$. The total area within which this portion is blocked out is just l^2. This then means that the *fraction, f,* of the total area that is obstructed is given by

$$f = \frac{\sigma n l^2 \Delta x}{l^2} = \sigma n \Delta x$$

Since the incident particles are striking the foil at completely random positions as measured on the scale of interatomic distances (even though the beam as a whole is limited in extent laterally, within perhaps a millimeter or so), the fraction of particles falling within a nuclear target area is identical with the fraction f given above. Thus if n' alpha particles approach the scattering foil and $\Delta n'$ of them undergo a nuclear scattering interaction of some kind, we have

$$f = \frac{\Delta n'}{n'} = \sigma n \Delta x \qquad (13\text{-}57)$$

Hence from a measurement of $\Delta n'/n'$, using a scattering foil of

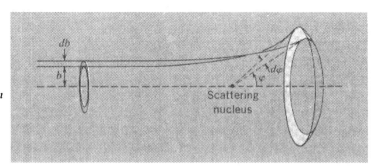

Fig. 13-37 Relation between scattering angles and impact parameters.

known thickness and composition, the effective cross section of an individual nucleus can be deduced.[1]

One often speaks of *partial* as well as total scattering cross sections. This concept is especially important in the analysis of "Rutherford scattering" and of many similar processes. The partial cross section has a very direct interpretation. Imagine that one is looking toward a scattering nucleus whose center is at the point O in Fig. 13-36(b). Then one can picture a ring contained between the impact parameters b and $b + db$. The area of this ring defines a partial cross section $d\sigma$:

$$d\sigma = 2\pi b \, db \tag{13-58}$$

Now Eq. (13-56) implies a unique connection between the value of b and the consequent scattering angle φ. This is indicated in Fig. 13-37. Since the partial cross section $d\sigma$ is completely defined by Eq. (13-58), it follows that the use of Eq. (13-56) leads to a specific prediction about the relative number of alpha particles scattered into angles between φ and $\varphi + d\varphi$. Let us make this quantitative.

Writing Eq. (13-56) in the form

$$b = \frac{kq_1q_2}{mv_0^2} \cot \frac{\varphi}{2}$$

we have

$$db = \frac{kq_1q_2}{2mv_0^2} \csc^2 \frac{\varphi}{2} \, d\varphi$$

---

[1] In this description, we have ignored the electrostatic "screening" provided by the electrons around a nucleus. Such screening drastically affects the scattering process for large impact parameters (comparable to an *atomic* radius).

LXI. *The Laws of Deflexion of α Particles through Large Angles*[*]. *By* Dr. H. Geiger *and* E. Marsden [†].

IN a former paper [‡] one of us has shown that in the passage of α particles through matter the deflexions are, on the average, small and of the order of a few degrees only. In the experiments a narrow pencil of α particles fell on a zinc-sulphide screen in vacuum, and the distribution of the scintillations on the screen was observed when different metal foils were placed in the path of the α particles. From the distribution obtained, the most probable angle of scattering could be deduced, and it was shown that the results could be explained on the assumption that the deflexion of a single α particle is the resultant of a large number of very small deflexions caused by the passage of the α particle through the successive individual atoms of the scattering substance.

[*] Communicated to *k. d.-k. Akad. d. Wiss. Wien*.
[†] Communicated by Prof. E. Rutherford, F.R.S.
[‡] H. Geiger, Roy. Soc. Proc. vol. lxxxiii. p. 492 (1910); vol. lxxxvi. p. 235 (1912).

In an earlier paper [*], however, we pointed out that α particles are sometimes turned through very large angles. This was made evident by the fact that when α particles fall on a metal plate, a small fraction of them, about 1/8000 in the case of platinum, appears to be diffusely reflected. This amount of reflection, although small, is, however, too large to be explained on the above simple theory of scattering. It is easy to calculate from the experimental data that the probability of a deflexion through an angle of 90° is vanishingly small, and of a different order to the value found experimentally.

Professor Rutherford [†] has recently developed a theory to account for the scattering of α particles through these large angles, the assumption being that the deflexions are the result of an intimate encounter of an α particle with a single atom of the matter traversed. In this theory an atom is supposed to consist of a strong positive or negative central charge concentrated within a sphere of less than about 3×10^{-12} cm. radius, and surrounded by electricity of the opposite sign distributed throughout the remainder of the atom of about 10^{-8} cm. radius. In considering the deflexion of an α particle directed against an atom, the main deflexion-effect can be supposed to be due to the central concentrated charge which will cause the α particle to describe an hyperbola with the centre of the atom as one focus.

The angle between the directions of the α particle before and after deflexion will depend on the perpendicular distance of the initial trajectory from the centre of the atom. The fraction of the α particles whose paths are sufficiently near to the centre of the atom will, however, be small, so that the probability of an α particle suffering a large deflexion of this nature will be correspondingly small. Thus, assuming a narrow pencil of α particles directed against a thin sheet of matter containing atoms distributed at random throughout its volume, if the scattered particles are counted by the scintillations they produce on a zinc-sulphide screen distance r from the point of incidence of the pencil in a direction making an angle ϕ with it, the number of α particles falling on unit area of the screen per second is deduced to be equal to

$$\frac{Q n t b^2 \operatorname{cosec}^4 \phi/2}{16 r^2},$$

where Q is the number of α particles per second in the

[*] H. Geiger and E. Marsden, Roy. Soc. Proc. vol. lxxxii. p. 495 (1909).
[†] E. Rutherford, Phil. Mag. vol. xxi. p. 669 (1911).

original pencil, n the number of atoms in unit volume of the material, and t the thickness of the foil. The quantity

$$b = \frac{2 N e E}{m u^2},$$

where Ne is the central charge of the atom, and m, E, and u are the respective mass, charge, and velocity of the α particle.

The number of deflected α particles is thus proportional to (1) $\operatorname{cosec}^4 \phi/2$, (2) thickness of scattering material t if the thickness is small, (3) the square of the central charge Ne of the atoms of the particular matter employed to scatter the particles, (4) the inverse fourth power of the velocity u of the incident α particles.

At the suggestion of Prof. Rutherford, we have carried out experiments to test the main conclusions of the above theory. The following points were investigated :—

(1) Variation with angle.
(2) Variation with thickness of scattering material.
(3) Variation with atomic weight of scattering material.
(4) Variation with velocity of incident α particles.
(5) The fraction of particles scattered through a definite angle.

The main difficulty of the experiments has arisen from the necessity of using a very intense and narrow source of α particles owing to the smallness of the scattering effect. All the measurements have been carried out by observing the scintillations due to the scattered α particles on a zinc-sulphide screen, and during the course of the experiments over 100,000 scintillations have been counted. It may be mentioned in anticipation that all the results of our investigation are in good agreement with the theoretical deductions of Prof. Rutherford, and afford strong evidence of the correctness of the underlying assumption that an atom contains a strong charge at the centre of dimensions, small compared with the diameter of the atom.

(1) *Variation of Scattering with Angle.*

We have already pointed out that to obtain measurable effects an intense pencil of α particles is required. It is further necessary that the path of the α particles should be in an evacuated chamber to avoid complications due to the absorption and scattering of the air. The apparatus used is shown in fig. 1, and mainly consisted of a strong cylindrical metal box B, which contained the source of α particles R, the scattering foil F, and a microscope M to which the zinc-sulphide screen S was rigidly attached. The box was fastened down to a graduated circular platform A, which could be rotated by means of a conical airtight joint C. By rotating the platform the box and microscope moved with it, whilst the scattering foil and radiating source remained in position, being attached to the tube T, which was fastened to the standard L. The box B was closed by the ground-glass plate P, and could be exhausted through the tube T.

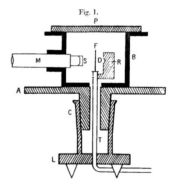

Fig. 1.

The source of α particles employed was similar to that used originally by Rutherford and Royds [*] in their experiments on the nature of the α particle. It consisted of a small thin-walled glass tube about 1 mm. in diameter, containing a large quantity of well purified radium emanation. The α particles emitted by the emanation and its active deposit could pass through the glass walls without much reduction of range. For these experiments the unhomogeneity of the source, due to the different α particles from the emanation, Ra A and Ra C, does not interfere with the application of the law of scattering with angle as deduced from the theory, as each group of α particles is scattered according to the same law.

By means of a diaphragm placed at D, a pencil of α particles was directed normally on to the scattering foil F. By

[*] E. Rutherford and T. Royds, Phil. Mag. vol. xvii. p. 281 (1909).

rotating the microscope the α particles scattered in different directions could be observed on the screen S. Although over 100 millicuries of radium emanation were available for the experiments, the smallness of the effect for the larger angles of deflexion necessitated short distances of screen and source from the scattering foil. In some experiments the distance between the source and scattering foil was 2·5 cm., and the screen moved in a circle of 1·6 cm. radius, while in other experiments these distances were increased. Observations were taken in various experiments for angles of deflexion from 5° to 150°. When measuring the scattering through large angles the zinc-sulphide screen had to be turned very near to the source, and the β and γ rays produced a considerable luminescence on it, thus making countings of the scintillations difficult. The effect of the β rays was reduced as far as possible by enclosing the source in a lead box shown shaded in the diagram. The amount of lead was, however, limited by considerations of the space taken up by it, and consequently observations could not be made for angles of deflexion between 150° and 180°.

In the investigation of the scattering through relatively small angles the distances of source and screen from the scattering foil were increased considerably in order to obtain beams of smaller solid angle.

TABLE VII.
Variation of Scattering with Velocity.

I. Number of sheets of mica.	II. Range R of α particles after leaving mica.	III. Relative values of $1/v^4$.	IV. Number N of scintillations per minute.	V. Nv^4.
0	5·5	1·0	24·7	25
1	4·76	1·21	29·0	24
2	4·05	1·50	33·4	22
3	3·32	1·91	44	23
4	2·51	2·84	81	28
5	1·84	4·32	101	23
6	1·04	9·22	255	28

TABLE II.
Variation of Scattering with Angle. (Collected results.)

I. Angle of deflexion, φ.	II. $\frac{1}{\sin^4 \phi/2}$	III. SILVER. Number of scintillations, N.	IV. $\frac{N}{\sin^4 \phi/2}$	V. GOLD. Number of scintillations, N.	VI. $\frac{N}{\sin^4 \phi/2}$
150	1·15	22·2	19·3	33·1	28·8
135	1·38	27·4	19·8	43·0	31·2
120	1·79	33·0	18·4	51·9	29·0
105	2·53	47·3	18·7	69·5	27·5
75	7·25	136	18·8	211	29·1
60	16·0	320	20·0	477	29·8
45	46·6	989	21·2	1435	30·8
37·5	93·7	1760	18·8	3300	35·3
30	223	5260	23·6	7800	35·0
22·5	690	20300	29·4	27300	39·6
15	3445	105400	30·6	132000	38·4
30	223	5·3	0·024	3·1	0·014
22·5	690	16·6	0·024	8·4	0·012
15	3445	93·0	0·027	48·2	0·014
10	17330	508	0·029	200	0·0115
7·5	54650	1710	0·031	607	0·011
5	276300	...	...	3320	0·012

Variation with Thickness of Material.

In investigating the variation of scattering with thickness of material, it seemed necessary to use a homogeneous source of α particles, for according to the theory the effect of the change of velocity with increasing thickness will be very appreciable for α particles of low velocity.

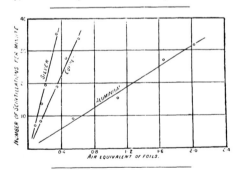

For all the metals examined the points lie on straight lines which pass through the origin. The experiments therefore prove that for small thicknesses of matter the scattering is proportional to the thickness.

Summary.

The experiments described in the foregoing paper were carried out to test a theory of the atom proposed by Prof. Rutherford, the main feature of which is that there exists at the centre of the atom an intense highly concentrated electrical charge. The verification is based on the laws of scattering which were deduced from this theory. The following relations have been verified experimentally:—

(1) The number of α particles emerging from a scattering foil at an angle φ with the original beam varies as $1/\sin^4 \phi/2$, when the α particles are counted on a definite area at a constant distance from the foil. This relation has been tested for angles varying from 5° to 150°, and over this range the number of α particles varied from 1 to 250,000 in good agreement with the theory.

(2) The number of α particles scattered in a definite direction is directly proportional to the thickness of the scattering foil for small thicknesses. For larger thicknesses the decrease of velocity of the α particles in the foil causes a somewhat more rapid increase in the amount of scattering.

(3) The scattering per atom of foils of different materials varies approximately as the square of the atomic weight. This relation was tested for foils of atomic weight from that of carbon to that of gold.

(4) The amount of scattering by a given foil is approximately proportional to the inverse fourth power of the velocity of the incident α particles. This relation was tested over a range of velocities such that the number of scattered particles varied as 1 : 10.

(5) Quantitative experiments show that the fraction of α particles of Ra C, which is scattered through an angle of 45° by a gold foil of 1 mm. air equivalent ($2 \cdot 1 \times 10^{-5}$ cm.), is $3 \cdot 7 \times 10^{-7}$ when the scattered particles are counted on a screen of 1 sq. mm. area placed at a distance of 1 cm. from the scattering foil. From this figure and the foregoing results, it can be calculated that the number of elementary charges composing the centre of the atom is equal to half the atomic weight.

(We have suppressed the negative sign that tells us that φ decreases as b increases.) Thus the partial cross section, as given by Eq. (13–58), becomes

$$d\sigma = \pi \left(\frac{kq_1q_2}{mv_0^2}\right)^2 \frac{\cos(\varphi/2)\, d\varphi}{\sin^3(\varphi/2)}$$

Now the scattering experiment is actually done by observing, as a function of φ, the number of alpha particles that enter a detector subtending a certain fixed solid angle at the place on the foil where the scattering processes are occurring. Thus a direct measure of the process is the amount of scattering per element of solid angle. This measure is provided by the so-called differential cross section, which is $d\sigma$ divided by the solid angle $d\Omega$ contained between the directions φ and $\varphi + d\varphi$. Since

$$d\Omega = 2\pi \sin\varphi\, d\varphi = 4\pi \sin\frac{\varphi}{2}\cos\frac{\varphi}{2}\, d\varphi$$

we have

$$\frac{d\sigma}{d\Omega} = \frac{1}{4}\left(\frac{kq_1q_2}{mv_0^2}\right)^2 \frac{1}{\sin^4(\varphi/2)} \tag{13-59}$$

This theoretical result for Coulomb scattering can then be compared with experimental observation, using an equation analogous to Eq. (13–57) but now limiting attention to scattering into a certain solid angle $\Delta\Omega$ at a scattering angle φ:

$$\Delta f = n\,\Delta x \left(\frac{d\sigma}{d\Omega}\right) \Delta\Omega \tag{13-60}$$

In Eq. (13–59), q_1 is the charge on the alpha particles, ($q_1 = 2e$, where e is the elementary charge) and q_2 ($= Ze$) is the charge on the scattering nucleus. mv_0^2 is just twice the kinetic energy of the alpha particles. The dependence of the observed scattering on Z, the atomic number of the scattering nucleus, on the energy of the alpha particles and on the angle of scattering φ agrees very well with the predictions of Eq. (13–59). One would expect deviations from the theory under two kinds of conditions:

1. If the impact parameter b is so *large* as to be a significant fraction of an atomic radius, the nuclear charge is partially shielded by the surrounding electron cloud, and the angle of scattering is correspondingly reduced, compared to that predicted by Eq. (13–56).

2. If the impact parameter is so *small* that the alpha particle comes within range of the specifically nuclear forces, one can no longer expect the scattering to conform to that calculated according to Coulomb's law for a point charge.

Since deviations of type 2 from the Rutherford scattering law [Eq. (13-59)] occur for those alpha particles which get in close to the scattering nucleus, they will become apparent in the backward scattering, especially for alpha particles of higher energy. From the observation of the energy of alpha particles for which this back scattering starts to depart from the Rutherford formula, one gets an upper limit for the size of the scattering nucleus. For the case of gold, one finds in this manner that the gold nucleus has a radius less than about 1.5×10^{-14} m ($= 15$ F). This may be compared to the radius of the gold atom which is about 1.5×10^{-10} m. The ratio of the radius of the atom to that of the nucleus is thus about 10^4, so that the volume of the atom is of the order of 10^{12} times that of its nucleus. The massive nucleus is thus concentrated into an extremely small fraction of the whole atomic volume, and the deflection of an alpha particle occurs as it passes through a region close to the nucleus, after penetrating the electrostatic shield provided by the atomic electrons.

AN HISTORICAL NOTE

Lord Rutherford (1871–1937), born in New Zealand, was an experimental physicist of remarkable skill. He received the Nobel prize in chemistry in 1908 and achieved the first experimental transmutation of matter in 1919 when he bombarded nitrogen with alpha particles, producing an isotope of oxygen. The steps that led to his discovery of the atomic nucleus are interestingly expressed in Rutherford's own words:

> Now I myself was very interested in the next stage, so I will give you it in some detail, and I would like to use this example to show how you often stumble upon facts by accident. In the early days I had observed the scattering of α-particles, and Dr. Geiger in my laboratory had examined it in detail. He found, in thin pieces of heavy metal, that the scattering was usually small, of the order of one degree. One day Geiger came to me and said, "Don't you think that young Marsden, whom I am training in radioactive methods, ought to begin a small research?"

Now I had thought that too, so I said, "Why not let him see if any α-particles can be scattered through a large angle?" I may tell you in confidence that I did not believe that they would be, since we knew that the α-particle was a very fast massive particle, with a great deal of energy, and you could show that if the scattering was due to the accumulated effect of a number of small scatterings the chance of an α-particle's being scattered backwards was very small. Then I remembered two or three days later Geiger coming to me in great excitement and saying, "We have been able to get some of the α-particles coming backwards..." It was quite the most incredible event that has ever happened to me in my life. It was almost as incredible as if you fired a 15-inch shell at a piece of tissue paper and it came back and hit you. On consideration I realized that this scattering backwards must be the result of a single collision, and when I made calculations I saw that it was impossible to get anything of that order of magnitude unless you took a system in which the greater part of the mass of the atom was concentrated in a minute nucleus. It was then that I had the idea of an atom with a minute massive centre carrying a charge. I worked out mathematically what laws the scattering should obey, and I found that the number of particles scattered through a given angle should be proportional to the thickness of the scattering foil, the square of the nuclear charge, and inversely proportional to the fourth power of the velocity. These deductions were later verified by Geiger and Marsden in a series of beautiful experiments.[1]

On pp. 612–613 we have reproduced some excerpts from the original paper by Geiger and Marsden. It is interesting to calculate the expected backward scattering ($> 90°$) on the basis of the most widely accepted atomic model of 1910. This was the Thomson model, in which the negative electrons were imagined to be distributed throughout a sphere of uniform positive charge of radius about 10^{-8} cm. A passing alpha particle could be deflected by the electrostatic repulsion of the positive charge, which constituted most of the mass of the atom.[2] The maximum deflection in a single encounter was quite small. However, "multiple scattering" from several atoms might occur

---

[1] From *Background to Modern Science* (ten lectures by various scientists at Cambridge University in 1936; J. Needham and W. Pagel, eds.), Cambridge University Press, England (1938).
[2] The electrons, as we mentioned earlier, are so light compared to the alpha particles that they would be simply brushed aside in a collision between the two.

in sufficiently thick foils, producing a net deflection which is large. For a gold foil 10^{-4} cm thick such as Geiger and Marsden used for some of their experiments, the Thomson theory predicted that the fraction of alpha particles scattered at angles greater than 90° would be about one out of every 10^{1000}! That is tantamount to saying that it would never happen. (Recall, for the purposes of comparison, that the total number of all the electrons, protons, and neutrons in all the galaxies of the observable universe is only about 10^{80}.) No wonder Rutherford was astonished when Geiger and Marsden observed for a foil of this thickness that approximately one out of every 10^4 alpha particles was deflected at angles greater than 90°.

PROBLEMS

13-1 The circular orbits under the action of a certain central force $F(r)$ are found all to have the same rate of sweeping out area by the radius vector, independent of the orbital radius. Determine how F varies with r.

13-2 In the Bohr model of the hydrogen atom an electron (mass m) moves in a circular orbit around an effectively stationary proton, under the central Coulomb force $F(r) = -ke^2/r^2$.

(a) Obtain an expression for the speed v of the electron as a function of r.

(b) Obtain an expression for the orbital angular momentum l as a function of r.

(c) Introduce Bohr's postulate (of the so-called "old quantum theory," now superseded) that the angular momentum in a circular orbit is equal to $nh/2\pi$, where h is Planck's constant. Obtain an expression for the permitted orbital radii.

(d) Calculate the potential energy of the system from the equation

$$U(r) = -\int_{\infty}^{r} F(r)\, dr$$

Hence find an expression for the total energy of the quantized system as a function of n.

(e) For the lowest energy state of the atom (corresponding to $n = 1$) calculate the numerical values of the orbital radius and the energy, measured in electron volts, needed to ionize the atom. ($k = 9 \times 10^9$ N-m^2/C^2; $e = 1.6 \times 10^{-19}$ C; $m = 9.1 \times 10^{-31}$ kg; $\hbar = h/2\pi = 1.05 \times 10^{-34}$ J-sec.)

13-3 A mass m is joined to a fixed point O by a string of length l.

Initially the string is slack and the mass is moving with constant speed v_0 along a straight line. At its closest approach the distance of the mass from O is h. When the mass reaches a distance l from O, the string becomes taut and the mass goes into a circular path around O. Find the ratio of the final kinetic energy of the mass to its initial kinetic energy. (Neglect any effects of gravity.) Where did the energy go?

13-4 A particle A, of mass m, is acted on by the gravitational force from a second particle, B, which remains fixed at the origin. Initially, when A is very far from B ($r = \infty$), A has a velocity v_0 directed along the line shown in the figure. The perpendicular distance between B and this line is D. The particle A is deflected from its initial course by B and moves along the trajectory shown in the figure. The shortest distance between this trajectory and B is found to be d. Deduce the mass of B in terms of the quantities given and the gravitational constant G.

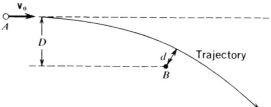

13-5 A particle of mass m moves in the field of a repulsive central force Am/r^3, where A is a constant. At a very large distance from the force center the particle has speed v_0 and its impact parameter is b. Show that the closest m comes to the center of force is given by

$$r_{\min} = (b^2 + A/v_0^2)^{1/2}$$

13-6 A nonrotating, spherical planet with no atmosphere has mass M and radius R. A particle is fired off from the surface with a speed equal to three quarters of the escape speed. By considering conservation of total energy and angular momentum, calculate the farthest distance that it reaches (measured from the center of the planet) if it is fired off (a) radially and (b) tangentially. Sketch the effective potential-energy curve, given by

$$U'(r) = -\frac{GMm}{r} + \frac{l^2}{2mr^2}$$

for case (b). Draw the line representing the total energy of the motion, and thus verify your result.

13-7 Imagine a spherical, nonrotating planet of mass M, radius R, that has no atmosphere. A satellite is fired from the surface of the planet with speed v_0 at 30° to the local vertical. In its subsequent orbit the satellite reaches a maximum distance of $5R/2$ from the center of the planet. Using the principles of conservation of energy and angular momentum, show that

$$v_0 = (5GM/4R)^{1/2}$$

13-8 A particle moves under the influence of a central *attractive* force, $-k/r^3$. At a very large (effectively infinite) distance away, it has a nonzero velocity that does *not* point toward the center. Construct the effective potential-energy diagram for the radial component of the motion. What conclusions can you draw about the dependence on r of the radial component of velocity?

13-9 A satellite in a circular orbit around the earth fires a small rocket. Without going into detailed calculations, consider how the orbit is changed according to whether the rocket is fired (a) forward; (b) backward; (c) toward the earth; and (d) perpendicular to the plane of the orbit.

13-10 Two spacecraft are coasting in exactly the same circular orbit around the earth, but one is a few hundred yards behind the other. An astronaut in the rear wants to throw a ham sandwich to his partner in the other craft. How can he do it? Qualitatively describe the various possible paths of transfer open to him. (This question was posed by Dr. Lee DuBridge in an after-dinner speech to the American Physical Society on April 27, 1960.)

13-11 The elliptical orbit of an earth satellite has major axis $2a$ and minor axis $2b$. The distance between the earth's center and the other focus is $2c$. The period is T.
  (a) Verify that $b = (a^2 - c^2)^{1/2}$.
  (b) Consider the satellite at perigee ($r_1 = a - c$) and apogee ($r_2 = a + c$). At these two points its velocity vector and its radius vector are at right angles. Verify that conservation of energy implies that

$$\tfrac{1}{2}mv_1^2 - \frac{GMm}{a-c} = \tfrac{1}{2}mv_2^2 - \frac{GMm}{a+c} = E$$

Verify also that conservation of angular momentum implies that

$$\frac{\pi ab}{T} = \tfrac{1}{2}(a-c)v_1 = \tfrac{1}{2}(a+c)v_2$$

  (c) From the above relationships, deduce the following results, corresponding to Eqs. (13-36) and (13-39) in the text:

$$T^2 = 4\pi^2 a^3/GM \quad \text{and} \quad E = -GMm/2a$$

13-12 A satellite of mass m is in an elliptical orbit about the earth. When the satellite is at its perigee, a distance R_0 from the center of the earth, it is traveling with a speed v_0. The mass of the earth, M, is much greater than m.

(a) If the length of the major axis of the elliptical orbit is $4R_0$, what is the speed of the satellite at its apogee (the maximum distance from the earth) in terms of G, M, and R_0?

(b) Show that the length of the minor axis of the elliptical orbit is $2\sqrt{3}\,R_0$, and find the period of the satellite in terms of v_0 and R_0.

13–13 A satellite of mass m is traveling at speed v_0 in a circular orbit of radius r_0 under the gravitational force of a fixed mass at O.

(a) Taking the potential energy to be zero at $r = \infty$, show that the total mechanical energy of the satellite is $-\tfrac{1}{2}mv_0^2$.

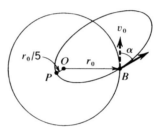

(b) At a certain point B in the orbit (see the figure) the direction of motion of the satellite is suddenly changed *without any change in the magnitude of the velocity*. As a result the satellite goes into an elliptic orbit. Its closest distance of approach to O (at point P) is now $r_0/5$. What is the speed of the satellite at P, expressed as a multiple of v_0?

(c) Through what angle α (see the figure) was the velocity of the satellite turned at the point B?

13–14 A small satellite is in a circular orbit of radius r_1 around the earth. The direction of the satellite's velocity is now changed, causing it to move in an elliptical orbit around the earth. The change in velocity is made in such a manner that the satellite loses half its orbital angular momentum, but its total energy remains unchanged. Calculate, in terms of r_1, the perigee and apogee distances of the new orbit (measured with respect to the earth's center).

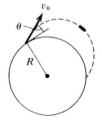

13–15 An experimental rocket is fired from Cape Kennedy with an initial speed v_0 and angle θ to the horizontal (see the figure). Neglecting air friction and the earth's rotational motion, calculate the maximum distance from the center of the earth that the rocket achieves in terms of the earth's mass and radius (M and R), the gravitational constant G, and θ and v_0.

13–16 A satellite of mass m is traveling in a perfectly circular orbit of radius r about the earth (mass M). An explosion breaks up the satellite into two equal fragments, each of mass $m/2$. Immediately after the explosion the two fragments have radial components of velocity equal to $v_0/2$, where v_0 is the orbital speed of the satellite prior to the explosion; in the reference frame of the satellite at the instant of the explosion the fragments appear to separate along the line joining the satellite to the center of the earth.

(a) In terms of G, M, m, and r, what are the energy *and* the angular momentum (with respect to the earth's center) of each fragment?

(b) Make a sketch showing the original circular orbit and the orbits of the two fragments. In making the sketch, use the fact that the major axis of the elliptic orbit of a satellite is inversely proportional to the total energy.

13-17 A spaceship is in an elliptical orbit around the earth. It has a certain amount of fuel for orbit alteration. Where in the orbit should this fuel be used to attain the greatest distance from earth? Do you notice any similarity between this problem and the one concerning a rocket ignited after falling down a chute (Problem 10-13)?

13-18 The commander of a spaceship that has shut down its engines and is coasting near a strange-appearing gas cloud notes that the ship is following a circular path that will lead directly *into* the cloud (see the figure). He also deduces from the ship's motion that its angular momentum with respect to the cloud is not changing. What attractive (central) force could account for such an orbit?

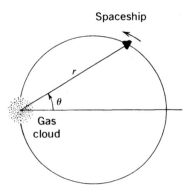

13-19 (a) Make an analysis of an earth-to-Mars orbit transfer similar to that carried out in the text for the transfer to Venus. Assume that earth and Mars are in circular orbits of radii 1 and 1.52 AU, respectively.

(b) In part (a), and in the discussion in the text, the gravitational fields of the planets are neglected. (The problem was taken to be simply that of shifting from one orbit to another, not from the surface of one planet to the surface of the other.) At what distance from the earth is the earth's field equal in magnitude to that of the sun? Similarly, at what distance from Mars is the sun's field equaled by that of the planet? Further, compare the work done against the sun's gravity in the transfer with that done against the earth's gravity, and with the energy gained from the gravitational field of Mars.

13-20 The problem of dropping a spacecraft into the sun from the earth's orbit with the application of minimum possible impulse (given to the spacecraft by firing a rocket engine) is not solved by firing the rocket in a direction opposite to the earth's orbital motion so as to reduce the velocity of the spacecraft to zero. A two-step process can accomplish the goal with a smaller rocket. Assume the initial orbit to be a circle of radius r_1 with the sun at the center (see the figure). By

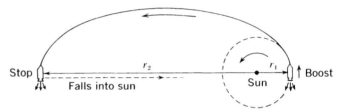

means of a brief rocket burn the spacecraft is speeded up tangentially in the direction of the orbit velocity, so that it assumes an elliptical orbit whose perihelion coincides with the firing point. At the aphelion of this orbit the spacecraft is given a backward impulse sufficient to reduce its space velocity to zero, so that it will subsequently fall into the sun. (As in the previous transfer problem, the effects of the earth's gravity are neglected.)

(a) For a given value of the aphelion distance, r_2 of the spacecraft, calculate the required increment of speed given to it at first firing.

(b) Find the speed of the spacecraft at its aphelion distance, and so find the sum of the speed increments that must be given to the spacecraft in the two steps to make it fall into the sun. This sum provides a measure of the total impulse that the rocket engine must be able to supply. Compare this sum with the speed of the spacecraft in its initial earth orbit for the case $r_2 = 10r_1$.

[*Note:* This problem is discussed by E. Feenberg, "Orbit to the Sun," *Am. J. Phys.*, **28**, 497 (1960).]

13-21 The sun loses mass at the rate of about 4×10^6 tons/sec. What change in the length of the year should this have produced within the span of recorded history ($\sim$5000 yr)? Note that the equation for circular motion can be employed (even though the earth spirals away from the sun) because the fractional yearly change in radius is so small. The other condition needed to describe the gradual shift is the over-all conservation of angular momentum about the CM of the system. (This problem was given in a simplified form as Problem 8-19.)

13-22 A particle of mass m moves about a massive center of force C, with the attraction given by $-f(r)\mathbf{e}_r$, where r is the position of the particle as measured from C. If the particle is also subjected to a retarding force $-\lambda \mathbf{v}$, and initially has angular momentum $\mathbf{L}_0$ about C, find its angular momentum as a function of time.

13-23 Consider a central force in a horizontal plane given by $\mathbf{F}(r) = -k\mathbf{r}$, where k is a constant. (This provides a good description, for example, of the pendulum encountered in the laboratory. Rarely is a pendulum physically confined to swing in only one vertical plane.)

(a) A particle of mass m is moving under the influence of such a force. Initially the particle has position vector $\mathbf{r}_0$ and velocity $\mathbf{v}_0$ as measured from the stationary force center. Set up a Cartesian coordinate system with the xy plane containing $\mathbf{r}_0$ and $\mathbf{v}_0$, and find the time dependence of the position (x, y) of the particle. Does the orbit correspond to any particular geometric curve? (Keep in mind the differences between this interaction and the gravitational problem.) What physical quantities are conserved?

(b) Suppose the particle is originally in a circular orbit of radius R. What is its orbital speed? If at some point its velocity is doubled, what will be the maximum value of r in its subsequent motion?

13-24 According to general relativity theory, the gravitational potential energy of a mass m orbiting about a mass M is modified by the addition of a term $-GMmC^2/c^2r^3$, where $C = r^2\, d\theta/dt$ and c is the speed of light. Thus the period of a circular orbit of radius r is slightly smaller than would be predicted by Newtonian theory.

(a) Show that the fractional change in the period of a circular orbit of radius r due to this relativistic term is $-(12\pi^2 r^2/c^2 T_0^2)$, where T_0 is the period predicted by Newtonian theory.

(b) Since, by Kepler's third law, we have $T_0^2 \sim r^3$, the effect of this relativistic correction is greatest for the planet closest to the sun, i.e., Mercury. Consider the effect of the relativistic term on the radial and angular periods, and see if you can thereby arrive at the famous result that the perihelion of Mercury's orbit precesses at the rate of about 43 seconds of arc per century. You may find it useful to refer back to Problem 8-20, which also deals with this question.

13-25 A beam of atoms traveling in the positive x direction and passing through a medium containing n particles per unit volume suffers an attenuation given by

$$\frac{dN(x)}{dx} = -AnN(x)$$

where A is the cross section for scattering of an atom in the beam by an atom of the medium. Therefore, if the beam contains N_0 atoms at $x = 0$, the number still traveling in the beam at x is just $N(x) = N_0 e^{-Anx}$

(a) Set up a simple model of beam attenuation that gives the results stated above.

(b) The graph summarizes a set of measurements of the attenuation of a beam of potassium atoms by argon gas at various pressures (the pressures are given in millimeters of mercury; the temperature is 0°C throughout). (These data are from the film "The Size of Atoms from an Atomic Beam Experiment," by J. G. King, Education Development Center, Newton, Mass., 1961.) Deduce the cross section for the scattering of a potassium atom by an argon atom. (1 cm^3 of a perfect gas at STP contains 2.7×10^{19} molecules.) Check whether the results for different values of the pressure agree.

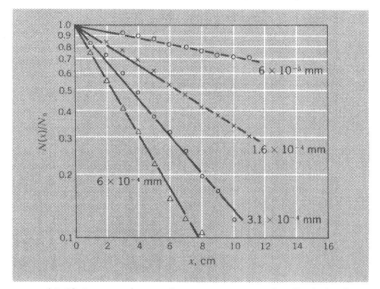

(c) If the potassium and argon atoms are visualized simply as hard spheres of radii r_K and r_A, respectively, what is implied about r_K and r_A by the result of part (b)?

13-26 (a) In the Rutherford scattering problem one can calculate a distance of closest approach d_0 for alpha particles of a given energy approaching a nucleus head on. Verify that d_0 is given by $d_0 = 2kq_1q_2/mv_0^2$.

(b) The force of repulsion between two protons, separated by 10^{-14} m, is 2.3 N. Use this to deduce the value of d_0 for alpha particles (charge 2e) of kinetic energy 5 MeV approaching nuclei of gold (charge 79e).

(c) By introducing d_0, the expression for the fraction of incident alpha particles scattered into $d\Omega$ at φ becomes

$$df = \frac{1}{16}(n\,\Delta s)\,d_0^2\,\frac{d\Omega}{\sin^4(\varphi/2)}$$

where n is the number of nuclei per unit volume and Δs is the length of the path through the foil. Putting $d\Omega = 2\pi \sin\varphi\,d\varphi$, show that the fraction of alpha particles scattered through angles $\geq \varphi_0$ is given by

$$f(\geq \varphi_0) = \frac{\pi}{4} n\,\Delta s\,d_0^2 \cot^2\frac{\varphi_0}{2}$$

(d) A foil of gold leaf 10^{-4} cm thick is bombarded with alpha particles of energy 5 MeV. Out of 1 million alpha particles, incident normally on the foil, how many would be deflected through 90° or more? (Density of gold = 1.9×10^4 kg/m^3; atomic weight = 197.)

In nature we have to deal, not with material points, but with material bodies of finite extent. But we may regard every body as composed of very many material points . . . . This reduces the question of the equations of motion of material bodies to that of the mechanics of systems of material points.

MAX PLANCK, *General Mechanics* (1916)

14

Extended systems and rotational dynamics

NEARLY ALL OUR discussion of dynamics so far has been limited to the translational motions of particles regarded as point masses. About the only exception has been our brief consideration of the vibration of a diatomic molecule (Chapter 10). But every particle has structure; it has a finite size and a greater or lesser degree of rigidity. The full description of the motion of any real physical object must include, in addition to the motion of its center of mass, a consideration of its rotation and other internal motions. In many instances one may regard a complex, extended object as being an assemblage of the ideal particles of basic mechanics. In this chapter we shall discuss a number of topics, touching upon what are in many respects widely different physical systems, yet having in common the feature that they involve the motions of two or more individual particles. We shall be devoting special attention to those physical systems in which particles interact strongly with one another and, in some instances, we shall treat the interactions as being so strong that the system is effectively rigid. The discussion will range from molecules to flywheels to galaxies; different though they may seem (and indeed are), they also have important properties in common. Most of this chapter will be concerned with rotational motion, but we shall begin by developing a couple of important general results that apply to the instantaneous motions of a collection of arbitrarily many particles engaged in any type of motion whatsoever.

MOMENTUM AND KINETIC ENERGY OF A MANY-PARTICLE SYSTEM

In Chapter 9 we analyzed the dynamics of two-particle systems as described in an arbitrary frame (the laboratory) and in the unique center-of-mass (CM) frame defined by the particles themselves. We saw how the introduction of the center of mass allows one to separate motions relative to the center of mass from bodily motions of the system as a whole. We shall now show that this possibility exists for any system of particles; it is a result that makes for very important simplifications in our analysis of complete objects of arbitrary shapes and sizes.

We shall suppose that our system is made up of particles of masses $m_1, m_2, m_3, \ldots$, located instantaneously at the positions $\mathbf{r}_1, \mathbf{r}_2, \mathbf{r}_3, \ldots$ and moving instantaneously with the velocities $\mathbf{v}_1, \mathbf{v}_2, \mathbf{v}_3, \ldots$. The position and velocity of the center of mass, C, are then defined by the following equations:

$$M\mathbf{r}_c = m_1\mathbf{r}_1 + m_2\mathbf{r}_2 + m_3\mathbf{r}_3 + \cdots$$
$$M\mathbf{v}_c = m_1\mathbf{v}_1 + m_2\mathbf{v}_2 + m_3\mathbf{v}_3 + \cdots$$

where

$$M = m_1 + m_2 + m_3 + \cdots$$

We can express these results more compactly as follows:

$$M\mathbf{r}_c = \sum_i m_i\mathbf{r}_i$$
$$M\mathbf{v}_c = \sum_i m_i\mathbf{v}_i$$
(14-1)

where the suffix i runs from 1 to N (N being the total number of particles in the system).

Consider now the statement of $\mathbf{F} = m\mathbf{a}$ as it applies to any one particle, i, in the system. In general it may be subjected to an external force, $\mathbf{F}_i$, and also to internal interactions from all the other particles of the system. We shall denote these latter by symbols such as $\mathbf{f}_{ik}$, to be read as the force exerted on particle i

by particle k.[1] Then the specific statement of Newton's law for particle i is as follows:

$$\mathbf{F}_i + \sum_{k \neq i} \mathbf{f}_{ik} = \frac{d}{dt}(m_i \mathbf{v}_i)$$

We can now proceed to write a similar equation for every other particle in the system and add them all up. When we do this, the right-hand side is, by Eq. (14-1), just the rate of change of the momentum of a single particle of mass M traveling at the center-of-mass velocity $\mathbf{v}_c$. What about the left-hand side? The first part of it is the sum ($\mathbf{F}_{ext}$) of all the external forces, regardless of which particles they are applied to. The second part—let us call it $\mathbf{f}_{int}$—is a double summation over all the interactions that can occur between the particles in pairs:

$$\mathbf{f}_{int} = \sum_i \sum_{k \neq i} \mathbf{f}_{ik}$$

Now if you consider what this summation entails, you will see that it can be broken down into a set of pairs of contributions of the type $\mathbf{f}_{ik} + \mathbf{f}_{ki}$. (It may help you to take the simplest specific case, $N = 3$, and write out all the terms in detail.) In other words, it is made up of a set of terms, each of which is the sum of the forces of action and reaction between two particles. Since, however, Newtonian mechanics has it as a basic tenet that the forces of action and reaction are equal and opposite, each one of these pairs gives zero, and it follows that the resultant of all the internal forces, $\mathbf{f}_{int}$, is itself zero. (This is, of course, an ancient piece of folk wisdom, as expressed colloquially in the statement "You can't pull yourself up by your bootstraps.") Thus, for any system of particles whatever, we have a statement of Newton's law exactly like that for a single particle of the total mass M:

$$\mathbf{F}_{ext} = \frac{d}{dt}(M\mathbf{v}_c) = M \frac{d\mathbf{v}_c}{dt} \tag{14-2}$$

Figure 14-1 shows an example of this result in action. The center of mass of a complex object, with innumerable internal interactions, follows a simple parabolic path under gravity.

The total kinetic energy of the system is also amenable to a

---

[1] Earlier, in dealing with two-particle systems only, we used the symbol F_{12} to denote the force exerted *by* particle 1 *on* particle 2. The revised definition is more convenient for our present purposes and should not lead to any confusion in the brief use that we shall be making of it.

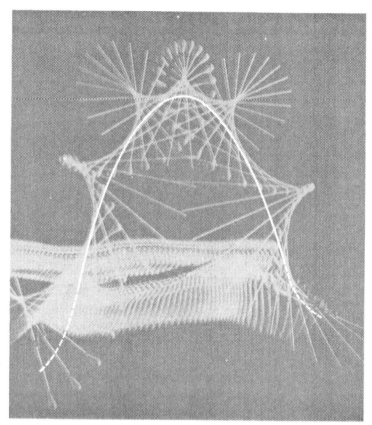

Fig. 14-1 The center of mass of a complicated object follows a simple parabolic path under the net gravitational force. Photograph by Prof. Harold E. Edgerton, M.I.T., of a drum majorette tossing a baton. Time between flashes was 1/60 sec. Dashed lines show path of CM before the baton was released and after it was caught again.

simple analysis. In this case we shall introduce the velocities $\mathbf{v}'_1, \mathbf{v}'_2, \mathbf{v}'_3, \ldots$ of the particles relative to the center of mass. Thus the velocity, $\mathbf{v}_i$, of any particle as measured in the laboratory can be written as $\mathbf{v}'_i + \mathbf{v}_c$. The kinetic energy K_i of this particle can thus be written as follows:

$$K_i = \tfrac{1}{2}m_i v_i^2 = \tfrac{1}{2}m_i(\mathbf{v}_i \cdot \mathbf{v}_i)$$
$$= \tfrac{1}{2}m_i(\mathbf{v}'_i + \mathbf{v}_c) \cdot (\mathbf{v}'_i + \mathbf{v}_c)$$
$$= \tfrac{1}{2}m_i(v'_i)^2 + m_i(\mathbf{v}'_i \cdot \mathbf{v}_c) + \tfrac{1}{2}m_i v_c^2$$

Thus we have

$$K_i = K'_i + (m_i \mathbf{v}'_i) \cdot \mathbf{v}_c + \tfrac{1}{2}m_i v_c^2$$

Let us now consider the result of summing the individual kinetic energies such as K_i for all the particles in the system. The first term on the right gives us the total kinetic energy, K', of all the particles relative to the center of mass. The last term gives us the kinetic energy of a particle of the total mass, M, moving with the speed v_c of the center of mass. And the *middle* term vanishes, because by the definition of the center of mass we have

$$\sum_i m_i \mathbf{r}'_i = 0$$

and hence

$$\left(\sum_i m_i \mathbf{v}'_i\right) \cdot \mathbf{v}_c = 0$$

Thus, for any system of particles, we can put

$$K = K' + \tfrac{1}{2} M v_c^2 \tag{14-3}$$

A very familiar example of such a system is a gas in a container. If the container is at rest in the laboratory the total momentum is zero (i.e., the laboratory frame and the CM frame coincide and $v_c = 0$) but the kinetic energy K' of the internal motion is large. Suppose we have 1 mole of gas at temperature $T\,°K$. It contains N molecules (N = Avogadro's number), each of mass m_0. The total kinetic energy of the internal motion is given by

$$K' = \tfrac{1}{2} m_0 (v_1'^2 + v_2'^2 + \cdots + v_N'^2)$$

Introducing the mean squared speed v_m^2, this can be written

$$K' = \tfrac{1}{2} N m_0 v_m^2 = \tfrac{1}{2} M v_m^2 \tag{14-4}$$

where M is the molecular weight.

From the kinetic theory of gases, however [see, for example, Eq. (9-39)], the pressure P of the gas, if it occupies a volume V, is given by

$$P = \tfrac{1}{3} n m_0 v_m^2 \quad \text{with } n = N/V \tag{14-5}$$

(n = number of molecules per unit volume, and we have assumed N molecules in a volume V.) From Eq. (14-5) we have

$$PV = \tfrac{1}{3} M v_m^2$$

and combining this with Eq. (14-4) gives us

$$K' = \tfrac{3}{2} PV = \tfrac{3}{2} RT$$

where R is the universal gas constant.[1] We know that 1 mole of

[1] $R = 8.32$ J/(mole-°K) $= 0.0821$ liter-atm/(mole-°K)

gas occupies a volume of 22.4 liters at a pressure of 1 atm ($= 1.013 \times 10^5$ N/m^2) at 0°C. Hence

$$K' = \tfrac{3}{2}(1.013 \times 10^5)(2.24 \times 10^{-2}) \approx 3.4 \times 10^3 \text{ J}$$

This is about equal to the kinetic energy of a 16-lb shot, as used in field events, traveling at nearly 70 mph! It is a good thing that molecular motions are random.

ANGULAR MOMENTUM

In Chapter 13 we recognized that the orbital angular momentum, **l**, of a particle with respect to a center of force is an important dynamical quantity. You will recall that **l** (the "moment of momentum") is defined through the following equation:

$$\mathbf{l} = \mathbf{r} \times (m\mathbf{v}) = \mathbf{r} \times \mathbf{p} \qquad (14\text{-}6)$$

Thus the actual magnitude of **l** depends on the particular choice of origin from which the position vector **r** of the particle is measured. If, as in the situations we considered, there is a well-defined fixed center of force, the appropriate choice of origin is clear. In general, however, the angular momentum of an individual moving particle is not a uniquely definable quantity. But as soon as one has two or more particles, or a single object that cannot be approximated as a point particle, it does become possible to speak unambiguously of the *internal* angular momentum of the complete system. Let us see how.

To introduce the discussion, consider first a very simple and specific situation. Two particles, of masses m_1 and m_2, are joined by a very light rigid bar that is pivoted at the center of mass, C, of the two particles [see Fig. 14–2(a)]. The system rotates with angular velocity ω about an axis through C perpendicular to the plane of the diagram. We shall calculate the total angular momentum of the two particles about a parallel axis through an arbitrary origin, O. With respect to O, the orbital angular momentum of m_1 is counterclockwise and that of m_2 is clockwise. To calculate the actual magnitude of the combined angular momentum, we can draw a line OA parallel to the line joining the masses. The velocity vectors are perpendicular to this line and intersect it at the points A and B. A line through C parallel to these velocity vectors intersects OA at a point D. Let $OD = d$. Then the total angular momentum of the two

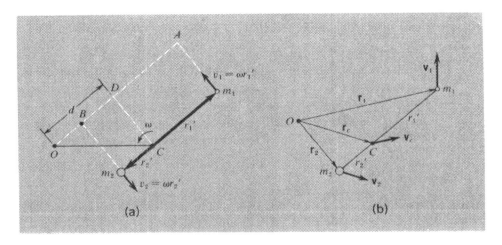

Fig. 14-2 (a) A rigid two-body system rotating about its center of mass, C. The angular momentum can be calculated with respect to C or to an arbitrary point, O. (b) Individual and center-of-mass motions in an arbitrary two-particle system.

particles with respect to O is given by the following expression:

$$L_O = m_1v_1(d + r_1') - m_2v_2(d - r_2')$$
$$= (m_1v_1 - m_2v_2)d + (m_1v_1r_1' + m_2v_2r_2')$$

where r_1' and r_2' are the distances of the particles from their common center of mass. The second term in parentheses is the total angular momentum, L_c, about the axis through C. The first term in parentheses is zero; this follows at once from the fact that we have taken C to be at rest and therefore we are in the zero-momentum frame (or we could write $v_1 = \omega r_1'$ and $v_2 = \omega r_2'$ and invoke the definition of the center of mass as such). Thus in this case we have $L_O = L_c$—the rotational angular momentum of the system has the same value about any axis parallel to the true rotational axis through C. Since $v_1 = \omega r_1'$ and $v_2 = \omega r_2'$, the magnitude of this total angular momentum is given by

$$L_c = (m_1r_1'^2 + m_2r_2'^2)\omega$$

Introducing the distance r between the particles, we have

$$r_1' = \frac{m_2}{m_1 + m_2} r \qquad r_2' = \frac{m_1}{m_1 + m_2} r$$

Substituting these values in the expression for L_c gives

$$L_c = \frac{m_1 m_2}{m_1 + m_2} r^2 \omega = \mu r^2 \omega \tag{14-7}$$

where $\mu \ [= m_1 m_2/(m_1 + m_2)]$ is the reduced mass of the system. Equation (14-7) is an important result for the angular momentum of a so-called "rigid rotator," and the quantity μr^2 is an example of what is called the *moment of inertia*, I, of a rigidly connected system. (We shall consider more complex cases later.) For such a system it is convenient to put

$$L_c = I\omega \tag{14-8}$$

where, in the present case, we have

$$I = \mu r^2 = m_1 r_1'^2 + m_2 r_2'^2$$

Let us now consider the more general case of a system of two particles moving with arbitrary velocities, as shown in Fig. 14-2(b). Again we shall refer the total angular momentum to an arbitrary origin O, and this time we make no assumption that the center of mass C is at rest relative to O. Instead we assume that it may have some velocity $\mathbf{v}_c$. We can always orient our diagram so that the origin O and the vector $\mathbf{r}$ from m_2 to m_1 lie in the plane of the paper, but the velocity vectors $\mathbf{v}_1$, $\mathbf{v}_2$, and $\mathbf{v}_c$ need not be confined to this plane. Let us now consider the total angular momentum defined by the vector sum of the contributions associated with m_1 and m_2 separately.

If the position vectors of m_1 and m_2 with respect to O are $\mathbf{r}_1$ and $\mathbf{r}_2$, as shown in the figure, we have

$$\mathbf{L}_O = \mathbf{r}_1 \times (m_1 \mathbf{v}_1) + \mathbf{r}_2 \times (m_2 \mathbf{v}_2)$$

Let us now introduce the positions and velocities of the particles relative to the center of mass:

$$\mathbf{r}_1 = \mathbf{r}_1' + \mathbf{r}_c \qquad \mathbf{r}_2 = \mathbf{r}_2' + \mathbf{r}_c$$
$$\mathbf{v}_1 = \mathbf{v}_1' + \mathbf{v}_c \qquad \mathbf{v}_2 = \mathbf{v}_2' + \mathbf{v}_c$$

Then we have

$$\mathbf{L}_O = (\mathbf{r}_1' + \mathbf{r}_c) \times m_1(\mathbf{v}_1' + \mathbf{v}_c) + (\mathbf{r}_2' + \mathbf{r}_c) \times m_2(\mathbf{v}_2' + \mathbf{v}_c)$$

This can be rearranged into a sum of four terms as follows:

$$\begin{aligned}\mathbf{L}_O = &\ (\mathbf{r}_1' \times m_1 \mathbf{v}_1' + \mathbf{r}_2' \times m_2 \mathbf{v}_2') \\ &+ (m_1 \mathbf{r}_1' + m_2 \mathbf{r}_2') \times \mathbf{v}_c \\ &+ \mathbf{r}_c \times (m_1 \mathbf{v}_1' + m_2 \mathbf{v}_2') \\ &+ \mathbf{r}_c \times (m_1 + m_2) \mathbf{v}_c\end{aligned}$$

Now it follows from the definition of the center of mass that the second and third of the above terms vanish, for we have

$$m_1 \mathbf{r}'_1 + m_2 \mathbf{r}'_2 = 0$$
$$m_1 \mathbf{v}'_1 + m_2 \mathbf{v}'_2 = 0$$

The first term in the expression for L_O is the combined angular momentum, L_c, of the particles about C, and so we have

$$\mathbf{L}_O = \mathbf{L}_c + \mathbf{r}_c \times M\mathbf{v}_c \tag{14-9}$$

where

$$\mathbf{L}_c = \mathbf{r}'_1 \times (m_1 \mathbf{v}'_1) + \mathbf{r}'_2 \times (m_2 \mathbf{v}'_2)$$

Thus the total angular momentum about O is the net angular momentum about the center of mass, plus the orbital angular momentum associated with the motion of the center of mass itself. If the particles have a rigid connection and so rotate as a unit about C, we can use Eq. (14-7) or (14-8) to give the explicit expression for the magnitude of L_c in Eq. (14-9).

A study of the above analysis will make it clear that equivalent results hold good for a system of arbitrarily many particles; thus in Eq. (14-9) we have a strong basis for the analysis of angular momentum in general. Notice in particular that if the center of mass of an arbitrary system of moving particles is at rest, then the total angular momentum has the same value, equal to L_c, about any point. The angular momentum of a rotating bicycle wheel, for instance (Fig. 14-3), has the same value about

Fig. 14-3 The rotational angular momentum of a bicycle wheel has the same value about the axis P of the pedal wheel as it has about its own axis through C.

a horizontal axis through the point P on the bicycle frame as it has about its own axis through C, because the wheel has no net translational velocity with respect to either point.

For any two-particle system, rigidly connected or not, the value of $\mathbf{L}_c$ can be conveniently expressed in terms of the relative coordinate $\mathbf{r}$ of the two masses and their relative velocity $\mathbf{v}_{\text{rel}}$. We take the expression for $\mathbf{L}_c$ in Eq. (14–9):

$$\mathbf{L}_c = \mathbf{r}'_1 \times (m_1 \mathbf{v}'_1) + \mathbf{r}'_2 \times (m_2 \mathbf{v}'_2)$$

and we first substitute for $\mathbf{r}'_1$ and $\mathbf{r}'_2$ in terms of $\mathbf{r}$. From the defining equations

$$m_1 \mathbf{r}'_1 + m_2 \mathbf{r}'_2 = 0$$
$$\mathbf{r}'_2 - \mathbf{r}'_1 = \mathbf{r}$$

we have

$$\mathbf{r}'_1 = -\frac{m_2}{m_1 + m_2}\mathbf{r} \qquad \mathbf{r}'_2 = \frac{m_1}{m_1 + m_2}\mathbf{r}$$

It then follows that $\mathbf{L}_c$ can be written as follows:

$$\mathbf{L}_c = \frac{m_1 m_2}{m_1 + m_2}(-\mathbf{r} \times \mathbf{v}'_1 + \mathbf{r} \times \mathbf{v}'_2)$$

i.e.,

$$\mathbf{L}_c = \mu \mathbf{r} \times (\mathbf{v}'_2 - \mathbf{v}'_1) = \mathbf{r} \times (\mu \mathbf{v}_{\text{rel}}) \qquad (14\text{–}10)$$

In the particular case of a rigid system rotating with angular velocity ω, we can further put

$$\mathbf{v}_{\text{rel}} = \boldsymbol{\omega} \times \mathbf{r}$$

If ω is perpendicular to $\mathbf{r}$ this reduces Eq. (14–10) to Eq. (14–7); otherwise the result is slightly more complicated.[1]

ANGULAR MOMENTUM AS A FUNDAMENTAL QUANTITY

The preceding analysis has established that two connected masses, regarded as a single system with mass and size, have what can

---

[1] We have, in fact, $\mathbf{L}_c = \mu \mathbf{r} \times (\boldsymbol{\omega} \times \mathbf{r})$. Now there is a general vector identity applying to any triple vector product: $\mathbf{A} \times (\mathbf{B} \times \mathbf{C}) = \mathbf{B}(\mathbf{A} \cdot \mathbf{C}) - \mathbf{C}(\mathbf{A} \cdot \mathbf{B})$. In the present case, with $\mathbf{A} = \mathbf{C} = \mathbf{r}$, $\mathbf{B} = \boldsymbol{\omega}$, this leads to the result $\mathbf{L}_c = \mu r^2 \boldsymbol{\omega} - \mu \mathbf{r}(\boldsymbol{\omega} \cdot \mathbf{r})$.

properly be described as an intrinsic angular momentum about the center of mass. Regardless of the actual motion of the CM, one can identify this rotational property of the system. But angular momentum took on a even more basic aspect when it was discovered, in the development of quantum mechanics, that there was a natural unit of angular momentum, equal to Planck's constant h divided by 2π:

$$\text{Basic unit of angular momentum} = \frac{h}{2\pi}$$
$$= 1.054 \times 10^{-34} \text{ kg-m}^2/\text{sec}$$

This is of course a very tiny unit, but it implies enormously high speeds of rotation in systems of atomic size. Let us consider one example. For many purposes a diatomic molecule, such as N_2, can be regarded as a rigid system such as we have discussed—two point masses a fixed distance apart. Nitrogen has two equal nuclei, each of mass about 2.3×10^{-26} kg, separated by about 1.1 Å $(= 1.1 \times 10^{-10}$ m). The moment of inertia [cf. Eq. (14–8)] is thus given by

$$I = 2m(s/2)^2$$
$$= 2(2.3 \times 10^{-26})(5.5 \times 10^{-11})^2 \text{ kg-m}^2$$
$$\approx 1.4 \times 10^{-46} \text{ kg/m}^2$$

If we put

$$I\omega = 1.054 \times 10^{-34} \text{ kg-m}^2/\text{sec}$$

then we find that

$$\omega \approx 7.5 \times 10^{11} \text{ sec}^{-1}$$

The frequency (rps) corresponding to this would be $\omega/2\pi$, or about 10^{11} sec^{-1}. Frequencies of this order are typical of molecules behaving as rigid rotators and can be studied through the techniques of microwave spectroscopy (the frequency just mentioned would correspond to a wavelength of 3 mm).

It may be noted that these rotational frequencies are far lower (by about two orders of magnitude) than the typical frequency of molecular vibration calculated at the end of Chapter 10. In actual molecules there is usually a complex admixture of vibrational and rotational motions, the latter providing a kind of fine detail superimposed on the former. In molecular spectroscopy one speaks of "rotation–vibration bands" arising in this way.

Even more fundamentally, it appears that all the elementary particles of the universe have an intrinsic angular momentum which is some integral multiple (including zero) of $h/4\pi$. In particular, our most familiar building blocks, nucleons and electrons, have just $h/4\pi$. At this level, however, the specification of what, if anything, is rotating becomes a moot question; one simply contents oneself with the fact of an intrinsic angular momentum that has the important property of being conserved in all the interactions and rearrangements of such particles. This conservation property of angular momentum in general is the subject of the next section.

Fig. 14-4 The conservation of angular momentum. A sequence showing a man making a backward somersault. These photographs were taken by Eadweard Muybridge, a pioneer of motion photography at the beginning of the 20th century. (From Eadweard Muybridge, The Human Figure in Motion, *Dover Publications, New York, 1955. Reprinted through permission of the publisher.)*

CONSERVATION OF ANGULAR MOMENTUM

Many interesting experiments can be performed that illustrate the important property that the total internal angular momentum of a system of particles is conserved if external influences are absent. Some of the qualitative demonstrations are no doubt familiar to you—the speeding up of a whirling ice skater, for example, or of an expert gymnast when he hunches his body after beginning a somersault (Fig. 14-4). A quite unskilled person (e.g., a professor) can do similar tricks if he sits on a freely pivoted stool, gets himself turning slowly with a couple of dumbbells held at arm's length, and then pulls the dumbbells inward [Fig. 14-5(a)].

The conservation of internal angular momentum holds good, whatever internal rearrangements of the system take place. Some particularly nice consequences of this conservation can be shown if one has a good flywheel—e.g., a bicycle wheel with an extra loading of lead around the rim. For example, one person can sit on a pivoted stool [Fig. 14-5(b)] and another person can hand him the wheel after it has been set spinning with angular momentum L_w as shown (corresponding to clockwise rotation about a vertical axis pointing upward). The person on the stool is not himself rotating, but the system, person + stool, has the

Fig. 14-5 Experiments on the conservation of angular momentum: (a) The person on the stool rotates faster if he pulls the dumbbells inward. (b) and (c) The man on the stool begins to rotate when he inverts the spinning wheel.

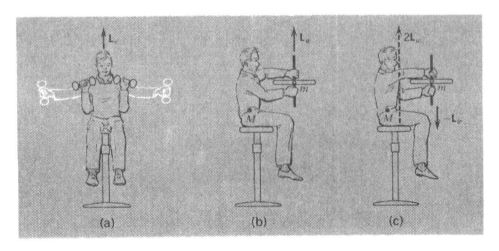

total internal angular momentum $\mathbf{L}_w$. If the wheel is now inverted its rotational angular momentum about its own center of mass is changed to $-\mathbf{L}_w$. It follows that the system of two masses, M (the person) and m (the wheel), must acquire a clockwise rotation with a total rotational angular momentum of $+2\mathbf{L}_w$ [Fig. 14–5(c)]. If the wheel in this new orientation is handed to the assistant, who inverts it and hands it back, the total angular momentum is raised to $3\mathbf{L}_w$. If the person on the stool again inverts the wheel, the general rotation of $M + m$ is raised to $5\mathbf{L}_w$. Thus angular momentum can be transferred back and forth in packets in such operations—although here we are going beyond the conservation of total angular momentum in a completely isolated system.

The formal proof of the conservation of angular momentum is not difficult. The total internal angular momentum (with respect to the center of mass or to any other point in the CM frame) is given, according to Eq. (14–9), by

$$\mathbf{L}_c = \sum \mathbf{r}'_i \times (m_i \mathbf{v}'_i) \tag{14-11}$$

Let us consider the variation of $\mathbf{L}_c$ with time. Differentiating, we have

$$\frac{d\mathbf{L}_c}{dt} = \sum \mathbf{v}'_i \times (m_i \mathbf{v}'_i) + \sum \mathbf{r}'_i \times (m_i \mathbf{a}'_i)$$

where $\mathbf{a}'_i$ is the acceleration of particle i relative to the CM. The first summation vanishes, because every product $\mathbf{v}'_i \times \mathbf{v}'_i$ is identically zero. In the second summation, we shall write $\mathbf{a}'_i$ as the vector difference between the true acceleration, $\mathbf{a}_i$, of particle i, as measured in an inertial frame, and the acceleration $\mathbf{a}_c$ ($= d\mathbf{v}_c/dt$) of the center of mass. (It is important to realize that $\mathbf{a}_c$ may exist even in a frame in which $\mathbf{v}_c$ is zero at some instant.) Thus we have

$$\frac{d\mathbf{L}_c}{dt} = \sum \mathbf{r}'_i \times m_i(\mathbf{a}_i - \mathbf{a}_c)$$
$$= \sum \mathbf{r}'_i \times (m_i \mathbf{a}_i) - \left(\sum m_i \mathbf{r}'_i\right) \times \mathbf{a}_c$$

However, by the definition of the CM, the summation in the second term is zero. The first term is the total torque about C of all the forces acting on the particles, because $m_i \mathbf{a}_i$ is the net force acting on any given particle. This force may be a combination of an external force $\mathbf{F}_i$ and a set of internal forces $\mathbf{f}_{ik}$:

thus we put

$$m_i \mathbf{a}_i = \mathbf{F}_i + \sum_k \mathbf{f}_{ik}$$

Substituting this statement of $m_i \mathbf{a}_i$ in the equation for $d\mathbf{L}_c/dt$ we therefore have

$$\frac{d\mathbf{L}_c}{dt} = \sum \mathbf{r}'_i \times \mathbf{F}_i + \sum_i \sum_k \mathbf{r}'_i \times \mathbf{f}_{ik}$$

Now, as in our earlier discussion of the total linear momentum of a system of particles, we can arrange the double summation involving internal forces into pairs of terms, in this case of the type

$$\mathbf{r}'_i \times \mathbf{f}_{ik} + \mathbf{r}'_k \times \mathbf{f}_{ki}$$

If, however, we can assume that the forces of interaction between any two particles are equal, opposite forces along the line joining them [Fig. 14-6(a)], then each such pair of torques adds up to zero, because each force has the same lever arm CD with respect to C. Thus we arrive at a very simple equation:

$$\frac{d\mathbf{L}_c}{dt} = \sum \mathbf{r}'_i \times \mathbf{F}_i = \sum \mathbf{M}'_i \qquad (14\text{--}12)$$

where $\mathbf{M}'_i$ is the torque exerted by $\mathbf{F}_i$ about the CM. Equation (14-12) is a very basic equation of rotational dynamics; we shall spell it out in words:

Regardless of any acceleration that the center of mass of a system of particles may have as a result of a net external force exerted on the system, the rate of change of internal angular momentum about the CM is equal to the total torque of the external forces about the CM.

In particular, therefore, if the net torque about the center of mass is zero, the internal angular momentum remains constant, whatever internal rearrangements may go on within the system.

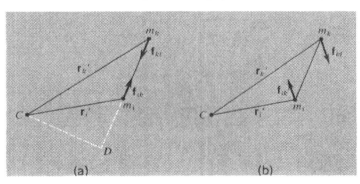

Fig. 14-6 (a) The equal and opposite internal forces between two particles can produce no torque about C if they are along the line joining the particles. (b) If the forces are not along the line joining the particles they produce a torque which is, however, nullified by other internal forces (see discussion in the text).

[Our derivation of Eq. (14–12) contains one weak link. This is the argument by which we conclude that the net torque of the internal interactions is zero. It is perfectly possible to imagine that the forces $\mathbf{f}_{ik}$ and $\mathbf{f}_{ki}$ are equal in magnitude and opposite in direction, thus conforming to Newton's third law, without having them act along the same line [see Fig. 14–6(b)]. In this case they would constitute a couple with a resultant torque about C or any other point. The vanishing of the net torque of the internal forces, and the consequent conservation of total angular momentum if external forces are absent, is however a result that holds good in general; it does not depend on the limited assumption that the forces of interaction act along the lines joining pairs of particles. A powerful theoretical argument in support of this proposition—but requiring virtually no mathematics at all—can be made on the basis of the uniformity and isotropy of space. It runs as follows.

Consider first the conservation of the total *linear* momentum of a system of particles. This holds good if the total potential energy, U, of the system remains unaffected by linear displacement, because the vanishing of **grad** U corresponds to the absence of any net force. If we know that external forces are absent, the *invariance* of U with respect to linear displacements is more or less axiomatic; it corresponds to our belief that absolute position along a line has no significance in physics. In an exactly similar way, we can argue that the total potential energy, U, associated with the internal forces of a system of particles is completely insensitive to a rotation through an arbitrary angle θ of the system as a whole. Now just as we can evaluate a force from a potential-energy function through relations of the type

$$F_r = -\frac{\partial U}{\partial x}$$

so we can evaluate torques through relations of the type

$$M_z = -\frac{\partial U}{\partial \theta} \tag{14-13}$$

where M_z is the torque about the axis z around which the rotation θ is imagined. Hence, if U is independent of θ, there can be no net torque, regardless of the detailed character of the internal interactions. Thus we can conclude that the conservation of total internal angular momentum of any isolated system must hold true in general.]

MOMENTS OF INERTIA OF EXTENDED OBJECTS

For an arbitrary system of particles, the intrinsic rotational angular momentum is defined by the equation

$$\mathbf{L}_c = \sum \mathbf{r}_i \times (m_i \mathbf{v}_i) \tag{14-14}$$

This is just a restatement of Eq. (14–11) except that we are now choosing an origin at the center of mass and using the unprimed symbols $\mathbf{r}_i$ and $\mathbf{v}_i$ to denote the position vectors and the velocities of the individual particles with respect to the CM. We shall now consider in detail the application of Eq. (14–14) to a system that is rotating as a unit with an angular velocity $\boldsymbol{\omega}$. In particular, we shall concern ourselves with the case of an object that has well-defined geometrical symmetry and is rotating about an axis of symmetry through the center of mass. For each particle in such an object, the linear velocity $\mathbf{v}_i$ is given by $\mathbf{v}_i = \boldsymbol{\omega} \times \mathbf{r}_i$, and hence the total angular momentum is given by

$$\mathbf{L}_c = \sum \mathbf{r}_i \times m_i(\boldsymbol{\omega} \times \mathbf{r}_i) \tag{14-15}$$

Let us consider the contribution to $\mathbf{L}_c$ from a pair of particles, i and j, of equal masses (because of the symmetry) situated symmetrically with respect to the axis of rotation as shown in Fig. 14–7(a). Particle i has a velocity $\boldsymbol{\omega} \times \mathbf{r}_i$ down into the plane of the diagram; the magnitude v_i of this velocity is given by

Fig. 14-7 (a) Contributions by two symmetrically placed particles to the net rotational angular momentum of a rigid object. (b) Main spin axis and two other principal axes of inertia of an object with well-defined symmetry.

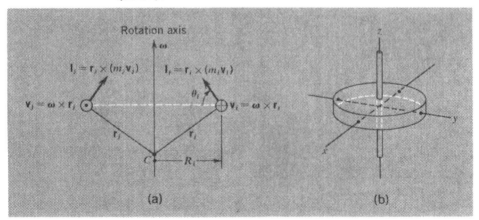

$$v_i = \omega r_i \sin \theta_i = \omega R_i$$

where R_i is the perpendicular distance of the particle from the axis. The angular momentum of this particle is then a vector $\mathbf{l}_i$ directed as shown in the diagram. Its magnitude is given by $|\mathbf{r}_i \times (m\mathbf{v}_i)|$, but since $\mathbf{v}_i$ is perpendicular to $\mathbf{r}_i$ itself, we have simply

$$l_i = m_i \omega R_i r_i$$

Now if we consider the angular momentum $\mathbf{l}_j$ of particle j, we see that it has the same magnitude as $\mathbf{l}_i$ but points in a different direction, in such a way that the components of $\mathbf{l}_i$ and $\mathbf{l}_j$ perpendicular to the rotation axis cancel, but their components parallel to the axis add. Thus, when we consider carrying out the summation represented by Eq. (14-15) for the complete system, we need take only the component of angular momentum parallel to $\boldsymbol{\omega}$ for each particle. The relevant contribution for particle i is thus given by $l_i \sin \theta_i$, and we have

$$l_i \sin \theta_i = m_i \omega R_i r_i \sin \theta_i = m_i R_i^2 \omega$$

It is clear that the total rotational angular momentum is then a vector in the direction of $\boldsymbol{\omega}$ itself, and we can put

$$\mathbf{L}_c = \left(\sum m_i R_i^2 \right) \boldsymbol{\omega} = I \boldsymbol{\omega} \qquad (14\text{-}16)$$

where

$$I = \sum m_i R_i^2 \qquad (14\text{-}17)$$

Equation (14-17) then defines the moment of inertia of the complete system about a given axis of symmetry.

If we consider an object with a well-defined symmetry, such as a flywheel [Fig. 14-7(b)] it is clear that the obvious axis—the normal rotational axis—may not be the only axis of symmetry that it possesses. If we label this normal rotational axis as z, then any other axis perpendicular to z and passing through the center of mass is also an axis of symmetry (although not of complete *rotational* symmetry) for the purposes of the kind of analysis we have presented above. Thus we recognize the possibility of defining, in addition to I_z, two other moments of inertia, I_x and I_y, about a pair of independent axes perpendicular to z and to one another. These three quantities (I_x, I_y, I_z) are known as the *principal moments of inertia* of the object. In the case illustrated, we would of course have $I_x = I_y$ ($\neq I_z$).

It is important to recognize that it is only when the axis of rotation, as defined by the direction of ω, coincides with an axis corresponding to a principal moment of inertia that the rotational angular momentum vector $\mathbf{L}_c$ is parallel to ω itself and can be expressed simply as $I\omega$. We shall, however, be concerned mostly with situations in which this is the case. [It is a remarkable fact that for any object at all, even if it has no kind of symmetry, one can still find a set of three orthogonal rotation axes for which $\mathbf{L}$ and ω are parallel.[1] The existence of these axes allows us to identify three principal moments of inertia for an object of completely arbitrary shape.]

For an object composed of so many particles that it is effectively continuous, we can write the moment of inertia about a given symmetry axis as an integral instead of a summation:

$$I = \int r^2 \, dm \qquad (14\text{-}18)$$

where dm is an element of mass situated at a perpendicular distance r from the axis through the center of mass.

Special cases

1. *Uniform ring.* (The axis is taken to be perpendicular to the plane of the ring.) If the mass of the ring is M and its radius is R [Fig. 14-8(a)] we have simply

$$(\text{Ring}) \qquad I = MR^2 \qquad (14\text{-}19)$$

2. *Uniform disk.* This can be regarded as a set of concentric rings, as in Fig. 14-8(b). An individual ring is made of the material lying between r and $r + dr$. Its area is thus $2\pi r \, dr$, and the area of the whole disk is πR^2. Hence the mass of the ring is given by

$$dm = \frac{2\pi r \, dr}{\pi R^2} M = \frac{2M}{R^2} r \, dr$$

Substituting this in Eq. (14-18) we have

$$(\text{Disk}) \qquad I = \frac{2M}{R^2} \int_0^R r^3 \, dr = \frac{MR^2}{2} \qquad (14\text{-}20)$$

[1]For a proof of this statement, see, for example, K. R. Symon, *Mechanics*, 2nd ed., Addison-Wesley, Reading, Mass., 1960.

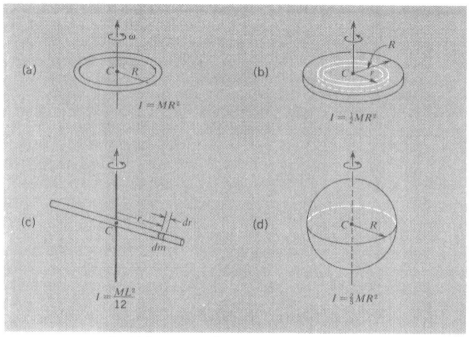

Fig. 14-8 Geometrically simple objects with exactly calculable moments of inertia: (a) ring, (b) disk, (c) bar or rod, and (d) solid sphere.

The same result applies to a long cylinder, regarded as a pile of disks. (The axis of rotation is the axis of the cylinder.)

3. *Uniform Bar.* (The axis is taken through the center, perpendicular to its length.) The length of the bar is L, so the radial distance r goes from 0 to $L/2$ on each side. If the bar is uniform, we have

$$dm = \frac{M}{L} dr$$

We evaluate $\int r^2 \, dm$ from $r = 0$ to $r = L/2$ and double it, to take account of both ends:

(Uniform bar or rod) $\quad I = \dfrac{2M}{L} \displaystyle\int_0^{L/2} r^2 \, dr = \dfrac{ML^2}{12}$ (14-21)

A result of exactly this same form holds also for the moment of inertia of a flat board of length L about an axis, perpendicular to the L dimension, that passes through the CM and lies in the plane of the board.

4. *Sphere.* [Any axis through the center. See Fig. 14-8(d).]

We shall quote this result without proof. (For a derivation, see almost any calculus text.)

$$\text{(Sphere)} \quad I = \tfrac{2}{5}MR^2 \quad (14\text{-}22)$$

It may be noted that in each case the moment of inertia is the mass of the object, times the square of a characteristic linear dimension, times a numerical coefficient not very different from 1. It is hard to be wildly wrong in estimating the moment of inertia of a body, even without detailed calculation. It is quite common practice to write the moment of inertia simply as the total mass times the square of a length k that is called the *radius of gyration* about the axis in question:

$$I = Mk^2 \quad (14\text{-}23)$$

Thus for the special cases considered above, we have the following values of the radius of gyration:

Object	Radius of gyration (k)
Ring	R
Disk	$R/\sqrt{2}$ $(= 0.707R)$
Bar (length L)	$L/2\sqrt{3}$ $(= 0.289L)$
Sphere	$R\sqrt{2/5}$ $(= 0.632R)$

TWO THEOREMS CONCERNING MOMENTS OF INERTIA

The calculation of moments of inertia can often be simplified with the help of two theorems that we shall now present. The first of them applies to any kind of object; the second is applicable only to objects that can be treated, to some approximation, as flat objects of negligible thickness.

The theorem of parallel axes

When we speak of "*the* moment of inertia" of an object we normally mean the moment of inertia about a symmetry axis drawn through the center of mass. There are, however, many situations in which the actual physical axis of rotation does not pass through the CM. In such cases one can make use of a theorem that directly relates the moment of inertia about the given axis to the moment of inertia about a parallel axis through the center of mass.

Figure 14-9(a) illustrates the situation. Suppose that the

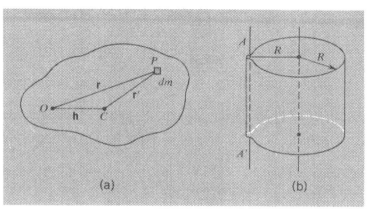

Fig. 14-9 (a) Diagram to show the basis of the parallel-axis theorem. (b) Use of the parallel-axis theorem makes easy such calculations as finding the moment of the cylinder about the hinge AA'.

given axis is perpendicular to the plane of the paper and passes through the point O. Imagine a parallel axis through C, and consider the object as being built up of a set of thin slices parallel to the plane of the paper. The vector distance from the axis through O to the axis through C is a constant, **h**. Within any one slice we can consider elements of mass such as dm at point P in the figure, a vector distance **r** from O and **r'** from C. We have

$$\mathbf{r} = \mathbf{h} + \mathbf{r'}$$

The contribution dI_O of dm to the moment of inertia about the axis through O is then given by

$$dI_O = r^2\, dm = (\mathbf{h} + \mathbf{r'}) \cdot (\mathbf{h} + \mathbf{r'})\, dm$$
$$= h^2\, dm + 2\mathbf{h} \cdot \mathbf{r'}\, dm + r'^2\, dm$$

The last term on the right is, however, just the contribution of dm to the moment of inertia about the parallel axis through C. Thus we have

$$dI_O = h^2\, dm + 2\mathbf{h} \cdot \mathbf{r'}\, dm + dI_c$$

We can now carry out the summation or integration of all such contributions, first within the slice and then over all the slices that build up to the complete object. This gives us

$$I_O = h^2 \int dm + 2\mathbf{h} \cdot \int \mathbf{r'}\, dm + I_c$$

By the definition of the center of mass, the middle term on the right is zero. The first term is simply the total mass, M, of the object times h^2. Thus we finally have the following result:

(Parallel-axis theorem) $I_O = I_c + Mh^2$ (14-24a)

If we choose to express I_c, according to Eq. (14-23), in terms of the mass M and the radius of gyration k, we can write Eq. (14-24a) in the alternative form:

(Parallel-axis theorem) $I_O = M(k^2 + h^2)$ (14-24b)

It may be seen from Eqs. (14-24a) and (14-24b) that the moment of inertia of an object about an arbitrary axis is always greater than its moment of inertia about a parallel axis through the CM.

Example. A cylinder of mass M and radius R is hinged about an axis AA' lying in its surface and running parallel to its axis, as shown in Fig. 14-9(b). What is its moment of inertia about AA'?

We know by Eq. (14-20) that the value of I_c for this case is $MR^2/2$. Hence by Eq. (14-24a), with $h = R$, we have

$$I_{AA'} = \tfrac{1}{2}MR^2 + MR^2 = \tfrac{3}{2}MR^2$$

Theorem of perpendicular axes

Suppose now that we have a flat object of arbitrary shape, cut out of thin sheet material [Fig. 14-10(a)]. Let us take an arbitrary origin O in the plane of the object, and a z axis perpendicular to it. Consider now the moment of inertia, I_z, of the object about the z axis. An element of mass dm, a distance r from the axis, makes the contribution $r^2\, dm$, and we have

$$I_z = \int r^2\, dm$$

However, since r lies in the xy plane, we have

$$r^2 = x^2 + y^2$$

Thus we can put

$$I_z = \int x^2\, dm + \int y^2\, dm$$

Since the object is flat, however, the first term on the right simply defines the moment of inertia I_y of the object about the y axis, and the second term correspondingly is I_x. Thus we have

(Flat objects) $I_z = I_x + I_y$ (14-25)

This is known as the perpendicular-axis theorem. Its usefulness

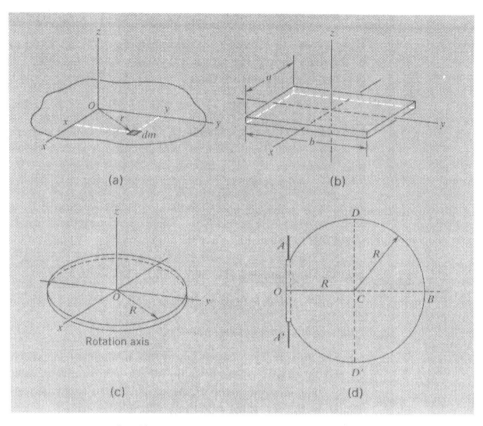

Fig. 14–10 (a) Diagram to show the basis of the perpendicular-axis theorem. (b) and (c) Flat objects to which the perpendicular-axis theorem can be usefully applied. (d) Calculation of the moment of inertia of a circular lid about the axis AA' exploits both the parallel-axis and the perpendicular-axis theorems.

can be illustrated by two different kinds of examples:

Example 1. An object is in the form of a uniform rectangular plate [Fig. 14–10(b)]. What is its moment of inertia about an axis through its center and perpendicular to its plane?

By Eq. (14–21) we have

$$I_x = \frac{Mb^2}{12} \quad I_y = \frac{Ma^2}{12}$$

It follows at once that

$$I_z = \frac{M(a^2 + b^2)}{12}$$

Example 2. A uniform disk has mass M and radius R. What is its moment of inertia about an axis along one of its own diameters, say the x axis in Fig. 14–10(c)?

This illustrates a more elegant use of the perpendicular-axis theorem. The direct calculation of the moment of inertia of a disk about a diameter would be quite awkward. We know, however, that the moment of inertia has the same value about any diameter. We also know, by Eq. (14–20), the moment of inertia I_z about the axis perpendicular to the disk through its center. Thus we can at once put

$$I_x + I_y = 2I_x = I_z = \frac{MR^2}{2}$$

Therefore,

$$I_x = \frac{MR^2}{4}$$

Finally, in Fig. 14–10(d), we show a situation in which we can exploit both of the above theorems. A circular disk (e.g., the lid of a cylindrical tank) is pivoted about an axis AA' that lies in the plane of the disk and is tangent to its periphery. What is the moment of inertia of the disk about AA'?

Beginning with Eq. (14–20) for the moment of inertia $MR^2/2$ about the axis through C perpendicular to the disk, we first use the perpendicular-axis theorem to deduce that the moment of inertia about the axis DD' is $MR^2/4$. We can then use the parallel-axis theorem, Eqs. (14–24a) and (14–24b), to deduce that the moment of inertia about AA' is given by

$$I_{AA'} = \frac{MR^2}{4} + MR^2 = \frac{5MR^2}{4}$$

Such a result is not worth memorizing for its own sake; the important thing is to realize that in these two theorems we have a powerful way of extending a few basic results, as represented for example by Eqs. (14–19) through (14–22), to handle a whole variety of more complicated situations.

KINETIC ENERGY OF ROTATING OBJECTS

A rotating system of course has kinetic energy by virtue of its rotation about its center of mass. Thanks to the general validity of Eq. (14–3) this can be calculated separately and added to any kinetic energy associated with motion of the center of mass itself. When a "rigid" object rotates with angular velocity ω about an axis, a particle within it of mass m, distance r from the axis, has a speed ωr and hence a kinetic energy $\frac{1}{2}m\omega^2 r^2$. The total kinetic energy of rotation is thus given by

$$K_{\text{rot}} = \tfrac{1}{2}\omega^2 \sum mr^2 = \tfrac{1}{2}I\omega^2 \tag{14-26}$$

Hence, if the CM has a speed v_c with respect to the laboratory, the total kinetic energy as measured in the laboratory is given by

$$K = \tfrac{1}{2}I\omega^2 + \tfrac{1}{2}Mv_c^2 \tag{14-27}$$

For an object that rolls along the ground, there will be a purely geometrical connection between v_c and ω (e.g., for a wheel of radius R, $v_c = \omega R$). In such a case the kinetic energy can be expressed in terms of ω (or v_c) alone.

If a round rigid object rolls down a slope, we can apply conservation of energy to calculate its acceleration. For example, if a solid cylinder rolls down a slope of angle θ [Fig. 14–11(a)], its total kinetic energy at any instant is given by putting $I = \frac{1}{2}MR^2$ and $\omega = v_c/R$ in Eq. (14-27):

$$K = \tfrac{1}{2}(\tfrac{1}{2}MR^2)\frac{v_c^2}{R^2} + \tfrac{1}{2}Mv_c^2$$
$$= \tfrac{3}{4}Mv_c^2$$

Let us, at this point, drop the suffix c on v_c and let the symbol v represent what we can properly call *the* translational speed of the rolling object. Then in a short space of time dt, the change of the total kinetic energy is given by

$$dK = \tfrac{3}{2}Mv\,dv$$

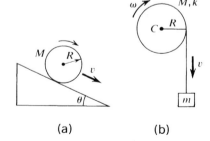

Fig. 14–11 (a) Disk or cylinder rolling down an inclined plane. (b) Mass descending at the end of a rope wrapped around a flywheel.

However, there is a corresponding change of gravitational potential energy. In time dt the CM of the cylinder travels a distance $v\,dt$ parallel to the slope, which means a (negative) change of vertical coordinate equal to $-v\,dt\,\sin\theta$. Hence

$$dU = -Mgv\,dt\,\sin\theta$$

Putting

$$dK + dU = 0$$

we have

$$\tfrac{3}{2}Mv\,dv - Mgv\,dt\,\sin\theta = 0$$

whence

$$a = \frac{dv}{dt} = \tfrac{2}{3}g\sin\theta$$

It is interesting that this result is independent of both the mass and the radius of the cylinder; only the fact that it is a solid cylinder is important. All solid *spheres* have some different characteristic acceleration in rolling down such a slope. (What is it?)

It may be noted that the need to create rotational kinetic energy causes the linear acceleration of a rolling object to be always less than the acceleration that the same object would have if it could simply slide, without friction, down the same slope. The rotational inertia acts, in effect, as a kind of brake on the motion. This inertial property can be exploited in the somewhat different context represented by Fig. 14–11(b). A mass m descends under gravity at the end of a rope that is wound around a flywheel mounted on a fixed axis through C. The flywheel has a moment of inertia $I\ (= Mk^2)$ and a radius R. The angular velocity ω of the wheel is equal to the linear velocity v of m divided by R. Thus the kinetic energy of the whole system is given by

$$\begin{aligned}K &= \tfrac{1}{2}mv^2 + \tfrac{1}{2}I\omega^2 \\ &= \tfrac{1}{2}mv^2 + \tfrac{1}{2}M\frac{k^2}{R^2}v^2 \\ &= \tfrac{1}{2}\left(m + M\frac{k^2}{R^2}\right)v^2\end{aligned}$$

The change of potential energy of m in descending a distance h

is, however, simply mgh. The speed v acquired by m in descending a distance h from rest is therefore given by

$$\tfrac{1}{2}m\left(1 + \frac{Mk^2}{mR^2}\right)v^2 = mgh$$

or

$$v^2 = 2\frac{g}{1 + (Mk^2/mR^2)}h = 2g'h$$

Thus the acceleration g' can be reduced to any desired fraction of g so as to produce a gentle, controlled descent.

Another aspect of rotational kinetic energy is that a large rotating object is an energy reservoir of possibly very large capacity. The use of flywheels as energy-storage devices in this sense is an important feature of all sorts of machines, giving to such systems a much improved stability with respect to sudden changes of load. One of the most impressive examples of the use of flywheels for energy storage is in the National Magnet Laboratory at M.I.T. There are two flywheels, each being an assembly of circular plates of steel, 190 in. in diameter, with a mass of 85 tons. They are part of a generating system for producing extremely strong magnetic fields. The normal speed of rotation of each flywheel is 390 rpm. From these figures we have (for each flywheel)

$$M = 7.7 \times 10^4 \text{ kg} \qquad R = 2.4 \text{ m}$$
$$I = \tfrac{1}{2}MR^2 \approx 2 \times 10^5 \text{ kg-m}^2$$
$$\omega = 2\pi \times 390/60 \approx 40 \text{ rad/sec}$$

Therefore,

$$K = \tfrac{1}{2}I\omega^2 \approx 1.6 \times 10^8 \text{ J}$$

When one of these flywheels is used as a power source, its speed of rotation can be lowered from 390 to 300 rpm in 5 sec. This means that about 40% of the stored energy can be drawn upon, at the rate of about 8% per second. The power output corresponding to this is close to 15 MW—enough, while it lasts, to equal the total rate of electrical energy consumption of a town of about 20,000 inhabitants.

ANGULAR MOMENTUM CONSERVATION AND KINETIC ENERGY

The conservation of rotational angular momentum has some

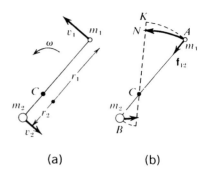

Fig. 14–12 (a) Pair of objects rotating at fixed distances about their center of mass. (b) Rotational displacements accompanied by a change of radial distance between the objects.

interesting implications for the total kinetic energy of a system that changes its shape or size. To take the simplest possible example, consider a system of two masses rotating about their center of mass [Fig. 14–12(a)]. We shall assume that the center of mass is stationary in our frame of reference, but in the absence of external forces it does not have to be defined by a fixed pivot or anything of that kind.

Suppose that the system is rotating with an angular velocity ω about an axis perpendicular to the plane of the diagram. Then if the distances of the masses from C are r_1 and r_2, respectively (with $m_1 r_1 = m_2 r_2$), we have

$$v_1 = \omega r_1 \qquad v_2 = \omega r_2$$

We can then proceed at once to write down expressions for the total angular momentum and the total kinetic energy:

$$L_c = (m_1 r_1^2 + m_2 r_2^2)\omega$$
$$K = \tfrac{1}{2}(m_1 r_1^2 + m_2 r_2^2)\omega^2$$

It will simplify things if we introduce the reduced mass μ [$= m_1 m_2/(m_1 + m_2)$], the relative separation r ($= r_1 + r_2$), and the moment of inertia I ($= \mu r^2$). We then have

$$L_c = \mu r^2 \omega = I\omega \ (= \text{const.})$$
$$K = \tfrac{1}{2}\mu r^2 \omega^2 = \tfrac{1}{2} I \omega^2$$

Combining these two equations, we arrive at the following result:

$$K = \frac{L_c^2}{2\mu r^2} = \frac{L_c^2}{2I} \qquad (14\text{–}28)$$

Consider now what happens if the particles draw closer together under some mutual interaction—e.g., the pull of a spring or elastic cord that connects them. The value of L_c remains constant, but r and the moment of inertia I decrease; hence the values of ω and K must increase. Where does the extra energy come from? Clearly, it has to be supplied through the work done by the internal forces that pull the masses together. One can feel this very directly if one does the experiment of sitting on a rotating stool with two weights at arm's length and then pulling the weights inward toward the axis [cf. Fig. 14–5(a)]. One very simple way of interpreting this from the standpoint of the rotating frame itself is to consider the small change of K associated with a small change of r. From Eq. (14–28) we have

$$\Delta K = -\frac{L_c^2}{\mu r^3} \Delta r$$

Substituting $L_c = \mu r^2 \omega$, this becomes

$$\Delta K = -\mu \omega^2 r \, \Delta r$$

But $\mu \omega^2 r$ is the magnitude of the centrifugal force that is trying (from the standpoint of the rotating frame) to make the masses fly apart. A force equal and opposite to this must be supplied to hold the masses at a constant separation r, and an amount of work equal to this force times the magnitude of the displacement Δr is needed to pull the masses toward one another. (Note that, in the case we have assumed, Δr is negative and so ΔK is positive.)

From the standpoint of a stationary observer, of course, the change of kinetic energy can be understood in terms of the fact that if r is changing [Fig. 14–12(b)], the radial forces of interaction, $\mathbf{f}_{12}$ and $\mathbf{f}_{21}$, have components along the paths of the masses (e.g., the curve AN for m_1) so that work is done, which would not be the case if the masses remained on circular arcs [e.g., AK in Fig. 14–12(b)] with the radial forces always perpendicular to the displacements and velocities.

The fact that the increase in rotational kinetic energy of a contracting system must come from the internal interactions places restrictions on the possibility that such contraction can occur at all in a particular case. If the increase that would be called for in the kinetic energy is greater than the work that could be supplied by the internal forces, then the contraction cannot take place. Especially interesting situations of this type may arise in the gravitational contraction and condensation of a slowly rotating galaxy or star. The mutual gravitational energy, U, of any system of particles is always negative and becomes more so as the linear dimensions of the system shrink and the particles of the system come closer together. (Remember that, for a two-particle system, $U = -Gm_1m_2/r$.) Thus, qualitatively at least, we recognize a source of the extra kinetic energy needed. There is more to it than this, however, because the magnitudes of the corresponding changes ΔK and ΔU, for a given change ΔR in the radius of the system, are not automatically equal. The relationship between them will depend on the magnitude of L_c and on the precise distribution of matter in the system. If ΔK would be larger than ΔU, the total energy would be required to increase, and this simply could not happen. If ΔK were less than ΔU, however, there would be a surplus of released gravitational energy that could be disposed of by developing random particle motions (heating) and by radiation of heat and light into space.

With the help of certain extreme simplifying assumptions the discussion can be made quantitative. Suppose, in particular, that the contraction occurs in such a way that the system merely undergoes a change of linear scale without altering the *relative* distributions of density or velocity. This means that if the linear dimensions shrink by a factor n, the density at distance r/n from the center is n^3 times the original density at r. The moment of inertia, I, is then simply proportional to the square of any characteristic dimension of the system—e.g., its outer radius R if it can be considered as being a spherical object with an identifiable boundary. Thus the kinetic energy K $(= L^2/2I)$ varies as $1/R^2$. The assumed uniform contraction also increases the potential energy $(\sim 1/r)$ between every pair of particles by the factor n. It follows that the total (negative) potential energy U varies as $1/R$. Under these assumptions, therefore, the total energy E of the system is given by an equation of the form

$$E(R) = \frac{A}{R^2} - \frac{B}{R} \qquad (14\text{-}29)$$

The constant A is given dimensionally, and perhaps even in order of magnitude, by the combination $L^2/2M$, where M is the total mass. Similarly, B is proportional to GM^2, where G is the universal gravitation constant. (The gravitational *self*-energy of a sphere of matter is of the order of $-GM^2/R$—see Problem 11-31.)

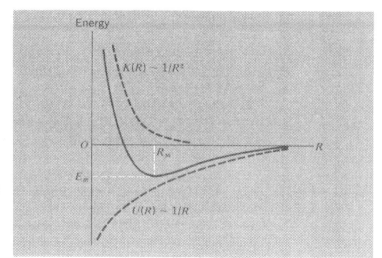

Fig. 14–13 Dependence of kinetic, potential, and total energy on radius for a rotating system held together by gravitational attraction.

The two contributions to E, and their sum, are shown graphically as functions of R in Fig. 14–13. The whole situation is very reminiscent of our discussion of energy diagrams in Chapters 10 and 11, and it is clear that we can calculate a radius R_m that corresponds to an equilibrium configuration of minimum energy:

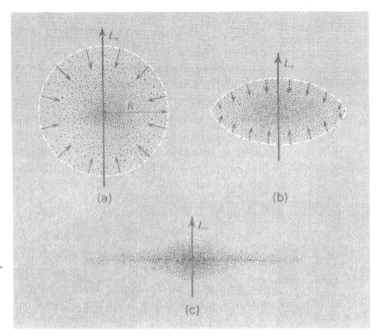

Fig. 14–14 Successive stages of contraction of a rotating gas cloud to form a disk-shaped galaxy.

Fig. 14-15 Example of a rotating galaxy, seen almost edge on (NGC 4565 in the constellation Coma Berenices). (Photograph from the Hale Observatories).

$$\frac{dU}{dR} = -\frac{2A}{R^3} + \frac{B}{R^2}$$

$$R_m = \frac{2A}{B}$$

If we put $A \approx L^2/2M$ and $B \approx GM^2$, we then have

$$R_m \approx \frac{L^2}{GM^3} \tag{14-30}$$

which would indicate the way in which the linear dimensions of similar galaxies might depend on the mass and the total angular momentum.

Figure 14-14 indicates the more probable trend of a contracting rotating mass. Since contraction parallel to the direction of $\mathbf{L}_c$ can take place without any increase of rotational kinetic energy, it is quite reasonable that this type of deformation can continue after a limit has been reached to the contraction radially inward toward the axis of rotation. One can certainly understand in these terms the progression through stages (a), (b), and (c) of Fig. 14-14, which might well result in the kind of galactic structure actually observed (Fig. 14-15).

TORSIONAL OSCILLATIONS AND RIGID PENDULUMS

One of the most valuable and widely used physical systems

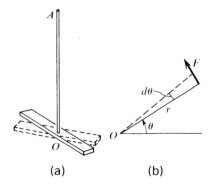

Fig. 14-16 (a) Simple torsional pendulum. (b) Diagram to indicate the work done in twisting a torsion fiber.

consists of a mass suspended from a wire, fiber, or other device that provides a torque in response to a twist [Fig. 14-16(a)]. The restoring torque comes from an elastic deformation of the suspension and, like the linear deformations discussed in connection with the linear harmonic oscillator problem in Chapter 7, such angular deformations usually result in a restoring effect proportional to the deformation.

These torsion devices are often used in static measurements —for example in ammeters, where a steady current passing through the instrument may be used (with the help of a permanent magnet) to produce a steady torque, leading to a steady deflection that is the meter reading of the current. But the free torsional oscillation of such a system is also of interest and importance. The analysis of this oscillatory motion is very conveniently made in terms of the constant total energy of the rotating system.

We shall suppose that the suspended mass has a moment of inertia I about the axis defined by the torsion fiber. Let the angle of deflection around this axis (we shall call it the z axis) be θ. Then the kinetic energy K is given by

$$K = \frac{1}{2} I \left(\frac{d\theta}{dt} \right)^2 \qquad (14\text{-}31)$$

The potential energy U is the work done in twisting the ends of the suspension through an angle θ relative to one another. If we let the fiber define a z axis, the restoring torque M_z, assumed proportional to θ, is given by

$$M_z = -c\theta \qquad (14\text{-}32)$$

where c is the *torsion constant* (measured in m-N/rad or dimensionally equivalent units).

It is not hard to guess that, by analogy with a stretched spring, a system that obeys Eq. (14-31) will lead to a storage of potential energy proportional to θ^2. To make this quantitative we need only consider one simple idea. Suppose we are looking down the axis of a torsion wire [Fig. 14-16(b)] to which is attached a lever arm of length r. A force F is applied at right angles to the end of the lever, just sufficient to overcome the torque M_z. This means that

$$rF = -M_z$$

Suppose that the angle of twist is increased by $d\theta$. Then the end of the lever moves through a distance $r\,d\theta$, and the work done by F is given by

$$dW = Fr\,d\theta = -M_z\,d\theta$$

Hence the total potential energy stored in the system, in going from its normal configuration to a twist θ, is given by

$$U(\theta) = \int_0^\theta c\theta\,d\theta = \tfrac{1}{2}c\theta^2 \tag{14-33}$$

Combining Eqs. (14-31) and (14-33), we see that the total mechanical energy of the system is given by

$$\tfrac{1}{2}I\left(\frac{d\theta}{dt}\right)^2 + \tfrac{1}{2}c\theta^2 = E \quad (=\text{const.}) \tag{14-34}$$

which is the familiar form of a harmonic-oscillator equation. The period will be given by

$$T = 2\pi \left(\frac{I}{c}\right)^{1/2} \tag{14-35}$$

It is worth noting, by the way, that the relation between M_z, as given by Eq. (14-32), and $U(\theta)$, as given by Eq. (14-33), satisfies the general relation between potential energy and torque that we cited earlier [Eq. (14-13)] in our discussion of angular-momentum conservation:

$$M_z = -\frac{\partial U}{\partial \theta}$$

Situations very similar to those of the torsional oscillator are represented by a so-called *rigid pendulum*—an arbitrary object free to swing about a horizontal axis, as shown in Fig.

Fig. 14-17 (a) Example of a rigid pendulum. (b) Period of oscillation of a rigid pendulum as a function of the distance h between the CM and the point of suspension.

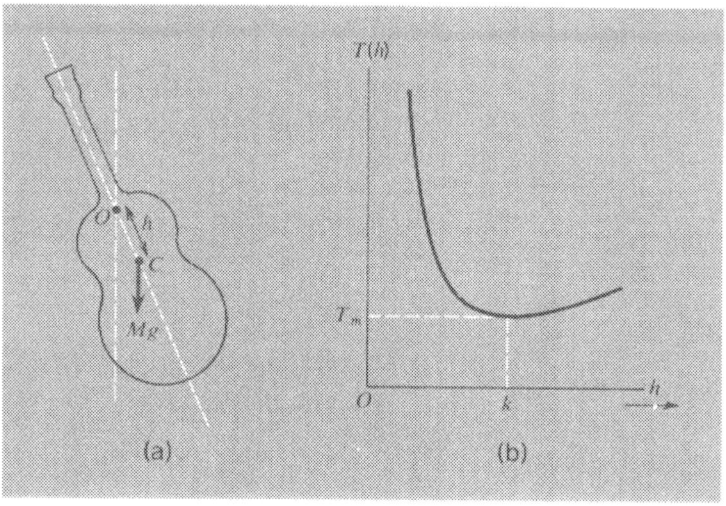

14-17(a). Let us suppose that this axis is a distance h from the center of mass. Normally, therefore, the center of mass C is a vertical distance h below the axis through O, but if the system is displaced through an angle θ, the CM moves along a circular arc of radius h. This causes the CM to rise through the distance $h(1 - \cos\theta)$; if θ is small the consequent increase of gravitational potential energy is given by

$$U(\theta) \approx \tfrac{1}{2} Mgh\theta^2 \qquad (14\text{-}36)$$

The *kinetic* energy of the system is equal to the kinetic energy associated with the linear velocity of the CM, plus the energy of rotation about the CM:

$$K = \tfrac{1}{2} M v_c^2 + \tfrac{1}{2} I_c \left(\frac{d\theta}{dt}\right)^2$$

[The term representing the rotational energy about C in this equation embodies an important feature. If the object has the angular velocity $d\theta/dt$ about its true axis of rotation through O, every point in it also has the angular velocity $d\theta/dt$ about a parallel axis through C, or through any other point for that matter. One can properly speak of *the* angular velocity of a rotating object without reference to a specific axis of rotation. Any line drawn on a rotating disk, for example, has the same rate of angular displacement as one of the radii.]

Returning now to the expression for K, we can put

$$v_c = h\frac{d\theta}{dt}$$

Thus we can write

$$K(\theta) = \tfrac{1}{2}(Mh^2 + I_c)\left(\frac{d\theta}{dt}\right)^2$$

or

$$K(\theta) = \tfrac{1}{2}I_0\left(\frac{d\theta}{dt}\right)^2$$

where I_O is the moment of inertia of the object about the axis through O, and the whole motion of every point in the object is expressed, as we know it can be, in terms of pure rotation about this axis.

For our present purposes, it is most illuminating to write I_c in the form Mk^2, where k is the radius of gyration. If we do this, we have

$$K(\theta) = \tfrac{1}{2}M(h^2 + k^2)\left(\frac{d\theta}{dt}\right)^2 \tag{14-37}$$

The equation of energy conservation, given by adding the results of Eqs. (14–36) and (14–37), is thus

$$\tfrac{1}{2}M(h^2 + k^2)\left(\frac{d\theta}{dt}\right)^2 + \tfrac{1}{2}Mgh\theta^2 = E \quad (= \text{const.}) \tag{14-38}$$

This defines simple harmonic vibrations with a period that depends in a systematic way on the distance h of the CM from the axis:

$$T(h) = 2\pi\left(\frac{h^2 + k^2}{gh}\right)^{1/2} \tag{14-39}$$

This period would become infinitely long for $h = 0$ (rotational axis passing through CM) and has a minimum value T_m for $h = k$. The over-all variation of T with h is as shown in Fig. 14–17(b).

For any given value of h, the period of oscillation corresponds to that of an "equivalent simple pendulum" of length l such that

$$l = \frac{h^2 + k^2}{h}$$

MOTION UNDER COMBINED FORCES AND TORQUES

Near the beginning of this chapter we developed the two results which, between them, provide the basis for analyzing the motion of extended objects under any circumstances. These results are as follows:

1. The rate of change of linear momentum is equal to the resultant external force. Expressed in terms of the motion of the center of mass, this statement becomes

$$\mathbf{F}_{net} = M\frac{d\mathbf{v}_c}{dt} = M\mathbf{a}_c \tag{14-40}$$

where M is the total mass.

2. The rate of change of angular momentum about the center of mass is equal to the resultant torque of the external forces about the CM:

$$\mathbf{M}_c = \frac{d}{dt}(\mathbf{L}_c) \tag{14-41}$$

In the present section we shall limit ourselves to cases in which both the torque and the angular motion are about an axis parallel to the axis of symmetry of the object [as defined in the discussion leading up to Eq. (14-16)]. This means that we can put $\mathbf{L}_c = I_c\boldsymbol{\omega}$, and Eq. (14-41) takes on the following special form:

$$(\text{Special case, } \mathbf{L}_c \parallel \boldsymbol{\omega}) \quad \mathbf{M}_c = I_c\frac{d\boldsymbol{\omega}}{dt} = I_c\boldsymbol{\alpha} \tag{14-42}$$

where $\boldsymbol{\alpha}$ is a vector representing the angular acceleration.

Let us at once consider a specific situation in which Eqs. (14-40) and (14-42) are applicable. An airplane is just touching down. When one of the landing wheels first makes contact with the ground, it has a large horizontal velocity v_0 but no angular motion; therefore, it is bound to skid at first. Anyone who has watched a plane landing will have seen the initial puff of smoke from burning rubber resulting from this violent skidding. After touchdown there are forces on the wheel applied at its two contacts with the external world—its axle and the place where the wheel touches the ground. The forces at the axle pass through the CM and so can exert no torque. The normal component, N, of the force of contact with the ground also passes through the CM; furthermore, since N is vertical, it does not affect the horizontal motion of the wheel. Thus we have two forces to consider in analyzing the forward motion of the wheel and its rotational motion. These are a force F, pushing the wheel forward at the axle, and a frictional force $\mathscr{F}$ acting backward, as shown in Fig. 14–18(a). By Eq. (14–40) we then have

$$F - \mathscr{F} = Ma_c$$

We can guess that a_c is negative, because the wheel remains attached to the plane (we hope) and the reaction force $-F$ applied by the wheel to the plane represents an unbalanced force that is acting to decelerate the plane as a whole. If we assume that a_c is constant, we then have

$$v_c = v_0 + a_c t \qquad (a_c < 0) \qquad (14\text{–}43)$$

When we look at the rotational component of the motion, we see that the only force that produces a torque about C is the frictional force $\mathscr{F}$. Furthermore, although $\mathscr{F}$ acts to slow down the linear motion, its torque is in such a direction as to speed up the angular motion. We have, in fact, by Eq. (14–42),

$$M_c = R\mathscr{F} = Mk^2 \frac{d\omega}{dt} = Mk^2 \alpha$$

where k is the radius of gyration of the wheel and R its actual radius. If we assume further that the angular acceleration α is constant, we have

$$\omega = \alpha t \qquad (14\text{–}44)$$

Fig. 14–18 (a) Forces and motions for a landing wheel of an airplane. (b) The velocity of any point on a wheel is the superposition of the linear motion of the center and the rotation about the center, as shown for an arbitrary point P and for the special cases represented by the top and bottom points, A and B.

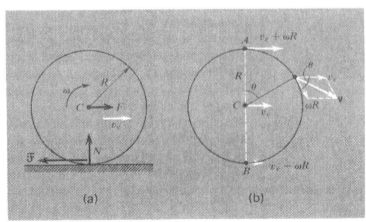

As long as the skidding goes on, Eqs. (14–43) and (14–44) operate separately to define the linear and angular velocities of the wheel. At any instant, the resultant velocity of any point on the rim of the wheel is the vector sum of the horizontal velocity v_c of C and a velocity of magnitude ωR along the tangent, as shown in Fig. 14–18(b). This then allows us to identify the condition for the skidding to stop. Skidding means the existence of relative motion between the ground and the point on the wheel that is instantaneously in contact with it. Since the ground defines our rest frame for this problem, the cessation of skidding requires that the velocity of the lowest point, B, on the wheel becomes zero. This velocity, v_B, is, however, the resultant of v_c forward and ωR backward. Thus we have

$$v_B = v_c - \omega R$$

Skidding therefore stops, and rolling begins, when we have

$$v_c = \omega R$$

Using Eqs. (14–43) and (14–44), we see that this occurs at a time t such that

$$v_0 + a_c t = \alpha R t$$

By using the dynamical equations that define the actual values of a_c and α, we can find the time t and hence the amount by which v_c has been reduced from its initial value v_0 at this instant. To solve the problem completely, we would also have to consider the linear deceleration imposed on the total mass of the plane by air resistance and by the forces of the type $-F$ due to all the wheels together.

The above problem is a valuable one because it does emphasize the separate consideration of linear and angular motions. On the other hand, it does not lend itself to well-defined calculations. A simpler problem of the same type is represented by a ball (e.g., a bowling ball) being projected horizontally along the floor [Fig. 14–19(a)]. In this case the frictional force alone provides both the linear deceleration and the angular acceleration. Furthermore, we can use our knowledge of the coefficient of friction and the weight of the ball to define the magnitude of $\mathcal{F}$:

$$\mathcal{F} = \mu N = \mu M g$$

The equations of motion then become the following:

Fig. 14-19 (a) Forces and motions for a ball projected along a floor. (b) Circular object accelerating down a slope. It may either roll or slip, depending on the angle of the slope and the coefficient of friction.

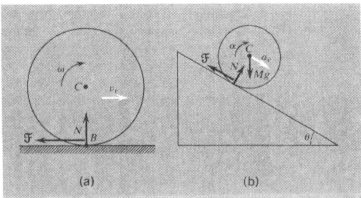

$$-\mu Mg = Ma_c$$
$$\mu MgR = Mk^2\alpha$$

Hence, if we assume that the ball starts out with horizontal velocity v_0 and no rotation, we have

$$v_c = v_o - \mu g t$$
$$\omega = \frac{\mu g R}{k^2} t$$

Rolling ($v_c = \omega R$) then begins at a time defined by the following equation:

$$v_o - \mu g t_{\text{roll}} = \mu g \frac{R^2}{k^2} t_{\text{roll}}$$

or

$$t_{\text{roll}} = \frac{v_o}{\mu g(1 + R^2/k^2)}$$

The linear velocity at this instant is therefore given by

$$v_{\text{roll}} = \frac{v_o}{1 + k^2/R^2}$$

Ideally, the ball would then continue to roll indefinitely at this speed.

If we can assume rolling from the outset, the calculations may be even more direct. Consider, for example, a problem such as that of a cylinder rolling down a slope, which we solved earlier by energy conservation. We shall now analyze it explicitly in terms of the forces and torques acting [see Fig. 14-19(b)]. We resolve the contact force at the plane into normal and tangential components, N and $\mathcal{F}$. Since there is no acceleration perpendicular to the plane, we have $N = Mg\cos\theta$. This defines a maximum available value of $\mathcal{F}$, equal to $\mu Mg\cos\theta$. At some point in the calculation we must check to see whether the assumption of rolling is consistent with this limitation. But to begin with, we proceed as though this were assured. The equations of rotation, and of linear acceleration along the plane, are as follows:

$$\mathcal{F}R = Mk^2\alpha$$
$$Mg\sin\theta - \mathcal{F} = Ma_c$$

In this case, however, we also have the purely geometrical connection between linear and angular accelerations if slipping is not to occur:

$$a_c = \alpha R$$

Solving the two dynamical equations with the help of this "equation of constraint," we at once find

$$a_c = \frac{g\sin\theta}{1 + k^2/R^2}$$

If we put $k^2 = R^2/2$ (corresponding to a uniform disk or cylinder) we arrive once again at the result obtained on page 653. Once we have the value of a_c or α demanded by this solution, we should check to see whether the associated value of $\mathcal{F}$ is under the permitted ceiling. If not, we must begin again with the value of $\mathcal{F}$ set equal to its maximum value μN, and calculate the linear and angular accelerations separately. Either way, however, the problem is soluble in definite terms.

One could easily multiply examples of this kind of analysis, but once one has clearly grasped the basic approach defined by Eqs. (14-40) and (14-42) the solution to any particular problem of this type should present no difficulties.

IMPULSIVE FORCES AND TORQUES

Anyone who has played a game that involves hitting a ball with a bat or a racket will have experienced the satisfying feeling that comes from hitting the ball just right, when it seems that the bat or racket is doing the work unaided. More often, perhaps, one is conscious of the sense of effort, or even a painful sting in the hands, when the ball does not make contact at the optimum point. These are both manifestations of the dynamical phenomena that occur when a rigid object (e.g., a bat) is subjected to a sudden impulse by the impact of the ball.

The simplest example of this kind of behavior is the application of an impulse to an object that is not restrained in any way. Imagine, for this purpose, an object, as shown in Fig. 14–20(a), resting on a frictionless horizontal surface and subjected to a force F for a time Δt along a line at a distance b from the center of mass, C. As a result of this the center of mass acquires a velocity v_c and the object acquires an angular velocity ω. Applying our basic equations [Eqs. (14–40) and (14–42)] we have

$$F\Delta t = Mv_c$$
$$bF\Delta t = Mk^2\omega$$

If we denote the magnitude of the impulse by Δp, representing the integral of F over time, whatever its precise variation, we have

$$v_c = \frac{\Delta p}{M}$$
$$\omega = \frac{b\Delta p}{Mk^2}$$

These equations, between them, determine the subsequent motion of every point in the object. The quantity $b\,\Delta p$ is the angular

Fig. 14–20 (a) An object completely free to recoil is given a sudden impulse by a force passing through the point A. (b) The object receives an impulse like that in (a) but is now forced to pivot about O. (c) A suspended baseball bat is struck by the ball. If the point of impact is correctly chosen, the bat pivots about O without producing any impulsive force there.

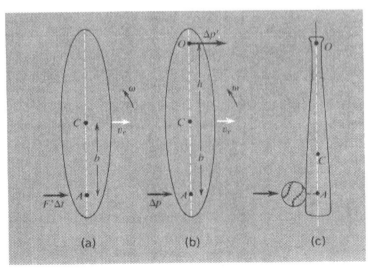

impulse about C due to the linear impulse Δp. And, just as a linear impulse has the dimensions of linear momentum, so an angular impulse has the dimensions of angular momentum.

The situation is changed considerably if the object is attached to a fixed axis passing through some point O, at a distance h from the CM [see Fig. 14–20(b)]. We must now assume that the impulse Δp due to F is accompanied by another impulse $\Delta p'$ applied to the object by the pivot. We now have

$$\Delta p + \Delta p' = Mv_c$$
$$b\,\Delta p - h\,\Delta p' = Mk^2\omega$$

If we take Δp and the values of M, k, and h as known, we have three quantities undetermined—v_c, ω, and $\Delta p'$—and so far only two equations. But we also have an equation of constraint; the point O must remain at rest. As Fig. 14–20(b) shows, we must have

$$v_O = v_c - \omega h$$

and this requires that

$$v_c = \omega h$$

If one solves these equations for $\Delta p'$ (do it!) one finds

$$\Delta p' = \frac{bh - k^2}{h^2 + k^2}\Delta p$$

Clearly $\Delta p'$ may be of either sign, depending on the precise values of b, k, and h. And we can now identify a condition that allows $\Delta p'$ to be zero; this means that even though the pivot is present at O, it is not called upon to supply any restraining force when the impulse Δp is applied, or in other words the point O automatically remains at rest. The necessary condition, from the above equation for $\Delta p'$, is

$$bh - k^2 = 0$$

or

$$b = \frac{k^2}{h}$$

This locates a point A on the axis of the object, a distance k^2/h below the center of mass. The distance l of A from the pivot point itself is given by

$$l = h + \frac{k^2}{h} = \frac{h^2 + k^2}{h}$$

If you refer back to Eq. (14–39) you will see that l is identical with the length of a simple pendulum having the same period as our object would have if allowed to swing about a horizontal axis through O.

We actually have here a good dynamical basis for locating the right point on a bat or racket at which the ball should be struck. This point is called the *center of percussion*. If one suspends a baseball bat, for example, on a long string [Fig. 14-20(c)], one can study directly the effect of making the impact with the ball occur at various distances on the far side of the center of mass from the region around O where the bat would normally be held. If CA is too large, the impact causes O to kick to the left (if the ball itself comes from the left, as shown). If CA is too small, then O kicks to the right. If CA has just the proper value (k^2/h) so that A is the center of percussion, the bat begins to rotate freely about O, which acts as a self-defined pivot. One could thus predict the optimum location of A first by a pendulum experiment and then see how things work in practice.

BACKGROUND TO GYROSCOPIC MOTION

Everybody is intrigued by gyroscopic devices, and probably everybody feels that their behavior somehow flies in the face of the usual rules of mechanics, even though intellectually one knows that this cannot be the case. It cannot be denied, however, that gyroscopic motions often seem surprising and bizarre, and this of course is the main source of their fascination.

The prime requirement for the appreciation and understanding of gyroscopic behavior is a full awareness of angular momentum as a vector. And with angular momentum as the central quantity, one must also learn to think primarily in terms of torques rather than forces, and in terms of angular rather than linear velocities and accelerations. Once one has achieved this, the phenomena fall far more readily into a rational pattern.

An effective, although rather empirical, way of demonstrating the vector character of angular momentum is to show that the rotational angular momenta of several separate flywheels can

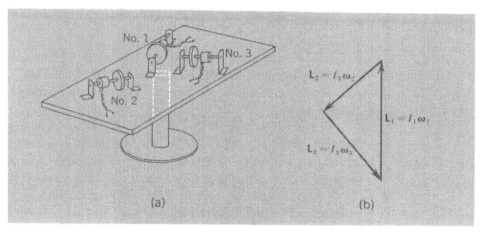

*Fig. 14-21 (a) An array of three motor-driven flywheels can be adjusted to have zero total spin momentum.
(b) The vector addition of the three angular momenta to give zero.*

be made to add up to zero.[1] This can be done by mounting the flywheels on a freely pivoted board [Fig. 14-21(a)]. Each flywheel has its own motor drive and can be run at a controlled and measurable rotation rate. The relative moments of inertia of the flywheels are also known. We shall take the case of three flywheels, as shown, set at different orientations. The system is statically balanced with respect to the pivot and has no preferred orientation if the flywheels are not rotating. If one of the flywheels is set spinning at high speed, the system acquires an easily recognizable gyroscopic stability; the board tends to preserve a given orientation. Setting a second flywheel into rotation does not change the situation qualitatively. But if now the third flywheel is set spinning, one can suddenly reach a condition in which the gyroscopic stability vanishes; the board will tend to flop over as if it simply had a few dead weights on it. It is then possible to verify that this condition is achieved when the individual angular momenta, as vectors, have a zero resultant as shown in Fig. 14-21(b). The values of these angular momenta are given by $L_1 = I_1\omega_1$, $L_2 = I_2\omega_2$, and $L_3 = I_3\omega_3$.

The above is a particularly simple situation to visualize, because we have individual flywheels, each with just one, readily identified contribution to its angular momentum. It is basic to

---

[1] The experiment described here is shown in the film "Angular Momentum—A Vector Quantity," by A. Lemonick, Education Development Center, Newton, Mass., 1962.

the analysis of gyroscopic problems, however, that we recognize the possibility that a single object may have simultaneous contributions, along different axes, to its total rotational angular momentum. An essential preliminary to this, which we have not needed to consider previously, is the full implication of angular velocity itself as a vector. If it is to be possible to speak of *the* angular velocity vector of a rotating object, then the instantaneous linear velocity, $\boldsymbol{\omega} \times \mathbf{r}$, of any point in the object must also be describable as the vector sum of the linear velocities due to component angular velocities ω_x, ω_y, and ω_z along orthogonal axes. The validity of this description also requires that, if only one component of $\boldsymbol{\omega}$—say ω_x—were present, the motion of the object would be pure rotation about the axis (x) in question.

To see how these ideas do work correctly in a simple specific case, consider the situation shown in Fig. 14–22. A uniform rectangular board is made to rotate with angular velocity ω about an axis (in the plane of the board) making an angle δ with an axis of x drawn parallel to one pair of edges. Consider the motion of a point P with coordinates described equivalently by (x, y) or (r, θ). The instantaneous velocity of P, as given by $\boldsymbol{\omega} \times \mathbf{r}$, is vertically upward from the plane of the diagram and its magnitude is given directly by the product of ω with the perpendicular distance PN from P to the rotation axis. Thus we have

$$v = \omega r \sin(\theta - \delta)$$
$$= \omega r \sin\theta \cos\delta - \omega r \cos\theta \sin\delta$$

However, by resolving $\boldsymbol{\omega}$ and $\mathbf{r}$ along the x and y axes, we have

$$\omega_x = \omega \cos\delta \qquad \omega_y = \omega \sin\delta$$
$$x = r \cos\theta \qquad y = r \sin\theta$$

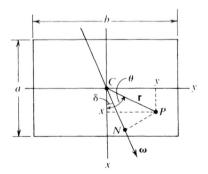

Fig. 14–22 *Angular velocity as a vector. The rotational dynamics of the rotating board can be completely analyzed in terms of the components of $\boldsymbol{\omega}$ along the x and y axes.*

This means that the expression for v can be rewritten as follows:

$$v = \omega_x y - \omega_y x$$

We see that, physically, this is exactly what we would get from the superposition of two separate rotational motions with angular velocites ω_x and ω_y about the x and y axes, respectively.

Let us proceed now to the angular momentum of the whole board about the axis of ω. The linear momentum of an element of mass dm at P is $v\,dm$, and its contribution to the angular momentum about ω is $v\,dm$ multiplied by the distance PN. Since v itself is equal to ω times PN, we see that the magnitude of the rotational angular momentum L_ω about the axis of ω is given by the following integral:

$$L_\omega = \int dm\,\omega r^2 \sin^2(\theta - \delta)$$

Expanding this by writing out the expression for $\sin^2(\theta - \delta)$ in full, we have

$$L_\omega = \omega \cos^2 \delta \int r^2 \sin^2 \theta\, dm + \omega \sin^2 \delta \int r^2 \cos^2 \theta\, dm$$
$$- 2\omega \sin \delta \cos \delta \int r^2 \sin \theta \cos \theta\, dm$$

That is,

$$L_\omega = \omega \cos^2 \delta \int y^2\, dm + \omega \sin^2 \delta \int x^2\, dm$$
$$- 2\omega \sin \delta \cos \delta \int xy\, dm$$

Now the first two integrals are the definitions of the moments of inertia, I_x and I_y, respectively, about the principal axes x and y. And the third integral (an example of what is called a *product of inertia*) vanishes, as one can see from the fact that for each element of mass dm at (x, y) there is an equal element at $(x, -y)$. Thus we have

$$L_\omega = I_x \omega \cos^2 \delta + I_y \omega \sin^2 \delta$$

Introducing the components ω_x and ω_y of ω, this can be written

$$L_\omega = I_x \omega_x \cos \delta + I_y \omega_y \sin \delta \tag{14-45}$$

This shows that the angular momentum about the axis of ω is precisely what we would get from the projections along ω of separate angular momenta $I_x\omega_x$ and $I_y\omega_y$ about the x and y axes. [In general, the vector combination of the angular momenta $I_x\omega_x$ and $I_y\omega_y$ will also have a component perpendicular to ω, and the *total* vector angular momentum of the board is the combination of L_ω with this other component. We shall not interrupt the present discussion to consider this other component of L. (See, however, Problem 14–30.)]

Finally, let us look at the total rotational kinetic energy. This is defined by

$$K = \int \tfrac{1}{2} v^2 \, dm$$

Putting $v = \omega r \sin(\theta - \delta)$, this becomes

$$K = \tfrac{1}{2}\omega^2 \int r^2 \sin^2(\theta - \delta) \, dm$$

We see that the expression for K involves precisely the same integral that appeared in the calculation of L_ω, and we therefore have

$$K = \tfrac{1}{2}(\omega^2 \cos^2 \delta)I_x + \tfrac{1}{2}(\omega^2 \sin^2 \delta)I_y$$

Substituting $\omega \cos \delta = \omega_x$, $\omega \sin \delta = \omega_y$, we thus arrive at the result

$$K = \tfrac{1}{2}I_x\omega_x^2 + \tfrac{1}{2}I_y\omega_y^2 \tag{14-46}$$

We see, then, (at least in the special case discussed above) how the dynamics of a rotating object is analyzable into separate contributions associated with component rotations about the principal axes.

Although we shall not take the matter any further here, it is not difficult to show that results of the form that we have developed are true in general.[1] The essential starting point is again the vector property of angular velocity, and we shall close this discussion with a few additional remarks about that. For a rigid object, pivoted at the CM or at some other fixed point, the motion of any given point in the object is confined to the surface of a sphere. Any given change in position can be produced by rotations about three chosen axes. It is, however,

---

[1]See, for example, K. R. Symon, *op. cit.* (see p. 645).

impossible to represent these finite angular *displacements* as rotation vectors with x, y, and z as axes, because (as we noted in Chapter 2) the resultant of two successive displacements depends on the order in which they are made [Fig. 14–23(a) shows a

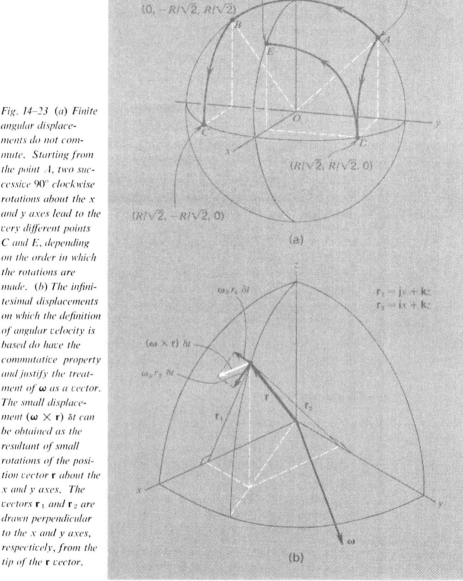

Fig. 14–23 (a) Finite angular displacements do not commute. Starting from the point A, two successive 90° clockwise rotations about the x and y axes lead to the very different points C and E, depending on the order in which the rotations are made. (b) The infinitesimal displacements on which the definition of angular velocity is based do have the commutative property and justify the treatment of $\boldsymbol{\omega}$ as a vector. The small displacement $(\boldsymbol{\omega} \times \mathbf{r})\, \delta t$ can be obtained as the resultant of small rotations of the position vector $\mathbf{r}$ about the x and y axes. The vectors $\mathbf{r}_1$ and $\mathbf{r}_2$ are drawn perpendicular to the x and y axes, respectively, from the tip of the $\mathbf{r}$ vector.

rather extreme example]. Thus there does not exist a unique angular displacement, characterized by an axis direction and a magnitude, that represents the combination of two or more individual rotations. However, the definition of angular *velocity*, like that of linear velocity, is based on infinitesimal displacements during a time δt that becomes zero in the limit. When one considers the combination of displacements of this type [Fig. 14-23(b)] one finds that the vector sum *is* unique, regardless of the order of addition, and can always be represented by a single infinitesimal rotation through the angle $\omega\, \delta t$, where ω is the vector sum of $\mathbf{i}\omega_x$, $\mathbf{j}\omega_y$, and $\mathbf{k}\omega_z$. (In our diagram we have assumed, for simplicity, that $\omega_z = 0$.) If we choose to spell out the calculation of the linear velocity of any point P in terms of components, we thus have

$$\mathbf{v} = (\mathbf{i}\omega_x + \mathbf{j}\omega_y + \mathbf{k}\omega_z) \times (\mathbf{i}x + \mathbf{j}y + \mathbf{k}z)$$

Using the relations $\mathbf{i} \times \mathbf{i} = 0$, $\mathbf{i} \times \mathbf{j} = -\mathbf{j} \times \mathbf{i} = \mathbf{k}$, and so on, it is easy to establish that this equation gives us

$$\mathbf{v} = \mathbf{i}(\omega_y z - \omega_z y) + \mathbf{j}(\omega_z x - \omega_x z) + \mathbf{k}(\omega_x y - \omega_y x)$$

It may be noted that the last term corresponds exactly to the value of $\mathbf{v}$ that we obtained in our special example of the rotating board (for which we took $\omega_z = 0$ and $z = 0$ everywhere). It is often convenient, as with any vector cross product expressed in orthogonal components, to write the general expression for $\mathbf{v}$ as a determinant:

$$\mathbf{v} = \begin{vmatrix} \mathbf{i} & \mathbf{j} & \mathbf{k} \\ \omega_x & \omega_y & \omega_z \\ x & y & z \end{vmatrix}$$

GYROSCOPE IN STEADY PRECESSION

A gyroscope is basically just a flywheel that is mounted so that it has three different possible axes of rotation, which can if desired be made orthogonal to one another. The first axis is the normal spin axis of the flywheel itself; the other two allow this spin axis to tilt in any direction. Figure 14-24 shows how this can be achieved by mounting the flywheel inside a pair of freely pivoted gimbal rings. Our concern will not be with the details of the arrangement but with the dynamics of the response of the

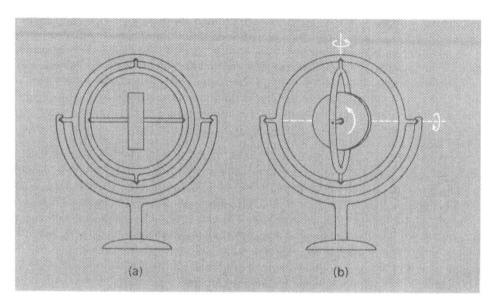

Fig. 14-24 (a) Gyroscope with its gimbal rings lying in the same plane. (b) Gyroscope with its gimbal rings perpendicular, showing the availability of three mutually perpendicular axes of rotation.

flywheel to torques of various kinds, granted that the complete latitude in angular position exists.

One of the simplest and most striking tricks that one can do with a gyroscope is to start it spinning about a horizontal axis and then set one end of its axis down on a pivot, as shown schematically in Fig. 14-25(a). The gyroscope then begins to *precess;* i.e., its axis OA, instead of slumping downward, proceeds to move around so that the extreme end A settles down (after some initial irregularities that we shall discuss later) into a horizontal circular path at constant speed. We can explain this as a direct consequence of the torque acting on the flywheel, but by way of introduction let us first consider a different situation that probably appears childishly simple. The flywheel is spinning with angular velocity ω about its axis [Fig. 14-25(b)] and a torque $\mathbf{M}$ is applied about that same axis for a time δt. Then of course the rotation rate simply speeds up or slows down. The general equation for the rate of change of angular momentum under the action of a torque [Eq. (14-41)] is

$$\mathbf{M}_c = \frac{d}{dt}(\mathbf{L}_c)$$

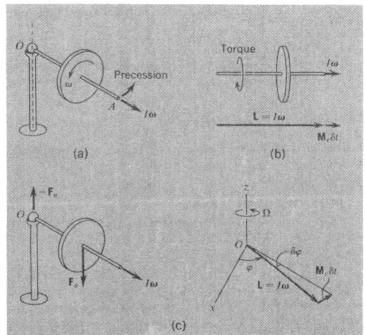

Fig. 14–25 (a) Gyroscope in steady precession. (b) Addition of angular momentum to a flywheel about the spin axis itself. (c) Addition of angular momentum to a flywheel at right angles to the initial spin axis, resulting in precession.

but it reduces in this case to the special form given in Eq. (14–42):

$$\mathbf{M}_c = I \frac{d\omega}{dt}$$

and we have

$$\mathbf{M}_c \, \delta t = \delta \mathbf{L} = I \, \delta \omega$$

The angular momentum vector simply gets longer or shorter without changing direction.

Consider now the circumstances in which precession occurs. The flywheel, supported at one end of its axis, is subjected to a vertical downward force $\mathbf{F}_g$ at its CM and an equal, opposite force at the pivot O [Fig. 14–25(c)]. These two forces constitute a couple whose axis is horizontal and at right angles to the spin axis. In a time δt the torque of this couple adds an amount of angular momentum $\mathbf{M}_c \, \delta t$ at right angles to $\mathbf{L}$. The result of this is to change the direction of $\mathbf{L}$ without changing its magnitude. If we denote by φ the angle between the axis of the flywheel and some standard horizontal reference axis (x) we have

$$M_c \, \delta t = L \, \delta \varphi$$

If the distance from the pivot O to the center of mass is l, we have

$$M_c = F_g l = Mgl$$

where M is the mass of the flywheel. Putting $L = I\omega$, we then find that

$$\Omega = \frac{d\varphi}{dt} = \frac{Mgl}{I\omega} \tag{14-47}$$

where Ω is the angular velocity of precession about the vertical (z) axis, as shown in Fig. 14-25(c). Equation (14-47) gives quantitative expression to the property of gyroscopic stability. The necessary condition is to make $I\omega$ large, so that the precessional angular velocity Ω is small under an applied torque and can, within certain limits, be made negligible in a practical gyroscopic system.

Notice the relationship between the directions of the various vectors involved here. The spin angular momentum is horizontal (a condition that we shall relax in a moment). The torque vector is also horizontal and is always perpendicular to **L**. The precession is described by an angular velocity vector **Ω** directed along the z axis, perpendicular to both **L** and **M**.

The above discussion shows that the steady precessional motion is a direct consequence of the basic vector relation between torque and the rate of change of angular momentum. One last remark may help to put the result in context. If we placed one end of the flywheel axis on the pivot O without giving the flywheel any initial spin, then the flywheel would of course fall down immediately, its center following a circular arc in a vertical plane. What we see here is again the addition of angular momentum about a horizontal axis by virtue of the gravitational torque, but in this case there is no pre-existing spin angular momentum to which it can be added, and the consequences appear (and are) very different.

MORE ABOUT PRECESSIONAL MOTION

The kind of steady precessional motion described in the last section is not confined to flywheels spinning with their axes horizontal. A gyroscope set down with its axis at an angle θ to the vertical [Fig. 14-26(a)] will quickly settle down into precession of just the kind that we have discussed. Its axis sweeps

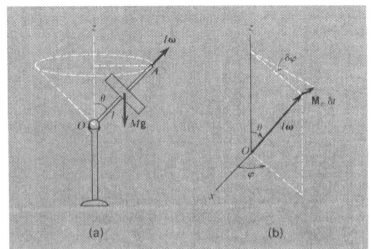

Fig. 14–26 (a) Steady precession with gyroscope axis at an arbitrary angle to the vertical. (b) Vector diagram to show the change of angular momentum in a short time δt in (a).

out a cone of semiangle θ, as indicated. Moreover, quantitative measurements would show that the time to complete one circuit of precession is almost independent of θ for a given value of the spin angular velocity ω and is always nearly equal to $2\pi/\Omega$, where Ω is calculated according to Eq. (14–47). This result is entirely reasonable when one considers that the gravitational torque for the inclination θ is reduced to $Mgl \sin \theta$, but that the amount of rotation about the z axis in time δt [see Fig. 14–26(b)] is equal to the angular impulse $M_c \, \delta t$ divided by the *projection* $I\omega \sin \theta$ of **L** onto a horizontal plane. Thus the quotient of these two quantities is independent of θ. For all its plausibility, however, this result is not quite correct. The reason is that the precessional motion itself implies a further contribution to the angular momentum, and this must be taken into account. We shall give a primitive argument to suggest what this entails.[1]

Consider the motion of a point P on the spin axis of the gyroscope [Fig. 14–27(a)]. Its instantaneous velocity is due to the precession alone and is of magnitude $\Omega r \sin \theta$, downward into the plane of the diagram. This same motion can be described alternatively in terms of a certain angular velocity ω_2 that the whole flywheel has about an axis perpendicular to its spin axis. All we need to do is to put

$$\omega_2 = \Omega \sin \theta$$

---

[1]The rest of this section can be omitted without loss of continuity.

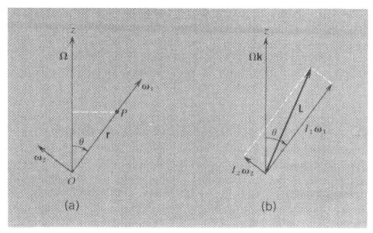

Fig. 14-27 (a) A precessing gyroscope has angular velocity about an axis perpendicular to its spin axis. (b) The total angular momentum vector of a precessing gyroscope is not parallel to the spin.

With this value of ω_2 we can describe the additional velocity, due to precession, of any point in the gyroscope. However, the direction of ω_2 corresponds to one of the principal axes of inertia of the flywheel. Thus we can write an expression for the total angular momentum in terms of the angular velocities ω_1 and ω_2, and the moments of inertia I_1 and I_2 for rotation about these axes:

$$\mathbf{L} = I_1\boldsymbol{\omega}_1 + I_2\boldsymbol{\omega}_2 \tag{14-48}$$

Now we know the way in which **L** is changing with time, for in steady precession the whole system turns with angular velocity $\boldsymbol{\Omega}$ about the vertical (z) axis. Thus we have the simple and very important equation

$$\frac{d\mathbf{L}}{dt} = \boldsymbol{\Omega} \times \mathbf{L} \tag{14-49}$$

corresponding to the rate of change of any vector that is rotating at the rate $\boldsymbol{\Omega}$. In the present case this therefore gives us

$$\frac{d\mathbf{L}}{dt} = I_1(\boldsymbol{\Omega} \times \boldsymbol{\omega}_1) + I_2(\boldsymbol{\Omega} \times \boldsymbol{\omega}_2)$$

Now if we look at the geometry of the situation [Fig. 14-27(a) and (b)], we see that $I_1\boldsymbol{\Omega} \times \boldsymbol{\omega}_1$ is a vector of length $I_1\Omega\omega_1 \sin\theta$ pointing down into the paper, and $I_2\boldsymbol{\Omega} \times \boldsymbol{\omega}_2$ is a vector of length $I_2\Omega\omega_2 \cos\theta$ pointing up from the paper. Together they form a horizontal vector, pointing down into the paper, of magnitude given by

$$\frac{dL}{dt} = I_1\omega_1\Omega \sin\theta - I_2\omega_2\Omega \cos\theta$$

However, we have seen that $\omega_2 = \Omega \sin\theta$. Thus we find that

$$\frac{dL}{dt} = I_1\omega_1\Omega \sin\theta - I_2\Omega^2 \sin\theta \cos\theta$$

Setting this rate of change of angular momentum equal to the gravitational torque in the same direction, we have

$$Mgl \sin\theta = I_1\omega_1\Omega \sin\theta - I_2\Omega^2 \sin\theta \cos\theta$$

or

$$Mgl = I_1\omega_1\Omega - I_2\Omega^2 \cos\theta \qquad (14\text{--}50)$$

For the particular case $\theta = \pi/2$ (spin axis horizontal) we reproduce the simple result expressed by Eq. (14–47) in the previous section. But we see that in general we are confronted with a quadratic equation leading to two different values of Ω. One of these represents a slow precession at a rate not very different from $Mgl/I_1\omega_1$, and this represents the normal situation. In principle, however, one can also obtain a fast precession, although it is difficult to start the gyroscope off with just the right motion to achieve it. In practice, as long as the gyroscope is operated in the usual way with a large value of ω_1, the possibility of the fast precession can be ignored.

GYROSCOPES IN NAVIGATION

We shall give here a very rudimentary discussion of the use of gyroscopes as devices for direction finding, guidance, and control. To introduce the subject, we shall describe another of the simple and initially surprising features of gyroscopic behavior. If a gyroscope, mounted in its gimbal rings, is set spinning about a horizontal axis, and the outer gimbal ring is rotated in a horizontal plane [Fig. 14–28(a)], then the inner gimbal ring, carrying the flywheel, tilts up out of the horizontal plane in which it, too, first lay, and after a few oscillations, damped out by friction, the axis of spin of the gyroscope settles down along a vertical direction.

This behavior can be qualitatively understood if one recognizes that the pivots on which the inner gimbal is mounted can supply torques about the vertical direction and about the spin

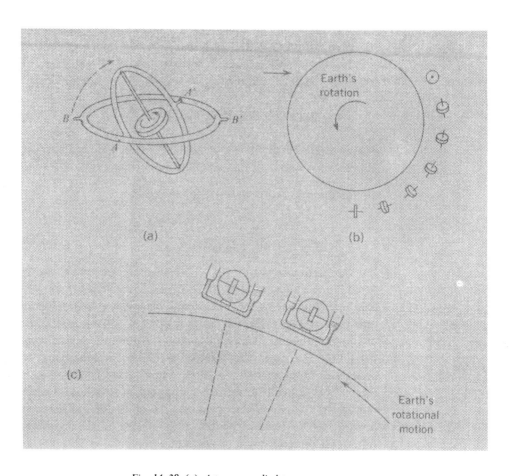

Fig. 14–28 (a) A torque applied to a gyroscope can force its spin axis to tilt. (b) Reorientation of a gyroscope through torque induced by the earth's rotational motion. (c) Principle of a gyroscopic compass. The earth's rotation causes a difference of levels in the U-tube attached to the gyroscope frame, and the resulting torque makes the gyroscope axis turn to align with a north–south direction.

axis of the flywheel, but not about the axis connecting the pivots themselves. This last fact, implying conservation of angular momentum about the axis AA', can be used to explain why the tilting begins. The attempt to rotate the spin axis of the gyroscope in a horizontal plane would, by itself, introduce a component of angular momentum along AA'. The tilting motion provides an equal and opposite angular momentum component that keeps the angular momentum about AA' equal to zero. The torques that *can* be supplied via the forces at A and A' are able to perform the main task of canceling out angular momentum about the axis BB' and creating angular momentum about the vertical axis, corresponding to the reorientation of the spin angular momentum $I\omega$ from horizontal to vertical. This process will continue until the axis of the flywheel is fully aligned with the axis of the torque.

Imagine now that a gyroscope, mounted as described above, were set spinning with its axis initially horizontal and in an east–west direction at a point on the equator [Fig. 14–28(b)]. Then the rotation of the earth itself would apply just such a torque to the gyroscope, about an axis parallel to the earth's axis of rotation. The gyroscope would therefore, in the course of time, pass through the succession of positions indicated in Fig. 14–28(b) until its spin axis was pointing true north, after which it would continue to act as a compass pointing in this direction.

In order to make a nonmagnetic compass of this type—a gyrocompass—for general navigational purposes, use is made of the gravitational torque that comes into existence if the gyroscope is unbalanced. This can be exploited in the way indicated schematically in Fig. 14–28(c). A U-tube containing liquid is clamped to the inner gimbal ring. (The outer gimbal ring is omitted from the diagram, but must be assumed present.) If one imagines an initial situation in which the axis of the gyroscope lies east–west, the preservation of this orientation would, through the earth's rotation, bring about a difference of levels in the two arms of the U-tube, because the new local horizontal makes an angle with the original one. This difference of levels results in a net gravitational torque parallel to the earth's axis, and the gyroscope reorients itself in the manner already described, turning within the outer gimbal ring, until the plane containing the U-tube and the spin axis of the gyroscope is parallel to the local north–south meridian.

Inertial guidance systems make use of very similar phenomena. Once a gyroscope has been set spinning about a chosen axis, any motion that would cause it to deviate from this direction calls into existence a torque that can be used as an error signal to control corrective maneuvers. The same techniques, in essence, can be used for stabilization of ships or vehicles against undesirable rolling motions.

Some of the most important applications of gyroscopic phenomena are in aircraft instruments. A free (perfectly balanced) gyroscope can give a constant indication of a chosen reference direction for checking the correct course, and a gravitationally loaded gyroscope can be used to identify the true vertical at all times and so define an artificial horizon against which the pilot can measure the angles of bank and climb.

ATOMS AND NUCLEI AS GYROSCOPES

We mentioned earlier the intrinsic "spin" angular momentum that is possessed by elementary particles. A large fraction of all atomic nuclei and neutral atoms also have such angular momentum, in amounts corresponding to simple multiples of the natural unit $h/2\pi$. This gives to these particles a natural gyroscopic stability. The spin angular momentum is, however, always accompanied by intrinsic magnetism; it is as if the atom or nucleus contains a tiny bar magnet pointing along the direction of its spin axis, and in the presence of a magnetic field this leads to steady precessional motion.

The basic phenomenon can be well described with the help of a simple classical model. Figure 14-29 portrays a spherical particle with its spin axis at angle θ to a magnetic field. If we imagine a bar magnet inside the particle, the north and south poles of this magnet experience forces in opposite directions parallel to **B**, as shown. These forces produce a torque with its axis pointing up out of the plane of the diagram. In the absence of the spin this torque would simply cause the magnet (after

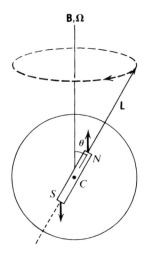

Fig. 14-29 Atoms and nuclei are like spinning magnets and precess in a magnetic field.

some preliminary oscillations) to align itself with the field, like a compass needle. The existence of an intrinsic angular momentum **L**, however, changes the situation and leads to precession in the direction indicated.

An expression for the rate of precession is easily derived if we picture the magnet as having poles of strength m separated by a distance l. The force F on each pole is equal to mB, so that the torque is given by

$$M_c = mBl \sin \theta$$

We can write the product ml as a single quantity, μ—the *magnetic dipole moment* of the particle. Thus we have

$$M_c = \mu B \sin \theta \tag{14-51a}$$

If we write the magnetic field and the magnetic moment as vectors, we can describe the torque completely by the equation

$$\mathbf{M}_c = \boldsymbol{\mu} \times \mathbf{B} \tag{14-51b}$$

Setting the torque equal to the rate of change of angular momentum due to a precessional angular velocity Ω, we then have

$$\mu B \sin \theta = \Omega L \sin \theta$$

or

$$\Omega = \frac{\mu B}{L} \tag{14-52}$$

The existence and the rate of this precession can be detected by picking up the tiny electrical signals that the rotating atomic or nuclear magnet can cause, by electromagnetic induction, in a coil placed nearby. Of course the signal from an individual particle is almost inconceivably small, but by using a sample containing vast numbers of identical particles, all precessing in the same way, the effect becomes measurable.[1] The detection of nuclear spin magnetism in this way is known as *nuclear magnetic induction* and was first studied by F. Bloch and E. M. Purcell (independently, and by different methods) in 1946. They shared the Nobel prize in 1952 for this work, and in his Nobel lecture

---

[1] We shall not go into the special techniques used to observe the precession. For an introductory account one cannot do better than to read the Nobel lectures by F. Bloch and E. M. Purcell in *Nobel Lectures, Physics*, 1942–1962, Elsevier, Amsterdam, 1964.

Purcell said: "I have not yet lost a feeling of wonder, and of delight, that this delicate motion should reside in all the ordinary things around us, revealing itself only to him who looks for it. I remember, in the winter of our first experiments... looking on snow with new eyes. There the snow lay around my doorstep... great heaps of protons quietly precessing in the earth's magnetic field."

(The actual precessional frequency of protons in the earth's field is about 2500 rps. The fundamental experiments were, however, done at a magnetic field about 10,000 times greater than this and hence at a correspondingly higher value of Ω.)

GYROSCOPIC MOTION IN TERMS OF F = ma[1]

Although the thoroughgoing use of angular quantities is certainly the most fruitful approach to gyroscopic phenomena, one may still wish to see how the motions relate back to a basic statement of Newton's law for individual particles. We shall therefore give here a simple example of an analysis in these terms.[2] Picture a primitive gyroscope made of four equal masses m, mounted symmetrically at distances r from a spindle, with their CM a distance l from the pivot point [Fig. 14-30(a)]. Consider the situation when one of the masses (No. 1) is at its highest point, traveling horizontally. The point C has a linear velocity $V\,(=\Omega l)$ due to the precession, and each mass has a velocity $v\,(=\omega r)$ relative to C. Mass 1 thus has, instantaneously, a velocity $v_1\,(= v - V)$ backward with respect to the precessional direction, and mass 3 has a velocity $v_3\,(= v + V)$ forward at this instant (we assume $v > V$). During a short time δt, the precession turns the axis of rotation through the angle $\Omega\,\delta t$, and this changes the directions of the velocities $\mathbf{v}_1$ and $\mathbf{v}_3$ in the manner shown in Fig. 14-30(b). Physically, this means that mass 1 has to be accelerated radially outward from the precession axis, and mass 3 has to be accelerated radially inward. The forces to supply these extra accelerations have to come from the supporting spokes, and if we consider the situation as it appears from the side (looking horizontally, perpendicular to the spin axis),

[1] This section can be omitted without loss of continuity.
[2] This analysis owes a debt to an article by F. W. Sears, *Am. J. Phys.*, **7**, 342 (1939). See also E. F. Barker, *Am. J. Phys.*, **28**, 808 (1960), and J. L. Snider, *Am. J. Phys.*, **33**, 847 (1965).

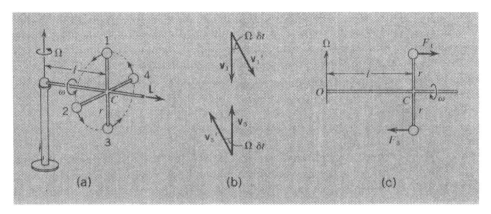

Fig. 14-30 (a) Simple gyroscope made of four concentrated masses. (b) Changes of the velocities $\mathbf{v}_1$ and $\mathbf{v}_3$ in a short time δt. (c) Forces needed to change the velocities of masses 1 and 3, corresponding to the precessional torque.

we see that the spokes must apply a net clockwise torque about C. Masses 2 and 4 do not require any such forces and torques because their instantaneous velocities, being vertical, are not reoriented by the precessional motion.

We have kept the above discussion rather qualitative, because as it stands it is not quite correct. If we used the diagrams of Fig. 14-30(b) to calculate the radial accelerations of the masses, we should have

$$a_1 = \Omega v_1 = \Omega(v - \Omega l) = \Omega v - \Omega^2 l$$
$$a_3 = -\Omega v_3 = -\Omega(v + \Omega l) = -\Omega v - \Omega^2 l$$

In both of these we recognize a centripetal acceleration (due to the precession) in a circle of radius l, which is as it should be, but the terms $\pm \Omega v$ do not represent all the rest of the story. The reason is somewhat subtle. What we have calculated is the change of a velocity vector that remains always at the top or bottom end of a vertical diameter of the rotor. However, the masses, by virtue of their rotation, move away from these positions during δt, and it turns out that this corresponds to a further contribution of $\pm \Omega v$ to the radial accelerations. (You will find a more rigorous discussion below if you are interested.) Thus we end up with the following corrected expressions:

$$F_1 = 2m\Omega v - m\Omega^2 l$$
$$F_3 = -2m\Omega v - m\Omega^2 l$$

You may recognize these as precisely the forces needed to balance the combined Coriolis and centrifugal forces for particles traveling horizontally at speed v in a frame rotating with angular velocity Ω about a vertical axis.

The net clockwise torque about the CM due to F_1 and F_3 combined is given by

$$M_c = 4m\Omega v r = 4mr^2\omega\Omega$$

We know, however, that $4mr^2$ is precisely the moment of inertia of the whole system of four masses about the spin axis. Thus we have $M_c = I\omega\Omega$, exactly as needed. Ultimately, as we know, this torque is supplied via the framework of the rotor by the gravitational forces acting on all four masses.

Formal analysis

Consider a particle of mass m at the position P shown in Fig. 14-31. Set up a system of axes x', y', and z' (with unit vectors $\mathbf{i'}$, $\mathbf{j'}$, and $\mathbf{k'}$) that rotate with the precessional angular velocity $\Omega\mathbf{k}$. The velocity $\mathbf{v'}$ of m in the precessing frame is given by

$$\mathbf{v'} = v\cos\theta\,\mathbf{i'} - v\sin\theta\,\mathbf{k}$$

Within this frame the motion of the particle is just around the circle of radius r and must be describable by the following statement of Newton's law:

$$\mathbf{F'} = -m\omega^2\mathbf{r}$$

However, the true force $\mathbf{F}$ acting on m, as measured in a sta-

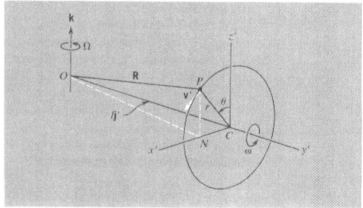

Fig. 14-31 Diagram for analysis of the forces acting on a particle P in a precessing gyroscope.

tionary frame, is related to $\mathbf{F}'$ by the following equation [cf. Eq. (12–13b)]:

$$\mathbf{F}' = \mathbf{F}_{true} + \mathbf{F}_{coriolis} + \mathbf{F}_{centrifugal}$$

This gives us

$$\mathbf{F}_{true} = -m\omega^2 \mathbf{r} + 2m\Omega \mathbf{k} \times \mathbf{v}' + m\Omega \mathbf{k} \times (\Omega \mathbf{k} \times \mathbf{R})$$

where

$$\mathbf{R} = l\mathbf{j}' + \mathbf{r} = l\mathbf{j}' + r\sin\theta \mathbf{i}' + r\cos\theta \mathbf{k}'$$

Now we have

$$\mathbf{k} \times \mathbf{v}' = v\cos\theta(\mathbf{k} \times \mathbf{i}') = v\cos\theta \mathbf{j}'$$
$$\mathbf{k} \times \mathbf{R} = l(\mathbf{k} \times \mathbf{j}') + r\sin\theta(\mathbf{k} \times \mathbf{i}') = -l\mathbf{i}' + r\sin\theta \mathbf{j}'$$

Substituting these above, we find

$$\mathbf{F}_{true} = -m\omega^2 \mathbf{r} + 2m\Omega v\cos\theta \mathbf{j}' + m\Omega \mathbf{k} \times (-\Omega l\mathbf{i}' + \Omega r\sin\theta \mathbf{j}')$$

or

$$\mathbf{F}_{true} = -m\omega^2 \mathbf{r} + 2m\Omega v\cos\theta \mathbf{j}' - m\Omega^2(l\mathbf{j}' + r\sin\theta \mathbf{i}') \quad (14\text{–}53)$$

In this expression for the total force we then recognize three distinct contributions:

1. The force needed to maintain the circular path of m around C.
2. The force corresponding to the Coriolis acceleration.
3. The force needed to give m the centripetal acceleration toward the precession axis when it is at any radial distance $ON (= l\mathbf{j}' + r\sin\theta \mathbf{i}')$ from that axis.

We can then substitute specific values of θ in Eq. (14–53) to correspond to the instantaneous positions of any set of particles comprising the spinning rotor. One can verify at once that putting $\theta = 0$ and π gives us two forces whose horizontal components are equal to F_1 and F_3, as quoted earlier.

NUTATION[1]

However convincing the analysis of gyroscopic precession may seem, one may still wonder how a gyroscope can possibly defy

[1] This section can be omitted without loss of continuity.

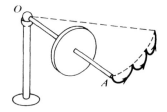

Fig. 14–32 Nutation of a precessing gyroscope.

gravity in the way that it appears to do. The answer is that this immunity *is* indeed only apparent. If a flywheel is set spinning about a horizontal axis, with both ends of its axle supported [Fig. 14–32], the first thing that happens if the support at one end (*A*) is removed is that this end does begin to fall vertically. Immediately thereafter, however, the precessional motion in a horizontal plane begins, and as this happens the falling motion slows down, until the point *A* is moving in a purely horizontal direction. It does not stay like this; what happens next is that the precession slows down and the end of the axle rises again, ideally to its initial level. This whole sequence is repeated over and over, as indicated in the figure. The process is called *nutation* (after the Latin word for "nodding") and always occurs unless a gyroscope is started out with the exact motion needed for steady precession. The net effect is that the center of gravity of the flywheel remains at an average level below its starting point, and one can guess that the gravitational potential energy so released provides what is needed for the kinetic energy of the precessional motion.

Formal analysis

We shall give a simple analysis of nutation for the type of situation discussed above, in which the support at one end of the axis of a flywheel is suddenly removed at $t = 0$. Figure 14–33(a) indicates the essential features as seen in a vertical plane containing the spin axis. The angle of precession of this plane about a vertical axis will be denoted by φ. The flywheel has a spin angular momentum $I_1\omega$; it has moments of inertia I_2 and I_3 (with $I_2 = I_3$) about the other two principal axes.

Since no torques act about a vertical axis, we have conservation of angular momentum about the axis Oz. The tilting downward of $I_1\omega$ would introduce a vertical component $I_1\omega\,\delta\theta$; this must therefore be compensated by an angular momentum $I_3\,\delta\Omega$ due to a change of the precessional velocity Ω. Thus

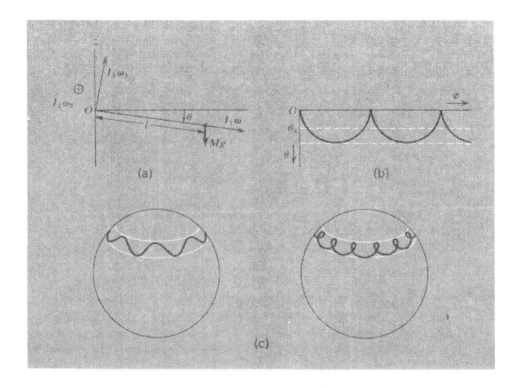

Fig. 14-33 (a) Intermediate stage in the nutation of a gyroscope whose spin axis is initially horizontal. (b) Relation of nutational angle (θ) to precessional angle (φ). (c) Examples of nutation with different initial conditions.

$$I_3\,\delta\Omega = I_1\omega\,\delta\theta$$

This can be integrated at once to give

$$\Omega = \frac{d\varphi}{dt} = \frac{I_1\omega}{I_3}\theta \tag{14-54}$$

(We put $d\varphi/dt = 0$ at $\theta = 0$.)

Next, let us consider energy conservation. A change of θ with time implies a kinetic energy of rotation about a horizontal axis perpendicular to the spin axis, and the precession means kinetic energy of rotation about a vertical axis. The release of gravitational potential energy is equal to $Mgl\,\delta\theta$. Thus we put

$$Mgl\,\delta\theta = \tfrac{1}{2}I_2\delta\left(\frac{d\theta}{dt}\right)^2 + \tfrac{1}{2}I_3\delta\left(\frac{d\varphi}{dt}\right)^2$$

Dividing throughout by δt and going to the limit $\delta t \to 0$, we have

$$Mgl\frac{d\theta}{dt} = I_2 \frac{d\theta}{dt}\frac{d^2\theta}{dt^2} + I_3 \frac{d\varphi}{dt}\frac{d^2\varphi}{dt^2}$$

Substituting for $d\varphi/dt$ from Eq. (14–54), we have

$$Mgl\frac{d\theta}{dt} = I_2 \frac{d\theta}{dt}\frac{d^2\theta}{dt^2} + \frac{(I_1\omega)^2}{I_3}\theta\frac{d\theta}{dt}$$

Canceling out $d\theta/dt$, and putting $I_2 = I_3$, this gives us

$$\frac{d^2\theta}{dt^2} = \frac{Mgl}{I_3} - \left(\frac{I_1\omega}{I_3}\right)^2 \theta$$

This is of the form

$$\frac{d^2\theta}{dt^2} = -k^2(\theta - \theta_0)$$

where $k = I_1\omega/I_3$ and $\theta_0 = MglI_3/(I_1\omega)^2$. The solution is a harmonic vibration about $\theta = \theta_0$:

$$\theta = \theta_0(1 - \cos kt)$$

Substituting this equation for θ into Eq. (14–54) and integrating, we find that

$$\varphi = \theta_0(kt - \sin kt)$$

The combination of these simultaneous variations of θ and φ is a cycloidal motion in angle, as shown in Fig. 14–33(b).

If the initial conditions are varied, different types of nutational motion may occur, as shown in Fig. 14–33(c), but they are all understandable in terms of the principles underlying the above analysis.

THE PRECESSION OF THE EQUINOXES

It is fitting that we should end this book with another of the great astronomical problems for which Newton first supplied the explanation. This is the slow precession of the earth, which behaves as a gyroscope under the torques due to the gravitational pulls of the sun and the moon.

The story begins with the ancient astronomers, who through their amazingly careful observations had discovered that the celestial sphere of "fixed" stars seems to be very gradually turn-

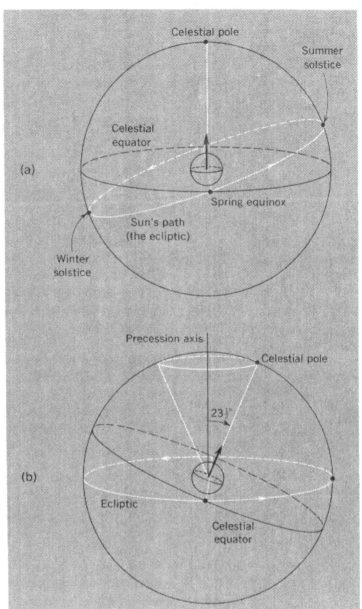

Fig. 14-34 Precession of the equinoxes as described (a) in terms of a movement of the ecliptic around the celestial sphere, and (b) in terms of the conical path traced out by the earth's spin axis with respect to a fixed axis perpendicular to the ecliptic.

ing from west to east with respect to a reference line defined by the intersection of the celestial equator with the ecliptic [Fig. 14-34(a)]. This reference line also defines the positions of the equinoxes, when the sun lies in the equatorial plane of the celestial sphere (and of the earth) so that day and night are of equal length over the whole earth. Since only relative motions are involved, the phenomenon could be described as a slow *westward* drift or precession of the equinoctial points themselves. It was the Greek astronomer Hipparchus, about 130 B.C., who discovered the phenomenon and reported its magnitude as about 36" of arc per year (the true value is close to 50"). He also recognized that the existence of this precession of the equinoxes allowed two different definitions of the year—either the time between spring equinoxes, bringing the sun back into the same positional relationship to the earth, or the slightly longer time (longer by about 20 min) for the sun to return to exactly the same place with respect to the fixed stars. These times are known as the *tropical year* and the *sidereal year*, respectively.

For Hipparchus the precession of the equinoxes was just an empirical fact, and so it remained until 1543, when Copernicus, in his *De Revolutionibus*, put forward the explanation—that the earth's axis, although it always keeps the same inclination ($66\frac{1}{2}°$) to the plane of the earth's orbit, nevertheless traces out a cone of semiangle $23\frac{1}{2}°$ with respect to the normal to this plane [Fig. 14-34(b)]. Copernicus also concluded that the average precessional rate is 50.2" per year, corresponding to a complete precessional period of just about 26,000 years. (The quoted number is excellent, but Copernicus was misled by some bad data into the false belief that the precessional rate is not constant.)

Thus by Newton's day the descriptive account of the precession of the equinoxes was well established, but its cause remained a mystery. Then, in the *Principia* (Book III: The System of the World, Proposition 39), Newton gave a quantitative dynamical explanation. It has something in common with the explanation of the tides (p. 531); the moon is the chief cause, but the sun also plays an important role, in just the same ratio as for tidal action [cf. Eq. (12-25)].

Figure 14-35 presents the basis for describing the phenomenon in modern terms, using the dynamics of gyroscopic precession. The earth's spin axis makes an angle θ to the normal to the plane of the ecliptic, so that the earth's equatorial bulge ($\Delta R/R_E \approx \frac{1}{300}$) is oriented unsymmetrically, as shown. Now the period of the precession is, as we have seen, immensely long; thus, from the standpoint of the earth, both the sun and the moon go through very many orbits within a time (e.g., 100 years) in which the direction of the spin axis hardly changes. This

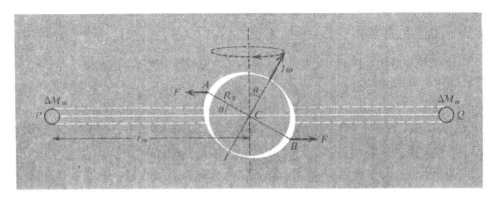

Fig. 14-35   Origin of precessional torque due to gravitational attraction between the earth's equatorial bulge and a ring representing the effective distribution of the moon's mass around its orbit.

means, in effect, that the mass represented by the sun or the moon is smeared out uniformly around its orbit as seen from the earth. In other words, the earth's gravitational environment is just like two rings of material; the one representing the moon is indicated in Fig. 14-35.

The origin of the precession is now clear. The earth's bulge in the vicinity of A experiences a net force toward the left of the diagram, and the bulge near B experiences an equal force to the right. Together these give a torque whose axis points upward from the plane of the diagram. Since the earth's steady rotation from west to east means a spin angular momentum $I\omega$ directed as shown, the result is a precession in which the tip of the spin angular momentum vector traces out a circular path from east to west.

What about the actual rate of precession? We shall indicate a very cavalier approach, just for the sake of coming up with an order of magnitude. (You will perhaps devise a much better calculation for yourself.) It is clear that the biggest effects are caused by interactions involving material close to the plane of Fig. 14-35—i.e., the plane that contains the spin axis and the normal to the ecliptic. If we consider an element of mass ΔM_E near A, acted on by equal elements of mass (ΔM_m) of the moon ring near P and Q, the force on ΔM_E is given by

$$F_m = \frac{G \Delta M_m \Delta M_E}{(r_m - R_E \cos \theta)^2} - \frac{G \Delta M_m \Delta M_E}{(r_m + R_E \cos \theta)^2}$$

or

$$F_m \approx \frac{4G\,\Delta M_m\,\Delta M_E R_E \cos\theta}{r_m^3} \tag{14-55a}$$

Let us arbitrarily say that an approximation to the net precessional force can be obtained by using in this equation the amounts of the earth's bulge and the moon's ring included within an angular range $\pm 45°$ of the center line (see plan view, Fig. 14-36). We then have $\Delta M_m = M_m/4$. To get an estimate of ΔM_E, we must calculate the volume of material contained in the bulge. Its thickness is about $R_E/300$, we have chosen its length equatorially to be a quarter of the circumference, i.e., $\pi R_E/2$, and now we must pick a value for its length from north to south. Let us try the range from $22\tfrac{1}{2}°$ south to $22\tfrac{1}{2}°$ north latitude; this takes us almost exactly from the plane of the ecliptic to the Tropic of Cancer and represents an eighth of the earth's circumference, or $\pi R_E/4$. Thus the volume, ΔV_E, of our patch would be given by

$$\Delta V_E \approx \frac{R_E}{300}\,\frac{\pi R_E}{2}\,\frac{\pi R_E}{4} = \frac{\pi^2 R_E^3}{2400}$$

The volume of the whole earth is $4\pi R_E^3/3$; hence we have

$$\frac{\Delta V_E}{V_E} \approx \frac{\pi^2}{2400}\,\frac{3}{4\pi} \approx 10^{-3}$$

Since the density of material near the earth's surface is only

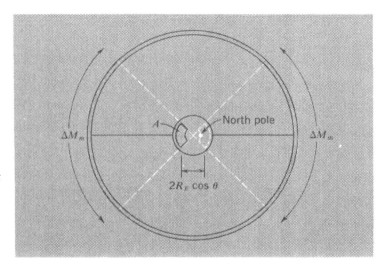

Fig. 14-36 Plan view to illustrate crude model of the precession-producing torque due to the moon acting on the earth's equatorial bulge.

about half of the mean density, we have thus defined a patch of material whose mass is given approximately by

$$\Delta M_E \approx 5 \times 10^{-4} M_E$$

That completes the most awkward (and most shaky) part of the calculation. If we substitute our values of ΔM_m and ΔM_E in Eq. (14–55a), we have

$$F_m \approx 5 \times 10^{-4} \frac{GM_m M_E R_E \cos\theta}{r_m^3} \qquad (14\text{–}55b)$$

Clearly a similar calculation will apply (or not!) to the earth–sun interaction, and so for the net torque-producing force we shall have

$$F \approx 5 \times 10^{-4} GM_E R_E \cos\theta \left(\frac{M_m}{r_m^3} + \frac{M_s}{r_s^3}\right) \qquad (14\text{–}56)$$

Although the magnitude of the numerical coefficient in F is quite doubtful, the *form* of the equation is correct. In particular, we see explicitly how the relative contributions of sun and moon are defined by the same combination of mass and distance as in the calculation of tide-producing forces.

Let us turn now to the calculation of the torque and the precessional velocity. For the torque, we have simply

$$M_c = 2R_E F \sin\theta$$

The moment of inertia, if we take the earth to be a uniform sphere, is $2M_E R_E^2/5$, and so the precessional equation is

$$2R_E F \sin\theta = \tfrac{2}{5} M_E R_E^2 \omega \Omega \sin\theta$$

or

$$\Omega = \frac{5F}{M_E R_E \omega}$$

with $\omega \approx 7 \times 10^{-5}$ sec^{-1}. Substituting for F from Eq. (14–56), we thus have

$$\Omega \approx \frac{2.5 \times 10^{-3} G \cos\theta}{\omega} \left(\frac{M_m}{r_m^3} + \frac{M_s}{r_s^3}\right)$$

$$\approx 2.5 \times 10^{-9} \left(\frac{M_m}{r_m^3} + \frac{M_s}{r_s^3}\right)$$

Now

$$M_m \approx 7 \times 10^{22} \text{ kg}, \quad r_m^3 \approx 6 \times 10^{25} \text{ m}^3$$
$$M_s \approx 2 \times 10^{30} \text{ kg}, \quad r_s^3 \approx 4 \times 10^{33} \text{ m}^3$$

Hence

$$\Omega \approx (2.5 \times 10^{-9})(1.7 \times 10^{-3}) \approx 4 \times 10^{-12} \text{ sec}^{-1}$$

and so

$$T = \frac{2\pi}{\Omega} \approx 1.5 \times 10^{12} \text{ sec} \approx 50{,}000 \text{ years}$$

We could try to trim this result a little—for example, we have somewhat underestimated the mean density of the earth's crust and (by treating the earth as a uniform sphere) we have somewhat overestimated the moment of inertia. Both of these would cause us to underestimate the precessional rate and obtain too large a value for the precessional period. But in view of the gross assumptions we have made elsewhere in the calculation, we should not set any great store on making small refinements of this type. The important thing is that, by quite simple means, we have verified that the precession of the equinoxes can indeed be understood in terms of Newtonian dynamical principles. But Newton got there first!

PROBLEMS

14-1 (a) Devise a criterion for whether there is external force acting on a system of two particles. Use this criterion on the following one-dimensional system. A particle of mass m is observed to follow the path

$$x(t) = A \sin(\omega t) + L + vt$$

The other particle, of mass M, follows

$$X(t) = B \sin(\omega t) + Vt$$

The different constants are arbitrary except that $mA = -MB$.

(b) Try it on the system with

$$x(t) = A \sin(\omega t) \quad \text{and} \quad X(t) = B \sin(\omega t + \varphi)$$

where A and B are related as before, and $\varphi \neq 0$.

14–2 Consider a system of three particles, each of mass m, which remain always in the same plane. The particles interact among themselves, always in a manner consistent with Newton's third law. If the particles A, B, and C have positions at various times as given in the table, determine whether any external forces are acting on the system.

Time	A	B	C
0	(1, 1)	(2, 2)	(3, 3)
1	(1, 0)	(0, 1)	(3, 3)
2	(0, 1)	(1, 2)	(2, 0)

14–3 Two skaters, each of mass 70 kg, skate at speeds of 4 m/sec in opposite directions along parallel lines 1.5 m apart. As they are about to pass one another they join hands and go into circular paths about their common center of mass.
  (a) What is their total angular momentum?
  (b) A third skater is skating at 2 m/sec along a line parallel to the initial directions of the other two and 6 m off to the side of the track of the nearer one. From his standpoint, what is the total angular momentum of the other two skaters as they rotate?

14–4 A molecule of carbon monoxide (CO) is moving along in a straight line with a kinetic energy equal to the value of kT at room temperature (k = Boltzmann's constant = 1.38×10^{-23} J/°K). The molecule is also rotating about its center of mass with a total angular momentum equal to $\hbar$ (= 1.05×10^{-34} J-sec). The internuclear distance in the CO molecule is 1.1 Å. Compare the kinetic energy of its rotational motion with its kinetic energy of translation. What does this result suggest about the ease or difficulty of exciting such rotational motion in a gas of CO molecules at room temperature?

14–5 A uniform disk of mass M and radius R is rotating freely about a vertical axis with initial angular velocity ω_0. Then sand is poured onto the disk in a thin stream so that it piles up on the disk at the radius r ($< R$). The sand is added at the constant rate μ (mass per unit time).
  (a) At what rate are the angular velocity and the rotational kinetic energy varying with time at a given instant?
  (b) After what length of time is the rotational kinetic energy reduced to half of its initial value? What has happened to this energy?

14–6 Two men, each of mass 100 kg, stand at opposite ends of the diameter of a rotating turntable of mass 200 kg and radius 3 m. Initially the turntable makes one revolution every 2 sec. The two men make their way to the middle of the turntable at equal rates.
  (a) Calculate the final rate of revolution and the factor by which

the kinetic energy of rotation has been increased.

(b) Analyze, at least qualitatively, the means by which the increase of rotational kinetic energy occurs.

(c) At what radial distance from the axis of rotation do the men experience the greatest centrifugal force as they make their way to the center?

14-7 Estimate the kinetic energy in a hurricane. Take the density of air as 1 kg/m³.

14-8 A useful way of calculating the approximate value of the moment of inertia of a continuous object is to consider the object as if it were built up of concentrated masses, and to calculate the value of $\sum mr^2$. As an example, take the case of a long uniform bar of mass M and length L (with its transverse dimensions much less than L). We know that its moment of inertia about one end is $ML^2/3$.

(a) The most primitive approximation is to consider the total mass M to be concentrated at the midpoint, distant $L/2$ from the end. You will not be surprised to find that this is a poor approximation.

(b) Next, treat the bar as being made up of two masses, each equal to $M/2$, at distances $L/4$ and $3L/4$ from one end.

(c) Examine the improvements obtained from finer subdivisions —e.g., 3 parts, 5 parts, 10 parts.

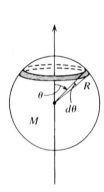

14-9 (a) Calculate the moment of inertia of a thin-walled spherical shell, of mass M and radius R, about an axis passing through its center. Consider the shell as a set of rings defined by the amounts of material lying within angular ranges $d\theta$ at the various angles θ to the axis (see the figure).

(b) Verify the formula $I = 2MR^2/5$ for the moment of inertia of a solid sphere of uniform density about an axis through its center. You can proceed just as in part (a), except that the system is to be regarded as a stack of circular disks instead of rings.

14-10 (a) Calculate the moment of inertia of a thin square plate about an axis through its center perpendicular to its plane. (Use the perpendicular-axis theorem.)

(b) Making appropriate use of the theorems of parallel and perpendicular axes, calculate the moment of inertia of a hollow cubical box about an axis passing through the centers of two opposite faces.

(c) Using the result of (a), deduce the moment of inertia of a uniform, solid cube about an axis passing through the midpoints of two opposite faces.

(d) For a cube of mass M and edge a, you should have obtained the result $Ma^2/6$. It is noteworthy that the moment of inertia has this same value about *any* axis passing through the center of the cube. See how far you can go toward verifying this result, perhaps by considering other special axes—e.g., an axis through diagonally opposite corners of the cube or an axis through the midpoints of opposite edges.

14-11 Refer to Fig. 11-19(a), which shows the variation of density with radial distance inside the earth. Using this graph, compare the moment of inertia of the earth about its axis with the moment of inertia of a sphere of the same mass and radius but of uniform density. You can do quite well by considering the earth to be made up of a central core and two thick concentric shells, each of approximately uniform density. The boundaries between these three regions correspond to the abrupt changes of density shown in the graph. [Alternatively, consider the earth as built up of three superposed solid spheres—a basic one, occupying the whole volume of the earth, with the density of the outermost region ($r > 0.54R_E$) and two other spheres with densities corresponding to the mean density differences between the successive regions.]

14-12 (a) A hoop of mass M and radius R *rolls* down a slope that makes an angle θ with the horizontal. This means that when the linear velocity of its center is v its angular velocity is v/R. Show that the kinetic energy of the rolling hoop is Mv^2.

(b) There is a traditional story about the camper–physicist who has a can of bouillon and a can of beans, but the labels have come off, so he lets them roll down a board to discover which is which. What would you expect to happen? Does the method work? (Try it!)

14-13 A skier is enjoying the mountain air while standing on a 30° snow slope when he suddenly notices a huge snowball rolling down at him. By the time he notices the ball, it is only 100 m away and is traveling at 25 m/sec. The skier gives himself a speed of 10 m/sec almost instantaneously and proceeds to accelerate down the slope at $g \sin \theta \, (= g/2)$. Does he get away? (Assume that the snowball has a constant acceleration corresponding to that of a sphere of given radius rolling, without slipping, down the slope. Assume that the moment of inertia of the snowball about an axis through its center is $2MR^2/5$.)

14-14 The preceding problem suggests another one. If an object is rolling down a slope, gathering material as it goes, how does its acceleration compare, in fact, with a similar object that is not adding material in this way? To give yourself a relatively straightforward situation to consider, take the case of a cylinder, rather than a sphere, that grows in size as it rolls. Make whatever assumptions seem reasonable about the way in which the rate of increase in radius depends on the existing radius, R, and on the instantaneous speed, v.

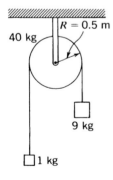

14-15 Two masses, of 9 kg and 1 kg, hang from the ends of a string that passes around a pulley of mass 40 kg and radius 0.5 m ($I = \frac{1}{2}MR^2$) as shown in the diagram. The system is released from rest and the 9-kg mass drops, starting the pulley rotating.

(a) What is the acceleration of the 9-kg mass?

(b) What is the angular velocity of the pulley after the 9-kg mass has dropped 2 m?

(c) What is the tension in the part of the string which is between the pulley and the 9-kg mass? Between the pulley and the 1-kg mass?

(d) If the coefficient of friction between the string and the pulley is 0.2, what is the least number of turns that the string must make around the pulley to prevent slipping? (Cf. Problem 5-14.)

14-16 An amusement park has a downhill racetrack in which the competitors ride down a 30° slope on small carts. Each cart has four wheels, each of mass 20 kg and diameter 1 m. The frame of each cart has a mass of 20 kg.

(a) What is the acceleration of a cart if its rider has a mass of 50 kg? (Assume that the moment of inertia of a wheel is given by $0.8MR^2$, where R is its radius.)

(b) If two riders, of masses 50 and 60 kg, respectively, start off simultaneously, what is the distance between them when the winner passes the finishing line 60 m down the slope?

14-17 A uniform rod of length $3b$ swings as a pendulum about a pivot a distance x from one end. For what value(s) of x does this pendulum have the same period as a simple pendulum of length $2b$?

14-18 (a) A piece of putty of mass m is stuck very near the rim of a uniform disk of mass $2m$ and radius R. The disk is set on edge on a table on which it can roll without slipping. The equilibrium position is obviously that in which the piece of putty is closest to the table. Find the period of small-amplitude oscillations about this position and the length of the equivalent simple pendulum.

(b) A circular hoop hangs over a nail on a wall. Find the period of its small-amplitude oscillations and the length of the equivalent simple pendulum.

(In these and similar problems, use the equation of conservation of energy as a starting point. The more complicated the system, the greater is the advantage that this method has over a direct application of Newton's law.)

14-19 A uniform cylinder of mass M and radius R can rotate about a shaft but is restrained by a spiral spring (like the balance wheel of your watch). When the cylinder is turned through an angle θ from its equilibrium position, the spring exerts a restoring torque M equal to $-c\theta$. Set up an equation for the angular oscillations of this system and find the period, T.

14-20 Assuming that you let your legs swing more or less like rigid pendulums, estimate the approximate time of one stride. Hence estimate your comfortable walking speed in miles per hour. How does it compare with your actual pace?

14-21 A torsion balance to measure the momentum of electrons consists of a rectangular vane of thin aluminum foil, 10 by 2 by 0.005 cm, attached to a very thin vertical fiber, as shown. The period of torsional oscillation is 20 sec, and the density of aluminum is 2.7 times that of water.

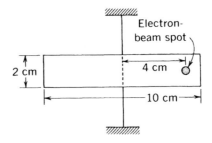

(a) What is the torsion constant of the suspension, in m-N/rad?

(b) What horizontal force, applied perpendicular to the surface of the vane at a point 3 cm from the axis, will produce an angular deflection of 10°?

(c) A beam of 1 mA of electrons accelerated through 500 V strikes the vane perpendicularly at a point 4 cm from the axis. What steady angular deflection is produced, assuming that the electrons are stopped in the vane?

14–22 The torsion constant of a wire or fiber of length l, and of circular cross section of radius a, is given by $c = E_s \pi a^4 / 2l$, where E_s is an elastic constant of the material known as the shear modulus, measured in N/m². The maximum load that can be supported by such a fiber is given by its cross-sectional area, πa^2, multiplied by the ultimate tensile strength of the material, also measured in N/m². For glass fibers the value of E_s is about 2.5×10^{10} N/m², and the ultimate tensile strength is about 10^9 N/m².

(a) Calculate the diameter of the thinnest glass fiber that can safely support two lead spheres, each of mass 20 g, in a gravity torsion balance. Allow a safety factor of about 3.

(b) If the spheres are at the ends of a light bar of length 20 cm, and the length of the suspending fiber is also 20 cm, what is the period of torsional oscillation of this system? (The measurement of this period is the practical way of inferring the torsion constant of the suspension.)

(c) What angular deflection of this system is produced by placing lead spheres of mass 2 kg with their centers 5 cm from the centers of the small suspended spheres? What linear displacement would this give in a spot of light reflected from a mirror on the torsion arm to a scale 5 m away? Compare this result with the figures used in Problem 5–3 on a Cavendish experiment.

14–23 A wheel of uniform thickness, of mass 10 kg and radius 10 cm, is driven by a motor through a belt (see the figure). The drive wheel on the motor is 2 cm in radius, and the motor is capable of delivering a torque of 5 m-N.

(a) Assuming that the belt does not slip on the wheel, how long does it take to accelerate the large wheel from rest up to 100 rpm?

(b) If the coefficient of friction between belt and wheel is

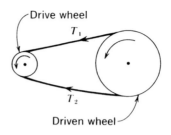

0.3, what are the tensions in the belt on the two sides of the wheel? (Assume that the belt touches the wheel over half its circumference.)

14-24 A possible scheme for stopping the rotation of a spacecraft of radius R is to let two small masses, m, swing out at the ends of strings of length l, which are attached to the spacecraft at the points P and P' (see the figure). Initially, the masses are held at the positions shown and are rotating with the body of the spacecraft. When the masses have swung out to their maximum distance, with the strings extending radially straight out, the ends P and P' of the strings are released from the spacecraft. For given values of m, R, and I (the moment of inertia of the spacecraft), what value of l will leave the spacecraft in a non-rotating state as a result of this operation? Apply the result to a spacecraft that can be regarded as a uniform disk of mass M and radius R. (Put in some numbers, too, maybe.)

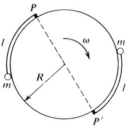

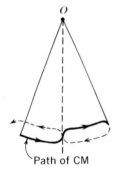

14-25 The technique of "pumping" a playground swing in order to increase the amplitude of its motion can be learned by example or (less easily) by trial and error. The mechanics of the procedure are not trivial. According to one model of the process, the pumping is taken to consist of a sudden elevation of the rider's center of mass at each passing of the vertical, or low point (the rider lifts and holds himself above the seat), and a subsequent return to resting on the swing seat at each turning point (see the figure). The support ropes are assumed to be always straight, and the *instantaneous* changes of effective length of the "pendulum" allow conservation of angular momentum to be applied not only to the low-point pumping motions but also to those at the turning points.

(a) Carry out the analysis as indicated above and show that increase of amplitude can be achieved. Note that the result agrees with the qualitative experience that any given amount of increase is more easily achieved as the amplitude increases.

(b) Consider in what ways the analysis indicated above may be imperfect. Also, how well does this idealized technique match the actual pumping method that children utilize every day? [The analysis suggested above may be found in an article by P. L. Tea, Jr., and H. Falk, "Pumping on a Swing," *Am. J. Phys.*, **36**, 1165 (1968).]

14-26 Two gear wheels, A and B, of radii R_A and R_B, and of moments of inertia I_A and I_B, respectively, are mounted on parallel shafts so that they are not quite in contact (see the figure). Both wheels can

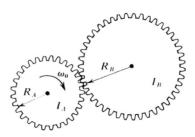

rotate completely freely on their shafts. Initially, A is rotating with angular velocity ω_0, and B is stationary. At a certain instant, one shaft is moved slightly so that the gear wheels engage. Find the resulting angular velocity of each in terms of the given quantities. (*Warning:* Do not be tempted into a glib use of angular momentum conservation. Consider the forces and torques resulting from the contact.)

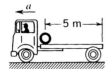

14-27 A section of steel pipe of large diameter and relatively thin wall is mounted as shown on a flat-bed truck. The driver of the truck, not realizing that the pipe has not been lashed in place, starts up the truck with a constant acceleration of 0.5 g. As a result, the pipe rolls backward (relative to the truck bed) without slipping, and falls to the ground. The length of the truck bed is 5 m.

(a) With what horizontal velocity does the pipe strike the ground?

(b) What is its angular velocity at this instant?

(c) How far does it skid before beginning to roll without slipping, if the coefficient of friction between pipe and ground is 0.3?

(d) What is its linear velocity when its motion changes to rolling without slipping?

14-28 (a) How far above the center of a billiard ball or pool ball should the ball be struck (horizontally) by the cue so that it will be sure to begin rolling without slipping?

(b) Analyze the consequences of striking the ball at the level of its center if the coefficient of friction between the ball and the table is μ.

14-29 A man kicks sharply at the bottom end of a vertical uniform post which is stuck in the ground so that 6 ft of it are above ground.

Unfortunately for him the post has rotted where it enters the ground and breaks off at this point. To appreciate why "unfortunately" is the appropriate word, consider the subsequent motion of the top end of the post.

14–30 Refer to page 673 for the discussion of a rectangular board rotating about an axis in its plane. Using the notation and the method of attack of that discussion, show that the angular momentum component L' about an axis in the plane of the board and perpendicular to ω is given by combining the resolved parts of $I_x\omega_x$ and $I_y\omega_y$ in this direction; i.e.,

$$L' = I_x\omega_x \sin \delta - I_y\omega_y \cos \delta$$

14–31 A flywheel in the form of a uniform disk of radius 5 cm is mounted on an axle that just fits along the diameter of a gimbal ring of diameter 12 cm. The flywheel is set rotating at 1000 rpm and the gimbal ring is supported at the point where one end of the axle meets it. Calculate the rate of precession in rpm.

14–32 In most cars the engine has its axis of rotation pointing fore and aft along the car. The gyroscopic properties of the engine when rotating at high speed are not negligible. Consider the tendency of this gyroscopic property to make the front end of the car rise or fall as the car follows a curve in the road. What about the corresponding effects for a car with its engine mounted transversely? Try to make some quantitative estimates of the importance of such effects. Consider whether a left-hand curve or a right-hand curve might involve the greater risk of losing control over the steering of the car.

14–33 See if you can pick up the challenge, given in the text, of making a more respectable calculation of the precession of the equinoxes.

Appendix

THE METRIC SYSTEM OF UNITS

ESTABLISHING a system of units and setting up standards for the quantitative measurement of length, mass, and time involves a great many practical problems, and also some philosophical ones (as discussed in Chapter 2). Primary standards are continually being revised and improved, with a trend toward incorporating atomic or nuclear phenomena that have the advantages of reproducibility and accessibility.

Mass

The standard kilogram, a platinum-iridium cylinder kept at Sèvres, France, is defined to have a mass of exactly 1 *kilogram*. Secondary mass standards are compared to this primary standard by using a beam balance.

Length

The current standard of length—the meter—is now (since October 1983) defined as the distance traveled by light in vacuum in a time of 1/299,792,458 seconds. This definition supersedes the previous definition (1960) in which the meter was defined as a certain number of wavelengths of light from a designated spectral source (^{86}Kr).

Time

The second has been (since 1967) defined as equal in duration to 9,192,631,770 cycles of vibration in an atomic clock, controlled by one of the characteristic frequencies associated with atoms of the isotope cesium-133.

Units of measurement which are definite multiples of the kilogram, meter, and second are denoted by prefixing the basic unit. A list of prefixes (and the multiple of the basic unit that each prefix represents) is given below.

Prefix	Abbreviation	Multiple
tera-	T	10^{12}
giga-	G	10^9
mega-	M	10^6
kilo-	k	10^3
centi-	c	10^{-2}
milli-	m	10^{-3}
micro-	μ	10^{-6}
nano-	n	10^{-9}
pico-	p	10^{-12}
femto-	f	10^{-15}
atto-	a	10^{-18}

In addition, special names have been given to particularly useful multiples of the meter:

$$10^{-15} \text{ m} = 1 \text{ fermi (F)}$$
$$10^{-10} \text{ m} = 1 \text{ angstrom (Å)}$$
$$10^{-6} \text{ m} = 1 \text{ micron } (\mu)$$

CONVERSION FACTORS

The values quoted have mostly been "rounded off" to three significant figures. For precise values, see a technical handbook or an advanced physics text.

Plane Angle   1 radian = 57.3° (57°20′) = $(1/2\pi)$ revolution
                        = 0.159 revolution

Solid Angle   1 steradian = $(1/4\pi)$ sphere = 0.0796 sphere

Length   1 in = 2.54 cm
                1 ft = 30.5 cm
                1 m = 39.37 in.
                1 km = 0.621 mile = 3281 ft

Length *(continued)*	1 mile = 5280 ft = 1609 m 1 astronomical unit = 1.49×10^8 km 1 light year = 9.46×10^{12} km 1 parsec = 3.08×10^{13} km
Mass, Weight, *and Force*	(where g has the standard value of 9.80665 m/sec^2) 1 newton = 10^5 dynes = 0.225 lb 1 kg mass weighs 2.2 lbs (or 9.8 newtons) 1 lb object has a mass of 0.4536 kilogram and weighs 4.448 newtons
Volume	1 liter is the volume of 1 kg of water at its maximum density (3.98°C and 1 atmosphere) = 1000.028 cm^3 1 cubic inch (1 in^3) = 16.4 cm^3 1 cubic foot = 0.0283 m^3 1 gallon (U.S.) = 231 in^3 1 gallon (U.K.) = 277 in^3
Time	(all values are mean solar) 1 year = 365.2 days = 3.16×10^7 sec 1 day = 8.64×10^4 sec

Speed	1 ft/sec = 0.305 m/sec = 1.10 km/hr
	         = 0.682 mile/hr
	1 mile/hr = 0.447 m/sec = 1.61 km/hr
	         = 1.47 ft/sec

Pressure	1 atmosphere = 76 cm of Hg
	            = 1.013×10^5 newtons/m^2
	1 mm of Hg = 1 Torr (or Tor)
	            = 133 newtons/m^2

Energy and
Work
	1 joule (1 newton-meter) = 0.738 ft-lb
	        = 6.24×10^{18} electron volts
	        = 0.239 calorie = 10^7 ergs
	1 electron volt = 1.60×10^{-19} joule
	Energy equivalent to ($E = mc^2$) $\begin{cases} 1 \text{ electron mass} = 0.511 \text{ Mev} \\ 1 \text{ proton mass} = 938.2 \text{ Mev} \\ 1 \text{ neutron mass} = 939.5 \text{ Mev} \\ 1 \text{ amu} = 931.1 \text{ Mev} \end{cases}$
	1 kilowatt-hour = 3.6×10^6 joules

GENERAL CONSTANTS

c	speed of light in vacuum = 3.00×10^8 m/sec
G	gravitational constant = 6.67×10^{-11} newton-meter^2/kg^2
N_0	Avogadro's number = 6.02×10^{23}/mole
h	Planck's constant = 6.63×10^{-34} joule-sec
$\hbar$	Planck's constant/2π = 1.05×10^{-34} joule-sec
e	electron charge = 1.60×10^{-19} coulomb
m_e	electron rest mass = 9.11×10^{-31} kg
M	nucleon rest mass (proton or neutron; their exact masses differ slightly) = 1.67×10^{-27} kg
M_p/m_e	(proton/electron) mass ratio = 1836 1 amu (1 atomic mass unit) = $\frac{1}{12}$ of mass of C^{12} = 1.66×10^{-27} kg

Note: For astronomical data, see pp. 34, 252, 290, 582.

Bibliography

IT HAS SEEMED useful to categorize this bibliography to some extent, although the divisions are not always clearcut. In particular, the profound involvement of the development of mechanics with observational astronomy, on the one hand, and with the whole rise of the scientific point of view, on the other, has suggested special attention to these areas as well as to a more conventional listing of textbooks of various types and levels. Since the teaching of mechanics is also so often a part of a more comprehensive introductory course, a short list of books written for this purpose is included. In a subject so vast there can be no claim to completeness; the author offers his apologies in advance for having omitted, through ignorance or inadvertence, many titles that should rightfully have been included.

SOME CLASSIC WORKS

Ptolemy, C., *The Almagest* (trans. R. C. Taliaferro), Encyclopaedia Britannica, Inc., Great Books, Vol. 16, pp. 1–478, Chicago, 1952.
  This book, incorporating Ptolemy's geocentric view of the universe, displays the marvelously detailed and exact knowledge of the heavens that had already been accumulated nearly 2000 years ago.

Copernicus, N., *On the Revolutions of the Heavenly Spheres* (1543) (trans. C. G. Wallis), Encyclopaedia Britannica, Inc., Great Books, Vol. 16, pp. 505–838, Chicago, 1952.
  The complete, documented account of the evidence and the reasoning by which Copernicus was led to the heliocentric picture of the universe.

Kepler, J., *Astronomia Nova* (Commentaries on the Motion of Mars) (1609), Johannes Kepler, Gesammelte Werke, Vol. 3, C. H. Beck, Munich, 1937.
  This book is the detailed account of Kepler's analysis of the

orbit of Mars. It contains the statement of the first two of Kepler's laws of planetary motion. (Unfortunately not available in English translation.)

———., *The Harmonies of the World* (1619) (trans. C. G. Wallis), Encyclopaedia Britannica, Inc., Great Books, Vol. 16, pp. 1009–1085, Chicago, 1952.

The scientific importance of this book is that it contains the statement of Kepler's third law. For the most part, however, it is a testimonial to Kepler's lifelong obsession with his quest for geometrical and numerical symmetries in the universe.

Galilei, G., *The Starry Messenger* (1610) (trans. Stillman Drake), in *Discoveries and Opinions of Galileo*, Doubleday Anchor, New York, 1957.

Galileo's own account of his first discoveries with his newly invented telescope. Includes samples of his observations on the moons of Jupiter.

———., *Dialogue Concerning the Two Chief World Systems* (1632) (trans. Stillman Drake), Univ. of California, Berkeley, California, 1953.

Galileo's lively dramatization of his advocacy of the Copernican system.

———., *Two New Sciences* (1638) (trans. H. Crew and A. DiSalvio) Dover, New York, 1952.

Chiefly of interest for Galileo's discussions of motion. It also presents some of his ideas on the constitution and strength of materials.

Newton, I., *Principia* (Mathematical Principles of Natural Philosophy) (1687). F. Cajori's revision of A. Motte's translation (1729). Univ. of California Press, Berkeley, California, 1960.

The extraordinary book that can still claim to be the greatest scientific work ever written. Anyone who studies it must surely echo the description of Newton (on his statue in the chapel of Trinity College, Cambridge) as the man "who surpassed the whole human race by his genius".

BIOGRAPHIES

Andrade, E. N. da C., *Sir Isaac Newton*, Doubleday Anchor, New York, 1958.

Armitage, A., *Copernicus*, Yoseloff, New York and London, 1957.

———., *John Kepler*, Roy, New York, 1966.

Caspar, M., *Kepler* (trans. C. D. Hellman), Abelard-Schuman, London and New York, 1959.

De Santillana, G., *The Crime of Galileo*, Univ. of Chicago Press, Chicago, 1955.

Fermi, L. and Bernardini, G., *Galileo and the Scientific Revolution*, Basic Books, New York, 1961.

Koestler, A., *The Watershed*, Doubleday Anchor, New York, 1960 (based on the major part of *The Sleepwalkers*, Macmillan, New York, 1959, by the same author).

Moore, P., *Isaac Newton*, Putnam, New York, 1958.

More, L. T., *Isaac Newton* (1934), Dover, New York, 1962.

North, J. D., *Isaac Newton*, Clarendon Press, Oxford, 1967.

HISTORICAL OR PHILOSOPHICAL IN EMPHASIS

Arons, A., *The Development of the Concepts of Physics*, Addison-Wesley, Reading, Massachusetts, 1965.

Ball, W. W. R., *An Essay on Newton's Principia*, Macmillan, London, 1893.
   A most interesting study of the *Principia* and its origins.

Colodny, R. (ed.), *Beyond the Edge of Certainty*, Prentice-Hall, Englewood Cliffs, New Jersey, 1965.
   A collection of essays, many of them concerned with the bases of classical mechanics.

Cooper, L. N., *An Introduction to the Meaning and Structure of Physics*, Harper and Row, New York, 1968.

Dijksterhuis, E. J., *The Mechanisation of the World Picture* (trans. C. Dikshoorn), Clarendon Press, Oxford, 1961.

Dugas, R., *A History of Mechanics* (trans. J. R. Maddox), Routledge and Kegan Paul, London, 1957.

Einstein, A. and Infeld, L., *The Evolution of Physics*, Cambridge Univ. Press, Cambridge, 1938.
   A survey, in simple, nonmathematical terms, that places the mechanical view in the context of physical theory generally.

Gillispie, C. G., *The Edge of Objectivity*, Princeton Univ. Press, Princeton, New Jersey, 1960.
   Subtitled "An Essay in the History of Scientific Ideas", this

graceful and scholarly book devotes several chapters to the development of mechanics.

Gold, T. (ed.), *The Nature of Time*, Cornell Univ. Press, Ithaca, New York, 1967.
 A report of a conference at which a number of distinguished scientists presented and discussed their ideas.

Hanson, N. R., *Patterns of Discovery*, Cambridge Univ. Press, Cambridge, 1958.
 The author, well versed in physics as well as philosophy, discusses how theories come into being. The book ends up with quantum theory but explores classical mechanics on the way. A provocative and unusual study.

Herivel, J. W., *The Background to Newton's Principia*, Clarendon Press, Oxford, 1965.
 Similar to Rouse Ball's book (v. sup.) but with the benefit of more recent scholarship.

Hesse, M., *Forces and Fields*, Nelson, London, 1961.
 A careful and scholarly study of the concept of "action at a distance."

Holton, G. and Roller, D. H. D., *Foundations of Modern Physical Science*, Addison-Wesley, Reading, Massachusetts, 1958.

Jammer, M., *Concepts of Force*, Harper Torchbooks, New York, 1962. *Concepts of Mass*, Harper Torchbooks, New York, 1964. *Concepts of Space*, Harper Torchbooks, New York, 1960.
 In these three monographs, the author discusses from a mainly philosophical standpoint the primary concepts of mechanics.

Kemble, H., *Physical Science, Its Structure and Development*, MIT Press, Cambridge, Massachusetts, 1966.
 A very good general survey.

Lindsay, R. B. and Margenau, H., *Foundations of Physics*, Dover, New York, 1957.
 A technical account, enriched by full discussion, of the basic theoretical concepts of physics.

Mach, E., *The Science of Mechanics* (9th ed., 1933) (trans. T. J. McCormack), Open Court Publishing Co., La Salle, Illinois, 1960.
 The first thorough critique (1st ed., 1883) of the bases of classical mechanics, and still well worth reading.

Magie, W. F., *A Source Book in Physics*, McGraw-Hill, New York, 1935.
 Conveniently accessible extracts from the writings of the pioneers.

Poincaré, H., *Science and Hypothesis* (1903), Dover, New York, 1952.
    A famous examination of fundamentals by one of the great men of nineteenth-century mathematical physics.

Reichenbach, H., *The Philosophy of Space and Time*, Dover, New York, 1957.
    Strongly oriented toward relativity theory (both special and general).

Schlegel, R., *Time and the Physical World*, Dover, New York, 1968.

Small, R., *An Account of the Astronomical Discoveries of Kepler* (1804), Univ. of Wisconsin Press, Madison, Wisconsin, 1963.
    A reprint of a detailed exposition of Kepler's researches. Though old, it has no modern counterpart.

Truesdell, C., *Essays in the History of Mechanics*, Springer-Verlag, New York, 1968.
    A collection of articles, illustrated with portraits and with reproductions from classic works on mechanics. Its erudite author (an applied mathematician) offers many pungent comments.

Whitrow, G. J., *The Natural Philosophy of Time*, Nelson, London, 1961.
    A lengthy essay by an expert in cosmology.

ASTRONOMICAL

Abell, G., *Exploration of the Universe*, Holt, Rinehart and Winston, New York, 1964.

Abetti, G., *History of Astronomy* (trans. B. B. Abetti), Abelard-Schuman, New York, 1952.

Baker, R. H. and Fredrick, L. W., *An Introduction to Astronomy* (7th ed.), Van Nostrand, Princeton, New Jersey, 1968.

Berry, A., *A Short History of Astronomy* (1898), Dover, New York, 1961.
    An excellent factual account.

Cohen, I. B., *The Birth of a New Physics*, Doubleday Anchor, New York, 1960.
    An excellent short, semi-popular account of the development from classical observational astronomy to the Newtonian synthesis.

Dreyer, J. L. E. (rev. W. H. Stahl), *A History of Astronomy from Thales to Kepler*, Dover, New York, 1953.

Duncan, J. C., *Astronomy* (5th ed.), Harper and Row, New York, 1955.

Hoyle, F., *Astronomy*, Macdonald, London, and Doubleday, New York, 1962.
An elegant and lavishly illustrated account of astronomy, with much fascinating documentation of its historical development.

Hubble, E., *The Realm of the Nebulae* (1936), Dover, New York, 1958.
The story of the discovery of the extra-Galactic universe, by the man who did most to elucidate the problem.

King, H. C., *Exploration of the Universe*, Signet Books, New York, 1964.
A most excellent short survey.

Krause, A., *Astronomy* (trans. M. and H. Seddon), Oliver and Boyd, Edinburgh and London, 1961.

Kuhn, T. S., *The Copernican Revolution*, Vintage Books, New York, 1959.

Lodge, O. J., *Pioneers of Science*, Macmillan, London, 1893.
Based on a set of lectures on the historical development of astronomy as related to Newtonian mechanics. Full of interesting details.

Motz, L. and Duveen, A., *Essentials of Astronomy*, Wadsworth, Belmont, California, 1966.

Pannekoek, A., *A History of Astronomy*, Allen and Unwin, London, 1961.

Turner, H. H., *Astronomical Discovery*, Edward Arnold, London, 1904.

SOME GENERAL TEXTBOOKS

Adair, R. K., *Concepts in Physics*, Academic Press, New York, 1969.

Atkins, K. R., *Physics*, Wiley, New York, 1965.

Borowitz, S. and Bornstein, L. A., *A Contemporary View of Elementary Physics*, McGraw-Hill, New York, 1968.

Feynman, R. P., Leighton, R. B., and Sands, M., *The Feynman Lectures in Physics*, Vol. I, Addison-Wesley, Reading, Massachusetts, 1963.

Ford, K. W., *Basic Physics*, Blaisdell, Waltham, Massachusetts, 1968.

Halliday, D. and Resnick, R., *Physics for Students of Science and Engineering*, Vol. I, Wiley, New York, 1966.

Miller, F., Jr., *College Physics*, Harcourt, Brace & World, New York, 1959.

P.S.S.C., *College Physics*, Heath, Lexington, Massachusetts, 1968.

Rogers, E. M., *Physics for the Inquiring Mind*, Princeton Univ. Press, Princeton, New Jersey, 1960.

Sears, F. W., and Zemansky, M. W., *University Physics*, Part I, Addison-Wesley, Reading, Massachusetts, 1963.

Weidner, R. T. and Sells, R. L., *Elementary Classical Physics*, Allyn and Bacon, Boston, Massachusetts, 1965.

MECHANICS TEXTS

Alonso, M. and Finn, E. J., *Mechanics* (Fundamental University Physics, Vol. I), Addison-Wesley, Reading, Massachusetts, 1967.

Arthur, W., and Fenster, S. K., *Mechanics*, Holt, Rinehart and Winston, New York, 1969.

Becker, R. A., *Introduction to Theoretical Mechanics*, McGraw-Hill, New York, 1954.

Bradbury, T. C., *Theoretical Mechanics*, Wiley, New York, 1968.

Bullen, K. E., *Introduction to the Theory of Mechanics* (7th ed.). Cambridge Univ. Press, Cambridge, 1965.

Cabannes, H., *General Mechanics* (2nd ed.), (trans. S. P. Sutera), Blaisdell, Waltham, Massachusetts, 1968.

Easthope, C. E., *Three-dimensional Dynamics* (2nd ed.), Butterworths, London, 1964.

Feather, N., *An Introduction to the Physics of Mass, Length and Time*, Edinburgh Univ. Press, Edinburgh, 1959.

Frank, N. H., *Introduction to Mechanics and Heat*, McGraw-Hill, New York, 1939.

Goldstein, H., *Classical Mechanics*, Addison-Wesley, Reading, Massachusetts, 1951.

Hill, R., *Principles of Dynamics*, Macmillan, New York, 1964.

Ingard, U., and Kraushaar, W. L., *Introduction to Mechanics, Matter and Waves*, Addison-Wesley, Reading, Massachusetts, 1960.

Kibble, T. W. B., *Classical Mechanics*, McGraw-Hill, New York, 1966.

Kittel, C., Knight, W. D., and Ruderman, M. A., *Mechanics* (Berkeley Physics Course, Vol. I), McGraw-Hill, New York, 1965.

Landau, L. D. and Lifshitz, E. M., *Mechanics*, (trans. J. B. Sykes and J. S. Bell), Pergamon, Oxford, 1960.

Lindsay, R. B., *Physical Mechanics*, Van Nostrand, Princeton, New Jersey, 1961.

Marion, J. B., *Classical Dynamics*, Academic Press, New York, 1965.

Maxwell, J. C., *Matter and Motion* (1877), Dover, New York, 1954.

McCuskey, S. W., *Introduction to Advanced Dynamics*, Addison-Wesley, Reading, Massachusetts, 1959.

Osgood, W. F., *Mechanics*, Macmillan, New York, 1937.

Pohl, R. W., *The Physical Principles of Mechanics and Acoustics*, Blackie, London and Glasgow, 1932.

Rutherford, D. E., *Classical Mechanics* (2nd ed.), Oliver and Boyd, Edinburgh and London, 1957.

Skinner, R., *Mechanics*, Blaisdell, Waltham, Massachusetts, 1969.

Slater, J. C., and Frank, N. H., *Mechanics*, McGraw-Hill, New York, 1947.

Sommerfeld, A., *Mechanics* (Lectures on Theoretical Physics, Vol. I) (trans. M. O. Stern), Academic Press, New York and London, 1964.

Symon, K. R., *Mechanics* (2nd ed.), Addison-Wesley, Reading, Massachusetts, 1960.

Taylor, E. F., *Introductory Mechanics*, Wiley, New York, 1963.

Young, H. D., *Fundamentals of Mechanics and Heat*, McGraw-Hill, New York, 1964.

ENGINEERING MECHANICS

Beer, F. P., and Johnston, E. R., Jr., *Vector Mechanics for Engineers* (2 vols.), McGraw-Hill, New York, 1962.

Chorlton, F., *Textbook of Dynamics*, Van Nostrand, Princeton, New Jersey, 1963.

Christie, D. E., *Vector Mechanics* (2nd ed.), McGraw-Hill, New York, 1964.

Halfman, R. L., *Dynamics: Particles, Rigid Bodies and Systems*, Vol. I, Addison-Wesley, Reading, Massachusetts, 1962.

Huang, T. C., *Engineering Mechanics*, Addison-Wesley, Reading, Massachusetts, 1967.

Meriam, J. L., *Dynamics*, Wiley, New York, 1966.

INDIVIDUAL TOPICS

Arnold, R. N., and Maunder, L., *Gyrodynamics*, Academic Press, New York and London, 1961.

Ball, K. J. and Osborne, G. F., *Space Vehicle Dynamics*, Clarendon Press, Oxford, 1967.

Blitzer, L. (ed.), *Kinematics and Dynamics of Satellite Orbits* (AAPT reprint volume), American Institute of Physics, New York, 1963.

Chalmers, B., *Energy*, Academic Press, New York and London, 1963.

Danby, J. M. A., *Fundamentals of Celestial Mechanics*, Macmillan, New York, 1962.

Darwin, G. H., *The Tides* (1898), Freeman, San Francisco and London, 1962.

Garland, G. D., *The Earth's Shape and Gravity*, Pergamon, Oxford, 1965.

Gray, A., *A Treatise on Gyrostatics and Rotational Motion* (1918), Dover, New York, 1959.

Greenhood, D., *Mapping*, Univ. of Chicago Press, Chicago, 1965.

Kisch, B., *Scales and Weights*, Yale Univ. Press, New Haven, Connecticut, 1965.

Macmillan, D. H., *Tides*, C. R. Books, London, 1966.

Mott-Smith, M., *The Concept of Energy Simply Explained*, Dover, New York, 1964.

Powles, J. B., *Particles and Their Interactions*, Addison-Wesley, Reading, Massachusetts, 1968.

Roy, A. E., *Foundations of Astrodynamics*, Macmillan, New York, 1965.

Ryabov, Y., *An Elementary Survey of Celestial Mechanics*, (trans. G. Yankovsky), Dover, New York, 1961.

Scarborough, J. B., *The Gyroscope: Theory and Applications*, Wiley (Interscience), New York, 1958.

Theobald, D. W., *The Concept of Energy*, E. and F. N. Spon, London, 1966.

Van der Kamp, P., *Elements of Astromechanics*, Freeman, San Francisco and London, 1964.

Vertregt, M., *Principles of Astronautics*, Elsevier, Amsterdam, London and New York, 1965.

Answers to problems

CHAPTER 1

1-2 (a) 10^{11}–10^{13} atoms; (b) 10^{50} atoms.
1-3 10^9 tons.
1-4 Probability of at least one ≈ 0.1.
1-5 (a) 10^{10} molecules/cm^3; (b) 5×10^{-4} cm.
1-6 6×10^{-5} cm^3.
1-7 (a) 3×10^{-9} H atoms/(m^3-year);
(b) 3×10^{15} H atoms/day or about 10^{-1} cm^3 at 1 Torr.
1-8 About 10^9 microorganisms; the number is comparable with the total human world population.
1-9 (a) $\bar{\rho} \approx 5 \times 10^{-11}$ kg/m^3; (b) $\bar{\rho}_{\text{belt}} \approx 2 \times 10^{-9}$ kg/m^3.
1-10 (a) 10^6 approx; (b) $p \approx 5 \times 10^{-15}$ atm $\approx 4 \times 10^{-12}$ mm Hg at $T = 300°$K.

CHAPTER 2

2-1 $x = 5.9$, $y = -15.8$, $r = 16.9$ (in miles).
2-3 (a) New York: $(x, y, z) = (1330, -4620, 4180)$ km;
Sydney: $(x, y, z) = (-4620, 2560, -3560)$ km;
(b) 12,100 km; (c) $s_{\min} = 16,000$ km.
2-4 (b) $(x, y, z) = R(\cos \alpha \cos \beta, \sin \beta, \sin \alpha \cos \beta)$.
2-6 $R_E \approx (25{,}000/2\pi) \approx 4000$ miles.
2-7 $R_U \approx 10^{41}$ fundamental length units; $T_U \approx 10^{41}$ fundamental time units.
2-10 (a) $\mathbf{r}(t) = Bt\mathbf{i} + AB^2t^2\mathbf{j}$; (b) $v = B[1 + (2ABt)^2]^{1/2}$.
2-11 $t_{AB}(x) = [(y_A{}^2 + x^2)^{1/2}/v_1] + \{[y_B{}^2 + (l - x)^2]^{1/2}/v_2\}$.

723

2-12 (a) $|v| = 20\sqrt{3}$ km/hr, directed 120° west of north;
(b) Minimum separation of 2.6 km occurs at about 13:17 hours.

2-13 Putting $T_0 = 2l/V$, the results are: (a) $T_0/(1 - v^2/V^2)$;
(b) $T_0/(1 - v^2/V^2)^{1/2}$;
(c) $T_0[1 - (v\sin\theta/V)^2]^{1/2}/(1 - v^2/V^2)$.

2-14 (b) The cutter reaches the ship a distance $Dv/(V^2 - v^2)^{1/2}$ down the coast from the port and a time $DV/v(V^2 - v^2)^{1/2}$ after leaving port.

2-15 (a) About 4 minutes.

2-17 (b) 3.8×10^5 km.

2-18 $\theta = 87°$ implies f (= sun's distance/moon's distance) = 19. Correct value of θ is 89°51′, corresponding to $f = 385$. For $\Delta\theta = \pm 0.1°$, $150 \leq f \leq 1300$.

CHAPTER 3

3-1 (b) $t_0 = (2l/g)^{1/2}$, $t = t_0$ (i.e. second object is not dropped until separation is l).

3-3 (a) $a = -2$ m/sec², $v_0 = 4$ m/sec; (c) $t = (2 \pm \sqrt{2})$ sec.

3-4 10 m.

3-5 (a) $v(25) \approx 35$ mph, $v(45) \approx 5$ mph, $v(65) \approx 40$ mph;
(c) $t_+ \approx 58$ sec, $t_- \approx 80$ sec.

3-6 (b) $x = a\tau(t - \tau/2)$;
(c) $v = 10$ m/sec, $a = 4.17$ m/sec², $\tau = 2.4$ sec. Distance to attain steady velocity $= 12$ m.

3-7 Overtaking car travels 775 ft approx; minimum distance ≈ 1523 ft ≈ 0.3 mile.

3-8 (a) $y_{\text{heaven}} \approx 1.9 \times 10^6$ km, $v_{\text{entry}} \approx 620$ km/sec (assuming 17.5 hr of daylight, which is the length of the longest day of the year in England, where Milton wrote).
(b) $y \approx 9 \times 10^4$ km, $v \approx 10$ km/sec.

3-9 (a) In first interval, $v_x = 4.3$ m/sec, $v_y = -2.7$ m/sec;
In second interval, $v_x = 4.9$ m/sec, $v_y = -2.1$ m/sec;
(b) $a_x = a_y = +30$ m/sec².

3-10 (b) $v \approx 380$ m/sec. This corresponds to the same value for $mv^2/2$ ($\sim T$) as for Cs and K in Fig. 3-10(b); $v_y \approx 1.3 \times 10^{-2}$ m/sec.

3-12 $2(1 + \sqrt{35})$ m ≈ 13.8 m behind thrower.

3-14 $v \approx 120$ ft/sec ≈ 84 mph.

3-15 (a) $\mathbf{r}(t = 8) - \mathbf{r}(t = 6) = -0.91\mathbf{i} + 2.80\mathbf{j}$,
where $\mathbf{r}(t = 0) = 2.5\mathbf{j}$; $\mathbf{r}(t = 2.5) = 2.5\mathbf{i}$.
(b) $\mathbf{v}(t = 4) = (\pi/2)[-\mathbf{i}\cos(\pi/5) - \mathbf{j}\sin(\pi/5)]$;
$\mathbf{a}(t = 4) = (\pi^2/10)[-\mathbf{i}\sin(\pi/5) + \mathbf{j}\cos(\pi/5)]$.

3-16 (a) 3×10^{-3} g; (b) 6×10^{-4} g; (c) 8×10^{21} g; (d) 40 g (all approx.).

3-17 (a) $a_x = (\ddot{r} - r\dot\theta^2)\cos\theta - (2\dot r\dot\theta + r\ddot\theta)\sin\theta$,
$a_y = (\ddot{r} - r\dot\theta^2)\sin\theta - (2\dot r\dot\theta + r\ddot\theta)\cos\theta$,
(b) $a_r = \ddot r - r\dot\theta^2$, $a_t = 2\dot r\dot\theta + r\ddot\theta$.
Each dot above a variable denotes a differentiation with respect to time; e.g. $\dot r = dr/dt$, $\ddot\theta = d^2\theta/dt^2$.

3-18 (a) $v = (\sqrt{3}/8)$ m/sec, $a = (5/8)$ m/sec²;
(c) $t = \frac{2}{5}[\pi - \cos^{-1}(4/5)]$ sec ≈ 1 sec.

CHAPTER 4

4-3 (a) $(\bar{x}, \bar{y}) = (1/W)[a(w_2 + w_3), b(w_3 + w_4)]$.

4-4 80¢.

4-6 In clockwise order beginning with the vertical rope, the tensions in each case are:
(a) W, $W/(\tan \varphi \cos \theta + \sin \theta)$, $W/(\tan \theta \cos \varphi + \sin \varphi)$;
(b) W, $W \cot \theta$, $W \csc \theta$.

4-7 (a) About 150 N. (b) The force on the car is about 2100 N; the distance it is moved by an additional 2 ft displacement of the midpoint is about 7 in.

4-9 (a) 100 lb.

4-10 (a) 3.5 ft;
(b) $F = 235$ lb in a direction 12° above the horizontal.

4-11 $2(8\sqrt{3} - 1)/3 \approx 8.6$ ft.

4-12 (a) 5 lb; anywhere along the left hand edge of the frame.

4-13 (a) $\theta = \cos^{-1}(r/R)$.

4-14 (a) $W/20$; (b) $(W - w)/20$.

4-15 $M = 25$m-N; $\cos(\mathbf{M} \wedge \mathbf{i}) = -\frac{16}{25}$, $\cos(\mathbf{M} \wedge \mathbf{j}) = +\frac{12}{25}$, $\cos(\mathbf{M} \wedge \mathbf{k}) = -\frac{3}{5}$.

CHAPTER 5

5-1 2.6×10^5 km $\approx \frac{2}{3} \times$ radius of moon's orbit around the earth.

5-2 3.5×10^{-4} seconds of arc. No.

5-3 (a) $G \approx 6.24 \times 10^{-11}$ N-m^2/kg^2; (b) 1.4% higher.

5-4 $(\Delta \theta)_A = (\Delta \theta)_B$.

5-5 (a) $F_C = 9.2 \times 10^{-8}$ N; $F_G = 4.1 \times 10^{-47}$ N;
(b) 2.4×10^6 km, slightly more than 6 times the distance from earth to moon.

5-6 650 kg.

5-7 about 5%.

5-8 10^{-18} approx.

5-9 $R \approx 10^{-13}$ m.

5-11 (a) $F_{VW} \approx 4 \times 10^{-2} F_C$; (b) $F_{VW} \approx 10^{-6} F_C$.

5-12 (a) 4×10^{-11} N; (b) 20 tons/in^2; 3×10^{-11} N.

5-14 (d) 350 lb, 1,230 lb, 4,330 lb, 15,200 lb.

5-15 $v \approx 0.4$ mm/sec.

CHAPTER 6

6-2 Yes—just barely.

6-3 (a) $F_{\text{avg}} \approx 2 \times 10^4$ N; (b) $h = 0.75$ m,

6-4 $x = (5\sqrt{2} - 7)$ m ≈ 0.07 m,
$v_x = (5\sqrt{2} - 3)$ m/sec ≈ 4.07 m/sec,
$y = [11 - 5(2\sqrt{3} - \sqrt{2})]$ m $\approx$
$v_y = [4 - 5(2\sqrt{3} - \sqrt{2})]$ m/sec ≈ -6.25 m/sec.

6-5 (a) $F_x = 0$, $F_y = -k$; (b) $F_x = -kx$, $F_y = -ky$;
(c) $F_x = -k$, $F_y = 0$.

6-6 (a) $F \geq 5 \times 10^{-5}$ N; (b) $\Delta p = 5 \times 10^{-5}$ N-sec;
(c) $W = 7.5 \times 10^{-7}$ joules.

6-7 (a) $F(t = 0) = 5$ N; (b) $F_{max} = \pm 10$ N.
6-8 (a) $F = -5.6 \times 10^3$ N, $F/mg = \frac{4}{7}$;
(b) $v \approx \sqrt{42}$ m/sec ≈ 15 mph.
6-9 (b) $F(t) = mA\alpha^2$ (**i** $\sin \alpha t +$ **j** $\cos \alpha t$); a mass attached to the rim of a rolling wheel.
6-10 (a) $v_{max} = (Tl/m)^{1/2}[1 - (mg/T)^2]^{1/2}$;
(b) $v_{max} = [(T - mg)l/m]^{1/2}$.

CHAPTER 7

7-1 $T_A = T_0 + 2ma$; $T_B = \frac{1}{2}T_0 + ma$.
7-2 (a) $a = 0.6$ m/sec^2; (b) $F_C = 3.2$ N.
7-3 (c) $a = [P(\cos \theta + \mu \sin \theta)/m] - \mu g$; (d) $\tan^{-1} \mu$.
7-4 (b) $T = m_1 m_2 g \sin \theta/(m_1 + m_2)$; $a_2 = m_2 g \sin \theta/(m_1 + m_2)$.
7-5 (a) $T_l = T_r = F/2$; (b) The dynamically important form of the relation is: $2a_P(t) = a_m(t) + a_M(t)$;
(c) $a_m = (F/2m) - g$; $a_M = (F/2M) - g$;
$a_P = -g + [(M + m)F/4Mm]$.
7-6 (a) $T_A = 2250$ N; $T_B = T_C = 1125$ N; (c) $F_C = 375$ N.
7-8 9 m/sec.
7-9 (a) $T(l) = F(1 - l/L)$, where l is the distance from the pulled end. Doing it vertically does not change the value of $T(l)$.
(b) $T(l) = [M + m(1 - l/L)]a$;
$T(l) = [M + m(1 - l/L)](a + g)$.
7-10 (a) $v = 2.466 \times 10^7$ m/sec; (b) $(e/m) = 1.75 \times 10^{11}$ coul/kg;
(c) Relativistic correction requires increasing the value obtained by 0.5%.
7-11 7 ft.
7-12 $\mu = 0.09$.
7-13 2800 volts.
7-14 $v_{max} = 3(g/l)^{1/2}/2\pi$.
7-15 $F = (0.98)\sqrt{2} \approx 1.4$ N in plane of motion and directed 45° above line to center from rim.
7-16 Begin to pull out at 1300 ft altitude.
7-17 (a) 12°; (b) 44 m/sec ≈ 97 mph.
7-19 (a) $t_1 = (mlT/F^2)^{1/2}$; (b) $a(t_1/2) = (T^2 + 16F^2)^{1/2}/4m$;
(c) $s = 3lT/2F$.
7-20 $v \approx 9 \times 10^7$ m/sec; $V_0 = 90$ million volts in Newtonian mechanics; relativistic correction increases V_0 by about 10%.
7-21 (a) $v_{min} = (gr/\mu)^{1/2}$; (b) $\varphi = \tan^{-1} \mu$;
(c) 9 m/sec = 20 mph; 31°.
7-22 (a) $r \leq 0.03$ mm; (b) $r \geq 2$ mm; (c) 9.5 m/sec; 4.6 m.
7-23 (a) 0.97μ; (b) 3 (negative); (c) 530.
7-24 $v_0' = v_0/4$; $v_1' = 2v_1 + (7/4)v_0$.
7-25 About 1 millisec.
7-26 50 cm/sec, $\frac{1}{10}$ sec.
7-27 (a) 200 N/m;
(b) $x(t) = 5 \cos 10t$ (x in cm, positive for extension, t in sec);
(c) $x(t) = 10 \sin 10t + 5 \cos 10t$ (x in cm, t in sec).
7-28 $\omega = (6g/5h)^{1/2}$.
7-29 $\omega = (\rho_0 g/\rho l)^{1/2}$.

7-30 (a) $2(T/mL)^{1/2}$; (b) $[TL/mD(L - D)]^{1/2}$.
7-31 (a) $gT^2/4\pi^2$; (b) $\mu gT^2/4\pi^2$.
7-32 (a) About 1.6 cm; (b) No ($a_{max} < g$).

CHAPTER 8

8-1 $T_s \approx 2.88 \times 10^{-4} T_E \approx 2\frac{1}{2}$ hr.
8-2 $5\sqrt{10}/4 \approx 4.0 < T_A/T_E < 3\sqrt{3} \approx 5.2$.
8-3 (a) $R - R_E \approx 1{,}700$ km; (b) $t \approx \frac{5}{11}$ hr ≈ 27 min $+ 20$ sec.
8-4 $R - R_J \approx 9 \times 10^7$ m $\approx 1.3 R_J$.
8-5 $v \approx 1.7 \times 10^3$ m/sec, $T \approx 6.5 \times 10^3$ sec ≈ 1 hr 50 min.
8-6 $\Delta r/r \approx 5.1 \times 10^{-6}$; or $\Delta r \approx 210$ m.
8-7 (a) $\Delta x \approx 10^{-4}$ m ≈ 0.1 mm;
    (b) $f \approx 1.35 \times 10^{-4}$ rev/sec ≈ 0.49 rev/hr.
8-8 $M_E \approx 9 \times 10^{24}$ kg, $\Delta\theta \approx 13\alpha$.
8-9 No, $\Delta g \approx 3.5 \times 10^{-2}$ mgal.
8-10 $T = (3\pi/\rho G)^{1/2}$; $T_{H_2O} \approx 1.2 \times 10^4$ sec.
8-11 $\rho_{sun} \approx 1.3 \times 10^3$ kg/m^3.
8-12 (a) $M \approx 10^5$ kg, $R = 1.9$ m; (b) No
8-15 (a) $T = (7.4 \times 10^5)(D^3/M)^{1/2}$; (b) $\frac{10}{11}$.
8-16 (a) $0.73 R_e$ from center of earth;
    (b) $\Delta T \approx (6.6 \times 10^{-3}) T_{month} \approx 4$ hours.
8-17 $M_{total} \approx 10^{11} M_{sun}$.
8-19 $\Delta T \approx 2.5 \times 10^{-3}$ sec.

CHAPTER 9

9-4 $v_3 = 28v_0/45$; $v_2 = v_0/15$.
9-5 $v = (v_0/4)(5 - 2\sqrt{2})^{1/2} \approx 0.37 v_0$,
    $\theta = \tan^{-1}[1/(2\sqrt{2} - 1)] \approx 29°$.
9-6 Two tosses in both cases.
9-7 $F_{av} = 36$ N.
9-8 $\Delta x = 2.35$ cm.
9-9 $v = 73$ ft/sec $= 22$ m/sec, $F = 350$ N ≈ 80 lb.
9-10 (a) $v_0 = [Mg/\pi R^2 \rho]^{1/2}$. (b) $v_0 = 15.5$ m/sec.
9-11 (a) $a = (\mu v_0/M_0) - g$; (b) $\mu = 7 \times 10^3$ kg/sec.
9-12 (b) $v_2 = v_0 \ln\left[\dfrac{n}{rn + (1 - r)}\right]$; (c) $n = \sqrt{N}$;
    (d) $v_{max} = 2 v_0 \ln\left[\dfrac{N^{1/2}}{rN^{1/2} + (1 - r)}\right]$;
    (e) $v = v_0 \ln\left[\dfrac{N}{rN + (1 - r)}\right]$.
9-14 (a) $n = 2$; (b) $F = 2A\rho v^2$; (c) $F = \pi r^2 \rho v^2$.
9-15 (a) $m_2/m_1 = 3$; (b) $v_{cm} = u_1/4$; (c) $\frac{3}{4}(\frac{1}{2} m_1 u_1^2)$;
    (d) $\frac{1}{4}(\frac{1}{2} m_1 u_1^2)$.
9-17 (c) $m_n = 0.12 m_0$, $v_n = 2.8 v_0$, $KE_n = 0.95 KE_0$.
9-18 (a) B; (b) $E_2 = 4E_0/5$, $E_8 = E_0/5$.
9-19 (a) implies $v_1 = 150$ mph; (b) $\frac{5}{8}$.

9-20 (a) $v_1 = \frac{3}{4}u$, $v_2 = \frac{1}{4}u$; (b) $E_0/8$ is lost.
9-21 (a) $M = 2m$, $v = u/\sqrt{3}$, $\theta = \pi/6$;
 (b) $\theta_m^* = 2\pi/3$, $\theta_M^* = 5\pi/3$.
9-23 Lab: $v = 0.48u$, $\theta = -34°$; CM, $v = (\sqrt{3}/6)u$, $\theta = -111°$.
9-24 (a) $M_n = (1.2 \pm 0.4)$ amu;
 (b) $v_{initial} = (3.1 \pm 0.4) \times 10^7$ m/sec.
9-25 (a) Proton; (b) Yes, within experimental error.
9-26 (a) $v_N^* = 12.7 \times 10^6$ m/sec, $v_{C-12}^* = 1.06 \times 10^6$ m/sec;
 (b) $v_N^{lab} = 10^7$ m/sec, $\theta = 132°$; (c) $n = 50$.
9-27 (b) $\theta = \tan^{-1}[(M - m)/(M + m)]^{1/2} = \frac{1}{2}\cos^{-1}(m/M)$.
9-28 (a) 2.01 Mev $\leq KE_p \leq$ 8.98 Mev; (b) $\theta_{max} \approx 68.5°$, $KE(\theta_{max}) = 0.25$ Mev.
9-29 (a) $d = mL/(M + m)$.

CHAPTER 10

10-1 $F_0^2 T^2/8m$.
10-2 (a) 10 m/sec; (b) $5\sqrt{6} \approx 12.2$ m/sec.
10-3 (a) $x(t) = F^3(t)/6C^2m$;
 (b) $x(t) = [v_0 F(t)/C][1 + (F^2(t)/6Cmv_0)]$.
10-4 (b) lever, l_2/l_1; inclined plane, cosec θ; pulley system, 6; crank, $R/(b - a)$.
10-5 (b) about 15 in.
10-6 (a) 5×10^4 N; (b) 5×10^5 J; (c) 2.5×10^5 J.
10-7 (a) $mu^2/2$; $m(u^2 + 2uv)/2$, where m is the mass of the ball.
10-8 (a) $F\omega R$.
10-9 (a) 1.95×10^6 ft lbs (= 1.33×10^6 J);
 (b) 118 min $\approx$ 2 hr; (c) 7.80×10^6 ft lbs.
10-12 (a) 2.4×10^{12} J; (b) 3.8×10^6 man-days $\approx 10^4$ man-years.
10-13 (a) $Um/M(M + m)g = H_0$;
 (b) $(2gh)^{1/2}$, $(2gh)^{1/2} + [2Um/M(M + m)]^{1/2}$;
 (c) $H_0 + 2(H_0 h)^{1/2}$, which is higher than in (a) by $2(H_0 h)^{1/2}$.
10-14 10^{-5} eV.
10-15 $(k_1/k_2)^{1/2}[(1 + 2mv^2 k_2/k_1^2)^{1/2} - 1]^{1/2}$.
10-17 $(gL/2)^{1/2}$.
10-18 (a) $(m_1 + m_2)g$.
10-19 (b) 4 J; (c) $x < 0$ m, $x > 2$ m.
10-20 About 39 sec.
10-22 (b) $r - (Mg/k)$;
 (c) $\Delta T = (\pi/\omega_0) + (2/\omega_0) \sin^{-1}\{1/[1 + (\omega_0 T_0/2)^2]^{1/2}\}$,
where T_0 = period of bouncing for perfectly rigid floor; $\omega_0 = \sqrt{k/M}$. Note that when $\omega_0 T_0 \gg 1$ (that is, when $H \gg Mg/k$), $\Delta T = \pi/\omega_0$; when $\omega_0 T_0 \ll 1$, $\Delta T = 2\pi/\omega_0$.
10-23 (b) $16/\pi^2$ J = 1.62 J.
10-24 (a) $U(x) = +(k_1 x^2/2) - (k_2 x^3/3)$; (b) $k_2 = (k_1/2b)$;
 (d) $v = (k_1 b^2/3m)^{1/2}$.
10-25 $(3mv^2/2\alpha)^{1/3}$.
10-26 $(2 + \sqrt{3})(mb^2/3U_0)^{1/2}$.
10-27 (b) $A = (k/8l_0^2)$.
10-30 (a) $(3mv^2/2) + (kL^2/8)$; (b) v; (c) $(3kL^2/8m)^{1/2}$;
 (d) $2\pi(2m/3k)^{1/2}$.

10-31 (a) 5.4 eV; (b) 1.9×10^{13} Hz.
10-32 (b) $r_0 = (mB/6A)^{1/(m-6)}$; (c) $D = (n-6)A/nr_0^6$.
10-33 (b) $r_0 = (nB/Ae^2)^{1/(n-1)}$; (c) $u = N_0(n-1)Ae^2/nr_0$;
(d) $u = 180$ kcal/mole.

CHAPTER 11

11-2 $a = g(2 - \sqrt{3})/6$ ($2m$ accelerates down the 30° slope);
$T = mg(1 + \sqrt{3})/3$.

11-4 (a) $(gl)^{1/2}$; (b) $2mg$; (c) $2(gl)^{1/2}/(n+1)$;
(d) $\cos\theta = 1 - (n-1)^2/2(n+1)^2$.

11-6 (a) $(l/\mu)[m/(M+m)]^2$; (b) $\mu_{\text{critical}} = 3m/2M$.

11-7 (a) $mg(5 + 2h/r)$ directed vertically upward;
(b) $g(3 + 2h/r)$.

11-8 (a) $(5gl)^{1/2}$; (b) $\cos^{-1}(2/3)$.

11-9 $(3 + \sqrt{3})$m.

11-10 (a) $\cos^{-1}(2/3)$; (b) $\cos^{-1}[(2 + v_0^2/gR)/3]$; (c) 5.8 m.

11-11 (a) $\theta = \cos^{-1}[(4 + \sqrt{3})/26]$;
(b) $\theta_{\text{buckling}} = \cos^{-1}[-2(r\cos\beta - L\cos\alpha)/3(L-r)]$;
(c) $\cos\alpha = (r/L)\cos\beta - (\sqrt{3}/2)[1 - (r/L)]$.

11-12 (a) 9 mm; (b) 5.4 sec.

11-13 It would lose nearly 2 sec per week.

11-14 (a) $q = 2L\sin(\theta/2)[mg\tan(\theta/2)/k]^{1/2}$.

11-16 (a) $M = 2\pi R^2 h \rho_s/[(\Delta T/T) + (h/R)]$; (b) 10 sec/day.

11-17 (a) For $0 < r < R$, $F(r) = 0$; For $R < r < 2R$,
$F(r) = GMm/r^2$; For $2R < r$, $F(r) = 3GMm/r^2$.
(b) $-2GMm/R$; (c) $2(GM/R)^{1/2}$.

11-18 (b) $T_{\text{tunnel}} = (3\pi/\rho G)^{1/2} = 1$ hr. 49 min;
(c) $T_{\text{satellite}} = T_{\text{tunnel}}$.

11-19 According to this criterion, the earth could retain all four of these gases; the moon, only N_2 and CO_2; Mars, all except H_2. It should be noted that an era of higher temperature could have resulted in the loss of heavier gases.

11-21 (a) $v_{\text{escape}} = (5GM/2R)^{1/2}$. (b) The launch point has a speed $0.12v_{\text{escape}}$ relative to the center of mass of the system, so the needed launch speed can be reduced.

11-22 (a) 8.4×10^8 J; (b) 2.7×10^9 J; (c) 5.8×10^9 J;
(d) The "hilltop" is about 0.9D from the earth, the kinetic energy needed at the earth's surface to overcome it is about 6.1×10^9 J.

11-23 (a) $F(z) = -2GMmz/(r^2 + z^2)^{3/2}$;
(b) $U(z) = -2GMm/(r^2 + z^2)^{1/2}$;
(c) For $z \gg r$, $F(z) \approx -2GMm/z^2$, $U(z) \approx -2GM/z$;
For $z \ll r$, $F(z) \approx -2GMmz/r^3$,
$U(z) = -2GM(1 - z^2/2r^2)/r$;
(d) $T_P = T_0/2\sqrt{2}$.

11-24 For the path given (straight line),
(a) $KE = k_1 a^2/2 + k_2 b^2/2$; (b) $KE = (k_1 + k_2)ab/2$.
The force of case (a) is the conservative force.

11-25 (a) $F_N(r) = [-\lambda e^{-r/r_0}/(r_0 r)][1 + (r_0/r)]$;
(b) $F_N(r = 1.4F) \approx -4.2 \times 10^3$ N,
$F_C(r = 1.4F) \approx 1.2 \times 10^2$ N;
(c) $F_N(r) = 10^{-2} F_N(r_0)$ for r just over 5F; the coulomb force is about 10 N there.

11-28 (a) $\theta = 45°$; (b) $\theta = \frac{1}{2}\tan^{-1}(-1/\mu)$; (c) $\theta \approx 50.8°$.
11-30 (a) $g = 2\pi G\sigma\{1 - [h/(R^2 + h^2)^{1/2}]\}$;
(b) For $R = 2h$, $g = 0.55(2\pi G\sigma)$; For $R = 5h$, $g = 0.80(2\pi G\sigma)$; For $R = 25h$, $g = 0.96(2\pi G\sigma)$;
(c) 0.08 sec/yr.

CHAPTER 12

12-1 $x = vt$, $y = -R\sin 2\pi nt$, $z = H + R\cos 2\pi nt$, where the observer's x axis has been taken along the direction of flight, where H is the (constant) altitude of the propeller hub, and where the sense of the propeller rotation is clockwise (as viewed by the pilot). The coordinate and time origin have been chosen for maximum simplicity.

12-2 (a) $r'(t) = (R^2 + D^2 - 2RD\cos\omega t)^{1/2}$;
$\theta'(t) = \tan^{-1}[y'(t)/x'(t)]$,
where $x'(t) = D\cos\omega t - R$, $y'(t) = -D\sin\omega t$.
(b) $dx'/dt = -\omega D\sin\omega t$; $dy'/dt = -\omega D\cos\omega t$;
$d^2x'/dt^2 = -\omega^2 D\cos\omega t$; $d^2y'/dt^2 = \omega^2 D\sin\omega t$.

12-3 $\tan^{-1}(a/g)$ (forward of the vertical).

12-4 (a) -1.5 m/sec^2 (relative to ground);
(b) $+5$ m/sec (relative to ground); (c) 33 m.

12-5 (a) 840 N; (b) 700 N; (c) 560 N; (d) 560 N, 700 N, 840 N;
(e) The elevator has an acceleration of 20/7 m/sec^2 directed upward. The direction of the elevator's motion is not determined.

12-6 $a_{max} = \mu g\cos\theta$.

12-7 (a) 0.5 m; (b) 0.25 m.

12-8 (b) $a_{slip} = 31g/8$.

12-9 Speed is 79 m/sec ($\approx$ 175 mph), 1180 m of runway is used ($\approx \frac{3}{4}$ mi.).

12-10 $a_{max} = 6.7 \times 10^4$ m/sec^2.

12-11 (a) $\tan^{-1}(a/g)$; the apple dropped in a straight line at an angle $\tan^{-1}(a/g)$ forward of "straight down". Thus, if $(h/d) < (a/g)$, it hit the floor; otherwise it hit the wall. (b) The balloon tilted *forward* at an angle $\tan^{-1}(a'/g)$ to the (upward) vertical.

12-12 (a) $a_{3m} = g/3$ downward; (b) $F_P = 2mg$.

12-13 (a) It is in equilibrium at $\tan^{-1}(a/g)$ back from vertically downward; $T = 2\pi[l/(a^2 + g^2)^{1/2}]^{1/2}$.
(b) It is in equilibrium when string is normal to track (i.e., θ back from vertical); $T = 2\pi[l/(g\cos\theta)]^{1/2}$.
(c) It is in equilibrium when string is vertically down, provided that $a < g$; $T = 2\pi[l/(g - a)]^{1/2}$.

12-14 500 lbs $\approx$ 2200 N.

12-15 (a) $\omega \geq [(a + g)/R]^{1/2}$; (b) $\omega \geq [(a^2 + g^2)^{1/2}/R]^{1/2}$.

12-16 (b) $\omega_{max} = (2kS_{max}/\rho)^{1/2}$;
(c) $\omega_{max} \approx 5.2 \times 10^3$ sec^{-1}, or about 50,000 rpm.

12-17 (a) $g_{eff} \approx 160,000 \times$ normal "g"; (b) $F = 9.5 \times 10^{-9}$ dynes;
(c) $v = 0.1$ mm/sec.

12-18 $15/\pi$ rpm.

12-19 $F_{net} = 3mg_R(\Delta R/R)$. (Note that F_{net} acts in the direction of the radial displacement ΔR.)

CHAPTER 13

13-1 $F \sim -1/r^3$.

13-2 (a) $v = (ke^2/mr)^{1/2}$; (b) $l = (ke^2mr)^{1/2}$;
(c) $r_n = n^2h^2/4\pi^2ke^2m$;
(d) $U(r_n) = -4\pi^2k^2e^4m/n^2h^2$;
thus $E(r_n) = -2\pi^2k^2e^4m/n^2h^2$;
(e) $r_1 \approx 0.5 \times 10^{-10}$ m ≈ 0.5 Å; $-E_1 \approx 14$ eV.

13-3 $KE_{\text{final}}/KE_{\text{initial}} = h^2/l^2$.

13-4 $m_B = v_0^2(D^2 - d^2)/2Gd$.

13-6 (a) $r_{\max} = 16R/7$; (b) $r_{\max} = 9R/7$.

13-12 (a) $v = v_0/3 = (GM/6R_0)^{1/2}$; (b) $T = 2\sqrt{3}(2\pi R_0/v_0)$.

13-13 (b) $v_P = 3v_0$; (c) $\alpha = \cos^{-1}(3/5)$.

13-14 $[1 \pm (\sqrt{3}/2)]r_1$.

13-15 $\{R/2(1-\alpha)\}\{1 + [1 - 4\alpha(1-\alpha)\cos^2\theta]^{1/2}\}$, where $\alpha = v_0^2R/2GM$.

13-16 (a) $E = -3GMm/16r$; $l = m(GMr)^{1/2}/2$.

13-18 $F \sim -1/r^5$.

13-19 (a) Using the notation of the text discussion, $v_1 = 1.1v_0 = 32.6$ km/sec; $T = (1.26)^{3/2}T_E \approx 1.4T_E$, so that the time of flight is 0.7 years, and $v_2 = 0.72v_0 \approx 21.3$ km/sec. Note that $v_0 = 29.6$ km/sec and $v_{\text{Mars}} = 24.0$ km/sec. (b) $g_E = g_{\text{sun}}$ at about 2×10^{-3} AU from earth; $g_{\text{Mars}} = g_{\text{sun}}$ at about 10^{-3} AU from Mars; work against sun $\approx 5 \times$ work against earth $\approx -25 \times$ work against Mars.

13-20 (a) $(r_{\text{ap}}/r_E) = [1 + (\Delta v_1/v_0)]^2/\{2 - [1 + (\Delta v_1/v_0)]^2\}$, where r_{ap} is the aphelion distance in AU, v_0 is the earth's orbital speed, and Δv_1 is the increment.
(b) $(v_{\text{ap}}/v_0) = [1 + (\Delta v_1/v_0)](r_E/r_{\text{ap}}) = \{2 - [1 + (\Delta v_1/v_0)]^2\}/[1 + (\Delta v_1/v_0)]$;
Thus, $[(\Delta v)_{\text{total needed}}/v_0] = (\Delta v_1/v_0) + (v_{\text{ap}}/v_0) = [1 - (\Delta v_1/v_0)]/[1 + (\Delta v_1/v_0)]$.

13-21 The year has lengthened about 20 millisec in 5000 years.

13-22 $L(t) = L_0 e^{-\lambda t/m}$.

13-23 (a) $x(t) = x_0 \cos \omega_0 t + (v_{x0}/\omega_0) \sin \omega_0 t$;
$y(t) = y_0 \cos \omega_0 t + (v_{y0}/\omega_0) \sin \omega_0 t$, where $\omega_0^2 = k/m$.
The orbit is an ellipse; total energy and angular momentum are conserved.
(b) $\omega_0 R$; $r_{\max} = 2R$.

13-25 (b) 1.7×10^{-14} cm^2; (c) $r_A + r_K = 7.4$ Å.

13-26 (b) $d_0 = 4.5 \times 10^{-12}$ m; (d) about 50.

CHAPTER 14

14-3 (a) ± 420 J-sec; the sign depends on the sense of the rotation
(b) ± 2310 J-sec if the third skater travels in same direction as the nearer one was initially skating; ± 1470 J-sec if the third skater travels in the opposite direction as the nearer one.

14-4 $KE_{\text{rot}} \approx 0.01\, KE_{\text{trans}}$.

14-5 (a) $d\omega/dt = -[\mu r^2 I_0/(I_0 + \mu r^2 t)^2]\omega_0$;
$dE_{\text{rot}}/dt = -[\mu r^2 I_0/(I_0 + \mu r^2 t)^2][I_0\omega_0^2/2]$, where $I_0 = MR^2/2$.
(b) $t = I_0/\mu r^2 = (M/\mu)(R^2/2r^2)$.

14-6 (a) $\omega_{\text{final}} = 3\omega_0$; rotational KE increased by factor 3.
(c) $R/\sqrt{6}$.

14-7 $10^{17\pm1}$ J.

14-8 In the answers below, I_n represents the value obtained by approximating the bar as n equal mass points.
(a) $I_1 = ML^2/4$; (b) $I_2 = 5ML^2/16$;
(c) $I_n = (M/n)\sum_{k=1}^{n}[(2k-1)L/(2n)]^2$
$= [ML^2/3][1 - (1/4n^2)]$.

14-9 (a) $I = 2MR^2/3$.

14-10 (a) $I = Ma^2/6$, plate of edge a, mass M;
(b) $I = 5Ma^2/18$, box of edge a, mass M;
(c) $I = Ma^2/6$, cube of edge a, mass M.

14-13 Yes.

14-15 (a) $4g/15$; (b) $8\sqrt{6}/3$ sec^{-1};
(c) 66 N; 12.7 N (using $g = 10$ m/sec^2);
(d) One full turn beyond the minimum half-turn.

14-16 (a) $0.35g = 3.4$ m/sec^2; (b) 6.6 m.

14-17 $x = b$.

14-18 (a) $T = 2\pi(3R/g)^{1/2}$; the equivalent simple pendulum has length $3R$.
(b) $T = 2\pi(2R/g)^{1/2}$; the equivalent simple pendulum has length $2R$.

14-19 $T = 2\pi R(M/2c)^{1/2}$.

14-21 (a) 2.2×10^{-8} m-N/rad; (b) 1.3×10^{-7} N; (c) 7.7°.

14-22 (a) 2.9×10^{-8} m-N/rad; (b) 12 min;
(c) 7.5×10^{-3} rad $= 0.43°$; 7.5 cm.

14-23 (a) $(\pi/150)$ sec ≈ 0.021 sec; (b) $T_1 \approx 409$ N; $T_2 \approx 159$ N.

14-24 $l = R[\sqrt{1 + (I/2mR^2)} - 1]$;
for $I = MR^2/2$; $l = R[\sqrt{1 + (M/4m)} - 1]$.

14-26 $\omega_{Af} = \omega_0/[1 + (I_B R_A^2/I_A R_B^2)]$; $\omega_{Bf} = -(R_A/R_B)\omega_{Af}$.

14-27 (a) 5 m/sec in the direction of the truck's motion;
(b) $\omega = -(5\text{ m/sec})/r$, where r is the radius of the pipe;
(c) $5/4\mu = 4.16$ m; (d) 0.

14-28 (a) $h = 2r/5$.

14-31 $108/\pi^2 \approx 11$ rev/min.

Index

Abetti, G., 717
Abraham, Z., 452
Acceleration, 85
 centripetal, 106, 200
 in circular motion, 106, 108
 invariance of, 175
 in polar coordinates, 108, 557
 radial, 108, 557
 related to force, 165
 transverse, 108, 200, 557
Accelerometers, 498, 501
Action and reaction
 in collisions, 313, 316, 320
 in jets and rockets, 324
 in static equilibrium, 123
Adam, 13
Adams, J. C., 292
Air resistance, 153, 214, 218
 and independence of motions, 225
Airy, G. B., 294
Almagest, 76
Alpha particle scattering, 604, 612
Anderson, O. L., 432
Angels, motive power of, 554
Ångström, A. J., 27
Angular momentum
 conservation of, 562, 639
 and kinetic energy, 654
 internal, 632, 641
 orbital, 560
 and centrifugal potential energy, 564
 quantum of, 637
 total, 633
 vector additivity of, 672, 679
Angular velocity, 107
 vector properties of, 673
Animals
 cruelty to, 104, 486
 equality of, 21
Aphelion, 577
Apogee, 577
Apple, *see* Moon
Approximations, 10, 12
Archimedes, 117, 133
Archimedes' principle, 512
Aristarchus, 75, 77, 78, 83, 275
Asteroids, 35
Astin, A. V., 64
Astronomical unit, 247, 279
Atomic mass unit (amu), 27
Atoms, 26
 free fall of, 98
 gyroscopic behavior of, 686
 masses of, 27
 (table), 27
 radii of, 28
 velocity distribution of, 100, 102
Austern, N., 181

Baade, W., 35
Baez, A., 585
Bagnold, R. A., 32
Balance, 117
Banking of curves, 199
Barker, E. F., 688
Barnard, E. E., 297
Barrère, M., 327
Bathtub vortex, 529
Beams, J. W., 513
Bell, E. T., 61
Berkeley, G., 542
Berry, A., 77, 296
Binary stars, 296
Bloch, F., 687
Bockelman, C. K., 404
Bode, J., 293
Bode's law, 293
Bounded orbits, 568
Boys, C. V., 155
Bragg, S. L., 327
Brahe, Tycho, 5, 14, 277, 578
Branson, H. R., 351
Broad, C. D., 114
Browne, C. P., 404
Bucherer, A. H., 169
Buechner, W. W., 404
Bullard, E. C., 452
Bursting speed (of rotating object), 208, 513
Butler, J. A. V., 29

Caesar, J., 39
Cartesian coordinates, 49
Cassini, G. D., 277
Catapult, 391
Cathode-ray tube, 195
Cavendish, H., 141
Cavendish experiment, 142, 154, 155
Cells, 31
Center of gravity, 132, 337
Center of mass, 296, 337
  and center of gravity, 132, 337
  kinetic energy of, 338, 631
  motion of, 353, 629
  See also CM frame
Center of percussion, 671
Central force
  conservative property of, 442
  definition of, 442

Central-force motion
  energy conservation in, 563
  and law of equal areas, 557, 584
  radial part of, 564
    effective potential-energy curves, 565, 570
  as two-body problem, 598
Centrifugal force, 507
Centrifugal potential energy, 564
Centrifuges, 511
Centripetal acceleration, 106, 200
Chadwick, J., 363
Circular motion
  energy conservation in, 426
  non-uniform, 108, 200
  uniform, 105
    acceleration in, 106
    of charged particle in magnetic field, 202
    dynamics of, 198
    relation to SHM, 233
Clemence, G. M., 66
Cloud chamber, electron paths in, 203
CM frame, 335
  and collision processes, 337, 341, 342, 345, 349
  kinetic energy in, 338, 339
Coe, L., 66
Collisions
  definition of, 351
  elastic and inelastic, 309
  elastic (perfectly elastic), 332, 342
  with energy storage, 400
  explosive, 347
  with external forces, 352
  first experiments on, 308
  and frames of reference, 331, 340, 342
  inelastic, 346
  invariance of KE changes, 334
  and kinetic energy, 333
  and momentum conservation, 308, 311
  nuclear, 342
  two-dimensional, 339
  and zero-momentum frame, 335, 341
Colodny, R. G., 180
Conservation
  of angular momentum, 562, 639
  of energy, *see* Energy
  of linear momentum, 308, 309, 312, 318
  of mass, 309

Conservative forces, 381, 423, 442, 457
Constraints, 425
Contact, 150, 151
Contact forces, 150, 152, 157
Conversion factors, 710, 711
Coordinate systems, 48
Coordinates
   Cartesian, 49, 50, 52
   oblique, 51
   orthogonal, 51, 56
   polar, 49, 50, 52
   rectangular, see Coordinates, Cartesian
Copernican system, 76, 247, 249
Copernicus, N., 77, 78, 249, 696
Coriolis, G., 514
Coriolis forces, 514
   and cyclones, 528
   and deviation of falling objects, 525
   and Foucault pendulum, 529
   in gyroscope, 691
Coulomb, C. A., 145
Coulomb forces, 145, 149, 151
Coulomb's law, 145
Crew, H., 162, 431
Cross product, 127
Cross section
   differential, 614
   partial, 611
   scattering, 609
Csikai, J., 357
Cyclotron, 204

DaVinci, L., 126
Day
   sidereal, 64, 82
   solar, 64, 82
De Revolutionibus, 77, 249
De Salvio, A., 162, 431
Descartes, R., 3, 306
Diatomic molecule
   rotation of, 637
   vibration of, 405
Dicke, R. H., 283, 546
Dirac, P. A. M., 65, 94
Displacement, 52, 53
   relative, 55
Dissipative forces, 210, 381, 470
DuBridge, L. A., 619
Du Mond, J. W. M., 221

Dust, 31
Dyne (def.), 119

Earth
   as gyroscope, 694
   mass of, 268, 302
   mean density of, 270
   radial variation of density, 452
   radius of, 259
   as rotating frame, 524
      deviation of falling objects, 525
      effect on g, 524
      and formation of cyclones, 528
      and Foucault pendulum, 529
Earth satellites, 265
   orbit decay of, 470
Earth-moon system
   gravitational potential of, 455
   and precession of equinoxes, 696
   and tide production, 532
Eddington, A. S., 163
Edgerton, H. E., 163, 630
Einstein, A., 8, 45, 161, 169, 178, 280, 299, 492, 506, 543
Eisenbud, L., 181
Electric field, 462, 467
Electric force, 145
   and motion of charged particles, 195, 467
   between neutral atoms, 149
Electrons, 24
   in combined electric and magnetic fields, 467
   in magnetic field, 206, 467
   trajectory in oscilloscope, 196
Ellipse, geometry of, 583
Elliptic orbits, see Orbits
Ellis, B., 180, 186
Energy, 367
   conservation of, 367, 377, 381, 425
   in central force motion, 563
   kinetic, see Kinetic energy
   potential, see Potential energy
   (table), 375
   units of, 373, 711
Energy diagrams, 382, 389
Eötvös, R., 281
Eötvös experiment, 282
Epicycle, 76, 77
Equal areas, law of, 557
Equant, 579

Equilibrant (of set of forces), 123
Equilibrium
  rotational, 120, 124, 131
  stable, 395
  static, 116, 119
  translational, 120
Equinoxes, precession of, 6, 694
Equipotentials, 463
Equivalence principle, 280, 506
Equivalent simple pendulum, 663
Eratosthenes, 79, 259
Eros, 278
  See also Venus, transit of
Escape velocities, 453
Estermann, I., 100, 101
Euclidean geometry, 59
Eve, 13
Explosive collisions, 347
Exponentials, 222
Extraneous roots, 93

Falk, H., 707
Faraday, M., 139
Feather, N., 66
Feenberg, E., 622
Fermat's principle, 81
Fermi, E., 26
Fermi (unit), 26
Feynman, R. P., 68, 554
Field(s), 461
  electric, 462, 467
  gravitational, 462, 473, 476
    of flat sheet, 477
    flux of, 473
    of sphere, 462, 476
Field lines, 462, 464
"Fixed stars" (reference frame), 47, 295, 540
Flamsteed, J., 277, 286, 291
Flux (of field), 473
Flywheels, 654
Force, 115
  central (def.), 442
  centrifugal, *see* Centrifugal force
  conservative, 381
    criteria for, 457
  electric, 145, 149, 151, 468
  electromagnetic, 139, 147
  of fluid jet, 324
  frictional, 152, 210

  gravitational, 139, 140
    conservative property of, 442
  inertial, *see* Inertial forces
  lines of, 462, 463
  magnetic, 146, 202, 205, 468
  nuclear, 139, 147, 156
  of particle stream, 321
  as rate of change of momentum, 315
  units of, 118, 171
  vector nature of, 119
  See also Forces
Forces
  contact, *see* Contact forces
  equilibrium of, 116
  polygon of, 121
  resolution of, 122
  vector combination of, 120, 168
Foucault, J. B. L., 529
Foucault pendulum, 529
Frame of reference, 46, 162
  and collisions, 331, 340, 342
  of "fixed stars," 47, 295, 540
  inertial, *see* Inertial frame
  linearly accelerated, 495
  and Newton's Second Law, 174
  rotating, 507
    the earth as, *see* Earth
    free fall in, 525
    inertial forces in, 510, 518, 523
Fraser, J. T., 66
Free fall, 95, 102
  and air resistance, 214
  of atoms, 98
  of neutrons, 100
  on rotating earth, 525
  and weightlessness, 285
Friction
  coefficient of, 153, 190
  dry, 152, 210
  fluid, 153, 211
    Newtonian theory of, 360
Frisch, D. H., 181

g, 96, 279
  altitude dependence of, 270
  as gravitational field, 462
  latitude dependence of, 284, 524
  local variations of, 272
G, 141, 268

Gal (unit), 272
Galaxies, 36
   contracting, dynamics of, 657
Galaxy
   mass of, 299
   rotation of, 298
Galilean transformation, 175, 334
Galilei, G., 84, 95, 112, 161, 162, 225, 240, 286, 288, 429, 494
Galle, J. G., 294
Gas
   internal KE of, 631
   pressure of, 354, 631
Gauss, K. F., 61
Gaussian surface, 474
Gauss's Law (or Theorem), 473
   applications of, 476
Geiger, H., 605, 612, 616
Geodesic, 299
Gillispie, C. G., 366
Gimbal rings, 677
Globular cluster, 143, 144
Gold, T., 66
Gravimeter, 272
Gravitation
   Einstein's theory of, 299
   law of, 5, 139, 140, 245, 256
   Newton's theory of, 5, 6, 245, 256
Gravitational forces, 139, 140
   conservative property of, 442
Gravity
   acceleration due to, *see* g
   equivalence to accelerated frame, 504
   force of, 119, 129
Gravity meter (gravimeter), 272
Great Pyramid of Gizeh, 414
Guye, C., 169
Gyrocompass, 685
Gyroscope, 677
   analyzed via $F = ma$, 688
   navigational use of, 683
   nutation of, 691
   in steady precession, 678, 681

Hadron, 147
Hafner, E. M., 66
Haldane, J. B. S., 11
Hall, E., 552
Halley, E., 255, 277, 597

Halley's comet, 597
Hanson, N. R., 546
Harmonic motion, *see* Simple harmonic motion
Harmonic oscillators, 226, 233, 393, 432, 437
Havens, W. W., 100
Heaven, altitude of, 111
Herivel, J., 559
Herodotus, 414
Herschel, J., 295
Herschel, W., 291, 295
Hipparchus, 75, 83, 260, 696
Hodograph, 103
Hooke, R., 227, 255, 272
Hooke's law, 227, 387
Hoyle, F., 36
Hubble, E., 36
Hubble's law, 16, 39
Hume, J. N. P., 519
Huygens, C., 255, 308, 331, 442, 479
Hyperbolic orbits, 604, 607
Hysteresis, 389

Iben, I., 452
Impact, laws of, 308
   See also Collisions
Impact parameter, 570, 607
Impulse, 173, 313, 369
Impulsive force, 315, 371, 669
Impulsive torque, 669
Independence of motions, 95, 194
   limitations to, 225
Inelastic collisions, 346
Inertia
   law of, 161, 494, 542
   moment of, *see* Moments of inertia
   origin of, 542
Inertial forces, 494, 497, 507, 518
   and Mach's principle, 545
Inertial frame(s), 163, 494
   dynamical equivalence of, 174, 497
   fundamental, 538
Inertial mass, 166, 280, 319, 543
   velocity dependence of, 169
Infeld, L., 492
Integrals of motion, 368
Invariance
   of energy changes, 334
   of Newton's law, 173

rotational, and torque, 642
translational, and force, 642
Inverse-square law
  electric, 145
    and α-particle scattering, 604
  gravitational, 5, 141, 245
    deduced from elliptic orbit, 583
    deduced from Kepler's Third Law, 254
    types of orbits, 600
Ivey, D. G., 519

Jeans, J. H., 39
Jeffery, G. B., 506
Jet propulsion, 324, 359
Johnson, N. B., 102
Jones, H. S., 295
Joule (unit), 173, 373
Jupiter (planet)
  mean density of, 286
  moons of, 278, 286, 288
    discovered by Galileo, 288
  orbit of, 248

Kant, I., 36
Kaufmann, W., 169
Kemble, E. C., 75
Kepler, J., 5, 253, 276, 577
Kepler's laws, 5, 7, 252
Kepler's Second Law, 5, 559
  implies central force, 559
Kepler's Third Law, 252
  explained by Newton's laws, 254
  and Jupiter's moons, 290
Kilogram (def.), 171
Kinetic energy (KE), 173
  in CM frame, 338
  in collisions, 333
  of many-particle system, 630
  of rolling object, 652
  of rotating objects, 651, 654, 675
  of two-body system, 338, 347, 398
King, J. G., 39, 149, 624
Kirchner, F., 236
Kirkpatrick, P., 4
Knot (unit), 67
Koestler, A., 582
Kramers, H. A., 62, 367

Laboratory frame, 337

Lagrange, J. L., 290
Laplace, P. S. de, 287
Lavanchy, C., 169
Lavoisier, A. L., 309
Lawrence, E. O., 204
Laws of motion, 162
Leibnitz, G. W. von, 368
Leighton, R. B., 68
Lemonick, A., 672
Length, units of, 63, 709, 710
Lesage, G. L., 304
Lever, law of, 117, 125, 133
LeVerrier, U. J., 292
Light, speed of, 67
Light-year, 35, 711
Lindsay, R. B., 366
Linear oscillator, 387
  as two-body problem, 397
Lines of force, 462, 463
Lippershey, H., 288
Locke, J., 583

Mach, E., 331, 542
Mach number, 67
Mach's principle, 543
Magnetic field, 205
Magnetic force, 146, 202, 205
  and motion of charged particles, 146, 202, 206, 467
Many-particle system
  angular momentum of, 643
  dynamics of, 629, 641
  kinetic energy of, 631
  momentum of, 628
Margenau, H., 366
Mars (planet)
  motion of, 74
  orbit of, 252, 581, 582
    analyzed by Kepler, 578
  parallax of, 277
Marsden, E., 605, 612, 616
Maskelyne, N., 302
Mass, 23, 164, 166
  additivity of, 166, 172
  conservation of, 309
  energy equivalence of, 376
  inertial, 166, 280, 319, 543
  scales of, 170, 319
  standards of, 171, 709

units of, 27, 171
velocity dependence of, 169
and weight, 279
Mass spectrograph, 206
Mechanical advantage, 412
Mercury (planet)
orbit of, 77, 246
precession of, 300, 305, 623
Meter (def.), 63
Metric system, 24, 709
Micron, 30
Milky Way, 22, 36
Millikan, R. A., 26, 218, 221
Milton, J., 111
MKS system, 24
Molecules, 28
diatomic
rotation of, 637
vibration of, 405, 409
Molière, 4
Moment, *see* Torque
Moment of momentum, *see* Angular momentum
Moments of inertia, 634, 644
principal, 644
special theorems, 647
Momentum, 166, 173, 310
angular, *see* Angular momentum
conservation of, 308, 309, 312, 318
rate of change of, 315, 321
vector character of, 310
Monkey-shooting demonstration, 104
viewed from monkey-frame, 497
Montgomery, D. J. X., 146
Moon
and apple, 256
distance to, 259
as a falling object, 256
and precession of equinoxes, 694
and tide production, 531
Morrison, P., 546
Morse, P. M., 409
Morse potential, 408
Motion, 3, 43
accelerated, 87, 165
under central force, *see* Central-force motion
of charged particles, 146, 195, 202, 205, 467

circular, *see* Circular motion
in conservative fields, 466
near earth's surface, 591
laws of, 162, 164
oscillatory, 395
relativity of, 46
against resistive forces, 210, 213
numerical solutions, 215
rotational, *see* Rotational motion
simple harmonic, *see* Simple harmonic motion
two-dimensional, 95, 102, 194
uniformly accelerated, 90, 91, 188
Muybridge, E., 638

Neptune, discovery of, 291, 294
Neutrino, 356
Neutrons, 24
elastic collisions of, 344, 363
free fall of, 100
Newman, J. R., 11, 61
Newton, I., 3, 21, 161, 166, 172, 256, 262, 281, 559
bucket experiment, 539
collision experiments, 308, 310, 314, 358
concepts of space and time, 43, 44
deduces $1/r^2$ law from ellipse, 583, 585
De Motu, 559
and law of equal areas, 559
on moon's motion, 256
Principia, 9, 18, 43, 46, 162, 172, 244, 254, 259, 269, 314, 538, 696
on proportionality of weight to mass, 281
quotations, 2, 7, 9, 20, 42, 43, 45, 46, 138, 160, 162, 244, 269
System of the World, 265, 295
theory of fluid resistance, 360
theory of precession of the equinoxes, 6, 694
theory of tides, 6, 533
and universal gravitation, 5, 6, 187, 245, 255, 265
Newton (unit), 118, 171
Newton's First Law, 162
See also Inertia, law of
Newton's Second Law, 166, 315
discussion of, 167, 173, 180
invariance of, 173, 177
and relativity, 174

simple applications of, 188
and time reversal, 178
Newton's Third Law, 314
limitations to, 316, 320
Nier, A. O., 207
Nuclear forces, 139, 147, 156
Nuclear reactions, dynamics of, 348
Nuclei, 25
gyroscopic behavior of, 686
and Rutherford scattering, 604, 609
Nucleons, 25
Numerical methods
in kinematics, 91
for harmonic oscillator, 229
for resisted motion, 215, 222
Nutation, 691

Occhialini, G. P. S., 342
Oil-drop experiment, 218, 240
Orbits
bounded, 568
calculated from initial conditions, 595
circular, 572
perturbed, 574
elliptic, 576, 577, 583
energy in, 589
families of, 589, 596
from force law, 600
hyperbolic, 604, 607
under inverse-square attraction, 577, 585, 600
shape versus energy, 604
parabolic, 604
planetary, 246, 249, 252, 577, 582, 583
precession of, 300, 305, 569, 623
unbounded, 569, 597, 604
Orders of magnitude, 10
Orwell, G., 21
Osgood, W. F., 511

Parabolic orbit, 604
Parabolic potential, 431
Parabolic trajectories, 95, 99, 103, 197, 496, 592
Parallax
of Mars, 277
of Venus, 278
Parallel-axis theorem, 647
Parallel forces, 125

Parasnis, D. S., 273
Particle, 21
properties of, 23
Paul, W., 323
Pedersen, K. O., 514
Pendulum
cycloidal, 442
energy conservation in, 427, 429
as harmonic oscillator, 437
period versus amplitude, 440
rigid, 434, 661
Percussion, center of, 671
Perigee, 577
Perihelion, 577
precession of, 300, 305, 623
Perpendicular-axis theorem, 649
Perrett, W., 506
Perturbation
of circular orbit, 574
of planetary orbits, 6, 287, 291, 295
Physical magnitudes (table), 11
Pisa, Tower of, 240
Planck, M., 627
Planetoids, 35
Planets, 33
data on (table), 34
motions of, 74, 82, 246
orbital radii, 247
orbits of, 246, 577
tabulated data, 16, 252, 582
periods of, 249
relative sizes of, 34
Pluto, discovery of, 6, 295
Polar coordinates, 49, 51
velocity and acceleration in, 556
Polya, G., 9
Polygon of forces, 121
Pope, A., 4
Potential, 463
Potential energy (PE), 377
effective, in central-force motion, 564, 573, 576
gradient of, 465
gravitational, 376, 384, 424, 431
scalar character of, 425, 446, 455
of spring, 387, 398
Pound (def.), 119
Powell, C. F., 342
Power, 373

Precession
  of atoms and nuclei, 686
  of equinoxes, 6, 694
  of gyroscope, 678
  of orbits, 569
  of perihelion of Mercury, 300, 305, 623
Pressure of a gas, 354
Principia, see Newton, *Principia*
Principle of equivalence, 280, 506
Product of inertia, 674
Proton-proton collisions, 342
Protons, 24
  magnetic deflection of, 146
Ptolemaic system, 76, 247
Ptolemy, C., 75, 78, 246
Pulleys, 130
Purcell, E. M., 687

Radius of gyration, 647
Rainwater, L. J., 100
Red Baron, 238
Reduced mass, 339
Reference frames, *see* Frame of reference
Reines, F., 358
Relative displacement, 55, 73
Relative motion, 46
Relative velocity, 72
Relativity
  Einstein's general theory, 280, 299, 506
  Einstein's special theory, 8, 45, 161, 169, 178
  Newtonian, 46, 173, 177, 497, 538
Resisted motion, 210, 213, 218
  growth and decay of, 221
  numerical analysis of, 215, 222
Resistive forces, 152, 153, 210, 360, 381, 470
Resultant (of set of forces), 123
Richer, J., 277
Rigid pendulum, 434, 661
Rocket
  principle of, 324
  thrust of, 325, 327
Rocket propulsion, 327
Roemer, C., 278
Rogers, C. W. C., 528
Rogers, E. M., 13, 74, 75, 290
Rolling objects, 652, 667
Rosebury, T., 40
Rosenfeld, J., 428

Rotating frame, *see* Frame of reference
Rotating objects
  fracture of, 208, 513
  kinetic energy of, 651, 654, 675
Rotational equilibrium, *see* Equilibrium
Rotational motion, 632, 639, 651
  combined with linear, 641, 664
Royds, T., 39
Runk, R. B., 432
Rutherford, E., 39, 187, 605, 615
Rutherford scattering, 604
  cross sections for, 611, 614

Sand, 31
Sandage, A., 36, 37
Sands, M., 68
Satellites
  of earth, 265
  of Jupiter, 286, 288, 290
  synchronous, 268
Saturn V rocket, 327
Scalar product, 57, 372, 424
Scattering of alpha-particles, 604, 612
Schlegel, R., 66
Schrader, E. F., 359
Schumacher, D. L., 66
Schwerdt, C. E., 30
Sciama, D. W., 546
Sears, F. W., 134, 688
Second (def.), 64, 65
Sellschop, J. P. F., 358
Shankland, R. S., 169
Shapiro, A. H., 529
Sidereal day, 64
Sidereal period, 250
Sidereal year, 696
Simple harmonic motion (SHM), 226, 231
  by energy method, 389, 393, 432
  geometrical representation, 232
  numerical solutions, 229
  in parabolic potential, 432
  related to circular motion, 233
Simple pendulum, 434
  as harmonic oscillator, 437
  isochronism of, 439
  See also Pendulum
Simpson, O. C., 100, 101
Sinden, F. W., 600
Slope of graph, 70

Small oscillations, 395
Smith, M. K., 441
Solar day, 64
Solar system, 34, 582
  acceleration of, 540
Solar year, 65
Space, 43
  curvature of, 59
  geometry of, 59
Space-time graphs, 66, 88
Special relativity, 8, 168, 169
Speed, 71
Sperduto, A., 404
Sphere
  gravitational attraction of, 261, 450
  moment of inertia of, 647
Spherical shell, gravitational effect of, 263, 446
Spring constant, 226, 387
  of diatomic molecule, 407
Sputnik I, 267
Squire, J., 5
Standards
  of length and time, 63, 709
  of mass, 171, 709
Stars
  apparent circular motions, 47
  binary, 296
  "fixed," 47, 295, 540
  globular clusters of, 143, 144
  orbits of, 296, 298, 540
  as particles, 22, 35
Static equilibrium, 115, 119, 124
Stern, O., 100, 101
Straight-line motion, 66, 85, 88
Strong interaction (nuclear), 147
Stull, J. L., 432
Sun
  distance to, 247, 275
  influence on tides, 537
  mass of, 274
  radial density variation of, 452
Superposition of forces, 168
Surface tension, 157
Sutton, G. P., 327
Sutton, O. G., 422
Svedberg, T., 513
Symon, K. R., 645, 675
Synchronous satellites, 268
Synodic period, 250

Szalay, A., 357

Tea, P. L., 707
Terminal speed, 212, 214
Thomas, G. B., 106
Thomson, J. J., 605
Threshold energy, 402
Tides, 531
  equilibrium theory of, 533
  height of, 535
Tilley, D., 365
Time, 45, 61, 66
  units of, 64
Time constant, 224
Time reversal and Newton's law, 178
Titius, J. D., 293
Tombaugh, C. W., 295
Torque, 125
  and change of angular momentum, 561, 641, 664
  vector character of, 126
Torques, vector addition of, 128
Torsion balance, 142, 145
Torsion constant, 154, 660
Torsional oscillations, 659
Trajectories
  helical, 204
  hyperbolic, 604, 607
  parabolic, 95, 103, 197, 496, 592
Transfer orbits (interplanetary), 592
Transit of Venus, 277
Trick
  clever and effective, 579
  neat (and valuable), 372
Turner, H. H., 294, 295

Ultracentrifuge, 513
Unbounded orbits, 569, 597, 604
Unit vectors, 49, 57, 58, 106
Units, 24, 709
  of force, 118
  of length, 63, 709, 710
  of mass, 27, 171
  of time, 64, 709, 711
Universal gravitation, 5, 6, 245, 256
  constant of, 141, 268
Uranus
  discovery of, 291
  perturbations of, 292

Van Atta, L. C., 404
Van der Waals, J., 149
Van der Waals forces, 149, 157
Vector product, *see* Vectors
Vectors, 48
  addition of, 53
    and properties of space, 59
  components of, 56
  cross product of, 127
  orthogonal, 57
  resolution of, 56
  scalar product of, 57, 372, 424
  subtraction of, 54
  vector product of, 127
Velocity, 67, 70
  angular, *see* Angular velocity
  instantaneous, 68
  in polar coordinates, 107, 556
  relative, 72
  unit of, 67
Velocity-time graphs, 88, 89, 90, 93
Venus
  motion of, 74
  orbit of, 77, 246
  parallax of, 278
  transfer orbit from earth, 593
  transit of, 277
Virus, 30
Vis viva, 334, 368
Von Laue, M., 2

Vulcan, 111

Wallis, J., 308
Walton, W. U., 501
Watt (unit), 374
Weak interaction (nuclear), 148
Weight, 129, 192, 279
  proportionality to mass, 279
Weightlessness, 130, 285
Weinstock, R., 181
Wessel, G., 323
Williams, R. C., 30
Wilson, C., 582
Wood, E. A., 503
Work, 173, 369
Wren, C., 308

Year
  sidereal, 696
  solar, 65
  tropical, 65, 696
Yukawa, H., 156, 485
Yukawa potential, 485

Zacharias, J. R., 139
Zeno, 87
Zero-momentum frame, 335
  See also CM frame
Zorn, J. C., 102